THE CUTTING EDGE

An Encyclopedia of Advanced Technologies

THE CUTTING EDGE

An Encyclopedia of Advanced Technologies

OXFORD
UNIVERSITY PRESS
2000

OXFORD
UNIVERSITY PRESS

Oxford   New York
Athens   Auckland   Bangkok   Bogotá   Buenos Aires   Calcutta
Cape Town   Chennai   Dar es Salaam   Dehli   Florence   Hong Kong   Istanbul
Karachi   Kuala Lumpur   Madrid   Melbourne   Mexico City   Mumbai
Nairobi   Paris   São Paulo   Singapore   Taipei   Tokyo   Toronto   Warsaw

and associated companies in

Berlin   Ibadan

Published by Oxford University Press, Inc.
198 Madison Avenue, New York, New York 10016
www.oup.com

Oxford is a registered trademark of Oxford University Press

Library of Congress Cataloging-in-Publication Data
The cutting edge : an encyclopedia of advanced technologies
p.   cm.
Includes bibliographical references and index.
1. Technology—Encyclopedias.
T9 .C96   2000   603—dc21   99-056125
ISBN 0-19-512899-0

Produced by The Moschovitis Group, Inc.
95 Madison Avenue, New York, New York 10016
www.mosgroup.com

Executive Editor: Valerie Tomaselli
Senior Editor: Jill Pope
Editorial Coordinator: Stephanie Schreiber
Design and Layout: Annemarie Redmond
Original Illustrations: Maruja Bedusa
Photo Researcher: Gillian Speeth, Picture This
Copyediting: Zeiders & Associates
Proofreading: Paul Scaramazza
Editorial Assistant: Renée Miller
Production Assistant: Yolanda Pluguez
Indexing Services: AEIOU, Inc.

Supplementary Web site at www.cuttingedge.oup.com

Cover Photo Credits *Clockwise from top left:* Corbis/Leif Skoogfors; Photo Researchers/Japan Air Lines; Courtesy of Sony
Electronics, Inc.; Corbis/Ed Eckstein; Courtesy of NASA; Corbis/Bettmann. *Center:* Courtesy of NASA.

9 8 7 6 5 4 3 2 1

Printed in the United States of America
on acid-free paper

CONTENTS

LIST OF ARTICLES

PREFACE

The Cutting Edge: An Encyclopedia of Advanced Technologies serves a crucial function for today's researcher. It explains cutting edge technologies in terms that the high school, college, and lay researcher can understand. It also places these technologies in context, outlining their historical development, current applications and limitations, and the issues and debates surrounding their current use and future implementation.

High technology is the general term used for any form of technology that uses the most advanced systems and techniques currently available. While this definition could encompass technological developments in any field, *The Cutting Edge* limits its focus to those technologies of most interest to a general audience. It includes technologies that impact our everyday lives and those that are frequently in the news, either because of the technological development itself or because of the societal issues surrounding it. With a few notable exceptions—technologies, such as SOLAR POWER SATELLITES, that we consider important to the future—the encyclopedia does not include technologies that are not now in use or that aid only pure research. Emphasis is placed on technologies associated with the home (such as Smart Homes), medicine and health care (such as Surgical Robotics and Telemedicine), and entertainment (such as Virtual Reality), as well as advanced forms of communications (Internet Protocol Telephony), transportation (Smart Highways), and information processing (Parallel Computing). The volume also includes entries in areas such as national defense, space and ocean exploration, biotechnology, materials science, and the environment.

The organization of *The Cutting Edge* is straightforward. The articles are in A to Z order, according to the name of the technology. All articles are structured in the same way, to offer ease of use and efficient comparison between technologies. Each article includes the following:

- Scientific and technological description of the technology (how it works)
- History of the technology (how it developed)
- Uses, effects, and limitations of the technology (what it has done and is capable of doing)
- Issues and debates arising from the technology

Original illustrations, included with each article, will help the reader comprehend the underlying principles or the structural design of the technologies. And photographs highlight significant historical figures and social aspects surrounding the technologies.

The articles end with comprehensive bibliographies that include print material as well as Internet sources. Extensive cross referencing is used throughout the articles, which leads the reader to related technologies included in the volume. The index offers a further means of locating topics of interest. And a supplementary Web site, located at www.cuttingedge.oup.com, offers links to the Internet sources cited in the articles; this site was designed to function as a gateway for further on-line research.

Advisory Board Members

William Allstetter is the Director of Science Communications for Columbia University's Health Sciences Division and the Executive Editor of the *Science & Technology Almanac* (Oryx Press, 1999). He has his journalism degree from Columbia University.

Joseph A. Angelo, Jr. is a consulting futurist, college professor, and retired Air Force officer. His interest include aerospace, energy, defense, and remote sensing. He received his Ph.D. in Nuclear Engineering from the University of Arizona in 1976.

Geoffrey C. Bowker is Associate Professor at the Graduate School of Library and Information Science at the University of Illinois at Urbana-Champaign. He is the author of *Science on the Run* (MIT Press, 1994) and *Sorting Things Out* (MIT Press, 1999). He has served as a department editor for the quarterly publication, *IEEE Annals of the History of Computing.*

Christos Moschovitis runs a strategic information technology consulting firm focused on providing high-end technology management expertise to small and medium-sized companies. He is co-author of the critically acclaimed *History of the Internet: 1843 to the Present* (ABC-CLIO, 1999).

Dr. Sal Restivo is Professor of Sociology and Science Studies at Rensselaer Polytechnic Institute in Troy, New York; and Special Professor of Mathematics Education at Nottingham University (U.K.). He is currently working on two books, *The Social Mind;* and *The Discovery of God.*

Dr. Tilli Tansey is a historian of modern medical sciences at the Wellcome Institute for the History of Medicine, London. She has also worked as a research neuroscientist. She is the co-editor of, among other works, *Ashes to Ashes: The History of Smoking and Health* (Rodopi, 1998).

Contributors

Joseph A. Angelo, Jr., a member of the advisory board, is also a contributor.

Paul Candon received a master's degree in psychology with a specialization in behavioral neuroscience from the State University of New York at Binghamton. He has written about recent developments in psychology for *Today's Science On File,* a monthly science news magazine.

Matthew Day trained as a molecular biologist, receiving a Ph.D. from Birkbeck College, London. Since then he has written for Facts On File on molecular biology and genetics. He currently works as an editor for upper level college texts.

Trevor Day is a lecturer and a writer of educational and popular science material. He has written more than 20 books, including *Oceans* (Facts On File, 1999) and *1001 Questions and Answers About The Human Body* (Random House, 1994). He has also written for the *Economist* and *Geographical.*

Christine Doane is Senior Editor at Advisor Media, Inc., a producer of publications for information technology professionals. She was formerly Associate Reviews Editor for *Windows Magazine,* where she wrote and edited articles on technology industry news, trends, and emerging products.

Philip Downey is a Canadian freelance science writer. He has worked at Harvard Medical School and has written for *Sky and Telescope* and *Today's Science On File.* He obtained his M.S. in science journalism from Boston University in 1998.

Mike Flynn is a freelance writer and editor based in London. He was a staff writer at the Science Museum in London for five years and has contributed to several award-winning books and CD-ROMs.

Kevin Manley is a systems software engineer based in Seattle, Washington. He has worked for several Internet companies including Doubleclick and Qpass, and has written for industry trade magazines such as *C++ Users Journal* and *Windows Tech Journal.*

Christina Roache has investigated a range of scientific subjects while working for the Massachusetts Institute of Technology's *Technology Review,* the American Association for the Advancement of Science, and PBS, among other organizations. She holds a Master's Degree in Science Journalism from Boston University.

Tamara Schuyler is a writer and editor of educational science material. She contributed to the Scientists, Mathematicians and Inventors volume of *Lives and Legacies: An Encyclopedia of People Who Changed the World* (Oryx Press, 1999) and co-authored *History of the Internet* (ABC-CLIO, 1999). She has also written extensively for the science-news magazine *Today's Science On File.*

Giles Sparrow is a freelance science writer and editor, trained at the University of London. He has contributed to the *Encyclopedia of Science in Action* (Macmillan, 1995), the *Eyewitness Science 2.0 CD-ROM* (Dorling Kindersley, 1997), and has recently worked as an editor on the *Encyclopedia of Space and the Universe* (Dorling Kindersley, 1999).

Roy C. Weatherford is Professor of Philosophy at the University of South Florida. He holds a Ph.D. in philosophy from Harvard University, and is the author of *World Peace* and the *Human Family* (Routledge, 1993), among other books.

Chris Woodford is a freelance science writer based in Hampshire, England. He has contributed to the *Encyclopedia of Inventors and Inventions* (Grolier), a 10-volume *Encyclopedia of Technology and Applied Science* (Marshall Cavendish), and the *Eyewitness Encyclopedia of Science CD-ROM* (Dorling Kindersley). He received an M.A. in Physical Sciences from Cambridge University, U.K.

AIRPLANE FUEL TECHNOLOGY

Recently, the rising costs of traditional fossil fuels, which are variations of normal gasoline, have led to the development of several new types of aircraft fuels. Some of these are already being used in mass-production commercial aircraft, while the building of prototype aircraft and engines capable of handling others is already under way. These new jet fuels help to address some of the major drawbacks relating to traditional fuels: expense, noise, and pollution.

Scientific and Technological Description

There is nothing about the design of an airplane jet engine that requires a specific type of fuel. A turbojet uses spinning blades at the front to suck in and compress air to 10 times normal atmospheric pressure. This air is then used to burn fuel, and the hot gases produced by the combustion naturally expand, forcing their way out of the exhaust, pushing the plane forward, and at the same time spinning the turbine that powers the blades at the air intakes. Other engine variants, including turbofans and turboprops, are scarcely more complicated, while the ramjet, which relies on airspeed alone to pressurize the air inside the engine, is simply a specially shaped open tube.

In all cases, the simplicity of the basic principle means that a jet can use any fuel capable of burning violently in air.

The more efficiently the fuel burns, the more power an engine can produce. Airplane fuels must overcome another important limitation as well—the fuel must not only burn efficiently but must also be reasonably light. Otherwise, the mass of a heavier fuel will counteract any additional power it produces.

All forms of gasoline are rich in hydrocarbon molecules, made up of hydrogen and carbon atoms. When burned, these hydrocarbons react with the oxygen in air to produce water vapor and carbon dioxide and monoxide. Jet A is by far the most common type of fuel used today. It is a type of gasoline, separated from crude oil in the normal refining process. Traditional jet A fuel has many disadvantages, however. Most important for the aviation industry, it is expensive to manufacture. In an attempt to hold down costs, several alternative fuels have been developed, the most successful of which so far is *synthetic gasoline,* made from shale oil.

Before the discovery of large crude oil deposits in the mid-19th century, oil was extracted from a far wider range of sources than are commonly used today, including whale blubber, pitch, and most important, the oily rock shale. To help avoid the economic problems caused by fluctuations in the crude oil price, shale oil is today once again being explored as a possible fuel source. When the rocks are crushed, ground, and separated from their oil, a form of syn-

Two Hydrogen Plane Designs

The demonstrator hydrogen-fueled aircraft currently under construction, based on the Fairchild Dornier 328 aircraft, will use external fuel tanks that are bolted onto its wings.

Another potential hydrogen airliner, a modified Airbus A310 aircraft, would place the fuel tank on top of the aircraft's fuselage (main body).

thetic crude oil can be created, which can then be distilled in the same way as normal oil, producing synthetic gasoline.

Another alternative fuel source under serious consideration is methane (CH_4). Methane is normally a gas, found in natural deposits, or produced as a by-product of the distillation of crude oil. However, when cooled it can be stored in liquid form and burns with a much higher energy efficiency per kilogram of fuel than gasoline.

Although synthetic gasoline and methane offer advantages in terms of reduced costs and increased fuel efficiency, one fuel is superior in every way to those in use today—hydrogen. At present, hydrogen is used primarily as a fuel in rockets. It burns with the highest energy efficiency of any fuel, generating nearly three times the energy of jet A fuel per kilogram. In addition, it is almost pollution free—the main product of hydrogen combustion is simply water vapor (with a very small proportion of nitrogen oxides).

Liquid hydrogen can be used as a fuel supply for a jet engine with only a few modifications to the engine itself, principally the addition of a heat-exchanger unit to heat the hydrogen just before it reaches the combustion chamber, converting it from a liquid to a gas. However, the safe storage of liquid hydrogen raises its own problems, and the costs of manufacturing liquid hydrogen are at present very high in comparison to those associated with conventional fuels.

Historical Development

No one really considered alternatives to jet A fuel until the 1970s, when the sudden world fuel crisis sent crude oil prices spiraling upward. The rising cost of fossil fuels coincided with a growing realization of the harm they were doing to the environment and led to increased pressure to find "green" fuels, fuels that come from renewable energy sources, such as the sun and wind, and have fewer adverse effects on the environment. Looking to the short term, fuel companies began to search for replacements for or supplements to the diminishing crude oil reserve, and this gave rise to renewed interest in shale oil extraction for synthetic gasoline. However, the longer-term view of alternative fuels must recognize that synthetic oil is another limited reserve and does nothing to solve the pollution issue. Most alternatives, including methane, are still hydrocarbon-based, so produce carbon monoxide and dioxide when burned.

Hydrogen, on the other hand, has been used as a fuel source since the beginning of the space age in the 1950s, and is an extremely clean fuel. The most efficient rockets carry liquid hydrogen and liquid oxygen in separate tanks, combining them together and igniting them in the rocket engine to produce forward thrust. The idea of using hydrogen in aircraft arose from two different quarters, appearing at first as a response to the energy crisis and growing ecological awareness of the 1970s. Speculative scientists supported hydrogen-fueled aircraft as just one of many hydrogen-based technologies that could be introduced to solve growing pollution problems. Commercial aerospace companies were more concerned with the alternative they offered to the rising costs of traditional fuels.

At present, the most promising development for hydrogen-powered flight is a collaboration involving the German corporations Daimler-Benz and Dornier, Russian aerospace company Tupolev, and Canadian engine manufacturers Pratt and Whitney. Work is currently under way to produce a hydrogen-powered "technology demonstrator," based on the small Dornier 328 commuter jet. According to current plans, this "cryoplane" (so called because of the extremely low, or cryogenic, temperature of the fuel, approximately -253°C) will be flying by 2000.

Meanwhile, the U.S. National Aeronautics and Space Administration (NASA) has seen a hydrogen-fueled aircraft as the solution to one of its own long-term goals, a single-stage spaceplane that is capable of hypersonic flight (travel at more than five times the speed of sound) to the edge of space and able to travel around the world in as little as two hours. Announced in 1987 as the National AeroSpace Plane (NASP), the initial development project soon ran into difficulties because of the huge range of technologies it tried to combine. Since 1995, NASA's research program has focused more on engine design, under the code name Hyper-X.

Spaceplanes will require an entirely new type of hybrid engine, called a *scramjet*, or *supersonic combustion ramjet* (a ramjet through which air passes at supersonic speeds). A hybrid scramjet uses oxygen from the atmosphere to burn its hydrogen fuel while in the atmosphere, then switches to internal liquid oxygen tanks to generate thrust like a rocket in space. A prototype Hyper-X vehicle, the X-43, is now undergoing a series of tests at NASA's Dryden Flight Research Center.

Uses, Effects, and Limitations

Although the basic principle of the jet engine remains the same whatever fuel is used, the different characteristics of different fuels require that some modifications to the basic jet engine design must be made. In addition, the storage problems posed by methane or hydrogen, with their larger volumes, could have a significant effect on the outward appearance of future aircraft. Internally, the most significant change that has to be made to a jet engine to allow for hydrogen burning is the insertion of a preheater stage. If an attempt was made to ignite the hydrogen straight from tanks in its liquid form, the result would be highly explosive. The heat of the engine must be used to boil the hydrogen on its way to the engine, converting it into a less explosive gas that will still generate a large thrust when combusted.

Another challenge is to ensure that the fuel is mixed thoroughly with the air so that it burns evenly. Traditional jets use fuel injection systems to pump the liquid fuel into the engine as a fine spray, but a gaseous fuel is more difficult to handle. For the first prototype aircraft, Pratt and Whitney have introduced a revolutionary new "micromix" injection system, injecting gas through thousands of tiny pores in the engine walls to create a large number of separate "flamelets," ensuring a thorough mix.

Although hydrogen-burning engines require these modifications, the use of hydrogen has another important advantage when it comes to supersonic and hypersonic flight. Because the gas is so combustible, flames spread through it almost instantaneously. This is significant because at very high speeds, it would be possible for a slower-igniting fuel to be pushed out of the back of the engine before it had time to combust. Using hydrogen fuel will also affect the overall shape and design of an aircraft. Most important, the volume occupied by liquid hydrogen is much larger than that filled by the equivalent amount of jet A fuel, even though hydrogen is considerably lighter. This means that any hydrogen-fueled aircraft will have very large hydrogen fuel tanks, and these tanks must be heavily insulated to keep the fuel below its boiling point (-253°C). One way of doing this would be to build a tank running above the fuselage, giving the aircraft a "bulge" without affecting its aerodynamic stability. The demonstrator currently under construction is likely to use bolted-on external tanks to avoid designing a completely new aircraft. NASA's spaceplane project will eventually store fuel in a hydrogen "slush," a mixture of solid and liquid hydrogen that takes up much less volume (although it must be kept at even colder temperatures, about -260°C). In the near future, however, methane may prove more popular as a compromise in terms of weight and volume.

One other application of liquid hydrogen fuel is as a coolant. All jet aircraft circulate fuel around their engines to absorb and carry away the heat they produce, and the very low temperatures of liquid hydrogen make it an ideal coolant, especially as the fuel needs to be preheated to a gas before use. A hypersonic hydrogen-powered aircraft would need cooling on all its surfaces to dissipate the heat built up from friction with the air (even current supersonic aircraft stretch considerably through heating while in flight). A fully developed hydrogen aircraft would probably incorporate a complex cooling system circulating fuel across all the surfaces most prone to heating.

Aviation is just one of many industries currently investigating hydrogen-based technology. Other applications include large-scale power generation, used by the automobile industry, and small-scale fuel cells that could generate electricity for domestic or business use at low costs (see ALTERNATIVE AUTOMOTIVE FUEL TECHNOLOGIES).

Issues and Debate

Although the by-products of hydrogen fuel are environmentally friendly, that is only one half of the equation. The other half is that hydrogen does not exist in its free form on Earth, so must be manufactured in some way. The method used for manufacture must also be environmentally sound if hydrogen fuel is to have a net positive effect on the environment. Perhaps the simplest method of hydrogen manufacture is by the electrolysis of water, splitting water molecules apart by passing an electric current through them. To avoid harm to the environment, the electricity used for hydrogen manufacture must come from a clean renewable resource. One other advantage of electrolysis is that it neatly balances the effects of hydrogen burning itself—water produced and oxygen absorbed during hydrogen combustion are countered by water used up and oxygen produced during electrolysis.

Ultimately, though, the adoption of hydrogen or other improved jet fuels will be spurred not by environmental but by commercial pressures. As with many other cutting-edge technologies, the speed of industrywide acceptance is the key to reducing costs. If the aircraft industry adopts hydrogen power with the same speed and enthusiasm with which it adopted the jet engine, the initial high prices of hypersonic flight in the new aircraft will fall rapidly. If one aircraft manufacturer or airline offers its customers the advantages of hydrogen-fueled flight, others will be forced to follow suit. Another way to encourage this is by levying taxes on environmentally harmful industries such as the gasoline industry. In the end, change will come about only by encouraging multinational companies to look to the world's long-term energy needs and seek alternatives to their current practices.
—*Giles Sparrow*

RELATED TOPICS

Alternative Automotive Fuel Technologies, Hydroelectric Power, Reusable Launch Vehicles, Solar Energy, Wind Energy

BIBLIOGRAPHY AND FURTHER RESEARCH

BOOKS

Brewer, G. Daniel. *Hydrogen Aircraft Technology*. Boca Raton, Fla.: CRC Press, 1991.

Cannon, James S., and Sharene L. Azimi, eds. *Harnessing Hydrogen: The Key to Sustainable Transportation*. New York: Inform, 1995.

Peschka, Walter, Edmund Wilhelm, and Ulrike Wilhelm. *Liquid Hydrogen: Fuel of the Future*. New York: Springer-Verlag, 1992.

Pohl, H.W. *Hydrogen and Other Alternative Fuels for Air and Ground Transportation*. New York: Wiley, 1996.

INTERNET RESOURCES

Hyper-X Hypersonic Experimental Vehicle
http://www.dfrc.nasa.gov/Projects/HyperX/index.html

National Hydrogen Association
http://www.ttcorp.com/nha/

U.S. Department of Energy Alternative Fuels Data Center
http://www.afdc.nrel.gov/

ALTERNATIVE AUTOMOTIVE FUEL TECHNOLOGIES

As sources of nonrenewable fossil fuels, such as oil and coal, become increasingly scarce, the automotive and energy industries have been forced to look elsewhere for the energy required

to power vehicles. Efforts to produce alternative automotive fuel technologies have concentrated mostly on ways of generating electrical power from the energy released by the chemical reaction of two or more substances. Notable exceptions, such as solar power, which relies on converting heat energy from the Sun into electricity, have also been considered but have been largely abandoned on the grounds that they are impractical.

The ultimate aim of many of those involved in developing alternative automotive fuel technologies is to produce a vehicle that can match the performance, cost, and convenience of a conventional gasoline-powered family car. Although it may be many years before such a vehicle is developed, ever-decreasing stocks of fossil fuels and political pressure to reduce toxic emissions from car engines are forcing automobile manufacturers to look seriously at workable alternative automotive fuel technologies.

Scientific and Technological Description

The most common power source for motor vehicles today is the internal combustion engine, in which gasoline is burned in a series of cylinders in order to produce large amounts of heat energy, which is then used to drive a set of pistons. The movement of the pistons rotates a crankshaft, one end of which is fitted with a heavy flywheel. Energy from the flywheel is then sent to the wheels of the vehicle via the transmission system.

There have been a number of attempts to produce non-gasoline engines that can run on fuels such as compressed natural gas (CNG) and dimethyl ether (DME), an alcohol-based fuel. Although these fuels have been able to supply the energy needed to drive a vehicle, the technology involved is essentially the same as that for ordinary engines. Moreover, bi-fuel vehicles, cars that can run on either CNG or diesel fuel, are already on the road. There have been several high-profile experiments with exotic-looking solar-powered cars, but solar power has largely been abandoned as an alternative means of powering motor vehicles. This is due to the fact that even the most efficient solar power cells cannot convert sufficient energy from the Sun's rays to drive anything more than the flimsiest of vehicles.

Electric Cars

A more practical option is the battery-powered electric car. Known as an EV (electric vehicle), the battery-powered car is fitted with an electric motor rather than a gasoline-burning internal combustion engine. Unlike a conventional car, where the energy needed to drive the vehicle is stored in the gas tank, the EV carries large, rechargeable battery

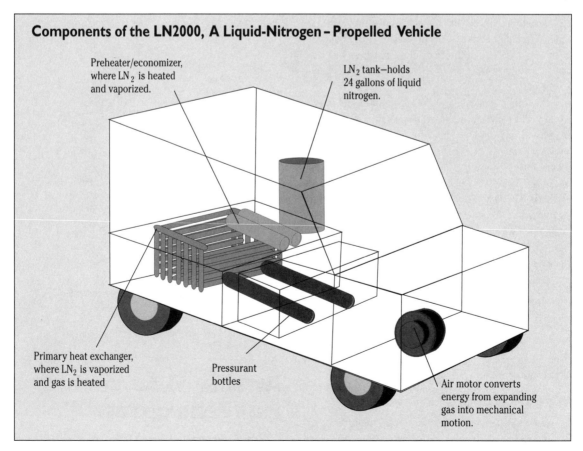

Components of the LN2000, A Liquid-Nitrogen–Propelled Vehicle

Preheater/economizer, where LN$_2$ is heated and vaporized.

LN$_2$ tank–holds 24 gallons of liquid nitrogen.

Primary heat exchanger, where LN$_2$ is vaporized and gas is heated

Pressurant bottles

Air motor converts energy from expanding gas into mechanical motion.

The LN2000, a converted mail delivery van, was developed by researchers at the University of Washington to run on liquid nitrogen fuel. (Courtesy of Aerospace and Energetics Research Program, University of Washington.)

packs, commonly placed in the trunk or built into the floor of the vehicle. The flow of electricity from the battery packs is regulated by the motor controller, which acts as the "brains" of the system. From the controller, the electricity passes into the motor, where electric power is converted into rotational mechanical power. This is then sent via an axle to drive the wheels of the vehicle. When the battery pack has been exhausted, it can be recharged with electricity from standard utility power lines.

Fuel Cells

An alternative to the rechargeable battery pack is offered by the hydrogen fuel cell. Unlike the battery pack, the hydrogen fuel cell does not need to be recharged from standard utility power lines. Power is harnessed from the energy produced when hydrogen and oxygen combine, a reaction that will continue to provide power for as long as there is an available supply of both elements. In a conventional battery, the active battery ingredients, such as nickel and cadmium, are incorporated into the battery's electrodes. Any chemical reaction will necessarily reduce the quantity of these active ingredients. This chemical alteration will eventually result in the ingredients becoming depleted to the point where no further useful power can be generated from them.

By contrast, the hydrogen in the fuel cell is supplied continuously to the anode from an external source while oxygen (usually from the surrounding air) is supplied to the cathode. The hydrogen at the anode dissociates (splits) into free electrons and positive hydrogen ions. The free electrons, now in the form of usable electric current, are conducted away through an external circuit before returning at the cathode, where they combine with the oxygen and the positive hydrogen ions to produce pure water and heat. The electricity produced at the anode is used to drive the vehicle. Water and heat are the harmless waste products of the reaction.

Because individual hydrogen fuel cells produce only about 0.6 volt of power, they must be stacked together to produce enough electrical power to drive a vehicle. Hydrogen is not the only fuel that can be used in such a cell. Methyl alcohol, hydrazine, and various simple hydrocarbons have also been used.

Liquid Nitrogen

A complete alternative to the fuel cell has also been developed—liquid nitrogen. Nitrogen has an extremely low boiling point. At temperatures above -195.8°C, it becomes a gas. As a result of this, when liquid nitrogen is stored in a pressure-controlled environment, such as a pressurized fuel tank, it has the potential to expand rapidly, releasing large amounts of energy in the process. This property has been exploited in order to drive a vehicle in much the same way as the steam from boiling water was used to drive a steam train. The rapidly expanding nitrogen gas pushes on pistons, which rotate a crankshaft (see figure).

To power a vehicle, liquid nitrogen is released from an insulated, pressurized storage tank and heated by the surrounding air. (On even the coldest days, liquid nitrogen will boil on exposure to surrounding temperatures.) This causes the gas to expand rapidly, releasing energy that is converted into mechanical motion by means of an "air" engine, which operates in much the same way as a steam engine. The only product of the rapid expansion of liquid nitrogen is nitrogen gas, the main element in air.

Historical Development

British scientist William R. Grove is credited with having created the world's first hydrogen fuel cell during the late 1830s. Then, as now, electric current was produced from the reaction of hydrogen and oxygen at the electrodes. Throughout the rest of the 19th century, further work was carried out in an effort to produce ever-larger and more stable amounts of electricity in this way, but none could compete with electric generators driven by steam or water power. Technology had improved, however, by the middle of the 20th century, and fuel cells capable of power generation on a large scale became a reality, driven in large part by the demands of various military and, later, space programs. Fuel cells were used during NASA's Apollo space program and are still in use on the various space shuttles.

As the 1960s came to an end there was a developing awareness of the fact that fossil fuels would eventually run out. In 1973, an embargo imposed by OPEC, the Organization of Petroleum Exporting Countries, heightened these concerns and led to an "energy crisis." That crisis fostered a search for energy alternatives that would last throughout the decade. This search combined with growing concerns about the amount of pollution being released into the atmosphere to spur the drive toward cleaner, more sustainable automotive fuel technologies.

During the 1970s and 1980s, many motor manufacturers began to invest in research into alternatives to the gasoline-powered internal combustion engine. This research has

General Motors' EV1, the first mass-produced electric car, went on sale in parts of California and Arizona in December 1996 (Photo Researchers/Will & Deni McIntyre).

produced a number of interesting vehicles, most of which are powered by electricity drawn from a conventional battery pack. At the same time, several university-based groups, such as groups at the Massachusetts Institute of Technology, University of Chicago, University of Washington, and University of Manchester Institute of Science and Technology (in the United Kingdom), began examining alternative ways of propelling vehicles, based largely on hydrogen fuel cell technology or on harnessing the energy of rapidly expanding liquid nitrogen. Many of the world's major automobile manufacturers are also committed to the search for alternatives to existing automotive fuel technologies and are actively attempting to produce feasible alternatives to existing technologies.

Uses, Effects, and Limitations

At present the most common alternative to the gasoline-powered car is the EV, or electric vehicle. Those produced by the major automobile manufacturers tend, at first glance, to resemble existing gasoline-powered models. On closer inspection, however, the EV is all but silent in operation and produces no toxic emissions. According to advocates of the EV, operating costs for the vehicle are approximately one-third those of gasoline-powered cars, depending on the time of the day the vehicle's batteries are recharged. Electricity tends to be cheaper during off-peak hours, typically 1:00 A.M. to 8:00 A.M. An additional advantage of the EV over conventional vehicles is the fact that the usual maintenance costs, such as oil changes, tune-ups, and exhaust replacements, are never incurred.

There are, however, a number of disadvantages to owning an EV. Most currently available EVs have an operating range of less than 100 miles on a single charge and a top speed of about 40 miles per hour. Even using special equipment to boost the power supplied by the standard utility power lines, it can still take up to eight hours to recharge the vehicle's battery pack. Clearly, this does not compare favorably with the time required to fill a tank with gasoline. Although faster methods are being developed, it is unlikely that the recharge time for EVs will ever match the refueling time of gasoline-powered cars.

Although researchers are working on increasing the top speeds of EVs, they are also trying to change motorists' perceptions of what they need from a car. The average speed of a car traveling in peak-time traffic in London, for example, is 11 miles per hour. A car designed for use in the city may not need to travel any faster than 40 mph. Hydrogen fuel cells, however, will run continuously, provided that the necessary hydrogen and oxygen are supplied. Unfortunately, the performance of vehicles powered in this way is often even worse than that of the standard EV. A very notable exception, however, is the NECAR 4 (New Electric Car) from DaimlerChrysler. Based on the existing Mercedes-Benz A-class compact car, the NECAR has achieved a top speed of

68 miles per hour and is able to travel over 250 miles before refueling, which can be done in as little as 15 minutes.

Liquid nitrogen–powered vehicles are still very much at the early stages of development and at present are not a serious alternative to the EV, being able to achieve speeds of little more than 20 miles per hour on an average fuel consumption of 5 gallons of liquid nitrogen per mile. Given time, however, the top speeds and efficiency of liquid nitrogen-powered vehicles are expected to improve.

Issues and Debate

Since the beginning of the industrial age, the human race has come to rely heavily on the energy released by burning nonrenewable fossil fuels. The automobile is a major consumer of fossil fuels and other natural resources. Any technologically advanced society will necessarily consume large quantities of oil, natural gas, and coal to maintain current standards of living. The industrial processes required to turn raw materials into cars produce almost as much pollution as do the resulting vehicles when in use on our roads. At present, motor vehicle manufacturing accounts for half the rubber, one-sixth of the aluminum, and at least one-seventh of the steel consumed in the United States every year.

To address this problem, national and local governments have introduced laws to control and ultimately reduce the toxic emissions from automobiles. The U.S. Environmental Protection Agency (EPA), taking its lead from the state of California (which has some of the most stringent emission control laws in the world), is currently proposing auto emission regulations for the period 2004–2010 that will promote the use of zero-emission vehicles. At the global level, in December 1997, representatives from more than 150 nations signed the Kyoto Protocol, a United Nations agreement to limit each nation's emissions of greenhouse gases, such as carbon dioxide, methane, and nitrous oxide. The signatory nations will have to find ways to lower their greenhouse gas emissions, which may include initiatives to develop alternative fuel technologies.

Even zero-emission vehicles, however, are not always as environmentally friendly as they might at first appear. Because the battery-powered EV must be recharged from standard utility power lines, it inevitably uses power that has been generated using polluting, nonrenewable fossil fuels. In addition, critics point out, advocates of the battery-powered EV often fail to take into account the fact that any significant increase in the demand for these vehicles will inevitably lead to an increase in the production of lead. (The batteries for a single conventional electric car contain around 1000 pounds of the noxious heavy metal.)

All the debate about the benefits or downsides of alternative automotive fuel technologies will come to nothing, however, if the alternative vehicles they are designed to power do not come into common use. Unfortunately, until such time as zero-emission vehicles can be shown to be comparable or superior in every way to conventional cars, it is unlikely that they will appeal to the average motorist.

—*Mike Flynn*

RELATED TOPICS
Airplane Fuel Technology, Solar Energy

BIBLIOGRAPHY AND FURTHER RESEARCH

BOOKS

Appleby, A.J., and F.R. Foulkes. *Fuel Cell Handbook*. New York: Van Nostrand Reinhold, 1989. Republished: Krieger Publishing Company, Melbourne, Fla., 1993.

Blomen, L.J., and M.N. Mugerwa, eds. *Fuel Cell Systems*. New York: Plenum Press, 1992.

Kordesch, Karl, and Gunter Simander. *Fuel Cells and Their Applications*. New York: VCH Publishers, 1996.

PERIODICALS

De Cicco, John, and Marc Ross. "Improving Automotive Efficiency." *Scientific American*, December 1994, 30–35.

Kaiser, Jocelyn. "Getting a Handle on Air Pollution's Tiny Killers." *Science*, April 4, 1997, 33.

Marshall, Eliot. "Slower Road for Clean-Car Program." *Science*, April 11, 1997, 194.

Nemecek, Sasha. "Bettering Batteries." *Scientific American*, November 1994, 86.

Sperling, Daniel. "The Case for Electric Vehicles." *Scientific American*, November 1996, 36–41.

Tickell, Oliver. "Clean Cars Will Power the Country." *New Scientist*, October 12, 1996, 42.

INTERNET RESOURCES

Alternatives to Gasoline-powered Engines
    http://www.washington.edu/alumni/columns/dec97/car2.html
DaimlerChrysler Presents First Drivable Fuel Cell Technology Car in the United States
    http://www.daimlerchrysler.de/index_e.htm?/news/top/t90317_e.htm
Electric Vehicle Association of the Americas' Home Page
    http://www.evaa.org/
Electric Vehicle Web
    http://www.mcclellan.af.mil/EM/EV/index.html
Information on Fuel Cells
    http://www.pipeline.com/~bkyaffe/altfuel/section/bibliodocs/h2faq.html
"The Ice Car Cometh" (*New Scientist* article on liquid–nitrogen fueled cars)
    http://www.newscientist.com/ns/970816/nsmog.html
University of Washington's Liquid Nitrogen Propelled Automobile
    http://www.aa.washington.edu:80/AERP/CRYOCAR/CryoCar.htm
U.S. Department of Energy's Office of Transportation Technologies
    http://www.ott.doe.gov/

ARTIFICIAL INTELLIGENCE

Artificial intelligence (AI) is a branch of computer science concerned with developing machines that can think, learn, and solve problems using approaches similar to those used by intelligent human beings.

The goals of AI research range from the pragmatic (producing computer-based systems that can solve one or more limited tasks using humanlike thinking and reasoning) to the ambitious (mimicking the human brain's general aptitude for a variety of tasks or developing an ultraintelligent machine [UIM]). AI has achieved considerable success in some fields, notably the development of expert systems (vast information databases trained to answer questions on a particular area of human knowledge). But the achievement of its wider goal, the development of UIMs that rival human intelligence, seems as far off today as when research began in the 1950s.

Scientific and Technological Description

AI research centers on methods of storing knowledge inside computer systems. Psychologists have long recognized the distinction between two types of human knowledge: procedural ("know how") and declarative ("know that"). *Procedural knowledge* is often a sequence of steps that have to be carried out to achieve a particular task, such as how to cross a road; *declarative knowledge* is a collection of facts and opinions, such as good places to cross particular roads. In computing, a program represents procedural knowledge and its data (information to be processed) represent declarative knowledge. AI researchers also make a distinction between procedural and declarative knowledge. For example, a chess-playing computer program needs a combination of procedural knowledge (how to move various pieces) and declarative knowledge (what different board configurations mean).

There are two principal approaches to representing and processing information in an AI system. One approach, known as *cognitivism*, *information processing*, or the *physical symbol system*, suggests that an artificial brain could work by manipulating symbols according to rules in the same way that an ordinary computer manipulates data according to a program. Using this approach, researchers do not worry about how symbols are stored or manipulated at the lowest level of an AI system, just as computer programmers do not worry about how electrons moving in transistors store their data; all that matters are the physical symbols (the objects or concepts to be stored or manipulated) and the operators (the rules used to manipulate the symbols) at the higher level. One example of representing knowledge using this approach is the *semantic network*, a complex map that describes relationships between objects or concepts and even the relationships between those relationships (see the figure). A semantic network is characterized by *nodes*, which represent different objects or concepts, and *links*, connecting the nodes, which indicate the relationships between the concepts.

Just as AI and computer science have led to other ways of understanding the human brain, so brain physiology has provided new inspiration for AI researchers and a second major approach to the problem of representing and processing information. This is known as *connectionism*, *parallel distributed processing* (PDP), or *neural networks* (see figure). In a neural network, different pieces of information can be processed in parallel by a large number of highly intercon-

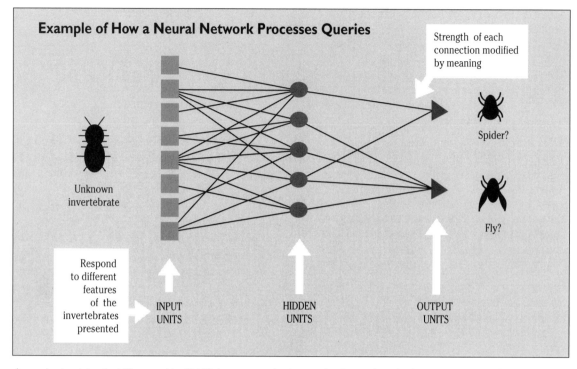

Example of How a Neural Network Processes Queries

Strength of each connection modified by meaning

Spider?

Fly?

Unknown invertebrate

Respond to different features of the invertebrates presented

INPUT UNITS

HIDDEN UNITS

OUTPUT UNITS

A neural network is a brainlike assembly of highly interconnected units or nodes that can be trained to recognize things if it is repeatedly presented with different examples. A neural network trained by examining hundreds or thousands of different spiders and flies builds up an idea of what constitutes a spider or a fly by adjusting the strengths of the connections between different units. Eventually, it can tell whether an unknown invertebrate is either a spider or a fly.

nected units that work simultaneously. They include *input units*, which recognize various features of the input, for example, information about an invertebrate that would help to determine whether it was a spider or a fly; *hidden units*, linked by connections of varying strength, which represent the knowledge in the network; and *output units*, which communicate the results—in this example, whether the unknown invertebrate was found to be a spider or a fly. These units are similar to the *neurons* (cells) in the human brain. Knowledge in a neural network does not exist in a single place, such as a ball in a box; instead, it is distributed throughout the network and its interconnections in the same way that memories are distributed throughout the network of neurons in the brain. Indeed, a neural network shares many of the properties of the human brain: It can learn, make mistakes, and suffer amnesia after lesions (partial brain damage).

Intelligent behavior involves not just storing information, but also knowing how to process it; in other words, it requires procedural as well as declarative knowledge. This includes understanding how a large problem or goal can be broken up into more manageable subproblems or subgoals, and how to search or narrow down the available knowledge to solve problems more quickly. There are two kinds of procedural knowledge: *algorithms*, fixed procedures to be followed to achieve predictable results, and *heuristics*, trial-and-error approaches to problem solving that are not guaranteed to produce results. An intelligent human being or an

AI system uses both algorithms and heuristics; a traditional (unintelligent) computer uses only algorithms. Procedural knowledge is often stored as a collection of IF . . . THEN . . . rules to be applied in a particular situation; this is sometimes known as a *rule-based system*.

Historical Development

The history of AI is usually traced to the publication of a scientific article in 1950 by British mathematician Alan Turing. Entitled "Computing Machinery and Intelligence," the article considered whether a machine could think and reformulated that question in a much more useful way: Can a machine be constructed that behaves in a way that is indistinguishable from the behavior of an intelligent human being? Thus was born the *Turing test of artificial intelligence*. In that test, an experimenter sits at a computer terminal connected to computers in two sealed rooms. In one room, a person sits at the terminal; in the other room, there is an artificially intelligent computer behaving like a person. The experimenter's job is to tell which room contains the human being and which contains the computer by asking questions directed at the occupant of each room. If the experimenter fails to tell the difference, reasoned Turing, the computer has demonstrated intelligence.

By 1954, the world's first artificial intelligence research group had been founded at Carnegie-Mellon University by Herbert Simon and Allen Newell. Two years later, a conference at Dartmouth University brought together electrical engineers, mathematicians, psychologists, and others interested in machine intelligence. That conference led John McCarthy, then of the Massachusetts Institute of Technology (MIT), to coin the term artificial intelligence.

Until the 1960s, AI research developed along two parallel lines. On the one hand, there was information processing, which involved manipulating symbols with rules, an approach advocated by Newell and Simon. On the other hand, there was an attempt to store and manipulate information using large networks of artificial neurons (similar to today's neural networks), which had been inspired by the work of neurophysiologist Walter McCulloch, logician Walter Pitt, and psychologist Donald Hebb in the 1940s. But in 1969, Marvin Minsky and Seymour Papert, two AI researchers at MIT, wrote a highly influential book, *Perceptrons*, effectively rejecting the idea that neural networks could solve simple logical problems. As a result, the information processing approach to AI triumphed as the major research strand until a resurgence of interest in neural networks around the mid-1980s.

Today, researchers such as Minsky are exploring the need to combine neural networks and information processing approaches to problem solving. They are also looking at the importance of AI systems acquiring generalized knowledge of the world akin to human "common sense." This means working out better ways for AI systems to interact with the human world; humans, after all, experience the world instead of having it preloaded at birth like a computer's operating system. At MIT, researchers are developing a robot known as COG that has sight, hearing, and touch, based on the assumption that developing human intelligence requires humanlike interactions with the world.

Uses, Effects, and Limitations

Expert systems (sometimes known as *knowledge-based systems*) have been the most successful applications of AI so far. They are effectively elaborate computer representations of knowledge extracted from a human expert arranged in a form accessible to later questioning. One of the first expert systems, MYCIN, was trained to diagnose blood diseases and suggest appropriate treatments using knowledge gained from physicians. Although it is simply a demonstration of AI, MYCIN is highly regarded and has been shown to reach the same diagnosis as that of a human doctor in 75 percent of cases. Other successful expert systems include DENDRAL, which can work out the structure of organic compounds from mass spectrometer results (a spectrometer measures the spectrum of energies emitted by a source), and PROSPECTOR, which uses geological data to predict the location of mineral ores. Expert systems have also been devised for use in economic forecasting and management.

Game playing is another area in which AI has been conspicuously successful. In the 1950s, AI researchers were confident that they would soon be able to develop machines advanced enough to compete with humans. By 1967, Arthur Samuel had devised a successful checkers program. But it was another 30 years before the IBM chess-playing computer Deep Blue beat grand master Gary Kasparov in a tournament in New York City.

Successful as they are, these systems are far from what we would describe as intelligent machines. Their main limitation is that they are specialized in only one task: The medical MYCIN program cannot beat Gary Kasparov at chess, and Deep Blue cannot diagnose infections. However good they are at their chosen task, they have no way of going beyond it; they lack what Marvin Minsky describes as encyclopedic knowledge of the world as a whole, and however good their specialized reasoning, they lack common sense. Furthermore, they have no understanding or insight into the decisions and conclusions they reach. MYCIN cannot explain why it reached a particular diagnosis; it can follow its thousands of IF . . . THEN . . . rules only according to information supplied to it.

Although AI has been successful in expert systems and game playing, progress has been slower in replicating other areas of human intelligence. Handwriting and speech recognition are now commonplace in palm-top computers, but computer understanding of natural (human) language is a more distant goal (see LANGUAGE RECOGNITION SOFTWARE). Machine vision (crude "eyesight" that allows computers to interpret images) also developed more slowly than other areas of AI until the renewal of interest in neural networks in the 1980s provided a fruitful new avenue for research.

Issues and Debate

Philosophers have wrestled with the problems of artificial intelligence ever since Alan Turing envisioned computing machines that could think. AI raises profound issues about exactly what constitutes intelligence and whether thinking machines can understand what they are thinking about. Some, such as U.S. philosopher John Searle, claim that just

Murray Campbell (left) and Joel Benjamin, two leaders of the IBM team that programmed Deep Blue for its chess match with Gary Kasparov, discuss strategy (Courtesy of IBM).

because computers can manipulate symbols in humanlike ways, it does not follow that they understand the meaning of the symbols or are displaying anything like human intelligence. A variation on this view is that digital computers (based on manipulation of the binary numbers 0 and 1) are so different from analog human brains (based on the transmission of chemical and electrical signals in any number of different ways) that they could never replicate them. In other words, AI systems lack human senses and bodies and do not interact with humans or other machines in a way that would allow them to be creative or cultural. Other researchers, such as Minsky, suggest that there is no reason why AI systems could not develop general knowledge, common sense, or culture, given time and exposure to a wide-enough range of humans or other AI systems.

But these issues are, if anything, the simplest part of the philosophical debate surrounding AI. Some experts, including Joseph Weizenbaum and Christopher Evans, have shown that artificial intelligence raises profound ethical, legal, and religious questions, whose starting point is the assumption that humans might eventually create an ultraintelligent machine. Would it be immoral to create a machine more intelligent than a human being? Now that there are expert systems capable of making medical diagnoses and recommending treatments, what happens when such a system advocates withdrawing treatment to a terminally ill patient or switching off a life-support machine? What happens when expert systems are used to analyze military plans and recommend bombing campaigns that result in the loss of human life? In other words, what happens when artificially intelligent machines have power over human lives? Conversely, if machines are granted the responsibilities of humans, should they also be granted human rights? Should a person who turns off the power to an artificially intelligent machine be charged with murder?

Some AI researchers dismiss these qualms as the kind of "technophobia" that seizes people whenever a new technology comes along. They suggest that AI could be harnessed in a limited, pragmatic way to improve the quality of human life, for example through better medical diagnosis or computer systems that interact more intelligently with humans. Others point out that technologies always come with responsibilities, and cite the case of the Hiroshima bombing at the end of World War II and the explosion at the Chernobyl nuclear plant in Ukraine in 1986 to show that scientists cannot always predict the consequences of the things they create.

—*Chris Woodford*

RELATED TOPICS
Artificial Life, Intelligent Agents, Parallel Computing, Robotics

BIBLIOGRAPHY AND FURTHER RESEARCH

BOOKS
Boden, Margaret. *Artificial Intelligence and Natural Man*. London: MIT Press, 1987.

Boden, Margaret, ed. *The Philosophy of Artificial Intelligence*. New York: Oxford University Press, 1990.

Cantoni, Virginio, Stefano Levaldi, and Vito Roberto, eds. *Artificial Vision: Image Description, Recognition, and Communication*. San Diego, Calif.: Academic Press, 1997.

Dennett, Daniel. *Consciousness Explained*. Boston: Little, Brown, 1991.

Hofstadter, Douglas. *Gödel, Escher, Bach: An Eternal Golden Braid*. New York: Basic Books, 1979.

McLelland, James, and David Rumelhart. *Parallel Distributed Processing*. Vol. 1. Cambridge, Mass.: MIT Press, 1986.

Minsky, Marvin. *The Society of Mind*. New York: Simon and Schuster, 1985.

Newborn, Monty. *Kasparov Versus Deep Blue: Computer Chess Comes of Age*. New York: Springer-Verlag, 1997.

Penrose, Roger. *The Emperor's New Mind*. Oxford: Oxford University Press, 1989.

Winston, Patrick. *Artificial Intelligence*. Reading, Mass.: Addison-Wesley, 1993.

PERIODICALS
Calvin, William. "The Emergence of Intelligence." *Scientific American*, October 1994, 100.

Cipra, Barry. "Artificial Intelligence: Will a Computer Checkmate a Chess Champion at Last?" *Science*, February 2, 1996, 599.

Milner, Peter. "The Mind and Donald O. Hebb." *Scientific American*, January 1993, 124.

Minsky, Marvin. "Will Robots Inherit the Earth?" *Scientific American*, October 1994, 108.

Pinker, Steven. "Out of the Minds of Babes." *Science*, January 1, 1999, 40.

"Profile: Humans Unite! Ben Schneiderman Wants to Make Computers into More Effective Tools—by Banishing Talk About Human Intelligence" (editorial). *Scientific American*, March 1999, 21.

Thomson, B., and W. Thomson. "Inside an Expert System." *BYTE*, April 1985, 315.

INTERNET RESOURCES
The Alan Turing Home Page
    http://www.turing.org.uk/turing
Carnegie-Mellon University AI Repository
    http://www.cs.cmu.edu/Groups/AI/html/repository.html
Marvin Minsky's Home Page
    http://www.ai.mit.edu/people/minsky/minsky.html
MIT AI Lab
    http://www.ai.mit.edu
An Online Timeline of Artificial Intelligence Milestones
    ftp.cs.cmu.edu/user/ai/pubs/faqs/ai/timeline.txt

ARTIFICIAL LIFE

Artificial life (A-life) research utilizes computer modeling and robotics to create new, lifelike forms that behave in ways similar to the actions of living organisms. With computers, scientists provide simulated environments in which digital life forms live, reproduce, and evolve. These forms can eventually show behavior that is adaptive to their environment. The emergent A-life populations display actions that resemble those of real-life populations, allowing scientists a better understanding of evolution, the origins of life, and the nature of learning and intelligence.

Researchers are also extending this technology into the field of ROBOTICS (see NANOTECHNOLOGY). A-life researchers believe that in the near future, a humanmade creature can be designed that will search for its own energy source, reproduce, and adapt in the real world, thus displaying the characteristics of a living organism. Some have raised concerns about the potentially disastrous implications of a self-replicating synthetic creature that cannot be unplugged.

Scientific and Technological Description

Artificial life refers to creating behaviors in computer programs and robots that mimic biology. A-life research is similar to, yet distinct from, the related field of ARTIFICIAL INTELLIGENCE, which utilizes a top-down approach to problem solving. With artificial intelligence, a particular task is clearly defined by the programmer, and the structure of the program is geared toward solving this specific problem. The extent of the program's knowledge is generally predetermined. For example, computer programs have been designed that skillfully play chess, with the ability to defeat any champion in the world, but the programs never learn to play checkers.

However, A-life programming involves a bottom-up approach. The researcher starts with a few simple, unconstrained rules that will govern an organism's behavior and

In 1988, after observing the flocking behavior of real birds, computer animator Craig Reynolds set animated birds, called "boids," in flight. He then set pillars in the boids' path to see how they would react. His program had not dictated the appropriate behavior for passing an obstacle, but his simulated birds divided briefly into separate flocks that rejoined after passing the obstacle. (Courtesy of Craig W. Reynolds, http://www.red.com/cwr/boids.html.)

allow the artificial organism's development and learning to emerge on its own. Emergent, unprogrammed behavior or phenomena are at the core of A-life research. A typical A-life computer program may start with a small number of instructions represented on the screen as insects. The insects will be programmed to find simulated food and to reproduce with other insects. At first, the insects will have a difficult time moving in the simulated environment, which may contain complicated trails and various obstacles. However, as the fittest creatures survive and reproduce (with some adaptations, which are not predetermined by the programmer and are thus considered mutations), a more-fit insect population emerges that is better able to navigate, find food, and reproduce. At times, the computer organisms display cooperative behaviors, helping one another survive. Some programmers have even witnessed the emergence of parasitic populations that feed off the host population's energy. After thousands of generations have passed, these insects are more capable of completing the specified task than is any organism created by the programmers. With these programs scientists have found that adaptations that occur over a span of 1000 years in nature may take place in 10 minutes in the computer lab.

Historical Development

Over the course of human history, philosophers, theologians, and scientists have attempted to define the parameters of life. These criteria have included, but have not been limited to, responsiveness, autonomy, evolution, a carbon-based structure, metabolism, and self-organization. However, one can easily find exceptions to most of these criteria (e.g., a mule is alive but cannot reproduce; parasites depend on a host to live). Therefore, some scientists have argued that artificial life—organisms that, although not carbon-structured, can seek and consume their own energy source, reproduce, and evolve—could be considered alive.

A-life originated in the late 1940s and early 1950s as the brainchild of John von Neumann, a Hungarian-born researcher who had one of the most fertile scientific minds of the 20th century. As well as working on the Manhattan Project in the construction of the first nuclear bomb during World War II, von Neumann was a pioneer in the early days of the computer age. He was first to develop the idea of a cellular automaton, a theoretical organism governed by the rules of logic. In von Neumann's theory, the hypothetical organism existed on a horizonless, infinitely extending checkerboard and covered about 200,000 squares, or cells. Each cell was in one of 29 states, and the combination of these cells in their respective states dictated the behavior of the organism through a set of logic-based rules. As each cell followed the rules, a large-scale change occurred: The organism self-replicated. The new organism then contained the same instructions for its own replication. In essence, von Neumann believed that many complex actions seen in nature were governed by less complex, logical rules. Von Neumann described his theory in a 1955 article in *Scientific American*. After his death in 1957, others continued mathematical work on cellular automata, but no large A-life advances were made until the advent of high-performance computing in the 1970s and 1980s.

The next big leap in A-life research came in 1979. Christopher Langton, now the editor of the journal *Artificial Life*, applied a variation of von Neumann's principles using an Apple computer, one of the first mass-market computers to be powerful enough to handle more advanced programming. First, he designed a single loop, a square with a short tail on one end, resembling the letter Q. Langton then wrote a set of rules that caused the Q to replicate: The tail thrusted outward and turned repeatedly until a new square with a tail formed. This daughter cell then contained instructions for its own replication. After many successive reproductions, some cells would be surrounded and could no longer reproduce. The process continued to form additional outer cells, and Langton noted that a complex pattern was forming, one that resembled the structure of a coral reef. Although Langton had not created something that was living, he showed that the forces of biology could be mimicked by rule-based machines. Other researchers have demonstrated that simple algorithmic formulas, called *L-systems*, when interpreted graphically, can produce complex forms such as ferns and trees. They found that certain rules controlled the plants' overall growth and others governed branching behavior.

As computers increased in power over the next decade, A-life research took important strides forward. The 1980s saw numerous investigators experimenting with computer programming and officially forming what is now the A-life field. Various researchers developed computer programs that simulate certain populations of creatures, typically insects. With these programs, researchers could develop "life" on the computer and, for the first time, manipulate its parameters. In addition, the 1980s witnessed the advent of COMPUTER VIRUSES. These are programs that self-replicate, spreading themselves through computer environments worldwide. Some observers consider computer viruses to be a form of A-life.

In 1988, computer animator Craig Reynolds made another important A-life discovery. By observing the flocking behavior of birds, he found that a simple set of rules controlled it. Flocking behavior, he realized, was governed by three rules: Stay together, maintain the speed of adjacent birds, and avoid collisions. With these rules he set animated birds in flight, witnessing a pattern of flocking that closely resembled that of real birds. He then set obstacles, such as pillars, in the birds' path to see what behavior the birds would display. Although his program had not dictated the appropriate behavior for passing an obstacle, he found that his simulated birds divided briefly into separate flocks that

John von Neumann, an early computer pioneer, was the first to develop the idea of a cellular automaton, a theoretical organism governed by the rules of logic (Corbis/Bettmann).

rejoined after passing the obstacle—in the same manner in which real birds behave (see figure, p.11).

Since 1987, A-life researchers have met annually at a conference in New Mexico to discuss advances in the field. Current research has extended A-life into the field of robotics. Although no one has yet developed a robotic creature instilled with the properties of A-life, many in the field consider this an inevitable future development.

Uses, Effects, and Limitations

Although some recent A-life research has involved robots, much of the research involves computer simulation. The primary benefit of A-life research has been the emergence of unprogrammed behavior that is better suited for a task than anything programmers can design. Computer animators and livestock caretakers have used these programs to predict the complex behaviors of animals. Moreover, variations of these programs have been used to create mathematical models of complex systems such as the stock market and intricate factory schedules. Programs have also allowed researchers insight into evolutionary processes. Programs such as Tierra and AntFarm have used logic-based rules to mimic the evolution of insect populations. Over the course of generations, the simulated insects display behaviors adaptive to their environment in a manner similar to that of real organisms, allowing researchers to analyze more closely the forces of evolution.

At times, A-life phenomena occur that are not found in nature. This has caused some biologists to broaden their view of what constitutes life. On Earth, only one example of an evolutionary process exists, with all living things sharing the same local historical events, biochemistry, genetic descent, and carbon-based structure. Because of computers, researchers are able to manipulate factors that one would be unable to control in the real world, thereby contemplating life-as-it-could-be. In this manner, biologists can better understand what exactly defines and controls living systems and how they operate. Researchers hope to use A-life to pinpoint the elusive laws of nature that govern biological systems.

Less theoretical and more practical applications of A-life research exist as well. Self-replicating programs have been modified and used by John Deere factories. The factories must schedule their workers so that the best scheduling combination is used out of 1 million possible combinations. The program begins by determining the productivity levels of a large number of scheduling possibilities. As the simulated schedules proceed from day to day, the less optimal schedules are discarded until at the end of a simulated month, the most efficient schedule for the month emerges. A-life algorithms have also been used by film studios to create realistic animation scenes. The Walt Disney Co. and Dream Works SKG have used algorithms to create the complex, animated movement of Huns charging through the snow, or masses of slaves moving about the desert. Companies that house large amounts of livestock have used A-life computer modeling to determine the best type of housing for confined animals.

Issues and Debate

The issue of A-life raises many philosophical and practical concerns. Although many people consider only organisms created in nature to be alive, a number of A-life researchers believe that they will soon develop small, robotic creatures that may fall in between bacteria and insects on the spectrum of life. But will such an organism be alive? If it can reproduce, evolve, and maintain itself in the natural world, many scientists feel that it will possess life as much as does any insect or animal. Clearly, further progress in A-life research will reawaken age-old philosophical debates over what constitutes life, forcing us to reevaluate and possibly revise current ideas. Some observers have wondered, for example, if organisms were to evolve whose intellect matches that of humans, would those creatures deserve civil rights?

A more immediate concern is the effect that a robotic A-life organism may have if such a creature is indeed developed and released into the real world. If the creature is truly autonomous (capable of reproduction, self-repair, searching for energy, etc.), humans may not be able to control it. Such a silicon-based organism may be able to find an ecological niche and evolve indefinitely, which would open up an enormous potential for ecosystem destruction. Humans could be directly in danger as well—even if the creatures were programmed specifically not to harm humans, a mutation could arise that allowed the creatures to override this specification.

Most A-life researchers believe that the development of an autonomous creature is not a question of "if," but a

question of "when." Yet despite the many concerns surrounding A-life research, most researchers continue with their work in the hope that its potential benefits will outweigh its possible drawbacks.

—Paul Candon

RELATED TOPICS
Artificial Intelligence, Computer Animation, Computer Viruses, Nanotechnology, Robotics

BIBLIOGRAPHY AND FURTHER RESEARCH

BOOKS

Boden, Margaret. *The Philosophy of Artificial Life*. New York: Oxford University Press, 1996.

Dyson, George. *Darwin Among the Machines: The Evolution of Global Intelligence*. Reading, Mass.: Addison-Wesley, 1997.

Kelly, Kevin. *Out of Control: The Rise of Neo-biological Civilization*. Reading, Mass.: Addison-Wesley, 1994.

Levy, Stephen. *Artificial Life: The Quest for a New Creation*. New York: Random House, 1992.

PERIODICALS

Dunn, Ashley. "The Cutting Edge: It's Alive! Well, Sort Of. Animation Studios, Manufacturing Firms, and Medicine Are Turning to the Once-Obscure Field of Artificial Life to Provide Solutions to Some of Their Toughest Problems." *Los Angeles Times*, July 6, 1998, 1.

Keeley, Brian L. "Artificial Life for Philosophers." *Philosophical Psychology*, June 1998, 251.

Stricklin, W.R., P. de Bourcier, J.Z. Zhou, and H.W. Gonyou. "Artificial Pigs in Space: Using Artificial Intelligence and Artificial Life Techniques to Design Animal Housing." *Journal of Animal Science*, October 1998, 2609.

Watts, J. M. "Computer-simulated Animals in Behavioral Research." *Journal of Animal Science*, October 1998, 2596.

INTERNET RESOURCES

Artificial Life Online
  http://alife.santafe.edu/

Boids: Background and Update, by Craig Reynolds
  http://hmt.com/cwr/boids.html

Scientific American article: "From Complexity to Perplexity: Can Science Achieve a Unified Theory of Complex Systems?"
  http://www.sciam.com/explorations/0695trends.html

Tierra Homepage
  http://www.hip.atr.co.jp/~ray/tierra/tierra.html

ARTIFICIAL ORGANS

Complete artificial organs for implantation in the human body have yet to be developed. However, more than a dozen different kinds of prostheses, ranging from artificial hip joints to electronic heart pacemakers, are used routinely to take over part of the function of an organ or a major structural element in the body.

Within the next three decades, bioartificial organs—part living tissue, part synthetic material—are likely to enter clinical use. Before that happens, major technical hurdles will have to be surmounted.

Scientific and Technological Description

An *organ* is a functional unit within the body comprised of several tissues: The heart, liver, kidney, and brain are exam-

Some Examples of Human Implants

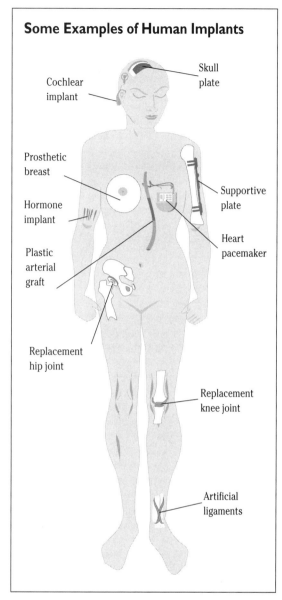

Skull plate

Cochlear implant

Prosthetic breast

Hormone implant

Plastic arterial graft

Supportive plate

Heart pacemaker

Replacement hip joint

Replacement knee joint

Artificial ligaments

ples. Millions of people each year suffer some loss of organ function as a result of accident, birth defect, or disease. In some cases, reparative surgery or drug treatment can make good or compensate for this damage (see figure). Since the 1960s, *organ transplantation*, the replacement of an organ by one donated from another person (living or dead), is used increasingly in cases where loss of organ function is life threatening and cannot be ameliorated in other ways (see ORGAN TRANSPLANTATION).

Two major limitations lie at the heart of organ transplantation as practiced currently. The first is the shortage of organ donors. Whether organs are obtained from a donor after death or from a living family member, there is a massive shortfall in organ supply compared to demand. In the United States alone, thousands of people die each year waiting for a transplanted kidney, lung, heart, or liver. According to the American Heart Association, in 1997 about 40,000

United States citizens needed a heart transplant; about 2300 obtained one. Second is the issue of *tissue rejection*. An organ transplanted from another biological source is normally sensed by the body to be foreign, and the body's immune system (the system of cells and chemical agents that defend the body against potentially harmful microbes or larger parasites) launches an attack against it. To control this, the organ recipient takes *immunosuppressant drugs*. Typically, such medication must be taken for the rest of the recipient's life. The possibility of manufacturing artificial organs that circumvent these difficulties is undoubtedly attractive. Robert Langer and Joseph P. Vacanti, two eminent tissue engineers, writing in *Scientific American* in 1995, predicted that "In the next three decades, medical science will move beyond the practice of transplantation and into the era of fabrication."

Considerable advances have already been achieved in the manufacture of artificial tissue. Artificial skin, blood, and tissue-engineered cartilage and bone are already available for clinical use or are at an advanced stage of clinical testing (see ARTIFICIAL TISSUE). These successes are for relatively simple tissues. Organ replacement is more complex because several tissues must be combined in a three-dimensional structure, normally of sufficient size that it requires an established blood supply to provide it with nutrients. In many cases, the artificial organ will need to be prepared for connection with the patient's nerves, blood vessels, and lymph channels if it is to function properly once in place.

Where artificial organ replacements have been most successful to date is in the development of prostheses or implants that take over part of the function of an organ or structural element within the body. A *prosthesis* is an artificial replacement for a part of the body. It may be functional or cosmetic, and may be internal, as in a heart valve, or external, as in a false arm. Hip joint replacement became a routine procedure in the 1960s, the same decade that electronic pacemakers (devices that maintain or control the heart's electrical rhythm) were introduced.

The production of artificial organs has entered a new phase with the advent of tissue engineering. Now, body tissues are beginning to be grown to order using cells from the recipient's own body. Within the next two or three decades, genetically engineered cells from sources outside the recipient could be grown to form replacement organs. To achieve this, ways must be found to overcome the problem of tissue rejection.

Historical Development

The development of artificial implants required major advances in several quite distinct fields: biomaterials science (the study of structural materials that occur within living organisms), surgery, antisepsis (the use of chemicals to counteract bacterial contamination), immunology (the study of immunity), and in some cases, microelectronics. The historical account given here centers on the hip joint and the heart pacemaker as two of the most successful implants, and then considers progress in developing a mechanical heart before considering the state of organ bioengineering.

In skeletal reconstruction, the most commonly used prosthesis is a hip joint replacement. The articulating surfaces of the hip joint may be damaged by physical accident (trauma), or more commonly by one of several forms of arthritis (joint inflammation). In either case, such damage is extremely painful and highly debilitating. In 1891, Theodore Gluck, a German surgeon, reputedly replaced a damaged hip with an ivory ball-and-socket joint. However, it was more than half a century before a truly workable hip joint replacement could be contemplated. A hip joint replacement has to be immensely strong, yet allow movement and at the same time be biologically inert (not harmful to biological processes or damaged by them). An artificial hip was achievable only after several major medical and technological hurdles had been overcome.

The articulating surfaces of a hip joint replacement need to be load bearing and yet very smooth, to minimize both friction and wear. The glue that holds component parts to bone must be long lasting and highly resilient, and any dangers of infection when the prosthesis is implanted should be minimal. Suitable biologically inert materials became available with the advent of acrylics and plastics technology in the 1930s and 1940s. Pioneers in artificial hip development included U.S. surgeon Marcus Smith-Petersen of Boston, Massachusetts, and the French brothers J. Judet and R. Judet. They made considerable progress in artificial hip design and surgical procedures, and their use of antibiotics minimized failures caused by bacterial infection. However, artificial joints still broke down or worked loose within a few years.

In the early 1960s, British orthopedic surgeon John Charnley developed an artificial joint and cement combination that was both smooth and long wearing. He eventually settled on high-density polyethylene for the socket in the pelvis and stainless steel for the replacement head of the femur. The high-grade steel, which is resistant to stress fractures from repeated high-impact activity over long periods, can last a patient's lifetime. The component parts were glued into place with methyl methacrylate, as volatile liquid that acts as an acrylic cement. Similar hip joint replacements are in common use today, with more than 1000 hip replacement operations performed each day around the world. The hip joint example serves to highlight some of the problems that implant developers must overcome. Similar and additional difficulties have confronted those who developed replacement knee, wrist, elbow, and shoulder joints—implants used in surgical procedures that are commonplace today.

The fruit electronic pacemaker was developed by the Swede Rune Elmqvist and implanted in 1960 by fellow Swede Ake Sening. Placed near the heart and connected to it by an insulated wire, the device takes over faulty functioning of the heart's own *pacemaker*, a region of heart tissue that stimulates the heart electrically to beat at a rate and depth to meet the body's demand for circulating blood. Longer-lasting lithium batteries to power pacemakers were introduced in 1973; in the 1980s, computerized programmable pacemakers were developed that were smaller, lasted longer, required less power, and were more responsive to the body's changing demands. Increasingly, heart pacemakers are able to record, store, and download diagnostic data on heart function as well as information about the operation of the pacemaker implant. By the early 1990s, about 100,000 heart pacemakers were being implanted annually in the United States alone.

Attempts to manufacture and implant an entire human heart have been made since the late 1960s but have met with rather limited success. In April 1969, U.S. heart surgeon Denton Cooley implanted an artificial heart in a patient who was dying of heart disease. Three days after implantation the mechanical device was replaced by a donor heart. The patient died soon after. In 1982, Robert Jarvik and a team at the University of Utah implanted a Jarvik Seven artificial heart in Barney Clark, a man with terminal heart disease. Clark lived for three more months but required intensive supervision, numerous clinical tests, and further surgery. The professional consensus is that a *total artificial heart* (TAH) can only be a temporary solution to a failing heart. It provides more time to find a heart donor but is not expected to serve as a permanent replacement.

Much greater success has been achieved with pumps that seek to assist the heart rather than to replace its function entirely. In a failing heart, the left ventricle, the heart chamber that supplies blood to most of the body, loses power. The *left ventricle assist device* (LVAD) was developed by U.S. cardiac specialist Michael DeBakey in the 1960s to assist just such a failing heart. By the early 1990s, smaller devices implanted by hospitals all over the world were keeping patients alive for months or even years before a donor heart became available. Initially about the size of a grapefruit, by the late 1990s, the pump was scaled down to thumbnail size and adopted NASA-style technology. It pushed blood around the body continuously using an Archimedes screw (a rotating screwthread blade) rather than a conventional pulse pump.

The LVAD is powered from an external energy source, typically, a battery pack on the body surface. Originally used as a stopgap measure for patients awaiting a heart transplant, some patients actually recovered while on LVAD and did not later require a heart transplant. In these patients, the rested heart cells of the left ventricle were seen to be recovering. If used early enough, the LVAD offers some the chance of recovery without need for a transplant. In those who go on to have a heart transplant, the LVAD may well improve their health before the operation and therefore increase the chance of a successful outcome.

Since the 1980s, some progress has been made in growing living tissue to implant in the body, as noted earlier. Work is currently under way to grow tissue implants—sometimes encased in plastic membrane, sometimes not—to replace some of the functions of the liver or the blood-glucose regulatory function of the pancreas. In the short to medium term, researchers see a bioartificial kidney or liver as a potential "bridge" organ that will support the function of a damaged organ while a living donor organ is being sought.

By spring 1999, Tony Atala and his associates at Children's Hospital, Boston, had grown an artificial human bladder in the laboratory. Six weeks after taking samples of bladder tissue, the team had grown a new bladder by applying cultured cells of two types, a few at a time, onto a bladder-shaped structure made of a polymer, a substance of high molecular mass constructed of repeating chemical units. In 1997, Atala's team had already shown that a tissue-engineered bladder could be transplanted and function in a dog. By the early 21st century, the procedure is likely to be tested on humans, thereby offering potential relief to those suffering severe bladder disease. Potentially, some 400 million people worldwide could benefit from such a transplant.

Uses, Effects, and Limitations

Implanting devices and artificial organs within the body is a highly challenging endeavor. First, the material used for construction must be biocompatible—it must not cause any untoward biochemical or biological effects. The materials used must be nontoxic and avoid causing inflammation or other adverse responses. This severely limits the choice of available materials. Second, the implant must, if possible, avoid stimulating attack by the body's immune system or must be able to withstand such an attack. Third, the implant needs to be as long lasting and reliable as possible, since in most cases replacement will require a surgical procedure. A sudden malfunction could be fatal.

In terms of materials, certain metal alloys, plastics, and ceramics have been found to meet all three criteria noted above. Extensive testing on animals over long periods of time is required before clinical testing on humans can take place. Quality control for implants—from development to final production—is a formidable undertaking. Despite such obstacles, great advances in organ construction have been achieved in the last four decades. The new wave of artificial organ development based on tissue engineering has to overcome such hurdles and more besides. Growing living tissues on an artificial scaffolding of nonliving material—in other words, growing bioartificial organs—requires the resolution of many problems. First, the cells used must be compatible with those of the donor. This means that the cells must either be grown from the recipient's own tissues, or must come from another source but lack the cell surface marker chemicals that identify the donor cells as "nonself." Currently, two approaches are being pursued: using embryonic stem cells (see FETAL TISSUE TRANSPLANTATION) or developing special cell lines that lack the special marker chemicals or have them masked in some way.

Second, the cells would need to be grown in large quantities, under safe conditions, and then encouraged to differentiate (become specialized) to form the various tissues within the organ. Initial findings suggest that tissues have the capacity to self-organize, but finding the chemical triggers that cause cells to differentiate as required to grow a particular organ is still many years away. Third, organs are three-dimensional structures and their tissues have specific biomechanical properties. Replacements need to share similar properties of strength and flexibility. Currently, polymer scaffolds are being used as templates on which to grow cells. By subjecting the cells to stresses and strains when they are cultured, they can be grown to mimic the mechanical properties of the tissue they are to replace.

Fourth, organs could be grown on demand, but ideally a stock of organs would be held in reserve, which would mean that techniques for organ preservation would need to be well advanced. Fifth, once transplanted, the artificially grown organ needs to be incorporated within the body in the same manner as its natural predecessor. Connections with blood vessels, nerves, and lymphatic vessels would need to be established. The organ would need to have the capacity to maintain itself, even to grow, but without becoming cancerous.

These hurdles are considerable, but the rewards—both financial and in terms of relieving human suffering—are immense. An artificial pancreas, for example, probably one of the simpler organs to recreate, could offer a cure for diabetes that would benefit millions of people worldwide.

Issues and Debate

The development of entirely artificial organs, as exemplified by the artificial heart, is a high-cost, long-term endeavor. The creation and implantation of an artificial heart is a "flagship" enterprise. It garners intense public interest and, almost regardless of the short-term outcome for the patient, is likely to boost the status of the surgeon involved. This can result in great pressure being applied to encourage patients to allow themselves to participate in helping to develop the technology. In any case, they are suffering from a terminal condition, which left untreated gives them only weeks or months to live.

Bioethics experts debate whether such patients can truly give informed consent to have novel procedures such as artificial heart implants performed on them. The surgeons informing patients of their options may also be those with the most to gain by carrying out the novel procedure. Pressure is being brought to bear by bioethics committees and other professional bodies to help ensure that such patients receive a balanced presentation of their options. Quality-of-life issues may outweigh the desire to have a potentially lifesaving operation, particularly where the risks are unknown and the quality of life after the operation, with exhaustive postoperative monitoring, is likely to be poor.

Despite slow progress in developing an effective mechanical heart, such projects continue to receive federal funding. Some U.S. bioethics experts question whether billions of dollars of public money should be invested in projects that are ultimately likely to benefit only those with the high level of medical insurance that will pay for a mechanical heart when the technology becomes available. Meanwhile, tissue engineering probably offers the best long-term solution to the organ transplant shortfall. In 30 years' time, "off-the-shelf" organs may be more commonplace than today's artificial hip joints and electronic heart pacemakers.

—*Trevor Day*

RELATED TOPICS

Artificial Tissue, Fetal Tissue Transplantation, In Vitro Fertilization, Organ Transplantation, Polymers, Tissue Transplantation

BIBLIOGRAPHY AND FURTHER RESEARCH

BOOKS

Caplan, Arthur L. *Am I My Brother's Keeper?* Bloomington, Ind.: Indiana University Press, 1997.

Kimball, Andrew. *The Human Body Shop*, 2nd ed. Washington, D.C.: Regnery Publishing, 1997.

Lanza, Robert P., Robert Langer, and William L. Chick, eds. *Principles of Tissue Engineering*. Austin, Texas: R. G. Landes and Academic Press, 1997.

Patrick, Charles W., Jr., Antonios G. Mikos, and Larry V. McIntire, eds. *Frontiers in Tissue Engineering*. Oxford: Pergamon Press, 1998.

Youngson, Robert. *The Surgery Book*. London: Century, 1993.

PERIODICALS

Doyle, Roger. "Health Care Costs." *Scientific American*, April 1999, 25.

Langer, Robert, and Joseph P. Vacanti. "Artificial Organs." *Scientific American*, September 1995, 130.

Langer, Robert S. and Joseph P. Vacanti. "Tissue Engineering: The Challenges Ahead." *Scientific American*, April 1999, 62.

Lysaght, Michael J., and Patrick Aebischer. "Encapsulated Cells as Therapy." *Scientific American*, April 1999, 52.

Mooney, David J., and Antonios G. Mikos. "Growing New Organs." *Scientific American*, April 1999, 38.

Morgan, Jeffrey R., and Martin L. Yarmush. "Tissue Engineering." *Science & Medicine*, November–December 1998, 6.

Pedersen, Roger A. "Embryonic Stem Cells for Medicine." *Scientific American*, April 1999, 44.

INTERNET RESOURCES

American Society for Artificial Organs
http://www.asaio.com/news.htm

Center for Biomedical Engineering, Massachusetts Institute of Technology
http://web.mit.edu/cbe/www/

International Society for Artificial Organs
http://www.strath.ac.uk/Departments/BioEng/isaotext.html

Pittsburgh Tissue Engineering Initiative
http://www.pittsburgh-tissue.net.html

The Society for Biomaterials
http://www.biomaterials.org/

ARTIFICIAL TISSUE

Artificial tissues are synthetic or partially synthetic substitutes to replace absent or malfunctioning tissues. Since the 1970s, the development of artificial tissues has shifted from the production of synthetic products based on plastics, metals, and other biocompatible materials, as in the case of silicone breast implants, to bioengineered tissues that commonly combine an artificial layer or framework in association with cultured living cells, as in some forms of skin replacement.

The emerging field of tissue engineering, in which living cells are combined with natural or artificial noncellular structures, draws upon numerous other disciplines, from materials science to molecular biology. The interdisciplinary nature of this technology means that advances are likely to arise in many and unexpected ways in this very active field of research and development.

Scientific and Technological Description

Artificial tissues, first developed in the 1970s, are designed to replace or augment missing or damaged human tissue, either temporarily or permanently. The first wave of artificial tissues, including breast implants and artificial ligaments, were based on the use of biocompatible materials such as silicone, plastics, and carbon fiber. Replacement heart valves made of combinations of metal, nylon, and rubber have been available since the 1960s. More recently, some replacement valves have incorporated pig or cow tissue. Artificial tissues seek to replace body cells that interact in a three-dimensional environment and themselves contain thousands of different chemicals. It is not surprising that in order to meet the body's stringent demands, researchers have resorted to employing the body's own cells in growing replacement tissue.

In the 1990s, the technology has shifted toward growing human tissues outside the body—often in tandem with artificial materials—and then incorporating these bioengineered tissues within the body. This newly emerging field, *tissue engineering*, draws upon a multiplicity of different specialties: biochemistry, biomaterials science, chemical and mechanical engineering, cell and molecular biology, physiology, and surgery. Ideally, the engineered tissue will become integrated with surrounding tissues within the patient and will effect a potentially permanent cure.

In tissue engineering, one of three approaches is usually adopted:

1. Growing human tissues outside the body for later implantation. Epidermal skin grafts, used as a treatment for burns, offer one example of this technique. Cells are removed from a healthy area of skin, separated by chemical treatment, and the epidermal (outer skin) cells are grown in sheets that are later transferred to the patient to cover damaged areas.

2. Implanting devices that induce the regeneration of appropriate local tissues. Such devices incorporate "signal" molecules that encourage local tissue growth. They

Artificial Tissue

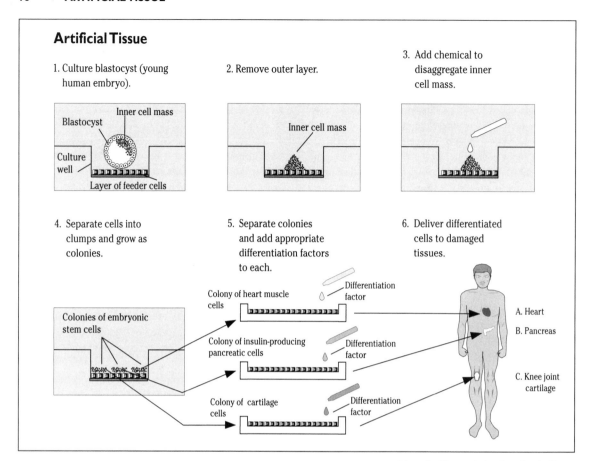

1. Culture blastocyst (young human embryo).

Blastocyst
Inner cell mass
Culture well
Layer of feeder cells

2. Remove outer layer.

Inner cell mass

3. Add chemical to disaggregate inner cell mass.

4. Separate cells into clumps and grow as colonies.

Colonies of embryonic stem cells

5. Separate colonies and add appropriate differentiation factors to each.

Colony of heart muscle cells
Differentiation factor

Colony of insulin-producing pancreatic cells
Differentiation factor

Colony of cartilage cells
Differentiation factor

6. Deliver differentiated cells to damaged tissues.

A. Heart
B. Pancreas
C. Knee joint cartilage

Adapted with permission from Laurie Grace.

also contain a three-dimensional polymer (a complex substance of large molecular size comprised of repeating chemical units) that forms a scaffolding (matrix) on which new tissue can grow.

3. Growing tissues outside the body within structural matrices (artificial scaffolding). Once grown, the sculpted tissue is placed within the body in the desired location. Cell-lined vascular (blood vessel) implants of this nature are being clinically tested.

The artificially induced regeneration of body parts is accelerating at such a pace that what is currently science fiction will soon be science fact. Currently, fabricated skin and connective tissues such as cartilage, bone, ligament, and tendon are finding surgical application in hospitals across the United States. Within the next few decades, tissue engineers are likely to tease out the processes by which undifferentiated (unspecialized) embryonic tissues change to become the differentiated (specialized) tissue of many major organs. Potentially, researchers will be able to customize tissue growth to replace specific tissues or even entire organs (see ARTIFICIAL ORGANS).

Historical Development

The beginnings of tissue culture—the technique that makes tissue engineering possible—can be traced back to experi-

ments by Ross G. Harrison at Yale University in 1907. He extracted nerve cells from an amphibian spinal cord and grew them in a culture of clotted tissue fluid in a warm, moist chamber. The nerve cells grew over several days. By 1913, Frenchman Alexis Carrel, working at the Rockefeller Institute in the United States, succeeded in culturing cells from birds and mammals (extracting the cells and growing them in a lab) for extended periods. He and his associates developed a strain of "immortal" cells, descended from cells obtained from the developing heart of a chick embryo, that was maintained between 1912 and 1942. By growing tissues in the laboratory in the longer term, biochemists and cell biologists could obtain a reliable stock of cells of a known type for experimentation.

Bacterial contamination was a major cause of failure of early tissue cultures. But with the advent of antibiotics (antibacterial drugs) in the 1940s, advances in tissue culture followed rapidly. By the mid-1970s, the blend of nutrients, salts, and antibiotics in conventional media began to be more finely tuned to the growth requirements of specific tissues. These included chemically defined media that contained one or more specific growth factors, hormones, and transferrin, a protein that carries iron into cells. By the late 1970s, all manner of human tissues—from heart tissue to skin cells—could be grown temporarily in culture. Typically, small tissue fragments, or explants, were seeded on the surface of plastic or glass containers suitably treated with growth media and containing components such as collagen that are found in the cell's normal tissue environment. Human skin cells, treated in this way, could be grown for several months in culture, dividing between 50 and 100 times during this time.

Two key discoveries have guided recent developments in tissue growth within the body, both involving the discovery of specific chemical factors that promote tissue growth *in situ* (in its natural environment). In 1965, Marshall R. Urist at the University of California at Los Angeles showed that implanted powdered bone would encourage the formation of new bony tissue. Since then, researchers have isolated bone morphogenetic proteins (BMPs), the chemicals that promote this effect. BMPs are now produced in large quantities. Clinical tests on people with poorly healing fractures, conducted by Creative BioMolecules, a Massachusetts company, show that artificially administered BMPs work as well as bone grafted from other parts of the body.

Another key discovery had to do with the blood vessels that feed nutrients to tissue. Any tissue more than a few millimeters thick normally requires its own blood vessels to supply it with nutrients for growth. Cancers, for example, require a rich blood supply to fuel their rapid expansion. In 1972, while searching for ways to combat the growth of cancerous cells, Judah Folkman at Harvard Medical School's Children's Hospital reasoned that if those specific molecules that inhibited blood vessel growth could be isolated, manufacturing and using them would be one way of fighting cancer. Finding chemicals that encourage blood vessel growth (angiogenesis) has been beneficial in the development of tissue engineering. Angiogenesis-stimulating molecules have since been isolated and manufactured.

Stimulating blood vessels or bone to grow in situ requires delivery of the appropriate growth-stimulating chemicals to the specific site and in the right amounts. Various approaches are being developed to achieve this. Several research centers in the United States are developing injectable biodegradable polymers that could fill missing areas of bone and be impregnated with BMPs. U.S. researchers David J. Mooney and Antonios G. Mikos are developing gelatinlike, water-filled polymers. These are used to encourage bone regeneration in tooth sockets following severe periodontal disease.

Uses, Effects, and Limitations

Synthetic tissues and tissue-engineered products must meet some of the same demanding requirements as drugs, but they must also fulfill many other criteria: in particular, biocompatibility (not being subject to attack by the body's immune system or by other cellular and biochemical phenomena), responsiveness to hormones (chemical messengers), and mechanical properties such as strength and flexibility. In the case of artificial skin, the healed wound needs to blend with the surrounding area and have texture and flexibility similar to those of adjacent skin.

The uses and limitations of artificial tissues are specific to particular applications, but two examples—skin and blood—serve to illustrate some of them. The U.S. company Integra LifeSciences manufactures one form of artificial skin that comes with a dermis (inner skin layer) comprised of a porous spongy matrix made of shark cartilage derivatives and collagen (fibrous connective tissue) obtained from cows. This combination forms a biodegradable matrix that encourages below-skin connective tissue and blood vessels to grow into it. The artificial dermis has a surface covering (artificial epidermis) comprised of silicone. Once underlying tissue has permeated the artificial dermis, and the matrix has degraded, the silicone is removed and replaced with the patient's own epidermal skin, which has been grown in tissue culture in the meantime. The use of artificial skin ensures that the final transplant of skin is much thinner than would otherwise be the case.

In another approach, Advanced Tissue Science (ATS), a company in La Jolla, California, has developed skin tissue grown from foreskin cells taken from newborns. These cells lack the distinguishing marker molecules that might trigger a recipient's immune system to reject it when transplanted. The tissue forms a temporary covering for burn patients. Organogenesis in Canton, Massachusetts, has devised similar skin tissue as a permanent graft for the treatment of skin ulcers. These artificial replacements are improvements on previous forms of treatment, which included the use of surgical dressings and skin tissues from cadavers. The new forms of artificial skin are likely to be cheaper, to shorten healing times, and to involve fewer complications.

As in the case of other donated tissues, blood is in short supply. In many parts of the world, blood is not donated and stored routinely. Even in those countries where it is, such as North America and Western Europe, demand commonly outstrips supply. Blood has to be screened for infections, carefully cross-matched between donor and recipient, and even then the blood can be stored for only a few weeks before it has to be discarded. A safe blood substitute would be invaluable.

Blood is a complex tissue and performs many functions, including wound healing and immune defense. It is the oxygen-carrying capacity of blood that has, to some extent, been duplicated in an artificial form. Two approaches have been adopted. Perfluorochemicals are inert substances that can dissolve about 50 times more oxygen than blood plasma (but many times less oxygen than real blood's oxygen carrier, hemoglobin). In one type of blood substitute, perfluorochemicals are mixed with fatty substances and are injected into the patient's blood as an emulsion. These blood substitutes can be manufactured in large amounts, and their purity easily assured, but their use requires the patient to breathe an oxygen-rich air mixture.

The second form of oxygen-carrying blood substitute utilizes hemoglobin. Instead of the hemoglobin being enclosed in cell membranes, as in real blood, the substance is free in suspension or solution. This has both benefits and disadvantages. One benefit is the avoidance of the need to match blood. The artificial substitute is not enclosed in a membrane (with real blood, the membrane proteins of red blood cells need to be cross-matched for compatibility between donor and recipient). Two disadvantages in using "naked" hemoglobin are that the chemical degrades much more quickly than when enclosed and that the naked chemical may be toxic to surrounding tissues. These difficulties are being addressed.

Currently, the implementation of tissue engineering in clinical use is limited largely to relatively simple tissues. At present, researchers are leaning toward culturing a patient's own cells and thereby avoiding tissue rejection problems, rather than implanting cells from other sources. As of early 1999, tissue engineering was being applied in many areas of medicine. In endocrinology (the study of the body's hormonal system) pancreas cells were being packaged in artificial membranes for implantation to treat diabetes. In rheumatology (the specialty concerned with joints, muscles, and connective tissues) U.S. surgeons were injecting tissue-cultured cartilage cells to repair damaged joints. In neurosurgery, artificial channels containing biologically active molecules were being tested to promote and guide nerve regeneration. Replacement breast tissue, cultured from smooth muscle elsewhere in the body and supported on a polymer scaffolding, was being developed as a potential alternative to silicone or saline implants.

Some researchers are investigating the potential of genetically engineering cells that lack the marker molecules that trigger immune reactions and so cause tissue rejection. Such cells would have the potential to form "universal donor" tissues. Embryonic stem cells (see FETAL TISSUE TRANSPLANTATION) hold the promise of providing undifferentiated tissue that could be encouraged to grow into a wide range of specialized tissues if nurtured in the correct physical and biochemical environment (see figure).

Issues and Debate

The medical need for tissue substitutes is great. As with whole organ substitutes (see ARTIFICIAL ORGANS, ORGAN TRANSPLANTATION, and TISSUE TRANSPLANTATION), this need arises for two main reasons. First, there is a great shortfall in tissues offered for transplantation. Second, unless grown from a person's own cells, tissues are likely to face rejection by the body's immune system. Artificial tissues could be of great benefit in helping meet the shortfall in the first instance and in circumventing biocompatibility problems in the second. They also have an important cosmetic role to play in resculpting body regions after surgery, as in the case of breast implants. Worldwide, more than 250,000 mastectomies (the surgical removal of part or all of a breast) are performed each year.

The stormy history of the use of silicone breast implants in the United States serves to illustrate some of the problems than can arise with even relatively simple forms of tissue implant. In 1992, the U.S. Food and Drug Administration (FDA) restricted the use of silicone gel–filled implants to women wishing to have breast reconstruction after surgical removal of a breast cancer. The FDA decision was a response to emerging concerns articulated by breast implant surgeons and their patients, including concern that contraction of fibrous tissue around the silicone implant was causing shape change, physical discomfort, and pain; concern about leakage of silicone into surrounding tissues; and the opinion of some specialists that autoimmune diseases (in which the immune system attacks the body's own cells) might be triggered by certain forms of implant.

By 1992, the silicone implant manufacturer Dow Corning had lost two major court cases to patients claiming that silicone implants had caused disease. Later claims against Dow Corning and other implant manufacturers formed part of a class action suit (where many claims are grouped in the same category to form part of a single legal action). By 1994, 20,000 lawsuits were filed against four silicone implant manufacturers, and a total of 480,000 claimants were vying for a portion of a massive $4.25 billion settlement.

As of the spring of 1999, there was still no widely accepted long-term study that showed that silicone implants cause autoimmune disease, the major allegation that triggered the court cases. Nevertheless, the cases have revealed that clinical testing by manufacturing companies was less than scrupulously thorough, a problem much less likely to apply to the testing of artificial tissues in the future. Breast implants in clinical use so far contain no living tissue.

The complexities of implant–host tissue interactions are likely to multiply once artificial tissues that contain living cells are in widespread use. Nevertheless, scientists and manufacturers in the tissue-engineering field are acutely aware of the need to develop implants that are not only safe but are also perceived as safe by the public. Tissue engineering is likely to benefit from advances in numerous other disciplines, from GENETIC ENGINEERING to NANOTECHNOLOGY (engineering of matter at a scale approaching that of individual atoms and molecules), and appears to have a very promising future.

—*Trevor Day*

RELATED TOPICS

Artificial Organs, Fetal Tissue Transplantation, In Vitro Fertilization, Nanotechnology, Organ Transplantation, Tissue Transplantation

BIBLIOGRAPHY AND FURTHER RESEARCH

BOOKS

Lanza, Robert P., Robert Langer, and William L. Chick, eds. *Principles of Tissue Engineering*. Austin, Texas: R.G. Landes and Academic Press, 1997.

Patrick, Charles W., Jr., Antonios G. Mikos, and Larry V. McIntire, eds. *Frontiers in Tissue Engineering.* Oxford: Pergamon Press, 1998.

PERIODICALS

Dickman, Steven, and Gabrielle Strobel. "Life in the Tissue Factory." *New Scientist*, March 11, 1995, 32.

Doyle, Roger. "Health Care Costs." *Scientific American*, April 1999, 25.

Langer, Robert, and Joseph P. Vacanti. "Artificial Organs." *Scientific American*, September 1995, 130.

Langer, Robert S., and Joseph P. Vacanti. "Tissue Engineering: The Challenges Ahead." *Scientific American*, April 1999, 62.

McCarthy, Michael. "Bio-Engineered Tissues Move Towards the Clinic." *Lancet*, August 17, 1996, 466.

Mooney, David J., and Antonios G. Mikos. "Growing New Organs." *Scientific American*, April 1999, 38.

Morgan, Jeffrey R., and Martin L. Yarmush. "Tissue Engineering." *Science and Medicine*, November/December 1998, 6.

Pedersen, Roger A. "Embryonic Stems Cells for Medicine." *Scientific American*, April 1999, 44.

Stix, Gary. "Growing a New Field." *Scientific American*, October 1997, 9.

INTERNET RESOURCES

Artificial Cells and Organs
http://www.physio.mcgill.ca/artcell/

Center for Biomedical Engineering, Massachusetts Institute of Technology
http://web.mit.edu/cbe/www/

Pittsburgh Tissue Engineering Initiative
http://www.pittsburgh-tissue.net.html

Prosthetic Heart Valve Information
http://www.csmc.edu/cvs/md/valve/default.htm

The Society for Biomaterials
http://www.biomaterials.org/

BIODIVERSITY PROSPECTING

Biodiversity prospecting, or bioprospecting, is the search for genetic and biochemical resources that can be developed into commercial products. Scientists collect plant and animal species and microorganisms, extract relevant compounds from them, and test them for medicinal or commercial value. Most of the products yielded from bioprospecting are pharma-ceuticals, but other types of products are discovered as well, such as environmentally friendly microbes that can replace harsh chemicals in industrial processes.

Although people have used natural medicines for most of history, controlled research into the medicinal value of plants and microorganisms did not occur until the 20th century. Plant-derived medicines now account for 25 percent of all prescription drugs in the United States and are used to treat a variety of ailments. Bioprospecting has advantages over other drug-development techniques such as COMBINATORIAL CHEMISTRY because the complex combinations of elements found in natural products would be extremely difficult to conceive in a lab. The use of bioprospecting by corporations in undeveloped countries or in regions occupied by indigenous people has raised concerns in the international community, however. Indigenous communities remain largely uncompensated for their contributions to profitable products.

Scientific and Technological Description

Bioprospecting involves the collection and study of biodiversity, the genes and species found in an environment. Scientists often travel to areas with a great amount of biodiversity (e.g., rain forests and coral reefs) to collect plants, microorganisms, or extracts from animal species. From these species, researchers remove relevant chemical compounds and then put them through screening procedures. Scientists design these chemical screens, or bioassays, to indicate if a chemical displays a specific trait that would work against a disease or perform other biologically relevant functions. If a chemical or extract shows potential usefulness in a bioassay, it can be further isolated and then tested in animals. If the compound is successful in animals, the scientists will test the chemical in humans, after which the scientists' affiliated pharmaceutical company can market the drug (see figure).

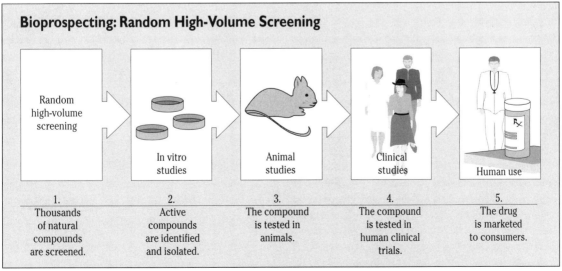

Bioprospecting: Random High-Volume Screening

Random high-volume screening	In vitro studies	Animal studies	Clinical studies	Human use
1. Thousands of natural compounds are screened.	2. Active compounds are identified and isolated.	3. The compound is tested in animals.	4. The compound is tested in human clinical trials.	5. The drug is marketed to consumers.

Adapted with permission from Shaman Pharmaceuticals.

To find effective chemical compounds among the great number of species in an environment, scientists use two methods. The first involves random, high-volume screening. A typical example of this technique can be seen in the recent development of a cholesterol-lowering drug by the U.S.-based pharmaceutical company Merck. Merck researchers wanted to find a natural compound that would lower cholesterol levels in humans. They looked at how the body processes cholesterol and found a point in the process at which they believed a compound could effectively alter cholesterol synthesis. They then designed a bioassay that would indicate if a substance affected this step. Thereafter, researchers screened thousands of natural compounds to find a substance that altered cholesterol synthesis. Eventually, a chemical from a soil sample from Spain proved effective. After further testing, Merck was able to market its cholesterol-lowering drug, now known as Mevacor.

The second major method involves tapping the knowledge of indigenous people. In recent years, some companies have begun to use the knowledge of local people to target specific plants and microbes for testing. Indigenous people tend to have traditional medical practices, using local plants and resources to treat specific illnesses. Many researchers have found that the traditional medical use of a plant is a strong indicator of its biological function. Shaman Pharmaceuticals, Inc., a U.S. research company, utilizes this screening technique. The company targets specific plant species for study after speaking with and observing indigenous healers. If the company finds three different indigenous communities using the same plant for similar purposes, its scientists decide to study that plant further. First, they test the relevant chemical in animals to evaluate its scientific validity. If the chemical proves effective, they further isolate it in laboratory tests and then evaluate the potency of the modified chemical in human clinical trials. Shaman claims to find useful compounds in half of its preliminary screening tests. This type of bioprospecting, with the help of indigenous knowledge, is 5000 times more likely to find a hit than simple random collection, according to the Rural Advancement Foundation International (RAFI), an international non-governmental organization.

Traditionally, bioprospecting involves the collection of a species, usually a plant, which is then dried and ground up. If more extract is needed, more samples are required as well. Newer technologies, however, need only to maintain plant cells in a test tube, where they are frozen and preserved. To do further research, scientists can later defrost the cells and grow more. Lab-based cell cultures can be up to 100 times less expensive to use than traditional plant samples. Lab-grown cells also allow for more flexibility in research. Plants tend to adapt on a minute-to-minute basis due to environmental changes. A plant specimen may show a different chemical makeup depending on the time of day it was picked, for example. However, researchers can expose lab-grown cells to changes in light and temperature as well as to different chemicals and hormones. In this manner, researchers can delineate all the chemical variations in a plant, thereby maximizing the chances of finding useful compounds.

Historical Development

Before the 19th century and the advent of organic chemistry, about 80 percent of all medicines originated from plant materials. The earliest known record of herbal medicine use dates back to 3000 B.C., when the Chinese used chaulmoogra oil from the fruit of the kalaw tree to treat leprosy. In Egypt, archeologists have excavated opium and castor oil seeds from tombs, indicating their medicinal use around 1500 B.C. However, beginning in the 20th century, scientists again began to discover the usefulness and marketability of organic products in both medical and industrial processes.

The modern history of bioprospecting dates to 1928, when a mold landed in a petri dish belonging to the British scientist Alexander Fleming. He noticed that the mold inhibited bacterial growth in the dish. This discovery eventually led to the development of the first antibiotic, penicillin, which became available in the 1940s. With this finding, researchers began studying many different microbial substances for possible medicinal purposes and eventually developed an array of antibiotic drugs.

In 1956, the U.S. National Cancer Institute started screening plants and animal extracts for potential cancer-fighting compounds. Before the program ended in 1981, researchers had screened over 35,000 species. Nonetheless, they failed to discover a significant number of effective compounds. A later study showed that the use of indigenous medical knowledge to help choose certain species would have doubled the project's success rate. At the time, however, the medicinal value of biodiversity seemed limited. In 1980, in fact, U.S. pharmaceutical companies spent no money on plant research.

By the mid-1980s, the destruction and disappearance of the world's rain forests, and the subsequent loss of plant and animal biodiversity, was becoming common knowledge. Experts informed pharmaceutical companies that the location—or loss—of each medicinally useful species represented $200 million in sales annually. Both vanishing sales potential and the success of certain plant-derived drugs such as the anticancer agent Taxol, which comes from the Pacific yew tree, spurred the pharmaceutical industry to invest more seriously in bioprospecting.

Recently, advances in molecular biology and related diagnostic screening techniques have made bioprospecting more cost-effective to perform. Companies are able to screen thousands of samples at a time, greatly increasing their chances of finding a "hit," a chemical that has any effect on a cell or protein. Widespread use of indigenous knowledge has also increased the probability of finding effective compounds. Currently, over 200 companies worldwide screen microorganisms, plants, and animal extracts for commercially useful compounds. Twenty-five percent of prescription medicines in the United States are derived from plant products or are synthesized to replicate the active ingredients in plants.

Uses, Effects, and Limitations

The main benefit of bioprospecting is the development of effective medicinal treatments. Plant-based drugs have been used to treat ailments ranging from headaches (aspirin, from the white willow tree) to Parkinson's disease (the compound L-dopa, from the velvet bean). The breadth of possible treatments is limited only by what researchers are capable of testing. Only 2 to 5 percent of flowering plants in the world have been tested for effectiveness against disease.

Microorganisms have even more varied uses. Companies have used microbes found in soil to flavor foods and vitamins. Derivatives of microorganisms have improved the way cheese is made and the manner in which paper is bleached. Often, environment-friendly microbes can replace toxic chemicals in the industrial process. Only 0.1 percent of all bacteria and fungi in the soil have been screened for their commercially useful compounds. Treatments for diseases such as AIDS (acquired immune deficiency syndrome) and cancer may lie untested among the world's biodiversity. In addition, some groups, such as Knowledge Recovery Foundation International, an environmental sciences research organization in New York City, have begun to develop plant-extract libraries. They collect and store plant extracts from around the world along with information about indigenous uses. The foundation plans to rent out the plant extracts to pharmaceutical companies; if a drug is developed, the foundation will require the company to pay a small royalty to the indigenous people from whom the sample came. With a plant-extract library, the world's biodiversity can be preserved for future medical use without the concern of losing species to extinction.

A method often used by pharmaceutical companies instead of bioprospecting is a technique called COMBINATORIAL CHEMISTRY, which allows scientists to combine diverse chemical groups together. With this technique, researchers can produce up to 40,000 compounds in a single experiment. Many view this as an effective method for discovering medically useful compounds, because researchers do not have to leave the lab to find, collect, and store biological material. However, as with random plant screening, the returns are often meager, and the compounds tested only infrequently show biological function. In addition, the combination of chemicals is limited by the imagination of the chemist. Combinatorial chemistry does not readily allow for the creation of complex and novel combinations such as those found in Taxol, which contains four intricately entwined carbon rings. Nonetheless, many researchers believe that the most promising future drugs will involve a blend of natural and synthetic products.

Issues and Debate

Prior to June 1992, many researchers considered biodiversity to be the "common heritage of humankind." Scientists were free to travel to areas rich with biodiversity and take home genetic resources and indigenous knowledge. Furthermore, they were able to develop and market derivative drugs or products without compensating the country of origin or the indigenous people who provided the resources. In fact, industrialized nations condemned any action taken to restrict access to biodiversity. However, at the United Nations' International Convention on Biological Diversity, held in 1992, over 150 nations signed an agreement recognizing the sovereign right of nations to protect their natural resources. As a result, nations can now create laws restricting access to their resources. The convention recognizes that indigenous people should share in the profits or patent rights if a company derives a useful product from their knowledge and resources. Nations may require informed consent and payment from researchers before allowing access to resources. In addition, the signed nations agreed to help conserve the world's biodiversity, with the industrialized nations supporting conservation economically in the poorer countries.

Originally, the United States refused to sign the agreement. The Clinton administration later reversed the decision, although the Senate has yet to approve it. The government at first felt that the agreement would undermine the interests of U.S. companies by restricting their ability to claim intellectual property rights on developed products. U.S. representatives claimed that companies and nations (or national institutions) should reach agreements on an individual basis rather than be guided by an international agreement. They pointed to a 1991 agreement between Merck and the Instituto Nacional de Bioversidad (INBio) of Costa Rica as an example. The institute agreed to provide Merck with extracts from plants, insects, and microorganisms. In return, Merck paid $1.14 million and gave training and technical assistance to the institute so that it could establish in-country drug research. In addition, Merck agreed to share any potential royalties from resulting products. Nonetheless, the agreement still came under fire for ignoring the rights of Costa Rica's indigenous people.

The guidelines for protecting the contributions of indigenous people under the 1992 agreement can be difficult to enforce, even within the countries that signed the agreement. Adherence rests mainly in the hands of the pharmaceutical companies and industrialized nations. Often, developing countries are in need of cash and eager to sell their resources without concern for greater, long-term control over their biodiversity. Also, governments of developing nations may be indifferent to or hostile toward the indigenous people living within their borders and may not be concerned with their legal rights. The indigenous people may not be familiar enough with international legal agreements and technical jargon to demand fair treatment. Furthermore, if numerous communities use a certain plant or species, proving from which group it originated may be difficult. Although such loopholes remain, pharmaceutical companies will probably continue to claim intellectual property rights on indigenous resources and knowledge.

—Paul Candon

RELATED TOPICS
Bioremediation and Phytoremediation, Combinatorial Chemistry

BIBLIOGRAPHY AND FURTHER RESEARCH

BOOKS
Joyce, Christopher. *Earthly Goods: Medicine-Hunting in the Rainforest.* Boston: Little, Brown, 1994.
World Resources Institute Staff. *Biodiversity Prospecting: Using Genetic Resources for Sustainable Development.* Washington, D.C.: World Resources Institute, 1993.

PERIODICALS
Dove, Alan. "Botanical Gardens Cope with Bioprospecting Loophole." *Science,* August 28, 1998, 281.
Eisner, Thomas, and Elizabeth Beiring. "Biotic Exploration Fund—Protecting Biodiversity Through Chemical Prospecting." *BioScience,* February 1994, 95.
Macilwain, Colin. "When Rhetoric Hits Reality in Debate on Bioprospecting." *Nature,* April 9, 1998, 535.
Makhubu, Lydia. "Bioprospecting in an African Context." *Science,* October 2, 1998, 41.
Posey, Darrell. "Protecting Indigenous Peoples' Rights to Biodiversity." *Environment,* October 1, 1996, 6.
Reid, Walter. "Bioprospecting: A Force for Sustainable Development." *Environment, Science, and Technology,* September 1993, 1730.
Seabrook, Charles. "Treasures from the Tropics: Georgia Scientists Are Prospecting—But Not for Gold. They're Hunting for Plants That May Hold the Key to Healing Human Ailments." *Atlanta Journal and Constitution,* February 7, 1999, C1.
Wickelgren, Ingrid. "Pay Dirt." *Popular Science,* February 1, 1996, 48.

INTERNET RESOURCES
Biodiversity Prospecting: Using Genetic Resources for Sustainable Development
  http://www.wri.org/wri/biodiv/bp-home.html
Bioprospecting/Biopiracy and Indigenous Peoples
  http://www.latinsynergy.org/bioprospecting.htm
Bioprospecting
  http://www.conservation.org/web/fieldact/c-c_prog/Econ/biopros.htm
Shaman Pharmaceuticals, Inc.
  http://www.shaman.com

BIOMASS ENERGY

Biomass energy refers to the usable energy derived from plant matter and animal wastes. In ecology, the term biomass typically refers to the total dry weight of all the living and once-living matter in a given area. In biomass-energy technology, various industrial processes harness the energy carried within agricultural by-products, lumber waste, "energy crops," and farm-animal wastes, and convert the energy into power for humans to use. Like sunlight, wind, and water, biomass is a renewable resource, and biomass energy creates less pollution than traditional methods of producing power, such as burning fossil fuels like coal and petroleum.

Advocates assert that biomass energy is a sustainable, relatively clean technology that warrants attention and financial backing. Yet some observers maintain that it is not worth developing because it will never contribute a significant portion of the power needed to meet people's energy needs. Many agree with the need to develop renewable energy sources such as biomass but have voiced concern about whether the benefits of developing the technology are likely to justify the costs.

Scientific and Technological Description

Biomass-energy technology involves the burning of carbohydrates in plant matter and animal waste (see figure). The Earth is continually bathed in SOLAR ENERGY. Plants store the Sun's energy directly, by using it to produce energy-rich substances, called carbohydrates, within their tissues. The energy is stored in chemical form, in the bonds between the carbon atoms that make up the skeletons of carbohydrates and the bonds between carbon atoms and hydrogen and oxygen atoms. When these bonds break, energy is released. One process that breaks these bonds is animal digestion, and this process enables humans and other herbivores to obtain energy from eating plants. The bonds are also broken when carbohydrates are burned. For example, when wood burns in a fireplace, the breaking of millions of chemical bonds produces energy in the form of heat.

Several sources of biomass are currently exploited for energy production. One source consists of the inedible parts of agricultural plants (such as stalks, leaves, and husks), which are abundant and typically otherwise wasted in landfills. Another source is waste from the lumber industry. Branches and debris are left on the forest floor following the cutting of trees, and waste wood accumulates in lumberyards as sawdust and scraps. These scraps are used to create biomass energy. *Energy crops,* crops that are grown specifically to be used for energy production, are a third source of biomass. A fourth source is animal waste, primarily the solid waste of plant-eating farm animals.

There are three main avenues for deriving power from the energy carried by biomass: direct combustion, gasification, and ethanol production.

Direct Combustion

Direct combustion, involving simply the burning of biomass, is the most common type of biomass technology. The heat produced by the burning process is captured to create high-pressure steam, which in turn is funneled into an industrial system where the steam powers generators that

Carbon Dioxide Exchange in Biomass-Energy Production

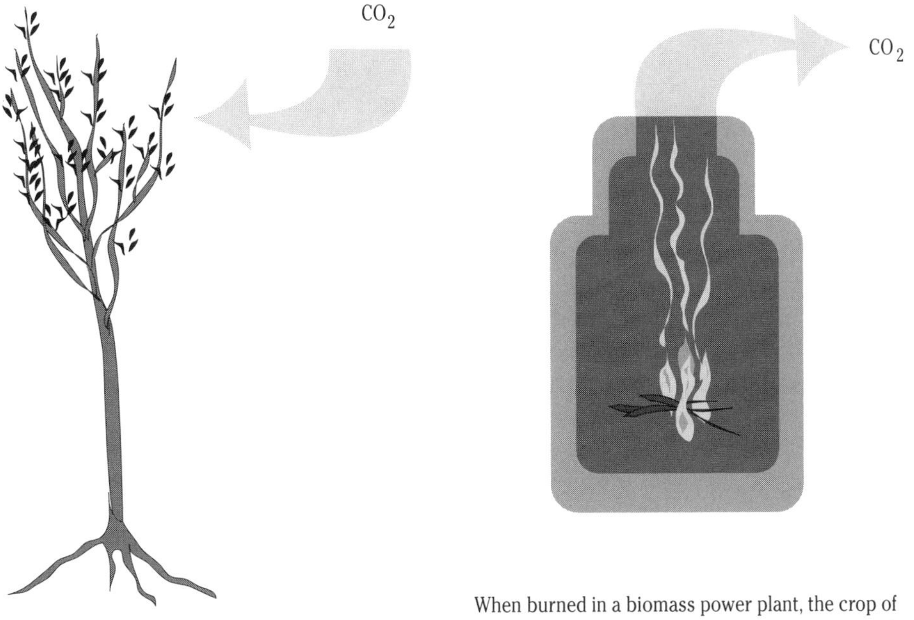

CO_2

CO_2

A crop of willow trees absorbs carbon dioxide from the atmosphere.

When burned in a biomass power plant, the crop of willows gives off the same amount of carbon dioxide that it absorbed while growing. Biomass energy causes no net increase in atmospheric carbon dioxide.

produce electricity. The same steps occur in a traditional power plant that burns coal to produce electricity.

Gasification

Gasification, a technology that is under development and not yet widely utilized, is the process of converting biomass into a gaseous fuel (similar to natural gas) consisting of combustible gases such as hydrogen, methane, and carbon monoxide. The process also yields noncombustible material, such as carbon dioxide, water, tar, and ash. The combustible gases are burned by direct combustion to create electricity or fed into machines that use the gas as a direct source of mechanization. Gasification is more efficient than direct combustion of biomass in that gasification converts a higher percentage of the biomass's original energy content into usable energy.

Ethanol Production

Ethanol is an alcohol that is added to some gasoline to produce a cleaner-burning fuel. Some biomass, particularly corn, potatoes, beets, sugarcane, and wheat, contains carbohydrates that can readily be converted into ethanol. In the production of ethanol from biomass, biomass is combined with yeast and heated, a process similar to the fermentation involved in making alcoholic beverages such as beer. The complex carbohydrates in the biomass break down into simpler substances, one of

which is ethanol. The ethanol is then isolated from the other substances.

Historical Development

The derivation of energy from biomass began hundreds of thousands of years ago, when human ancestors built fires to warm themselves and cook food. Throughout human history, people have used biomass, typically wood, to produce heat in campfires, fireplaces, and wood stoves. The development of a large-scale technology to use biomass to produce electricity and fuel arose nearly 1700 years later, in the late 20th century, as a potential way to replace coal as an energy source. Coal, which itself consists of plant matter heated and compressed over millions of years by geological processes, has a much shorter history than wood as a producer of energy for humans. Around the year 300, the Chinese began to replace wood with coal in the making of cast iron. The first mention of coal in a European text was in a book by the explorer Marco Polo written in 1298, but it was not until the 17th century that the practice of burning coal was widely adopted in Europe. As the need for electricity and mechanization increased during the Industrial Revolution in the mid-19th century, coal began to be burned in abundance. Coal provides more energy per volume of material than wood.

By the mid-20th century, people had recognized that burning coal damages the environment by releasing pollutants. Individuals and organizations in industrialized coun-

Sorghum, a so-called energy crop, is grown both to be burned as fuel and for use in making ethanol, an alcohol added to gasoline to produce a cleaner-burning fuel (Courtesy of U.S. Energy Department).

tries began to seek ways to reduce the impact of burning coal and other polluting and nonrenewable resources such as oil and natural gas. The notion to develop a technology and industry based on the burning of biomass emerged in the 1970s, along with the start of other alternative energy technologies, in response to public concern over the country's dependence on foreign-oil imports. The first biomass-fueled generators in the United States were built and operated in the late 1970s; other pioneering work in biomass technology took place in Sweden. In the 1980s, the U.S. government reduced its support of renewable energy resources, and Sweden maintained the lead in advanced biomass-energy research and implementation. In the 1990s, in the northeastern United States and in other places, a few amibitious projects have gotten under way for the cultivation of biomass crops.

Uses, Effects, and Limitations

In the United States in 1998, biomass-derived energy constituted 4 percent of the energy consumed by the country. Advocates claim that biomass technology is sufficient to support 20 percent of the country's energy needs. In the United Kingdom also, biomass is estimated to be capable of providing 20 percent of the nation's energy. The governments of numerous industrialized countries have dedicated conservative amounts of money to the development of biomass technology.

In the late 1990s, the planting of hybrid willow trees as dedicated energy crops stood out as one of the most promising biomass endeavors. In Sweden, Scotland, and New York State, cooperation among government, industry, and researchers led to the establishment of power plants fueled by fast-growing, fruitless willows.

The arguments in favor of using biomass to generate electricity revolve primarily around environmental issues. Biomass represents an alternative to traditional methods of generating power that harm the environment. Burning coal releases sulfur and nitrogen compounds into the air, causing acid rain and polluting aquatic ecosystems. Burning coal also emits excess carbon dioxide, a gas that contributes to global warming by breaking down the protective ozone layer in Earth's upper atmosphere. In addition to releasing carbon dioxide, burning petroleum also releases carbon monoxide, which pollutes the air. Moreover, coal and petroleum are limited resources; geological processes do not generate them as quickly as humans deplete them.

Biomass offers partial solutions to these environmental problems. Biomass technology releases no excess amount of carbon dioxide into the atmosphere. As plants grow, they absorb carbon dioxide from the air around them to use in cellular processes such as photosynthesis. When plants are burned at biomass power stations, the same amount of carbon dioxide that they absorbed during their growth is released into the atmosphere. Although some sulfur and nitrogen compounds may be released by the burning of certain types of biomass, the amount is less than that released by the burning of coal.

Ethanol derived from biomass is added to gasoline in some U.S. cities, such as Denver and Phoenix, where smog is a continual problem. The disadvantage of ethanol as a fuel is that a gallon of ethanol contains only two-thirds the energy of a gallon of gasoline. However, cutting gasoline to produce a mixture of 90 percent gasoline and 10 percent ethanol reduces the amount of carbon monoxide, carbon dioxide, and sulfur dioxide emitted by the mixed fuel by more than 15 percent. Some automobile research is devoted to developing an efficient engine fueled entirely by ethanol (see ALTERNATIVE AUTOMOTIVE FUEL TECHNOLOGIES). Biomass technology also benefits the environment by reducing the pressure on landfills; agricultural and lumber wastes burned for energy would otherwise have ended up further expanding the planet's garbage dumps.

Issues and Debate

There is little controversy in the biomass-energy industry itself, and few people would argue that investigating clean, renewable energy resources is an undesirable endeavor. Skeptics have noted, however, that before continuing to develop biomass technology, it is important that governments and societies evaluate the costs involved and whether the technology entails any adverse environmental effects.

Because biomass energy is a new technology, the cost of operating a biomass-powered plant has not been lowered to the level at which coal and natural gas plants are run. But strides are being made in that direction, and advocates believe that cost will not be a limiting factor for biomass energy by 2010. Skeptics argue that costs will never be as low as those for coal, because biomass carries less energy than coal, volume for volume, and because biomass has to be collected from diverse locations and often must be processed to some degree before burning. The coal that fuels a given plant, on the other hand, is typically removed from at most a few coal mines and requires little modification before it can be burned.

Some observers wonder whether demand for biomass will cause problems if the technology becomes more widespread. Perhaps, they suggest, if energy crops such as willow are lucrative, farmers might replace food crops with energy crops. Biomass advocates respond that it is highly unlikely that the value of energy crops will ever exceed that of food crops.

Biomass critics also point out that the current high demand for wood already results in the clearing of vast swaths of forest across the globe. It has been suggested that if biomass energy were considered to be yet another use for lumber, even more trees would disappear. However, chopping down virgin forest to feed power plants is not among the avenues being considered by biomass technology advocates. The only materials from the lumber industry that biomass technologists say they want to use are the scraps otherwise wasted.

By the account of many authorities, the outlook for biomass technology is positive. Biomass will probably supply a larger share of the energy consumed by the human population in the future.

—*Tamara Schuyler*

RELATED TOPICS

Alternative Automotive Fuel Technologies, Biomining, Bioremediation and Phytoremediation, Geothermal Energy, Hydroelectric Power, Nuclear Energy, Solar Energy, Wave and Ocean Energy, Wind Energy

BIBLIOGRAPHY AND FURTHER RESEARCH

BOOKS

Berger, John J. *Charging Ahead: The Business of Renewable Energy and What It Means for America*. Berkeley, Calif.: University of California Press, 1998.

Hagan, Essel B., and Charles Wereko-Brobby. *Biomass Conversion and Technology*. New York: Wiley, 1996.

International Energy Agency. *Key Issues in Developing Renewables*. Paris: Organization for Economic Coordination and Development (OECD), 1997.

Klass, Donald L. *Biomass for Renewable Energy, Fuels, and Chemicals*. San Diego, Calif.: Academic Press, 1998.

Saha, Badal C., and Jonathan Woodward, eds. *Fuels and Chemicals from Biomass*. Washington, D.C.: American Chemical Society, 1997.

White, James C., ed. *Global Energy Strategies: Living with Restricted Greenhouse Gas Emissions*. New York: Plenum Press, 1993.

Zoebelein, Hans. *Dictionary of Renewable Resources*. New York: VCH, 1997.

PERIODICALS

Beardsley, Tim. "Bring Me a Shrubbery." *Scientific American*, November 1996, 20.

Donovan, Christine, and Jeffrey E. Fehrs. "Wood Alternatives." *Independent Energy*, October 1998, 44.

Fitzgerald, Mark C. "Renewable Energy Today and Tomorrow." *The World & I*, March 1999, 150.

Halbouty, Michel J. "Using All Energy Resources a Matter of National Survival." *Houston Chronicle*, September 13, 1998, 4.

"The State of U.S. Renewable Power." *Mother Earth News*, February/March 1999, 16.

INTERNET RESOURCES

Biomass
http://www.nrel.gov/research/industrial_tech/biomass.html

Biomass Energy Research Association
http://www.crada.com/bera/about.html

Biopower Home Page
http://www.eren.doe.gov/biopower/

EREC Brief: Biomass Energy Information Sources
http://www.eren.doe.gov/consumerinfo/refbriefs/t316.html

BIOMINING

Biomining uses bacteria to mine and extract metals and minerals such as copper, iron, gold, phosphate, and uranium from the earth in mining and environmental cleanup. Certain bacteria have the ability to release these materials from rock and mineral deposits, making them reclaimable by traditional mining methods.

Mining with bacteria is a truly simple method of extracting valuable metals and minerals from ore. The costs to set up a biomining reactor are low and the technology is quite simple. Although still in its infancy as a technique, biomining has proven to be an environmentally safe and extremely efficient mining approach that is likely to come into wider use in the 21st century.

Scientific and Technological Description

Hydrometallurgy, the science of extracting metal from solutions, has been practiced for thousands of years. *Biomining* (sometimes called *biohydrometallurgy*) applies bacteria to assist. Certain bacteria of the genera *Thiobacillus* and *Leptospirillum* have the ability to use sulfur as their means of generating energy. (Humans and animals use oxygen.) These bacteria can live only in rocky environments and are often found in wastewater from mines. As the bacteria use sulfur, they convert the metals that are locked in rocks and minerals into water-soluble ionized forms. *Ionization* is the addition or removal of electrons from a neutrally charged molecule. Ions are usually more water soluble than are neutral metals. The metal ions in the solution, including iron, copper, uranium, and gold, are easily converted into pure metals by centuries-old industrial processes.

The key to this extraction method is the ability of the bacterium *Thiobacillus ferrooxidans* to take sulfur and, in the presence of oxygen and iron, convert it into sulfuric acid. This causes an increase in the acidity of the surrounding water, reducing the pH value (a measure of acidity) to 1.5. Such a strong acidity helps catalyze conversion of the metals in rocks and minerals into soluble ions. When the solution surrounding these bacteria is drained off and an electric current is run through it, the metal precipitates out of the solution and can be extracted in its pure form. Biomining is essentially an open pit of ground-up ore, doused with mild sulfuric acid, with a pipe to collect the runoff.

Typical Biomining Operation

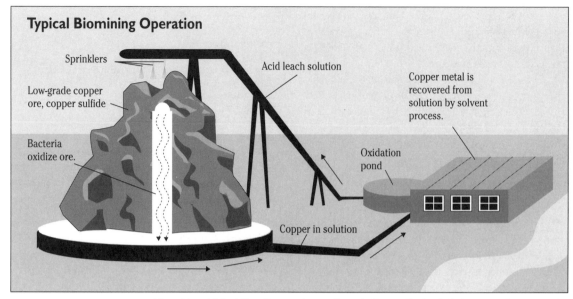

Sprinklers

Acid leach solution

Low-grade copper
ore, copper sulfide

Copper metal is
recovered from
solution by solvent
process.

Bacteria
oxidize ore.

Oxidation
pond

Copper in solution

Adapted from Michael Sims, Bioremediation: Nature's Cleanup Tool, *DIANE Publishing Company, 1993.*

Miners often mine the earth for high-grade copper in open-pit mines. The ores recovered from mines are "raw" rocks, a mixture of many metals and minerals. High-grade ores contain high concentrations of easily extractable copper. Digging into and tearing up the earth also generates tons of low-grade ores that are much more difficult and expensive to process and extract copper from. Formerly, these ores would be tossed aside as waste products. Now, with the aid of bacteria, miners can get the copper out of this low-grade ore with relative ease and high efficiency (see figure).

The refiners grind the low-grade copper ores to a powder and spread it out over large fields called *dumps* or *heaps*. A leach solution of water and sulfuric acid is poured over the ore. The solution trickles through the ore, which stimulates *Thiobacillus* to grow and convert the copper ores into water-soluble copper sulfate. Iron and sulfur molecules, an essential part of the reaction, are commonly found in all sorts of raw ores. When they are bonded together, they form the very common mineral pyrite, also known as fool's gold. The sulfuric acid reacts with the iron and sulfur and the bacteria use them to produce ferric iron and sulfate ions, both of which catalyze the conversion of ore-bound copper to its soluble form. This reaction also increases the acidity of the solution, which stimulates *Thiobacillus* to continue growing and producing acid. After the solution is drained out and collected from the dumps, adding metallic iron causes a reaction that creates soluble iron sulfate and metallic copper. The sulfuric acid produced by the bacteria is also recovered, and it can be used again on another heap.

Gold is another metal that can be recovered from low-grade ores by bacteria. (Ironically, pyrite is an ore that also contains small amounts of gold.) Gold in ores is usually bound to sulfur and is very hard to release, so traditional extraction methods are time consuming, expensive, and harmful to the environment. One such method involves heating the ores to drive out the sulfur, which is released into the environment, where it may contribute to acid rain formation. Another technique is to treat the ores with oxygen in a giant vat under tremendous pressure, which is terribly expensive. But it is fairly simple to grind the ores into powder, let the bacteria do their work, and collect the gold ions in the runoff solution. Gold is recovered from the solution by adding cyanide.

Phosphate can also be mined by bacteria. Phosphate is a mineral, not a metal, composed of a phosphorus molecule attached to four oxygen molecules. A mineral is inorganic (not carbon-based) but not a metal. Phosphate is the world's second most common fertilizer (after nitrogen) and is also used in soaps and detergents. More than 6 million tons of phosphate is mined yearly in the United States. Traditional phosphate mining is accomplished by burning raw ores, which yields solid phosphorus; the other method is to add sulfuric acid to the ore, which produces phosphoric acid and gypsum, a relatively useless rock. Both methods harm the environment. Researchers have found that the bacteria *Pseudomonas cepacia* E-37 and *Erwinia herbicola* convert glucose, a sugar, into two different acids that can dissolve the phosphate molecules out of ores. Most important, the bacteria do this at room temperature. Utilizing a process that does not involve sulfuric acid greatly benefits the environment. This phosphate-mining process is still in development and has yet to take place on a large scale.

Historical Development

Mining has always been a cornerstone of human civilization. One need only look at the historical eras labeled the Bronze Age, Iron Age, and so on, to see the importance of metals in human history. In the last 1000 years, little has changed in the field of mining. There are two basic ways to get ores from a mine. The first is to tunnel into the earth, following the paths of the richest ores, called *seams*. The ores can be broken apart by explosives or manual labor and removed from the tunnels. The second way is to strip mine, churning up the Earth's surface with explosives and massive earthmoving equipment to obtain thousands of tons of mixed high- and low-grade ores. What strip mining lacks in finding high-quality ores it makes up for in volume, rendering the process economically feasible.

Once the raw ores have been taken from a mine, the metals are purified by a process called *leaching*, which extracts metals from rocks into aqueous (water-based) solutions, or by *smelting*, which melts rocks to separate, purify, and collect the metals locked inside. Biomining is one facet of leaching, since it assists in converting metals into water-soluble forms but does not convert the ions into their pure metallic forms.

Two thousand years ago the Romans observed blue water running out of a Spanish copper mine. The blue color indicated that copper ions were in the water. The Romans never knew how the copper got into the water, but they recognized that copper was there and were able to extract it. It was not until 1957 that S. Zimmerley of the Kennecott Copper Company discovered that the bacterium *Thiobacillus ferrooxidans*, found in copper mines in Colorado, was responsible for this and other conversions. Many scientists had previously thought that the leaching of copper into water was an entirely chemical process that did not rely on bacteria. Since then, the intentional use of bacteria for mining has grown steadily in the mining industry. The most successful sector of the industry has clearly been copper; presently, around 30 percent of copper worldwide is produced with the aid of bacteria. Other sectors have had varying amounts of success: Prospectors for gold have had some luck, while uranium biomining has come and gone in cycles, and phosphate and cobalt are entering the development stage.

Uses, Effects, and Limitations

Like any bacterium, *Thiobacillus* requires certain growth conditions. It grows best in the acidic pH range 1.3 to 4.5. Although it can grow at any temperature from 10 to 37°C, it thrives in the range 30 to 35°C. It can grow with or without oxygen; however, only with oxygen can it perform any of the reactions useful to mining. In either case, it can grow only in an environment with rocks and minerals that provide acidity: iron and sulfur.

All the reactions performed by these bacteria generate heat. Therefore, as the bacteria work in the ores, the surrounding temperature will begin to rise. Eventually, it will rise to the point where the bacteria are killed off by the heat, or slow down and stop growing and metabolizing until the temperature falls back to normal ranges. Some scientists are looking into using *Archaebacteria*, a group of bacteria that can survive at elevated temperatures, for mining purposes. Others are looking at genetically engineering the bac-

Biomining offers an alternative to strip-mining, such as the operation shown above in this area northwest of Knoxville, Tennessee. Companies are beginning to rely on biomining to oxidize lower-grade ores that would not be feasible to extract using traditional methods (Photo Researchers/Hank Morgan).

teria in use today so that they become more efficient at reactions they already effect, or by modifying the temperatures and pH values at which they can thrive.

Many mines contain other metals that can be poisonous to bacteria, such as mercury, arsenic, and cadmium, in addition to gold and copper deposits. (Any acid other than sulfuric acid can also kill the bacteria.) Any of these metals can inhibit the growth of, or kill, the bacteria and bring extraction of useful metals to a standstill. Researchers are trying to overcome this by selectively breeding bacterial strains that can resist metals; many are also looking at methods of genetically engineering the bacteria, perhaps by transferring genes from other bacteria that can resist heavy metals. It will not be an easy task. Little is known about the basics of *Thiobacillus* genetics when compared to such well-studied bacteria as *Escherichia coli*.

Coal miners are also looking at using bacteria to remove sulfur from coal. The world's supplies of high-grade coal, which has a low sulfur content and burns relatively cleanly, are quickly disappearing. Sulfur released into the atmosphere by burning coal is a major contributor to acid rain. *Thiobacillus* can remove sulfur from low-grade coal, but unfortunately, the process takes months, whereas coal refiners need it to happen in days.

Issues and Debate

Biomining offers solutions to some serious economic and environmental problems associated with traditional forms of mining. The major benefit to using bacteria in the mining process comes from their ability to extract metals from ores cheaply and cleanly. Many of the ores being mined today are of low quality, as the higher-grade ores found near the Earth's surface are being depleted. Copper miners ran into this problem decades ago, as rich ores became increasingly rare. Gold miners began seeing this in the 1980s—the rich high-grade gold ores near the surface of the planet had all been oxidized (a type of ionization involving the loss of electrons) by bacteria, sunlight, or water and were easy to find and manipulate. Strip-mining for high-grade ores also produces lower grades of ores, which are usually bound to sulfur and are so difficult and expensive to extract in a way that makes economic sense but is still environmentally friendly that in the past companies usually just discarded them. Now, companies are beginning to rely on bacteria to oxidize lower-grade ores that would not be feasible to extract using traditional methods.

Some researchers believe that biomining is feasible without previous strip mining. Instead of creating a giant strip mine, they propose, the leach solution can be injected into underground deposits and later pumped out and transported to a refinery, where it can be treated. Biomining may also have political implications. Some nations are concerned about the ability of *Thiobacillus* to extract uranium from ore, just as it does copper and gold. News reports have stated that India is looking at using *Thiobacillus* to mine uranium for use in nuclear weapons. The Indian ores discovered to date are not plentiful enough to fuel a nuclear weapons program, but using bacteria to extract fully all the available uranium could conceivably produce enough for such a program. Declared nuclear powers such as the United States and the United Kingdom fear that India might use the uranium to develop nuclear weapons against its longtime rival and neighbor, Pakistan.

In addition to biomining's environmental benefits and low setup costs, bacteria usually recover more metal than do more traditional industrial processes. In copper mining, the differences are only a few percentage points, but when those points are multiplied by thousands or millions of tons of ore, bacteria are clearly better miners than are refineries and smelters. Since biomining is such a recent field, the technology is still quite primitive. In the coming decades many researchers will be looking for other bacteria that can oxidize metals. They will also be looking to improve the bacteria they already have, by searching for the most efficient bacterial strains, and by using techniques of genetic engineering to enhance and modify bacteria's current abilities.

—*Philip Downey*

RELATED TOPICS
Bioremediation and Phytoremediation

BIBLIOGRAPHY AND FURTHER RESEARCH

BOOKS
Tortora, Gerard J., Berdell R. Funke, and Christine L. Case. *Microbiology: An Introduction*, 4th ed. Redwood City, Calif.: Benjamin-Cummings, 1992.
Upland, P. *Bioremediation: Nature's Cleanup Tool*. Upland, Pa.: Diane Publications, 1993.

PERIODICALS
Cook, William J. "The Little Bugs That Dig for Gold." *U.S. News & World Report*, April 17, 1989, 62.
"Mining with Microbes." *Science News*, April 16, 1990, 236.
Moffat, Anne Simon. "Microbial Mining Boosts the Environment, Bottom Line." *Science*, May 6, 1994, 778.

INTERNET RESOURCES
Heavy Metal Bioleaching and Stabilization of Municipal Sludges
http://strategis.ic.gc.ca/SSG/es30674e.html
The Nuclear Bug
http://www.newscientist.com/ns/980711/nnuclearbug.html
Thiobacillus ferrooxidans
http://www.mines.edu/fs_home/jhoran/ch126/thiobaci.htm

BIOREMEDIATION AND PHYTOREMEDIATION

Human activities unleash enormous quantities of toxic substances, such as fuel and industrial chemicals, into the environment each year. These substances persist in soil and water, contaminating valuable resources and altering the chemistry and biology of natural systems. One pioneering cleanup technology of the 1990s is the use of bacteria and plants to degrade or stabilize toxic compounds present in soil and water. The use of bacteria is called bioremediation and that of plants is called phytoremediation.

Bioremediation and phytoremediation technologies are often preferred over traditional restoration methods, because they typically cost less and are less invasive. Their development has not been hindered by opposition or controversy. However, some scientists have noted that in certain instances bioremediation involves a significant modification of the microbiology of a contaminated site. This modification may have unforeseen ecological consequences.

Scientific and Technological Description

Bioremediation

Bioremediation takes advantage of the diversity of bacterial biology. To grow, all bacteria need a source of energy and a source of carbon to build cellular components, but different types of bacteria obtain these essentials in varying ways. Some bacteria use sunlight as their energy source, just as plants do. Others harness energy directly from chemicals that are present in their environment, such as hydrogen sulfide and ammonia. Like plants, certain bacteria obtain carbon from the carbon dioxide that is in the surrounding air, water, soil, or rock. But other bacteria rely on more complex substances as their source of carbon; indeed, some bacteria in this category can gather all the carbon they need from almost any carbon-based substance, including substances that are toxic to most other organisms.

To extract useful carbon, bacteria consume carbon-based substances and break them down into smaller constituents. When a carbon-based substance is broken down by an organism (a process called metabolism), the substance is transformed, and the products of this process have different chemical characteristics than those of the original substance. In bioremediation, bacteria break down toxic carbon-based compounds such as petroleum oil. The bacteria use the carbon they need, and the by-products of the metabolic process are typically carbon dioxide, water, and various simple, nontoxic carbon compounds that can persist harmlessly in the environment (see figure).

The most common applications of bioremediation involve the breakdown of carbon-based, or organic, compounds, as described above. But in some cases, bacteria detoxify harmful inorganic (non-carbon-based) compounds, such as nitrate, a common agricultural waste. Detoxification, which consists of transforming the original compound into nonreactive, stable compounds, occurs as a result of the various biochemical reactions that proceed during the growth cycle of a bacterial cell.

Phytoremediation

Plants, too, can aid in environmental decontamination. There are two main types of phytoremediation; in one, the presence of plants simply enhances the efficiency of bioremediation taking place naturally in the soil surrounding the plants' roots. In the other type, plants themselves either transform toxic compounds into benign ones or absorb toxic compounds and store them. The layer of soil in which plants' roots grow often contains more oxygen than does plant-free soil, because plant roots serve as a site of gas exchange between plant and soil. Oxygen is important in the metabolism of many bacteria. Consequently, there is often a higher concentration of bacteria around plant roots than in deeper soil. Thus if toxin-degrading bacteria are living in a given area of shallow soil, their numbers may be increased by the presence of plants. A greater number of toxin-degrading bacteria yields a higher rate of bioremediation.

Some plants' roots readily absorb metals such as nickel, zinc, chromium, and copper, which are toxic when present in certain forms and in high concentrations. Industrial processes release these and other harmful metals into the environment. Ragweed, mustard, broccoli, cabbage, and sunflowers are among the plants that have been shown to take up large quantities of metals through their roots from the surrounding soil. Some plants transform metals into a stable form. Others simply store the metals in their tissues.

Bioremediation and phytoremediation technologies consist of various techniques that harness, and in some cases supplement, the natural biological capabilities of bacteria and plants. The techniques are outlined in detail following a short historical review.

Historical Development

The development of bioremediation as a technology began in the early 1970s as a direct response to the increasing frequency of accidental leaks of gasoline and other toxic but biodegradable fuels. (The term *biodegradable* refers to any substance that can be broken down by an organism. Substances that no organism can break down, such as most plastics, are called *nonbiodegradable*.) Biodegradable fuels consist of carbon-based compounds called petroleum hydrocarbons, which are harmful to humans and other animals. In many cases, fuel leaks were large enough that they could not be contained, and fuel soaked into the ground and contaminated water supplies.

Scientists had known since the 1940s that some bacteria are able to degrade petroleum hydrocarbons, but in the 1970s it had only recently been suggested that underground water systems might host such bacteria. The problem of fuel contamination motivated vigorous investigation into the question of whether indigenous bacteria could help remove petroleum hydrocarbons from water supplies. One of the first research projects in this field centered around a broken gasoline pipeline in Whitemarsh Township, Pennsylvania. In 1971 several hundred thousand

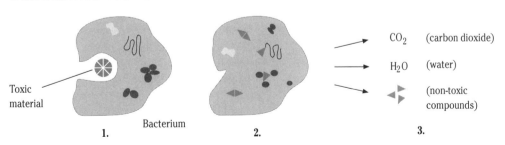

Bacterial Breakdown of Toxic Material

CO$_2$ (carbon dioxide)

H$_2$O   (water)

(non-toxic compounds)

Toxic material

Bacterium

1.   2.   3.

1. Bacteria engulf or absorb toxic material from their surroundings.

2. Metabolic processes break down the toxic material within the bacteria. Some parts of the toxic material are used as building blocks for cellular components.

3. The by-products of this process are CO$_2$, H$_2$O, and non-toxic compounds.

gallons of gasoline spilled into the surrounding environment, and some gasoline reached the local municipal water-supply well. Biologists studying the site in the aftermath of the disaster discovered that the water supply contained hydrocarbon-degrading bacteria. Their investigations also showed that the addition of excess oxygen, nitrogen, and phosphate to these bacteria (in the laboratory) increased the rate of bacterial growth. These results led to field studies showing that delivering oxygen and nutrients directly to water supplies could assist petroleum-cleanup efforts by stimulating bacteria to grow.

In 1974, engineer Richard Raymond of Suntech Oil, Inc., was issued a patent for the design of a method to introduce oxygen and nutrients to underground water systems. Raymond's design involved injection wells through which the added compounds enter a water system, and pumping wells, which circulate the compounds through a water system. Variations on Raymond's original design were used to help remove petroleum from water supplies throughout the United States during the 1970s and 1980s. These were the first large-scale applications of bioremediation.

In the 1980s, careful monitoring of the spread of petroleum leaks led to the discovery that in some cases bioremediation occurs efficiently without any help from humans. A large study published in 1991 showed that more than 7000 of California's municipal water wells were protected from petroleum contamination by natural unenhanced bioremediation, also called *intrinsic bioremediation*. Intrinsic bioremediation has since been considered an effective restoration process, although it is always accompanied by rigorous monitoring to evaluate success and determine whether human intervention is desirable.

The 1980s and 1990s saw the development of bioremediation methods applicable to contaminated soil. Phytoremediation entered the picture in the 1990s, as restoration research began to focus on plants as decontamination vessels. Fungi represent the newest candidate in the biological restoration field; research is now under way to determine fungal decontamination capabilities.

Uses, Effects, and Limitations

Soil and water are the two principal targets of bioremediation technology. Fuel, industrial chemicals, and agricultural products enter the environment either accidentally via spills or deliberately via regulated emissions from factories. Uncontained toxic material soaks into soil and eventually contaminates the underground water system. Toxic substances modify the chemistry of soil and reduce its capacity to sustain plant life; in water, toxins threaten the health of people and wildlife. Bioremediation has become a crucial player in efforts to render soil and water toxin-free.

Bioremediation may be applied to contaminated material *in situ* ("in place") or *ex situ* ("out of place"). In situ applications involve treating the material in its original location,

and ex situ applications involve removing the material from its original location, treating it, and redepositing it. Water-system bioremediation typically involves in situ rather than ex situ applications, because contaminated water is in a continual state of motion as it flows through a system. Water enters and leaves the system at various sites and in fluctuating volumes depending on aboveground factors such as precipitation, humidity, and air temperature. It would be virtually impossible to isolate and remove the contaminated water from such a dynamic system. Intrinsic bioremediation, in which degradation takes place naturally within indigenous bacteria without assistance from humans, is a widespread and effective in situ technique for cleaning water supplies. A second in situ application for water is *biostimulation*, which involves the delivery of oxygen and nutrients to the water supply to provide optimal conditions for bacterial growth.

Bioremediation of soil includes ex situ as well as in situ methods because soil is more readily accessible than underground water sources, and because soil can be more easily contained and moved. Ex situ applications for soil decontamination include *bioreactors* and *land farming*. Bioreactors are large vessels in which soil is mixed with microorganisms that can degrade the contaminant in question; bioreactors might be chosen when the contaminated soil does not naturally host the appropriate bacteria. (The application of nonindigenous bacteria to a soil sample is called *bioaugmentation*.) Land farming involves excavating and spreading the contaminated soil in a thin layer over a large area and introducing either indigenous or foreign bacteria. When the soil is land farmed, it is exposed to more oxygen as a result of the spreading process, and the soil is easier to monitor because more of it is accessible from above ground.

In situ methods of soil bioremediation include various types of biostimulation. *Passive venting*, in which channels are constructed for air to flow easily from the surface through deep layers of soil, is one way to deliver oxygen to soil; alternatively, oxygen and nutrients may be actively introduced to soil by pump systems tailored to the chemical and physical conditions of the contaminated area.

Although implementation of phytoremediation is in its infancy, several successful applications are already widely practiced. Toxin-degrading grasses and trees have been planted along the borders of crops to transform toxic pesticides and fertilizers or to enhance intrinsic bioremediation of those substances. The land surrounding garbage dumps has been protected from waste leakage by the planting of toxin-eating vegetation around dump perimeters. Plants have been installed in carefully engineered phytoremediation systems along the edges of wetlands to catch harmful compounds that would otherwise enter the sensitive ecosystems from rainfall and river runoff.

Bioremediation and phytoremediation have several advantages over traditional methods of cleaning toxic waste out of soil and water. Traditional methods include removing damaged soil and burying it in a toxic-waste dump, an expensive procedure that simply moves the toxic material to a new location where it can still leak into the surrounding environment. The dig-and-dump method also upsets the ecology of the contaminated site considerably, as it typically involves removing plants and enormous amounts of soil, both of which play important roles in the health of the entire local ecosystem.

A traditional method of cleaning water systems is to flush the system with acids or other chemicals that either break down or neutralize toxins. Other chemicals bind to toxins and carry them out of the system. The downside of this flush technique is that it kills indigenous bacteria and other microorganisms that are beneficial to the system; it may take years for populations of these organisms to become reestablished.

Biological remediation techniques, particularly in situ applications, cause fewer side effects than traditional methods, leaving the contaminated site with more of its natural components intact. They are also typically less expensive than traditional methods. There exist a few limitations of biological remediation. The restoration process may take much longer than traditional cleanups, because it relies on the natural growth and life cycles of living organisms. For the same reason, it is harder to control biological remediation. Bioaugmentation of water systems presents another problem. The addition of foreign bacteria to water is often unsuccessful, because the foreign bacteria are unable to compete with the native bacteria for oxygen and nutrients and thus disappear before they have a chance to clean the water.

Issues and Debate

Only a few controversial issues related to the development of biological remediation technology have been raised. According to some observers, it is important that biological remediation methods not divert environmental engineers' attention from developing emergency remedies. Biological remediation is not appropriate for containing disastrous accidents quickly, because it does not take effect immediately. In addition, some critics warn that biological remediation has the potential to serve as a justification for increasing the allowable amount of industrial waste. Environmentalists worry that industrial authorities might argue that the amount of waste generated is unimportant, as long as bioremediation offers a natural, efficient, and harmless way to clean up those wastes. Finally, some scientists have noted that the complexity of the underground ecosystem and its interaction with the rest of the biosphere is not well understood. They point out that introducing nonnative bacteria and plants to large areas of soil and water might alter the biological and physical relationships within the ecosystem in unanticipated ways.

—*Tamara Schuyler*

RELATED TOPICS
Biomining, Genetic Engineering, Remote Sensing, Robotics

BIBLIOGRAPHY AND FURTHER RESEARCH

BOOKS
Baker, K. H., and D. S. Herson, eds. *Bioremediation*. New York: McGraw-Hill, 1994.
Barry, K. R. *Practical Environmental Bioremediation*. Boca Raton, Fla.: Lewis Publishers, 1992.
Crawford, Ronald L., and Don L. Crawford, eds. *Bioremediation: Principles and Applications*. New York: Cambridge University Press, 1996.
de Serres, Frederick J., and Arthur D. Bloom, eds. *Ecotoxicity and Human Health: A Biological Approach to Environmental Remediation*. Boca Raton, Fla.: CRC/Lewis, 1996.
Eweis, Juana B. *Bioremediation Principles*. Boston: WCB/McGraw-Hill, 1998.
Sayler, Gary S., John Sanseverino, and Kimberly L. Davis, eds. *Biotechnology in the Sustainable Environment*. New York: Plenum Press, 1997.

PERIODICALS
Coghlan, Andy. "How Plants Guzzle Heavy Metals." *New Scientist*, February 17, 1996, 17.
Johnson, Dan. "Flowers That Fight Pollution." *Futurist*, April 1999, 6.
Johnson, Tom. "Bioremediation Is Effective for Contaminated Site Cleanup." *Capital District Business Review*, September 28, 1998, 31.
Mountain, Stewart, and Mario Verdibello. "Oil in the Soil." *Civil Engineering*, November 1998, 52.
"Plants Recruit Oil-Detoxifying Microbes." *Science News*, August 5, 1995, 84.
Riggle, David, and Kevin Gray. "Using Plants to Purify Wastewater." *BioCycle*, January 1999, 40.

INTERNET RESOURCES
Bioremediation Resources
    http://www.nalusda.gov/bic/Biorem/biorem.htm
Environment Canada's Bioremediation Resources on the Internet
    http://gw2.cciw.ca/internet/bioremediation/
EPA's Hazardous Waste Cleanup Information Site
    http://www.clu-in.org/
Phytoremediation
    http://www.engg.ksu.edu/HSRC/phytorem/
U.S. Geological Survey's Bioremediation Site
    http://h2o.usgs.gov/public/wid/html/bioremed.html
Using Vegetation to Enhance In Situ Bioremediation
    http://www.engg.ksu.edu/HSRC/phytorem/vegenhance.html
Other Bioremediation Web Sites
    http://www.lbl.gov/NABIR/help/help2.html

BIOSENSORS

Biosensors are instruments that use biological material to detect the presence or amount of a chemical compound within a complex solution or substance. Biosensor technology couples highly specific biological interactions with electrochemistry (the study of the electrical properties of chemical reactions) and physics to yield quick, accurate detection devices.

Biosensors are used in medical, environmental, chemical, and food-manufacturing industries, to detect compounds such as glucose in blood, pesticides in water, and harmful bacteria in prepackaged food. Tiny microchip-based biosensors may one day be implanted in the human body to perform

Reaction Pathway in a Theoretical Biosensor

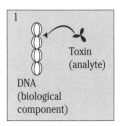

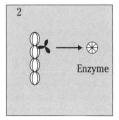

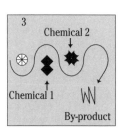

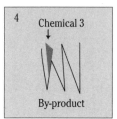

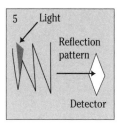

1. The analyte binds to the DNA in a biosensor.

2. The bound analyte triggers the DNA to instruct the production of an enzyme.

3. The enzyme catalyzes a reaction between two chemicals. There is a reaction by-product.

4. The by-product binds to another chemical.

5. The resulting compound reflects light in a detectable pattern. The biosensor's transducer translates the light pattern into a readable signal.

automatically analyses, such as blood tests, that must now be performed outside the body.

Scientific and Technological Description

Living organisms are the sum of millions of chemical and physical processes that take place among their constituent cells, molecules, and atoms. These chemical and physical processes include movements of charged particles, changes in molecular structure, and reactions that yield chemical by-products. Biosensors work by detecting and quantifying these processes.

Biosensors have two main parts, a biological component and a transducer. The biological component may consist of microorganisms, enzymes (proteins that catalyze, or jump start, a chemical reaction), DNA (deoxyribonucleic acid, the genetic material within living cells), antibodies (components of the immune system), or various other elements of biological origin. The biological component in a biosensor serves to detect a target compound, which is called the *analyte*. For example, in the case of a biosensor that measures the level of a toxin in a water sample, the toxin is the analyte. Biosensors are designed so that the chosen biological component will have a detectable response to the presence of the target analyte.

The *transducer* of a biosensor is a device that translates the response of the biological component into information that scientists can interpret. This information can be in the form of an electrical signal, an acoustic (sound-based) signal, or an optical (light-based) signal. An electrical signal, measured as a change in current or voltage, is produced when a flow of charged particles results from interaction between the biological component and the analyte. An acoustic signal is produced when this interaction causes a change in the mass of the biological component; for example, the analyte may bind directly to the biological component, increasing its mass. The acoustic signal is measured as a change in the frequency of a sound wave that is sent through the biological component. An optical signal is produced when the interaction between biological component and analyte causes a change in the way that light behaves with respect to the biological component. The component may begin or stop fluorescing (glowing), or it may absorb or reflect a different amount of light.

The path from the original biological interaction to the transducer's signal may be direct, or it may consist of numerous steps. What follows is a theoretical multistep path in a biosensor with DNA as the biological component and a toxin as the analyte. The toxin is a substance that binds to the DNA at a particular site and activates a particular gene. When the gene is activated, it directs the manufacture of an enzyme, which itself catalyzes a reaction between two chemicals. The reaction has a chemical by-product, which binds to another compound and causes it to change shape and reflect light at a different wavelength. The bound compound is isolated, and its new reflection pattern produces an optical signal translated by the transducer. The design of such a biosensor would be highly specific and complex.

Miniaturization in biosensor technology has resulted in the development of devices dubbed microbiosensors. A microbiosensor is built on a silicon-based microchip. The biological element is typically an enzyme, and the transducer operates by translating a flow of charged particles into a digital signal. Some microbiosensors provide electricity to drive a reaction between the biological component and analyte that would not otherwise take place.

One advantage of microbiosensors over their larger counterparts is that several distinct devices can be integrated in one chip, allowing simultaneous measurements of several analytes within a single small sample. In addition, microbiosensors are small and may therefore readily be transported to testing sites, are inexpensive, and may be mass produced. In some cases, microbiosensors are able to detect and quantify much smaller amounts of analyte than are larger biosensors. Some biosensors detect simply whether an analyte is present within a substance. Others measure the amount of analyte at a particular location within a substance or the concentration of analyte throughout the substance.

Historical Development

Enzymes, which served as the biological component in the earliest biosensors, were discovered in 1897 by German biologist Eduard Buchner. The discovery emerged from studies of fermentation (the breakdown of sugars by bacteria and yeast that is used to produce bread, alcohol, and other products). At the time it was widely believed that fermentation required the participation of live yeast cells. Skeptical of this, Buchner ground up yeast cells with sand and extracted a cell-free mixture. Added to a sugar solution, this mixture caused fermentation just as live yeast cells would have. Buchner concluded that there was a substance in yeast cells responsible for activating the fermentation process, and he called the substance *zymase*. We now call the family of such substances, which catalyze biochemical reactions, *enzymes*.

The nature of enzymes remained a mystery until 1926, when J. B. Sumner succeeded in crystallizing the enzyme urease. He isolated samples of the enzyme in solid crystal form and was thus able to analyze the enzyme's structure. His experiment suggested that enzymes were proteins. Biologist John Northrop extended the suggestion by crystallizing numerous enzymes that play roles in animal digestion. The understanding of enzymes as proteins allowed enzymes to be isolated from biological material and studied; medical research in particular focused on enzymes.

Before enzymes could be joined with electrical devices to become biosensing agents, the science of electrochemistry had to mature. Electrochemistry is the study of the exchange of charged particles between chemical compounds and the resulting electrical properties of chemical reactions. Although much work had been done in electrochemistry by the end of the 19th century, a crucial step had yet to be taken that would allow scientists to use electrochemical methods to identify and quantify the constituents of a chemical solution. To do this, scientists had to be able to translate a readable signal, such as the electrical potential of an electrochemical system, into information about the compounds in the system. In the 1920s, Czech chemist Jaroslav Heyrovsky provided the answer; he elucidated the relationship between a system's electrical potential and the chemical reactions taking place within it. Scientists used Heyrovsky's work as a basis for further research into electrochemical systems as sensing instruments.

In 1962, the first working biosensor was introduced by L. C. Clark and C. Lyons at a meeting of the New York Academy of Sciences. The device was designed to measure glucose levels in blood. Its biological component was the enzyme glucose oxidase, which catalyzes a reaction in which glucose is converted into gluconic acid and hydrogen peroxide. This reaction removes oxygen from the surrounding solution. Clark and Lyon's biosensor detected and measured this drop in oxygen as an electrical signal and translated it into a direct quantification of the concentration of glucose in the blood sample.

Clark continued to work in the pioneering biosensor field, and in 1974 a glucose biosensor of his design (the Model 23 YSI analyzer), built by the Yellow Springs Instrument Company, appeared on the market. Glucose biosensors are primarily for diabetics, who must monitor their glucose level to remain healthy. Biosensor development blossomed and slowly moved into non-medical applications such as food and environmental analysis. As scientists' understanding of molecular biology increased, the pool of candidates for biological components widened to include cells, DNA, and antibodies. Medical uses continue to dominate biosensor technology; in the mid-1990s, 80 percent of the biosensor market belonged to glucose analysis specifically.

Uses, Effects, and Limitations

Biosensors serve in numerous applications. Detecting and quantifying both wanted and unwanted substances is a common procedure in medicine, food manufacturing, various chemistry-based industries, and environmental monitoring. In addition to checking the level of glucose in diabetics' blood, medical uses of biosensors include detection of the following: viruses (such as influenza), poisonous contaminants or drugs in blood, protein in urine, genetic disease, and pregnancy. One food-manufacturing use of biosensors is detection of harmful microorganisms, such as *Salmonella* bacteria, in both fresh and prepackaged meat and dairy products. Other applications related to food include monitoring levels of various compounds during the fermentation process, and testing for compounds that indicate flavor, nutrition, and shelf life of a product. Manufacturers of agricultural and industrial chemicals use biosensors to check the chemical makeup of solutions at various stages of the processes that yield their products. In environmental engineering, biosensors are used to test water supplies for the presence of microorganisms and chemical contamination. Also, biosensors may be used to check the content of industrial wastes.

Several factors make biosensors more desirable than traditional detection methods that use chemical and physical means to signal the presence of an analyte. One example of a traditional detection technique is chromatography, in which a mixture is passed over or through a liquid or solid substance; molecules within the mixture will adhere to the substance more or less strongly, depending on their chemical and physical properties. Typically, there is some visible way to identify the analyte when it adheres to the substance. Biological sensing material, in many cases, can be more specific and faster than chemical or physical sensors. Unlike some detection methods, biosensors do not always require that a testing sample be removed from the substance that is being tested. The removal and preparation of a sample for traditional analysis can alter the chemical makeup of a substance, yielding a faulty analysis. Also, sample preparation takes time, and the original substance may have changed during the time between sample removal and test result. Biosensors are capable of delivering results that concur with the actual state of the tested substance. Another benefit of biosensors is that they are often reversible and reusable. For example, if the biological component is altered during detection of the analyte, a controlled reverse reaction may return the biological component to its original state, ready to perform detection again.

Some traditional detection methods require extensive equipment; biosensors are typically compact and may be taken into the field for on-site testing. Such portability is desirable for most environmental monitoring, so that samples do not have to be transported to a laboratory. Some traditional sensing techniques involve using toxic substances as part of the detection procedure. Biosensors therefore reduce the production of toxic wastes by a small amount.

Biosensor technology is limited by several weaknesses. Some biological components are either expensive or not readily available, and for some analytes, there is no known biological component that can serve as a detection device. A major drawback of biosensors is that most biological components are unstable and can easily or quickly undergo changes that impede their function. Efforts to overcome this roadblock include the development of stable synthetic enzymes, which some scientists call *synzymes*. Synzymes might also solve a similar problem, the creation of disruptive by-products during reactions taking place within a biosensor.

Issues and Debate

In general, biosensors are widely regarded as a beneficial, benign development in sensing technology; few controversies related to biosensors have been brought forward to date. One issue that might engender opponents in the future, however, concerns implantable microbiosensors. The small size of microbiosensors makes them suitable for implantation into humans and other animals. Because microbiosensors give an electronic signal, the detection results they deliver could be checked remotely. Science fiction writers have imagined people being implanted with microchips for a variety of purposes. Future implantable microbiosensors might perform convenient analyses, such as monitoring glucose levels without removal of blood samples. But some observers worry that such microbiosensors might be capable of being implanted, without a person's knowledge or consent, for detecting private information such as the presence of genetic disease.

—Tamara Schuyler

RELATED TOPICS
Genetic Engineering, Light-Activated Drugs, Supercritical Fluids

BIBLIOGRAPHY AND FURTHER RESEARCH

BOOKS
Blum, Loïc J., and Pierre R. Coulet, eds. *Biosensor Principles and Applications*. New York: Marcel Dekker, 1991.
Bockris, J. O., and Amulya K. N. Reddy. *Modern Electrochemistry*. New York: Plenum Press, 1998.
Chaplin, M.F., and C. Bucke. *Enzyme Technology*. New York: Cambridge University Press, 1990.
Scott, A. O., ed. *Biosensors for Food Analysis*. Cambridge: Royal Society of Chemistry, 1998.
Turner, Anthony P.F., Isao Karube, and George S. Wilson. *Biosensors: Fundamentals and Applications*. New York: Oxford University Press, 1987.

PERIODICALS
Chase, Victor D. "Building Better Biosensors." *Appliance Manufacturer*, February 1999, 8.
Hartman, Nile F. "Reducing Food Poisoning Deaths." *Resource*, August 1997, 11.
Sikorski, Robert, and Richard Peters. "Lactamase Live!" *Science*, January 16, 1998, 412.
Weaver, Tara. "Biosensor Detects Chemical Residues." *Agricultural Research*, November 1998, 14.

INTERNET RESOURCES
American Chemical Society's Environmental Biosensors Status Report
http://www.westminster.edu/Acad/chem/faculty/ttw/chem612/environmentalbisensors/environmentalbisensor.htm
Massachusetts Institute of Technology Biology Hypertextbook: Enzyme Biochemistry Chapter
http://esg-www.mit.edu:8001/esgbio/eb/ebdir.html
Quantech Biosensors Site
http://www.biosensor.com/

CD-ROMs

CD-ROMs are compact disks (CDs) with read-only memory (ROM) used to store vast amounts of computer data (approximately 450 times as much as can be stored on a 3.5-inch floppy computer disk). There are numerous kinds of CD-ROMs, and they are used for a variety of different purposes, such as distributing software and running computer games.

Although CD-ROMs have revolutionized computing, they are not without their problems. It takes much longer to retrieve data from a CD-ROM than from an ordinary hard drive; CD-ROM capacity (however large it seems) is still limited; and although the majority of desktop computers are shipped with CD-ROM drives, most of these are read-only devices (they cannot write data to CD-ROMs). Although CD-ROMs were greeted with great excitement when they were launched in the early 1990s, they are widely expected to be replaced by DIGITAL VIDEO TECHNOLOGY and the Internet.

Scientific and Technological Description

A CD is a very thin disk of aluminum between two protective layers of plastic. Traditionally, computers store and process data (information) in binary code, which is a method of representing decimal numbers using only the

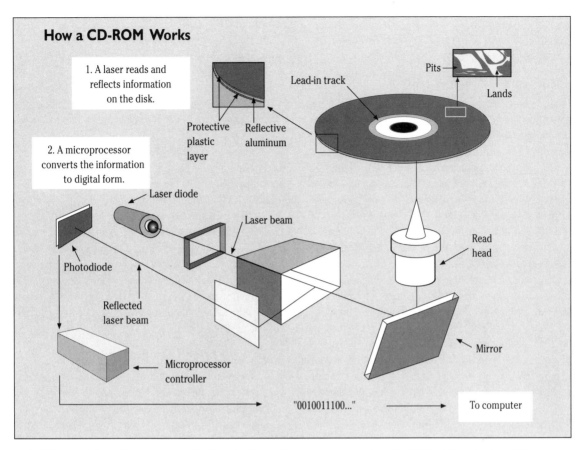

How a CD-ROM Works

1. A laser reads and reflects information on the disk.

2. A microprocessor converts the information to digital form.

Protective plastic layer

Reflective aluminum

Lead-in track

Pits

Lands

Laser diode

Laser beam

Photodiode

Reflected laser beam

Read head

Mirror

Microprocessor controller

"0010011100..."

To computer

A CD-ROM stores information as a pattern of hollows (pits) in a reflective aluminum surface (lands). The information is read by a scanning laser beam, which reflects back from the lands but not from the pits. A microprocessor in a CD-ROM player interprets the pattern of pits and lands into digital (numerical) form that can be understood by a computer.

numbers zero and one. Binary information is stored on a CD using a pattern of *pits* (tiny holes) in the aluminum; the start or end of a pit represents a one, whereas a *land* (a flat area of aluminum where there is no pit) represents a zero. The pits and lands are arranged in a long, continuous spiral that runs clockwise from the center of the CD to the rim.

Unlike traditional methods of storing computer data, which used magnetism, CD-ROMs work using optical technology (i.e., light). The pattern of pits and lands is read using a precisely focused laser beam that scans across the surface. Lands reflect the beam straight back, because they are part of the undisturbed aluminum surface; pits, on the other hand, scatter the laser beam and reflect it only partially. A light-sensitive device called a photodiode mounted next to the laser detects a full or partial reflected beam, and a microprocessor in the CD drive works out from this the pattern of zeros and ones stored on the disk (see figure).

The laser can scan backward and forward to access data stored in different parts of the disk. In fact, the lead-in track (the part of the CD closest to the center) contains a table of contents of what is on the CD and where it is stored. However, moving the laser to the place where the data are

stored takes time, perhaps 150 milliseconds (0.15 second), or 20 times as long as a hard drive takes to locate data.

Although CDs are popularly known as a read-only medium, recordable CDs became widely available from 1998 onward. CD-R (CD-recordable) disks are also known as CD-WORMs (write once read many), and are instantly recognizable from their highly reflective gold coloring. The process of recording a CD is known as *burning*, because a laser creates the pattern of pits by melting a special recording layer (between the gold layer and the plastic base) made of an organic dye such as cyanine. This process can happen only once, so once written, a CD-R cannot be rewritten.

CD-RW (CD-rewritable) goes a step further, enabling data to be written and rewritten thousands of times. CD-RWs use a different type of CD, whose recording layer consists of an alloy (fused mixture of metals) of silver, indium, antimony, and tellurium. The alloy can exist in one of two states, either as a regular crystal structure or as an amorphous (structured in no special way) noncrystalline solid. In a blank CD-RW, the recording layer is entirely in the crystalline form. Data are recorded using a laser that heats a precise area of the crystal structure above its melting temperature and turns it into the amorphous form. Data are stored in this way until the recording layer is heated again, which transforms the alloy back into a regular crystalline structure, erasing the data and allowing other data to be recorded.

Historical Development

Compact disks first appeared in 1983 with the announcement of compact disk digital audio (CD-DA) or music CDs by the Sony and Philips corporations. Music CDs are recorded according to a set of standards known as the Red Book, which ensures compatibility among different manufacturers. CD-DA rapidly evolved into other formats that could be used to store data as well as music. The Yellow Book is a set of general standards for CD-ROMs. Other standards include the Orange Book, which defines how CD-R (WORM) disks should be recorded, and the White Book, which refers specifically to video CD-ROMs.

Some corporations have found the original CD-ROM standards too restricting and have introduced alternative standards of their own. For example, the Green Book is concerned with CD-I (CD-interactive). This type of CD-ROM was launched by Philips in 1991 as a sophisticated form of multimedia, which included CD-quality stereo sound and full-motion video. It was designed to run on 16-bit processors (computer chips that process 16 binary digits at a time).

Another CD-ROM standard known as 3DO was launched in 1993 by the U.S. computer games company Electronic Arts. As the name suggests, it was designed specifically to handle the fast three-dimensional motion and lifelike animation that computer game players demand. It was the first standard for 32-bit processors (chips twice as powerful as 16-bit processors) and used a technique known as RISC (reduced instruction set computing), which makes computers work faster by reducing the number of instructions they have to understand. A refinement of 3DO known

Harvard University Professor Dr. Kwame Appiah prepares a demonstration of the Encarta Africana Multimedia CD-ROM in February 1999. Appiah and fellow professor Dr. Henry Louis Gates (center) co-edited the CD-ROM encyclopedia of black history and culture. Because CD-ROMs can store a wide variety of data types in digital form, text, graphics, animation, video, and sound can all be combined to create a multimedia product (AP/Bebeto Matthews).

as 3DO M2, introduced in 1995, could handle even more sophisticated graphics, thanks to its use of 64-bit processors (four times as powerful as the 16-bit processors used in early CD-I machines). Games manufacturers such as Nintendo and Sega have developed their own proprietary formats. Another proprietary format is Kodak's Photo-CD system, which enables high-resolution photographs to be stored in digital format on CD-ROM.

Other developments of CD-ROM technology have improved the speed and capacity of CDs. A typical CD can store about 74 minutes of music or 650 megabytes (MB; 1 million bytes) of data, which is roughly 450 times as much as can be stored on a 3.5-inch disk. The first CD-ROM drives transferred data to the computer's memory at the rate of about 150 kilobytes per second (kB/s). Today's CD-ROM drives are described as so many times this rate; for example, a 4X CD-ROM drive transfers data at 600 kB/s. The latest CD-ROMs, known as DVDs (digital video disks or digital versatile disks), can store up to 4.7 gigabytes (GB; 1 billion bytes or 1000 megabytes) of data.

Uses, Effects, and Limitations

It is no exaggeration to say that CD-ROMs have revolutionized computing. Before their invention, computer games were much less sophisticated, software was distributed on cumbersome 3.5-inch disks, and there was no such thing as multimedia. Computer games have benefited from CD-ROMs in two principal ways. CD-ROMs can hold 650 MB of data, which enables games to store complex backgrounds, data, or sound on a CD and load it into a computer's memory only when it is needed, which has made games much more sophisticated. In addition, the animation, video, and sound capabilities of CD-ROMs have transformed simple computer graphics into a primitive kind of VIRTUAL REALITY.

Software is also much cheaper to distribute and easier to install from CD-ROMs than from 3.5-inch disks. Large programs such as the Windows operating system used to be distributed on a stack of 3.5-inch disks, which had to be installed one at a time. A single CD-ROM is big enough to hold the entire essential Windows software; installing it is a matter of a few mouse clicks. Because CD-ROMs are so inexpensive to manufacture, it is now commonplace for computer magazines to attach to their covers CDs containing free trial software. But it is in the development of multimedia that CD-ROMs have had their greatest impact. Throughout the 1990s, innovative publishers have used the power of CD-ROMs to produce educational and games software that combines text, graphics, animation, video, and sound. The *Encyclopaedia Britannica*, which used to be available only as a shelf full of heavy paper volumes, can now be bought on a single CD-ROM for about $150. Multimedia CD-ROMs have also played a key role in the explosive growth of the multimedia Internet. Had CD-ROMs not spurred the development of computers that can handle three-dimensional graphics, animation, sound, and video, the Internet might still be restricted to text and simple images.

The success of CD-ROMs can largely be attributed to their high capacity for storing a wide variety of data types in digital (numerical) form. Unlike analog methods of recording (such as vinyl records), the data in digital recordings are unaffected by slight scratches and blemishes. CDs are uniquely forgiving; coffee spilled on them can be washed off. However, CD-ROMs are not without disadvantages. The plethora of largely incompatible formats (from CD-I to 3DO and CD-RW to Photo-CD) continues to confuse people; most CD-ROM formats can be played on only one type of player. Although the capacity of CD-ROMs seemed huge when they were first launched, the format struggles to keep up with the ever-increasing size of computer hard disks (measured in megabytes when CD-ROMs were invented, but now measured in gigabytes).

Issues and Debate

Unlike many other cutting-edge technologies, CD-ROMs have no major drawbacks. The main problem with them so far has been the poor compatibility between the various CD-ROM formats, which means that different CD-ROM readers sometimes have to be purchased for different applications. But unless manufacturers had been willing to risk improved versions of basic CD-ROM standards such as 3DO, CD-ROMs might not have progressed at all. Compatibility is less of a problem with newer formats, such as CD-R. A CD-R disk can record data from a variety of other formats, including CD-DA, CD-ROM, and CD-I, but not from the proprietary formats of computer game manufacturers. CD-I and 3DO can also view photographs stored using Kodak's Photo-CD system, and most computer CD-ROM drives can play audio CDs.

Recordable CD-ROMs (CD-R and CD-RW) have raised copyright issues, because in theory they permit large-scale illegal duplication of audio and software CDs. It was opposition from the music business that slowed the spread of recordable CD technology at first, but that technology is now becoming widespread as a result of antipiracy initiatives. First, blank audio CD-R disks are more expensive than their equivalent in computer disks; part of that price is distributed to recording artists in the form of a royalty to compensate for the inevitable piracy of their material. Also, audio CD-R players contain devices that prevent copies from being made of other copies; in other words, they ensure that a certain number of original disks must always be purchased. In another antipiracy technique, CDs and DVDs are coated with a polymer that gradually darkens, restricting the number of times they can be played.

Probably the most significant area of debate concerns whether CDs are the ultimate method of information storage. This seems unlikely, for a number of reasons. In a dynamic world, CD-ROMs become out of date almost as soon as they are created; the World Wide Web (WWW) has already proved to be a much more versatile means of distributing up-to-date information. But even as static methods of storage, CD-ROMs already seem dated. DVDs and DVD-ROMs can now store about seven times more information than CD-ROMs by using laser beams of shorter wavelength, which enable the pits and lands to be reduced to less than half of their standard size. Scientists are already speculating that computers of the future might store binary information by changing the state of individual atoms in a solid material. That new approach, known as *quantum computing* because it utilizes the principles of quantum mechanics, would enable the storage of unprecedented amounts of information, making CD-ROMs seem as outdated and cumbersome as chiseled tablets of stone seem today.

—Chris Woodford

RELATED TOPICS
Digital Video Technology, High-Fidelity Audio

BIBLIOGRAPHY AND FURTHER RESEARCH

BOOKS
Cunningham, Steve, and Judson Rosebush. *Electronic Publishing on CD-ROM: Authoring, Development, and Distribution*. Huntsville, Ala.: The CD Info Company, 1996.
Hoffos, Signe. *CD-I Designer's Guide*. London: McGraw-Hlll, 1992.
Holsinger, Erik. *How Multimedia Works*. Emeryville, Calif.: Ziff-Davis, 1994.
Multimedia: The Complete Guide. London: Dorling Kindersley, 1998.
Pivovarnick, John. *The Complete Idiot's Guide to CD-ROM*. Indianapolis, Ind.: Alpha Books, 1994.

PERIODICALS
Andrews, Dave. "CD-ROM Weds the Web." *BYTE*, November 1996, 32.
Bell, Alan. "Next Generation Compact Discs." *Scientific American*, July 1996, 28.
Fox, Barry. "This Message Will Self Destruct." *New Scientist*, November 15, 1997, 7.
Ghanadan, Hamid. "Making Multimedia." *Science*, September 18, 1998, 1814.
Gunshor, Robert, and Arto Nurmikko. "Blue-Laser CD Technology." *Scientific American*, July 1996, 34.
Halfhill, Tom. "CDs for the Gigabyte Era." *BYTE*, October 1996, 139
O'Malley, Chris. "Finally! Recordable CDs." *Popular Science*, June 1998, 51.
O'Malley, Chris. "The Gig Is Up." *Popular Science*, February 1999, 68.

INTERNET RESOURCES
Hardware Central: Multimedia Basics
http://hardwarecentral.com/hardware/multimedia/mmbasics/
Online & CD-ROM Review Magazine
http://www.learned.co.uk/olr/index.asp
Special Interest Group on CD/DVD Technology
http://www.sigcat.org/

CLONING

Cloning refers to the creation of an organism whose genetic makeup is identical to that of another organism. Plant cloning is widespread in agriculture, and amphibian cloning is common among researchers who investigate animal development. However, the technology of cloning mammals is in its infancy. Scientists hope that once it has become financially feasible, cloning will become an efficient means of producing farm animals that can serve a wide variety of functions. General cloning research also promises to offer insights into human development and disease.

Cloning moved into the public spotlight in 1997, when sensational media coverage of Dolly, the first mammal cloned from an adult cell, stirred up questions of where cloning technology would lead. The most contentious issue surrounding cloning is the possibility that advances in the field will lead to the ability to clone human adults, which many people believe would be unethical.

Scientific and Technological Description

Cloning concerns the biology of development, the processes through which an organism proceeds on the way from fertilization to the adult stage. A single fertilized egg cell holds the potential to become a fully formed organism; it begins this process by dividing into two cells, four cells, eight cells, and so on. Every new cell contains the same genetic information (the same set of genes) as that of the original egg cell. As division proceeds, the cells of the resulting embryo begin to differentiate or become specialized with respect to function. Some cells are destined to become brain cells, others will be part of the liver, others form the skin, and so on. Although all the cells contain copies of the same genes, during differentiation each cell type is programmed to activate only some of its genes. For example, in brain cells, the genes that are involved in producing *neurotransmitters* (chemicals that participate in the function of the nervous system) are turned on, whereas in liver cells these genes are turned off. Experiments have shown that a differentiated cell can actually revert back to an undifferentiated state. That is, a cell can recover the potential to activate any of its genes, even those it has been programmed to turn off. Recent cloning takes advantage of this process of *reversibility*.

The basic technique involved in the cloning of animals, called *nuclear transfer*, involves the transfer of one cell's nucleus into another cell (see figure). This represents a transfer of genetic material, as the nucleus of a cell contains most of its genetic material. Nuclear transfer thus involves two cells: a *donor cell* (the one whose genetic blueprint is to be replicated in the clone), and an *egg cell* (the one that serves as the originator of the development process). First, the egg cell's nucleus is removed, a procedure called *enucleation*. Second, the donor cell is forced into the gap zero (G0) stage, during which the cell is inactive but not dead. If the donor cell is not in this dormant state, it will not be accepted by the egg cell. Next, the donor cell's nucleus is removed and transplanted into the egg cell; alternatively, the two eggs are placed next to each other and given an electric jolt that causes them to fuse. Finally, the egg cell, outfitted with a new nucleus, is triggered to begin development, just as if it were a fertilized egg. But instead of having genetic material consisting of genes from both of its parents (as a fertilized egg cell would have), the egg cell in a nuclear transfer experiment has genetic material from just one source: the donor cell.

Two types of nuclear transfer are used in cloning. One, developed at the Roslin Institute, located several miles outside Edinburgh, Scotland, is known as the *Roslin technique*. In this procedure, the donor cell is isolated from the donor animal and allowed to divide in a culture medium in the laboratory. Many copies of the donor cell result, and this can be useful if scientists want to alter the genetic

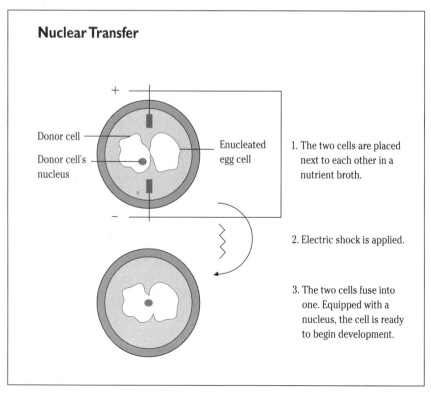

Nuclear Transfer

Donor cell

Donor cell's nucleus

Enucleated egg cell

1. The two cells are placed next to each other in a nutrient broth.

2. Electric shock is applied.

3. The two cells fuse into one. Equipped with a nucleus, the cell is ready to begin development.

During nuclear transfer, one way to move the donor cell's egg into the egg cell is by fusing the two cells with an electric shock.

material of the donor before proceeding. Then the experimenter coaxes the donor cells into the G0 stage by placing them in a substance that contains only a few nutrients; the cells stay alive but become inactive. After the egg cell has received the donor nucleus and begun dividing, the embryo is transplanted into a surrogate mother's uterus to continue development.

The second type of nuclear transfer is the *Honolulu technique*, developed at the University of Hawaii at Honolulu. To avoid the problem of having to force the donor cell into the G0 stage, scientists in Hawaii used cumulus cells as donors. Cumulus cells, which surround and nurture the egg cell before its release from the ovary, are almost always in the G0 stage. Unlike the Roslin technique, the Honolulu technique involves no culturing of the donor cell. The donor cell's nucleus is transplanted into the egg cell immediately upon the donor cell's extraction from the donor animal. The remaining steps of the Honolulu technique are the same as described for the Roslin technique.

Historical Development

People have long been fascinated by animal development. As early as the 4th century B.C., philosopher and naturalist Aristotle recorded observations concerning the development of animal embryos. But developmental biology emerged as a distinct field of research only at the beginning of the 20th century. German biologist Hans Spemann performed numerous experiments on amphibian embryos starting in the early 20th century that formed the basis of modern developmental biology. He won a Nobel prize in 1935 for his work. In 1918 Spemann split a two-celled salamander embryo by lassoing the cells apart with a human baby's hair. The two cells developed independently into a normal embryo. This experiment was the first to disprove an earlier hypothesis stating that the genetic material in each cell diminishes as the cells divide in a developing embryo. It suggested instead that each cell carries the genetic material necessary for developing into a complete embryo.

In 1928, Spemann conducted the first nuclear transfer experiment. Working again with salamander embryos, he placed the nucleus of one embryonic cell into another that had had its nucleus removed. The cell with the transplanted nucleus grew into a normal embryo. Spemann published the results of this and similar experiments in 1938 in the book *Embryonic Development and Induction*. In it, he called for experiments to test the feasibility of transferring the nucleus of an adult cell (rather than an undifferentiated embryonic cell) into another cell to create a genetic replica of an adult animal. Many scientists believed that this would not be possible because they thought that differentiation was not reversible.

In 1958, F.E. Steward of Cornell University grew whole carrot plants from a differentiated carrot root cell. Giving a name to the emerging technology, British biologist J. B. S. Haldane coined the term *clone* in a 1963 speech about the future of biological technologies. In 1984, the first cloning of a mammal took place. Danish scientist Steen Willadsen cloned a sheep from an undifferentiated embryo cell. Two years later, he cloned a cow from a one-week-old differentiated embryo cell. His work influenced Ian Wilmut and Keith Campbell, both of the Roslin Institute, to attempt to clone animals from adult cells. After years of refining the Roslin technique of nuclear transfer, Wilmut and Campbell cloned two sheep (named Megan and Morag) from differentiated embryo cells in 1995. The event was largely overlooked by the press. But two years later, in 1997, when they announced the 1996 birth of Dolly, the world's first mammal cloned from an adult cell, the world turned its attention to their work. Dolly's birth initiated discussions of human cloning within the scientific community and the public arena.

In 1998, Ryuzo Yanagimachi and Teruhiko Wakayama of the University of Hawaii cloned 50 mice using the Honolulu technique of nuclear transfer, which they developed. This technique is more efficient than the Roslin technique and is likely to be the one used in most future animal-cloning research. By the end of 1998, several cows had also been cloned successfully.

Uses, Effects, and Limitations

Scientists and advocates of cloning research envision numerous ways in which cloning might benefit society if it becomes financially feasible on a large scale. Cloned farm animals might serve several purposes. One is the more rapid spread of beneficial traits among herds. Currently, farmers use IN VITRO FERTILIZATION (IVF) to generate larger numbers of elite animals (the largest milk producers, the providers of the most tender meat or the softest wool, etc.). IVF, in which the joining of the chosen egg cell and sperm sample takes place in a test tube, allows farmers to direct which animals will be the parents of a new generation. However, there is no guarantee that mixing the mother's

In January 1998, Dr. Richard Seed spoke to reporters outside his home in Riverside, Illinois. Seed made headlines in late 1997 when he announced plans to start a fertility clinic that would begin work on cloning human beings (AP/Charles Bennett).

and father's genes will give rise to offspring that have the same beneficial traits as those of the parents. Cloning, on the other hand, would deliver offspring genetically identical to the donor animal.

Cloning may also be useful in creating animals that secrete valuable proteins in their milk. Some human diseases can be treated with doses of particular proteins, but the traditional methods of producing these proteins are problematic. Recently, scientists have begun to modify the genetic material of cows, sheep, and goats such that the animals secrete such therapeutic human proteins in their milk. Combining cloning with GENETIC ENGINEERING could produce protein-secreting animals more efficiently than do current methods. The first genetically modified cloned sheep (named Polly and announced by the Roslin Institute in 1997) secretes a human protein in her milk.

Cloning may yield animals that can provide organs for transplantation into humans. Genetically engineered pigs whose tissue is compatible with human blood are being researched as sources of transplantable organs. Cloning such pigs might be a quick means of relieving the current shortage of transplantable organs (see ORGAN TRANSPLANTATION). Animals produced by cloning offer an unprecedented opportunity for scientists to study developmental and aging processes. The typical adult cell has already undergone millions of replications, during which small mistakes in the cells' genetic material, called *mutations*, occur. Animals cloned from adult cells inherit the mistakes that have accumulated in the adult donor cell. Because mutations contribute to general aging phenomena and to an increased risk of cancer, studying these animals might give scientists insights into how genetic mistakes cause these conditions.

Although these visions of cloning's future are optimistic, they will not be realized immediately because the technology is still inefficient and expensive. To produce Dolly, scientists at the Roslin Institute performed 277 nuclear transfers and only 29 of these eggs developed normally during the first week of divisions. Of those, only one resulted in the live birth of a normal lamb (Dolly). Some died before birth or soon after, and some of those that died had developmental abnormalities. Although the Honolulu technique has a higher success rate than the Roslin technique (about three normal births for every 100 nuclear transfers), it is still costly. John Woolliams, a scientist at the Roslin Institute, predicts that it will be at least 10 years, and possibly 20, before the technology has been improved enough to allow the creation of entire herds of cloned livestock.

Issues and Debate

Most of the debate concerning cloning technology relates to the idea of cloning humans. Some observers feel that if cloning research continues, it will inevitably lead to the ability to clone human adults. In fact, American scientist Richard Seed has publicly stated that he has initiated research into cloning humans and intends to be the first to succeed. He and his supporters consider human cloning to be an ethical future solution for couples who are unable to conceive a child. Many people, including many scientists, believe that cloning humans would be unethical. Some view cloning as an inappropriate modification of the natural process of evolution. Some imagine that if human cloning (combined with genetic engineering) were available, it would lead to the ability to shape the physical characteristics of children and the ability to select children of a particular gender. Others point out that since scientists have had so little opportunity to observe the growth and development of animal clones, there is no way to be sure that cloned humans would have normal developmental processes throughout their lives. Still other opponents have noted that children cloned from a parent might experience emotional difficulties upon learning that they are genetic duplicates of one of their parents.

According to advocates, some of the remaining concerns raised by opponents of human cloning are based on misconceptions. For example, some people have claimed that cloning is objectionable because it is akin to "playing God." Advocates respond that this is not the case; cloning does not create life but reproduces life from existing life. Also, some opponents have argued that a clone would not be a normal human being and that clones might be construed as beings that could be owned by noncloned humans. Advocates refute this belief, stating that clones would be considered normal humans with the same rights as all others, just as identical twins, who have identical genetic material, are considered normal humans. Finally, some critics foretell the creation of unconscious clones to provide organs for transplantation. Scientists respond by asserting that not only would this type of experiment be unethical, it would also be impossible, because the nature and physical location of consciousness are not currently known.

In response to the controversy that followed the announcement of Dolly, some countries took official action regarding human cloning. President Clinton requested that human-cloning research not be undertaken until further studies reveal more of the issues involved. The U.S. government has not passed a law banning research involving human cloning, although the government does not fund research involving human embryonic tissue. Numerous European governments have signed a treaty calling for a moratorium on human-cloning research. The United Kingdom and Germany did not sign this treaty, although both nations have their own restrictions on research uses of human embryos.

There are also controversial issues surrounding animal cloning. Critics worry that cloning will reduce genetic diversity. Advocates agree that protecting genetic diversity should be attended to closely if cloning becomes widespread. Also, critics have pointed out the danger of transmitting viruses from animals to humans through therapeutic milk from cloned animals. A final objection concerns expense; some critics believe that the eventual benefits of cloning could not justify the huge cost of cloning research.

—Tamara Schuyler

RELATED TOPICS
DNA Fingerprinting, Gene Therapy, Genetic Engineering, Genetic Testing, In Vitro Fertilization, Organ Transplantation

BIBLIOGRAPHY AND FURTHER RESEARCH

BOOKS
Cole-Turner, Ronald, ed. *Human Cloning: Religious Responses.* Louisville, Ky.: Westminster John Knox Press, 1997.
Foote, Robert H. *Artificial Insemination to Cloning: Tracing 50 Years of Research.* Ithaca, N.Y.: Cornell University Press, 1998.
Kass, Leon, and James Q. Wilson. *The Ethics of Human Cloning.* Washington, D.C.: AEI Press, 1998.
Kolata, Gina Bari. Clone: *The Road to Dolly, and the Path Ahead.* New York: William Morrow, 1998.
Nussbaum, Martha C., and Cass R. Sunstein, eds. *Clones and Clones: Facts and Fantasies About Human Cloning.* New York: W.W. Norton, 1998.
Pence, Gregory E., ed. *Flesh of My Flesh: The Ethics of Cloning Humans: A Reader.* Lanham, Md.: Rowman & Littlefield, 1998.
Silver, Lee M. *Remaking Eden: Cloning and Beyond in a Brave New World.* New York: Avon Books, 1997.

PERIODICALS
Eisenberg, Leon. "Would Cloned Humans Really Be Like Sheep?" *New England Journal of Medicine.* February 11, 1999, 471.
Gillis, Justin. "Cows and Clones on a Virginia Farm; Where Animals Make Drugs and Gene Research Goes to the Frontier." *Washington Post,* February 28, 1999, A1.
Klotzko, Arlene Judith. "We Can Rebuild . . . " *New Scientist,* February 27, 1999, 52.
Stone, Richard. "Cloning the Woolly Mammoth." *Discover,* April 1999, 56.

INTERNET RESOURCES
Conceiving a Clone
    http://library.advanced.org/24355/home.html
New Scientist Planet Science/Cloning Report: The Latest News and Opinions on Cloning and Related Issues
    http://www.newscientist.com/nsplus/insight/clone/clone.html
Roslin Institute Online: Information on Cloning and Nuclear Transfer
    http://www2.ri.bbsrc.ac.uk/library/research/cloning/cloning.html

COMBINATORIAL CHEMISTRY

Combinatorial chemistry is a method of rapidly creating thousands of different chemicals. Using robots and computer automation, thousands of different compounds can be synthesized overnight. These compounds are stored in "libraries" and can be analyzed later at leisure.

Pharmaceutical and agrochemical companies are the largest users of combinatorial chemistry. They use it to generate thousands of chemicals for future use as drugs, pesticides, and herbicides. Synthesizing new chemicals used to be the slowest part of the drug-discovery process. Combinatorial chemistry has speeded the process so much that other researchers are scrambling to develop new ways of sorting and testing all of the newly created molecules that have resulted.

Scientific and Technological Description

Combinatorial chemistry quickly creates thousands of different chemical compounds through experiments that are precisely controlled by robots and computers. The technique involves starting with simple, well-understood molecules, and slowly adding chemicals to them to generate new molecules. The idea behind combinatorial chemistry is that small, random, step-by-step additions to a molecule will eventually yield new, interesting, and useful ones.

Stepwise additions of new chemicals encounter a part of the original molecule and add something to it; or the added chemical may encounter a new, growing branch of the molecule and add to that. In either case, the original molecule starts growing and may quickly acquire many branches and extensions. This process can be repeated many times in hundreds of test tubes, creating thousands of new chemicals.

The most popular method of combinatorial chemistry is *parallel synthesis* (see figure). A typical combinatorial experiment would involve a rack of test tubes arrayed in eight rows and 12 columns, forming 96 individual *wells*. Along each row, researchers add one specific chemical. Small amounts of each chemical have previously been attached to plastic resin beads. Keeping the molecules on beads rather than in a solution allows researchers to separate and purify compounds more easily at the end of the experiment. It is much easier to trap the beads and free the molecule from the bead's surface than to extract and purify the chemicals from a solution.

Next, an array of robotic arms holding chemical injectors is placed above the rack of test tubes. A computer controls this array as it moves around over the test tubes, adding chemicals. The additions are carefully timed and are recorded by a computer. This allows the researchers to go back and determine the process that created the final molecule.

To begin the chemical reactions, the first injector goes down a column and adds one chemical to the eight individual wells. The result is that the same chemical is added to the eight different molecules. In the next column, a second injector goes down the column and injects a different chemical, which reacts with the original molecule. The process is repeated across all 12 columns of the array, with a different chemical being added in each column. The end result is 96 (8 x 12) new molecules. The plastic beads are then rinsed with a mild solution to clear out any unreacted molecules, molecules that didn't react with the ones on the beads.

If desired, the process can be repeated, and another set of injectors containing different molecules can travel over the array, creating new molecules by reacting with the products of the first round. This would create an even larger variety of molecules. When the scientists decide that the experiment is over, the newly produced molecules, called a

A Common Combinatorial Chemical Technique: Parallel Synthesis

STEP 1
Start with a molded plastic sheet of wells (called a microtitre plate) that have been partly filled with a solution containing inert polystyrene beads (black circles). Typical microtitre plates have 8 rows and 12 columns, for a total of 96 wells; this example illustrates only the upper left corner of the plate and expanded view of one well.

STEP 2
Add the first set of molecules – the A class (squares) – to the beads, putting the A1 molecules into the wells of the first row, the A2 molecules into the wells of the second row, and so on. After one set of molecules has been added, filter the material in every well to eliminate unreacted chemicals (those unattached to beads).

STEP 3
Add the second set of molecules – the B class (triangles) – down the columns, with B1 in the first column, B2 in the second, and so on. Filter away unreacted chemicals as in step 2.

STEP 4
Once the 96-member library has been produced, detach the final structures from the beads so that the compounds can be screened for biological activity.

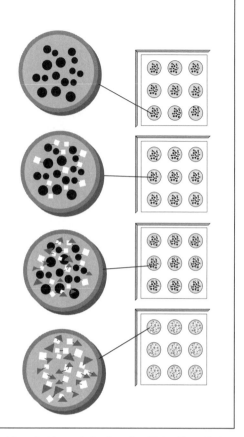

Adapted with permission from Jared Schneidman Design.

library, are still attached to their microscopic plastic resin beads, and are fished out of the wells. Chemists can then release the chemicals from the beads and use the new molecules in other tests.

Another variation of the combinatorial technique is *split-and-mix synthesis* (see figure). Instead of starting with just a single chemical in each well, researchers use three or four chemicals. It is important to keep track of all the different reactions that take place, so each bead carries only one chemical on it before it is mixed with other beads into a single well. Now, instead of starting with one pure chemical per well, chemists may begin with three or four to a well, all of which are mounted on their separate beads. After that, the split-and-mix method is just like the parallel technique. Because of the difficulties created in keeping track of so many more chemicals, not to mention trying to separate them after the experiment, most companies prefer to use the simpler parallel synthesis method.

Several drug and pharmaceutical companies have developed ways of tracking all the changes that can happen to a molecule during a combinatorial experiment. One option is to use beads that contain memory chips and miniature radio transmitters at their core. These components can be coated with resin. During the experiment radio signals are sent to the chips, which record what is being added to the molecule within which they are embedded. If researchers find a useful chemical, the chip and its memory act like the bar codes used to identify products in department and grocery stores. Comparing its number with computer records gives a history of the molecule's creation.

Other companies track additions directly into a computer database, giving them a rough idea of the reactions that occur. Researchers can also add chemical marker molecules to the beads, which attach to the resin but not to the other molecules, during the reaction and therefore track the modifications by noting which bead carries which set of markers. When researchers find that a chemical they have created has useful activities, they can trace backward to find exactly how it was made. They can then quickly identify its structure at the molecular level using other tests and assays.

The new compounds can then be tested for biological activity. A chemical that has any effect on a cell or protein is called a *hit*. Chemists make in-depth studies of hits and try to take them to the next stage of development, *leads*. Leads are chemicals that chemists believe most likely to lead to successful drugs, pesticides, and so on, and are the entities in which companies may eventually invest tens of millions of dollars. Hits can also be refined by putting them through a second combinatorial process, or they can be modified by other techniques.

Historical Development

New compounds are created by modifying older ones, a successful technique for discovering new molecules. Combinatorial chemistry is merely a method of automating and vastly increasing the speed at which this happens.

Penicillin was the first antibiotic to be discovered, in 1928. Since World War II, penicillin has been modified and improved many times to produce new antibiotics. The new antibiotics have had improved activities and properties, including improved safety in humans, stronger killing ability against bacteria, and a broadening of the range of bacteria that can be killed. Other branches of chemistry, especially those involving plastics, vinyls, and other POLYMERS, have also benefited from chemists making small refinements to tweak the performance of already understood molecules. Following the synthesis of Bakelite, the first completely synthetic polymer, in 1909, chemists have gone on to create other polymers: nylon, Styrofoam, and Teflon, to name a few.

Combinatorial chemistry was developed in the 1980s. Molecular biologists had had many successes in the 1970s creating long strands of DNA (deoxyribonucleic acid), the molecule that makes up the genes of all living things. DNA is composed of only four different molecules, called *bases*, which can be strung together in any order into a long nonbranching strand. These strands of DNA can easily be connected into strands measuring in the millions of bases. Advances in computers and robotics in the 1980s and 1990s have increased the speed at which sequences can be made and deciphered.

In the mid-1980s, chemists decided to see if the techniques of molecular biology could be extended to chemistry. H. Mario Geysen used combinatorial chemistry for rapid generation of short stretches of proteins, called *peptides*, for use in identifying the short sections of large proteins to which antibodies bind. Other chemists found that there were few problems in applying the theories that underpinned the creation of strands of DNA to chemistry, and the field of combinatorial chemistry was born. Since then almost every large company involved in pharmaceuticals and agrochemistry has gotten involved in combinatorial chemistry by partnering with specialized biotechnology firms that produce combinatorial libraries, developing their own in-house combinatorial projects or buying successful combinatorial firms.

Uses, Effects, and Limitations

Combinatorial chemistry is undertaken primarily by drug and pharmaceutical companies, which must search through millions of natural compounds that are difficult to purify and analyze, as well as synthetic compounds that take time to create in their own labs. Combinatorial chemistry can speed up this process by a factor of 1000. Instead of a team of chemists synthesizing four or five new molecules each month, one

Pharmaceutical companies are among the largest users of combinatorial chemistry. They use it to generate thousands of chemicals, which may be used in over-the-counter remedies and prescription drugs (Photo Researchers/David R. Frazier).

chemist using combinatorial chemistry can make thousands of chemical compounds in a couple of days. The companies' goal is to find a hit, a biologically active compound that can hopefully be developed into a useful drug.

Combinatorial chemistry is also used by chemists searching for agriculturally useful compounds, such as herbicides and pesticides. Peter G. Schultz and other researchers have also used combinatorial chemistry to discover new SUPERCONDUCTORS, materials that conduct electricity perfectly without wasting any as heat. Combinatorial chemistry techniques have also yielded fluorescent molecules for use in light-emitting diodes (LEDs). LEDs emit light when electricity passes through them; in alarm clocks, for example, they glow but do not illuminate.

Although combinatorial chemistry can quickly extend and modify chemicals we already know and understand, it usually doesn't create novel structures. Without these new structures, some scientists believe that we will create only small modifications of drugs we already have. This has both benefits and drawbacks.

Taxol, a well-known breast cancer drug, is an example of a molecule that could never have been discovered in a laboratory using combinatorial or any other technique of randomly generating molecules. It is a tremendously complex natural product that was discovered in yew trees in the western United States. Chemists have only recently learned to synthesize it in the laboratory.

Extending the activities of drugs through small modifications has a long and distinguished history in chemistry, however. This has happened in every field of applied chemistry, including plastics, pharmaceuticals, ceramics, and metals. A recent example is Crixivan, a potent drug against HIV (human immunodeficiency virus, which causes AIDS), which was created by modifying a previously known and understood heart drug.

In the drug industry, the creation of new molecules used to be the most difficult and slowest part of the drug-discovery process. Combinatorial chemistry has changed that, and now the slowest stages of drug discovery are testing molecules for biological activity and testing them in clinical trials with humans. Clinical trials are controlled studies in which one group of people is treated with a new drug while another group receives a placebo medication, to measure the new drug's effectiveness. The agrochemical industry is equally interested in combinatorial chemistry. Its members are using the technique to search for pesticides, herbicides, fertilizers, and other agriculturally useful molecules. Combinatorial chemistry has also been used for the development of interesting metals and alloys, although that is a much smaller field. Combinatorial chemistry has so many uses that some observers have gone so far as to call the process a "rain forest surrogate," an allusion to the millions of potentially useful chemicals that are thought to reside in the plants and trees of the world's rain forests.

Issues and Debate

Although combinatorial chemistry holds out the promise of locating many useful drugs more quickly, the technique has yet to deliver substantial results. Drug companies have spent millions of dollars on combinatorial chemistry but so far have seen little return on their investment. No drug developed through combinatorial chemistry has yet received approval to be sold on the market. A few drugs developed through combinatorial chemistry are now involved in clinical trials: drugs for treating pain, asthma, and the central nervous system, for example. There may be more; pharmaceutical firms don't always like to reveal their development techniques. Researchers have also developed novel antibiotics, which are desperately needed in this era of growing resistance to antibacterial drugs. One drug developed for the treatment of migraine headaches took only two years to develop and get into human clinical trials—no small feat considering that most drugs usually take at least five years to reach that stage.

Although a company can easily produce thousands or millions of chemicals, sorting through them is very difficult. Throwing millions of chemicals at thousands of biological targets yields billions of potential combinations. Assessing these combinations is now one of the most time-consuming stages of drug development. Many companies are expecting this to change in the next decade as automated high-throughput screening tests are developed. Today, most lab testing is still done by human technicians. New, faster techniques to automate and speed testing are expected to rely as heavily on computers and robotic controls as combinatorial chemistry does now.

Since combinatorial chemistry is done primarily by large pharmaceutical and agrochemical companies and smaller biotechnology firms, most of the information gleaned and chemicals produced by these techniques are considered proprietary information and are not released to the public. In that way, information about new chemical compounds, and therefore new potential drugs, is treated differently than information gathered about the human genetic sequence, for example, which may also lead to medical breakthroughs. As part of a cooperative effort to map the genetic sequence of the entire human genome, geneticists release to the public the sequences of genes they study. Discoveries in both areas can lead to the developments of treatments to improve human health. But the fact that so much potential profit hangs in the balance when a new drug is being developed raises questions about whether the public interest is taken into account adequately in the drug development process.

Combinatorial chemistry has made the economics of the pharmaceutical industry more complex by adding new layers of profit-making entities to the established hierarchy. Smaller biotech companies are making money by charging larger firms for access to their libraries of chemicals. Some of the more successful biotech companies have in fact been acquired by large pharmaceutical firms. Such investments are a clear sign that combinatorial chemistry is here to stay.

—Philip Downey

RELATED TOPICS
Bioprospecting, Polymers, Rational Drug Design, Superconductors

BIBLIOGRAPHY AND FURTHER RESEARCH

BOOKS
Terrett, Nicholas K. *Combinatorial Chemistry*. Oxford: Oxford University Press, 1998.
Wilson, Stephen R., and Anthony W. Czarnik. *Combinatorial Chemistry: Synthesis and Application*. New York: Wiley, 1997.

PERIODICALS
Plunkett, Matthew L., and Jonathan Ellman. "Combinatorial Chemistry and New Drugs." *Scientific American*, April 1997, 68.
Persidis, Aris. "Combinatorial Chemistry." *Nature Biotechnology*, July 1998, 691.
Persidis, Aris. "Combinatorial Chemistry." *Nature Biotechnology*, April 1997, 391.
Persidis, Aris. "High-Throughput Screening." *Nature Biotechnology*, May 1998, 488.

INTERNET RESOURCES
Combinatorial Chemistry
    http://pubs.acs.org/hotartcl/cenear/970224/comb.html
Combinatorial Chemistry: A Strategy for the Future
    http://www.netsci.org/Science/Combichem/feature02.html
"Drug, Biotech Firms Beginning to Embrace Combinatorial Chemistry"
    http://www.the-scientist.library.upenn.edu/yr1996/may/combo_960513.html
Planet Science: Combinatorial Chemistry
    http://www.newscientist.com/nsplus/insight/future/balasubramanian.html
RepliGen R&D: Combinatorial Chemistry
    http://www.repligen.com/rnd_chem.html

COMPOSITE MATERIALS

A composite is a material that combines two or more substances in such a way that the new material borrows qualities from each of them, resulting in a material that is an improvement over either of the originals. Some of the best known composites are those used in the aerospace and automobile industries, which often combine lightness with strength. However, other composites can be made that have, for example, specific electrical or heat-conducting properties.

Although composites have been used for many centuries, recent improvements in materials science have led to a boom in their importance to modern society. Today, composites are widely used, especially as lightweight, strong materials in air, sea, and road vehicles. In the future, new composites may even be able to adapt to changes in their environment (see SMART MATERIALS).

Scientific and Technological Description

A huge range of different materials can be classed as composites according to the definition given above, ranging from biological tissues, such as muscle, to building components such as steel-reinforced concrete girders. However, the study of composites is usually limited to artificial materials, those that are manufactured by mixing substances together at an intimate molecular level. Because the different components of a composite are chosen for their different properties, they usually have different physical structures. Nearly all composites with practical uses today consist of fibers or "filaments" of one material embedded in a three-dimensional *matrix* of another material.

The most frequently used filaments are carbon fibers—long, thin strands of carbon atoms strongly bonded in narrow chains. They combine immense strength and resistance to tension forces with flexibility in other directions. *Whiskers*, another type of carbon fiber filament, which are shorter and have proportionately thicker fibers, have more rigidity but less tensile strength. Carbon fibers are manufactured by a technique called *pyrolysis*. This involves the charring of long chainlike molecules called POLYMERS, which are themselves composed of other atoms grouped around a central chain of carbon. By heating some *polymers* in a reduced supply of oxygen, the outer atoms can be burned off, leaving just the carbon behind. Other types of filaments are glass- or metal-based.

Filaments themselves are impractical for most applications apart from creating strong yarns or ropes. The matrix in a composite binds them together, adds bulk to the material, and often provides other useful properties such as thermal stability (preventing the composite from expanding when heated), and electrical insulation. The matrix is what turns the fibers into a workable material (see *Uses, Effects,*

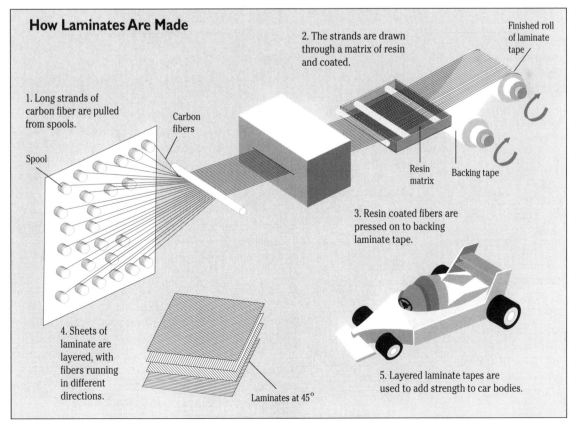

How Laminates Are Made

1. Long strands of carbon fiber are pulled from spools.

Spool

Carbon fibers

2. The strands are drawn through a matrix of resin and coated.

Finished roll of laminate tape

Resin matrix

Backing tape

3. Resin coated fibers are pressed on to backing laminate tape.

4. Sheets of laminate are layered, with fibers running in different directions.

Laminates at 45°

5. Layered laminate tapes are used to add strength to car bodies.

Adapted from Durant et al., Encyclopedia of Science in Action, *Macmillan Publishers Ltd., 1995.*

and Limitations). Several different kinds of material can provide a matrix for composites. The most widespread are polymers, but metals and ceramics can also be used.

Polymer–Matrix Composites

In polymer–matrix composites, the most common type, the fibers are embedded in a matrix of polymer molecules, forming a plasticlike material. Like other types of plastic, polymer–matrix composites (PMCs) can be either *thermoplastic* (capable of being heated and reshaped many times) or *thermosetting* (with a shape permanently fixed after their initial cooling). Thermoplastic composites have obvious advantages, but thermosetting composites are currently easier to manufacture, and a wider range of different materials of that type has been produced. In both cases the composite itself is embedded in the polymer matrix during the manufacturing process.

One of the most common forms of PMCs manufactured is the *laminate*. A laminate is a type of composite in which fibers or filaments of a strong structural material (e.g., carbon fibers) are sandwiched between layers of a supporting board, which acts as the matrix, giving the composite structural support. Often, laminates are made by laying filaments in parallel along each layer, creating a material that is extremely strong in one direction but flexible in others. These laminates have the advantage of being pliable, so they can be bent over a form, creating the curved shapes often required for aircraft, boat, or automobile bodies. The overall structural strength can then be built up by laying additional layers onto the form, each with their filaments aligned in different directions. As well as large structural boards, laminate tapes can be manufactured with the same properties. Careful layering of laminate tapes (either by hand, or increasingly by computer-controlled robot) is used to build up smooth corners on, for instance, car bodies.

Another type of laminate is made by chopping the filaments into strands and then dropping them onto the bottom layer of board at random. The varying directions in which the filaments lie in the completed laminate layer give it considerable strength in all directions, although not as great as that of multilayered laminates.

A more advanced method of composite manufacture is a process called *filament winding*. In this technique, a computer-controlled winding machine spins out very long lengths of filament, wrapping them around and around a precast form, such as that of a monocoque—a racing car body—until it is eventually completely cocooned in filament. The cocoon is then "cured" by heating to high temperatures in an oven. The filaments bond together, producing an extremely light, yet strong shell.

Metal–Matrix Composites

Polymer–matrix composites are by far the most widespread type in use today, but they are not suitable for every application, principally because they will melt and lose their structural rigidity at relatively low temperatures. Metal–matrix composites are an intermediate type that can withstand higher temperatures. They are manufactured by casting a mixture of fibers and molten matrix material together in a mold, in a manner similar to that of some polymer-matrix composites. They have advantages, including increased toughness, but their principal disadvantage is the increased weight of the metal.

Ceramic–Matrix Composites

At the most extreme temperatures, metal–matrix composites will also melt, so a third class of materials has been developed. Ceramic matrix composites have fibers embedded in a ceramic compound that remains stable at very high temperatures. Because ceramics cannot be melted and re-formed around the filaments, these composites have to be manufactured by *sintering*, a process in which the filaments are mixed with ground-up ceramic and slurry (ceramic dust mixed with a suitable liquid). When heated, the ceramic particles can reach a semimolten state in which they fuse around the filaments. The one major problem with ceramic composites is that they are brittle.

Carbon–Carbon Composites

One final type of high-performance composite that is sometimes used in extremely high temperature applications is the carbon–carbon composite. These are made by embedding carbon fibers in a matrix also made of carbon.

Historical Development

The advantages of combining two materials in a composite have been recognized throughout human history, stretching back into prehistory. Probably the first application of a composite material was seen in the deadly "composite bow," invented as early as 3000 B.C. This weapon combined the elasticity of a wooden "stave" with the stiffness and strength of a horn reinforcement grafted onto the rear of the bow. Composite bows remained one of the most important battlefield weapons until the arrival of firearms. Early in the Iron Age (from 500 B.C. onward), blacksmiths learned how to combine different types of metal to improve properties of weight, strength, and sharpness.

One of the most important early composites was concrete, invented by the Romans around the 2nd century B.C. Concrete is a mixture of cement (made by the Romans using lime, sand, water, and volcanic ash) with fragments of stone. The cement acts as a binding agent, and the stone provides strength. The invention of concrete allowed the construction of spectacular buildings such as the Pantheon in Rome (built under the Emperor Hadrian, ca. A.D. 118), with its domed roof 142 feet (43.2 meters) across.

Although other composite materials continued to be used throughout the Middle Ages and through the Industrial

Revolution, the development of modern composite technology began after World War II, with the race between the United States and then-Soviet Union to develop rocket launch vehicles for military and space uses. The efficiency of a rocket is crucially dependent on its weight, so there was a sudden drive to develop new lightweight materials that would still be capable of withstanding the extreme stresses and high temperatures involved in spaceflight. One of the first lightweight composites developed was fiberglass, a type of plastic reinforced with short glass fibers, which are brittle but extremely strong.

The new materials that emerged from the space race soon found other applications, which drove the development of new composites. The aircraft industry in particular embraced composites for use at first in military aircraft and later in passenger jets. Until this time, the major structural materials used in aircraft parts had been alloys of metals such as aluminum and titanium. Alloys contain intermixed atoms of two or more metals, combining the properties of both to produce a strong and comparatively light material. However, they could not compete against the arrival of early composites such as fiberglass.

The first automobile with composite shell components was the 1953 General Motors Corvette, which had a fiberglass body. Previously, most car bodies had been made of metal. Within a very few years, composites overwhelmed the motor industry, transforming car design in the process, allowing improved streamlining and higher fuel efficiency. Similarly, composites were soon applied to water transport, particularly racing yachts and speedboats, where their reduced weight allowed boats to travel faster with the same power behind them.

Uses, Effects, and Limitations

Composites today are widely used in a huge range of applications, although transportation, as mentioned above, is still one of the most important. Aircraft, in particular, use different types of composites in different areas; they may have a fuselage (main body) made from a polymer–matrix composite, overlaid with a heat-resistant metal–matrix skin, and have many engine parts made from ceramic–matrix composites.

Another important area for the use of PMC materials is that of leisure and sporting goods, such as tennis rackets and fishing rods. Metal–matrix composites are used in other applications as varied as high-quality cutting tools and *heat sinks* for cooling electronic circuits (a heat sink is a heat-absorbing block attached to a densely packed silicon chip to drain off heat that could otherwise impair the chip's function). As well as being used to create materials with purely mechanical properties, composites can be used to create materials with other specific requirements. These include electrical properties, rates of thermal expansion, and in the case of composites for medical use (such as prostheses), compatibility with organic tissue (see Artificial Tissue; Artificial Organs).

The most important step in utilizing composites is turning them from laboratory samples into useful materials capable of being worked, shaped, and cut into useful objects. Metal–and–ceramic-matrix composites are usually produced in the component shapes that will eventually be required. PMCs in particular can be produced in other forms, ranging from *extrusion* (pulling out) of long cylinders and beams of composite (a method similar to that used for wiremaking), to casting in molds ranging in size from small components to entire boats and wing forms.

Issues and Debate

Composite materials have had a major effect on modern life, largely through their positive effects on transportation. The development of lightweight components for use in the aerospace industry has been a major factor in the spread of reasonably priced high-speed air travel. Although many composites are based on polymers and therefore derived from the oil industry, the benefits of manufacturing composites, and the fuel efficiency savings they create, more than outweigh the cost of the original resources. In addition, the use of composites helps preserve stocks of other expensive metals that were previously used in the aerospace industry.

In the future, new types of composites are likely to have even more of an effect on our lives. A new generation of smart materials is being developed that will combine components with different qualities, enabling them to react in specific ways to changes in their surroundings. Examples include piezoelectric materials, embedded with crystals that change their shape and stiffen when an electric current is passed through them, and electrorheostatic fluids, which solidify when electricity is applied to them. The most complex smart materials of all will have microelectronic sensors embedded in them to measure changing conditions and trigger specific responses in the material. As with many cutting-edge technologies, only a few applications of these advanced materials can be foreseen at present. But as the materials themselves become available, they will soon find uses that we cannot now imagine. Investment in further scientific research on composite materials may well result in societal benefits— whether of an economic, timesaving, or aesthetic nature.

—Giles Sparrow

RELATED TOPICS
Artificial Organs, Artificial Tissue, Polymers,
    Smart Materials, Space-Based Materials Processing

BIBLIOGRAPHY AND FURTHER READING

BOOKS
Chawla, Krishnan K. *Composite Materials: Science and Engineering.*
    New York: Springer-Verlag, 1998.
Vigo, Tyrone L., and Albin F. Turbak. *High-Tech Fibrous Materials.*
    Washington, D.C. : American Chemical Society, 1991.

PERIODICALS
Gibbs, W. Wayt. "Smart Materials." *Scientific American,* May 1996.

Rogers, Craig A. "Intelligent Materials." *Scientific American*, September 1995.

INTERNET RESOURCES
USAF Advanced Composites Program Office
http://www.mcclellan.af.mil/MLS/acpob.html

COMPUTER ANIMATION

Computer animation is animation that is generated in whole or in part by computer rather than being done entirely by hand. Computer animation is used to make logos that fly across television screens, virtual houses that architects can "walk" through, and detailed maps of the human brain. It enhances and is beginning to replace traditional animation, and it has revolutionized the entertainment industry.

Scientific and Technological Description

Animation works for the same reason that live-action film does—if individual images are shown at high speed, the human brain ignores the breaks between the images and sees uninterrupted motion. But for the illusion to work, the images must be shown very quickly. Film images are usually shown at a rate of 24 images per second. This means that 86,400 images are needed to make an hour of film. With traditional animation, each of those 86,400 images has to be drawn or modeled (out of clay or using puppets, for example), set up, and shot individually. Animators have figured out many laborsaving techniques, such as reusing a sequence of film or a particular set of drawings, but producing a piece of traditional animation still takes a lot of people and a very long time, because animators must create and shoot each image individually.

Enter computer animation. Computers have greatly sped the production of traditional-looking two-dimensional animated films. For example, a computer animator who wants to make a film starring a frog could create a simple line drawing of the frog in several ways. He or she could draw it directly onto the computer, using a stylus and pad (a hand-held computerized pen and electronic sketch pad), a mouse, or even a programming language. Or the animator could simply draw it on a piece of paper and scan the image into the computer.

Coloring the frog on the computer is much easier than coloring it by hand. Indeed, painting software has largely replaced the traditional method of coloring animated film, which required a battalion of workers who hand-tinted every image. With a computer, the animator has to color only one drawing of the frog, because the computer can automatically assign the selected colors to every other drawing of it.

But the advantage of using a computer really comes when the frog moves—at 24 images a second. In a traditional animation studio, there is usually one head animator and a bevy of assistants. The head animator would draw what were called *keyframes*, the images that showed where the frog started its movement and where it ended up. The assistants were kept busy drawing the innumerable in-between images that showed the frog actually moving, a job called *tweening*. The computer can take over the tedious tweening process, sometimes called *interpolation* when performed by computer. With computer animation,

Modeling and Rendering in Computer Animation

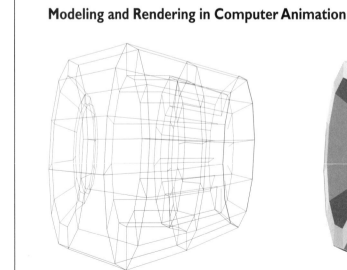

1. 3-D polygonal modeling of a prototype engine casement

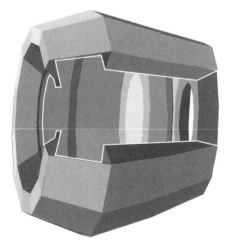

2. 3-D rendered model showing cutaway to interior details

Two computer-generated models of a prototype engine casement are shown above. The model on the left was created using polygons. The one on the right shows what the polygonal model looks like after it has been rendered. (Adapted from Ted Boardman and Jeremy Hubbell, Inside 3D Studio Max 2: Modeling and Materials, *New Riders Publishers, 1998.)*

the animator draws the keyframes, and the computer draws everything else.

Three-Dimensional Animation

With three-dimensional animation, computers are used to create animation that would be all but impossible to make with traditional techniques. The creation of three-dimensional computer animation can be broken down into three steps: modeling, building on a computer the basic shapes that will appear in the animation; rendering, adding detail, such as surface texture, color, and lighting effects; and motion, adding movement to the piece.

Modeling. Most COMPUTER MODELING programs come with a library of primitive three-dimensional shapes: spheres, cubes, columns, and the like. These shapes can be made larger or smaller, and usually can be twisted or otherwise distorted. Other shapes can be made by spinning, taking a two-dimensional drawing and rotating it around an axis to create a three-dimensional shape. Animators also look to real life, the preferred source for animation related to medicine or engineering. The animator can use a variety of actual body or machinery scans or probes to read the dimensions of an object and record them into the computer. The same object can be represented on the computer in a number of different ways. Most commonly, an object is represented as being only a surface that is made up by a variety of polygons (see figure). Polygons are easy to edit and manipulate, and they can be arranged to represent many different shapes. But polygons may create problems for animators, as they consume a large amount of computer memory (see *Uses, Effects, and Limitations*). Objects can also be represented with primitive shapes, such as spheres and cubes. For example, an animator wanting to make a camera could join a rectangle and a column.

A more recent development in modeling software is tools based on particles, which are computer-generated clouds of specks or dots. The size of a particle can be defined, as can the relation between the particles. Small particles with lots of space between them can be used to create dust or fog. Larger particles that cling together strongly make a virtual viscous goo. If this goo is attached by an animator to, say, a small human skeleton, the result can be a realistic-looking chubby baby. This is an exciting development for animators because making rounded, organic shapes such as babies is hard to do using polygons and primitive shapes.

Rendering. Rendering is a general term that means adding detail to a three-dimensional scene. Three-dimensional scenes can be colored and surfaces given texture at the rendering stage. Rendering software usually has an array of lighting options. There are several different methods that a computer can use to represent how light hits a scene, but the general rule of thumb is: The greater the realism, the greater the time it takes for the computer to render it.

One example of this is a highly realistic lighting technique called *ray tracing*. In ray tracing, the computer traces each virtual ray of light until it dissipates or leaves the scene. Since the ray is behaving very much like an actual ray of light, the room's lighting will look very realistic. But since the computer has to calculate the various paths of all the rays of light in the scene, the room will take a long time to render.

Motion. Motion in three-dimensional animation can be based on a kinematic or animator-controlled system, or on a dynamic or computer-controlled system. *Kinematic systems*, in which the animator controls every aspect of movement, are both more common and more traditional. In a kinematic system, an animator who wants to show a ball bouncing on a floor can control how fast the ball falls, how much the impact distorts the ball, and how quickly and in which direction the ball flies after each bounce. Because the animator controls all the variables, the movement can be very unrealistic—the ball can gain speed after each bounce, for example. In a *dynamic system* the animator sets certain parameters—the strength of gravity, the softness of the ball, and the shape of the floor in this example—and the computer determines how objects will move. Dynamic systems generally create very realistic-looking movement and are useful for creating simulations.

Historical Development

The first animated film was produced in 1906 by the American cartoonist J. S. Blackton. Called *The Humorous Phases of Funny Faces*, the film featured a drawing of a face on an easel that changed expressions. Blackton created this effect by stopping the camera after shooting a few frames and switching the drawing. But the drawing-on-paper method of animation was short-lived, because if a scene was complex, the animator had to draw the background hundreds of times. During the 1910s, U.S. filmmakers developed *cel animation*, in which moving characters and objects are painted onto transparent pieces of celluloid. The pieces of celluloid are filmed on plates of glass suspended above a background, with the camera positioned above the entire apparatus. In this way, only the parts of a scene that move have to be painted hundreds of times. Cel animation is still the traditional method of creating two-dimensional animation.

Traditional three-dimensional animation, which uses puppets, was first used by Russian filmmaker Ladislas Starevich in 1911. Starevich's idea spread quickly, with three-dimensional animated films being produced in the United States as early as 1917. His method of creating animation by shooting one frame of film, then moving the puppet slightly and shooting the next, is basical-

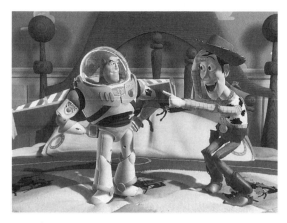

In 1995, Pixar Animation Studios released Toy Story, *the first completely computer-animated feature film (Photofest).*

ly the same method as that used today among traditional animators.

The first computer-animated films were produced by scientists at Bell Laboratories in the early 1960s. Computer animation was developed as a tool for engineers, who could use it to visualize automobile or aerospace designs. During the 1960s, scientists at Bell Laboratories created many of the basic techniques of computer animation, such as polygon modeling. Programmer Kenneth Knowlton created the first computer animation language, and A. Michael Noll, who as a child was fascinated by a three-dimensional viewer, developed a variety of ways to represent three- and even four-dimensional objects using computer animation. Other companies quickly jumped on board. In the late 1960s, the Mathematical Applications Group Inc. (better known as MAGI) took software it had developed to determine how nuclear radiation would move through space and began to modify it to create ray tracing. At about the same time, General Electric used computer animation to create the first flight simulator for NASA.

As computers became more powerful in the 1970s, programmers began to refine computer animation. A variety of programs were developed to smooth edges and to create realistic lighting effects without using the memory-intensive technique of ray tracing. The first practical computer-animation systems were manufactured and sold in the mid-1970s. As computer animation became more accessible and better looking in the 1970s, film producers began to use it to create special effects in live-action films. Movies became the proving ground for new techniques in computer animation, both because studios had the money to spend on expensive new technology and because the publicity a company received for creating effects on a successful movie drew investors and clients from all fields.

In the mid-1980s, several new companies devoted to computer animation were founded, most notably Pixar, a spin-off of famed director George Lucas's Lucasfilm. These companies focused on making three-dimensional animated films created entirely on computer rather than just creating computer-generated effects for live-action films. Pixar's *Tin Toy* brought the technology to public notice when it won the 1989 Academy Award for best animated short film. Almost all two-dimensional animated films released in the 1990s used computer animation in some form, and computer-generated special effects became more common in live-action films as the decade progressed. In 1995, Pixar's *Toy Story*, the first completely computer-animated feature film, wowed audiences and demonstrated the unique visual impact of three-dimensional animation. The successful *Toy Story* was followed in 1998 by *Antz* and *A Bug's Life*, produced entirely on computers, which were both commercial and critical successes.

Uses, Effects, and Limitations

Although computer animation is used most visibly in the entertainment industry, it has had a significant impact on many fields. Computer animation is used to create highly realistic flight simulations and other forms of VIRTUAL REALITY. Certain types of medical scans use computer animation to create dynamic models of a patient's organs. Architects create models of buildings using computer animation techniques, then render the scenes to determine how the building will look in different light. Computer animation is still commonly used by engineers for its original purpose, the testing in virtual space of design prototypes.

But three-dimensional computer animation pushes the envelope technologically. Creating elaborate or lengthy pieces of three-dimensional animation on a computer requires a tremendous amount of memory and computing power. This has strictly limited the types of techniques an animator who is not associated with a large research facility can use. Some "new" computer animation techniques, such as ray tracing, required so much computing power that they were inaccessible to most animators for decades after they were originally developed.

Animators have also run into problems with modeling software. The use of polygons in modeling, for example, creates problems because they consume a lot of memory. The computer has to remember the shape of each polygon and its place in relation to the other polygons. An animator can save memory by cutting down the number of polygons and making each individual polygon larger. But a lower polygon count makes curved objects look blocky and unnatural.

Some of these problems can be avoided by using primitive shapes. Primitive shapes that come premade with the modeling software have smooth surfaces and use very little memory. But objects built from primitive shapes cannot be manipulated as easily as objects built other ways—they cannot be cut in half, for example. Rendering, too, can be problematic, as it is especially complex for a computer to perform. It can take days for even a high-powered specialized computer to render a highly detailed and realistic-looking animated short film.

Issues and Debate

Computer-generated motion can be aesthetically unsatisfying. Computers create mathematically based motion that tends to be unnaturally smooth and fluid, like that of a satellite traveling through space. A computer animator can address this problem by drawing many more keyframes or by writing software that will create more realistic motion (or by buying the software, if it exists). The animator can also look to the source: real life. With the right equipment, the animator can take a film or video recording of an actual moving object and map the animated image over it. Professional animators sometimes hire actors to wear sensor-laden suits and walk or dance around. Those movements are recorded and used as the basis for an animated character's motions. The problem with these methods is that they are expensive and time consuming.

But in general, computer animation saves time, labor, and most important, money. Although traditional animation may seem a simple affair that requires only a pencil and paper, in reality professional-quality traditionally animated films of any substantial length require armies of workers. For decades, a handful of well-financed U.S. companies such as Disney and Warner Brothers Studios produced nearly all commercially available animation. Computer animation has helped open the field to smaller companies and non-U.S. producers.

Without a doubt, computer animation has changed the way that professional animators do their jobs. Many of the low-level hand-tinting and tweening jobs in the industry have vanished, and computer literacy is more likely to be required of aspiring animators. These changes have hardly been bemoaned by many in the animation industry, however—hand-tinting and tweening were notoriously tedious jobs. Part of the reason that computer animation's impact on the industry has been viewed so positively is that animators have not generally lost their jobs to computers. Animation is used much more often now that computers have made it cheaper and more realistic looking. Movies and even television commercials now routinely feature computer-generated effects, and computer animation is used in fields such as engineering that almost never relied on traditional animation. This means more jobs for animators, especially if they are computer literate.

Computer animation is also making animation more accessible to nonprofessionals. Although still fairly expensive, equipment and software now exist that allow inexperienced home users to model, render, and set into motion three-dimensional animated scenes. Some of these programs include many sophisticated options that were once available only to professionals. Although animation has hardly become a quick and easy way for regular people to communicate ideas, computer animation is making that more of a possibility.

—*Mary Barr Sisson*

RELATED TOPICS
Digital Video Technology, Virtual Reality

BIBLIOGRAPHY AND FURTHER RESEARCH

BOOKS

Auzenne, Valliere Richard. *The Visualization Quest: A History of Computer Animation*. Rutherford, N.J.: Fairleigh Dickinson University Press, 1994.

Hoffer, Thomas. *Animation: A Reference Guide*. Westport, Conn.: Greenwood Press, 1981.

MacNicol, Gregory. *Desktop Computer Animation: A Guide to Low-Cost Computer Animation*. London: Focal Press, 1992.

Mealing, Stuart, et al. *Principles of Modelling and Rendering: 3D Studio*. Lisse, The Netherlands: Swets en Zeitlinger Publishers, 1998.

Street, Rita. *Computer Animation: A Whole New World*. Gloucester, Mass.: Rockport Publishers, 1998.

Taylor, Richard. *Encyclopedia of Animation Techniques*. London: Focal Press, 1996.

PERIODICALS

Greene, Donna. "Charles Machover: A Computer Pioneer Takes a Look Back." *New York Times*, May 25, 1997, WC3.

Hodgins, Jessica K. "Animating Human Motion." *Scientific American*, March 1998, 64–69.

Salerno, Heather. "Watershed Moment for an Effects Breakthrough: Arc Second's Hardware Measured Up in 'Titanic.'" *Washington Post*, March 16, 1998, F5.

Squires, Sally. "Million-Dollar Images." *Washington Post*, November 6, 1990, Z12.

Thomas, C., et al. "Four-Dimensional Imaging: Computer Visualization of 3D Movements in Living Specimens." *Science*, August 2, 1996, 603.

INTERNET RESOURCES

A Beginner's Guide to Computer Animation
http://www.acs.ucalgary.ca/~vlwalker/animation.html

Animation Now & Then
http://www.myhouse.com/anim/index.htm

DV Live: Digital Video Resources Online
http://www.dv.com

Pixar
http://www.pixar.com

3-D Animation Workshop
http://www.webreference.com/3d/

COMPUTER MODELING

A computer model is a detailed representation of part of the real world that can be used to make forecasts or predictions or to test how that part of the world would respond to a variety of external factors or forces. Computer models are used to forecast the world's weather, predict the ups and downs of the stock market, design and test new products, and construct walk-in visual simulations of the world that are known as VIRTUAL REALITY.

Computer models have proved to be a highly effective way of testing theories about the world, but their use has often been controversial. Models of the world's economy have predicted overpopulation, famine, and economic collapse;

weather models have predicted global warming, sea-level rise, and the disappearance of low-lying islands. Although computer models can provide answers to the world's biggest problems, those answers must be based on the assumptions and data fed into them, and they may not be the ones that people want to hear.

Scientific and Technological Description

The heart of a computer model is a set of algebraic equations that describe how different aspects of the particular part of the world to be modeled (known as the *system*) relate to one another. For example, a computer model of how a golf ball flies through the air could use the equations of motion from physics to relate a number of variable quantities (distance traveled, velocity, and time) to a number of constant quantities (acceleration due to gravity and initial velocity) to work out that the ball follows a curved path known as a parabola. This simplified model ignores air resistance, wind, how the ball is hit, and any obstacles in the flight path. It includes only three variable quantities (usually just called *variables*); by comparison, a more complex model, such as a climate model, might contain up to 1 million variables (see figure).

Several distinct stages are involved in creating a computer model. The first and most important stage is to decide exactly what features the model should include and what assumptions can be made to simplify the task of constructing it. Getting these initial assumptions wrong can produce misleading results. Early models of Earth's ozone layer (a thin layer of the upper atmosphere that screens out the Sun's harmful ultraviolet rays) failed to predict that a hole was forming due to humankind's use of aerosol propellants and other chemicals. Only when the effects of ice particles at Earth's poles (which hasten the ozone problem) were included did the models begin to show the hole correctly.

Once the essence of a model has been captured as a set of mathematical equations, the next step is to convert these into a computer program. Large-scale computer models such as those used in weather forecasting require billions of individual calculations to be performed each second on vast amounts of very similar data. "Number-crunching" SUPERCOMPUTERS are often used for this purpose, but may still take hours or days to come up with results.

When the model has been constructed and loaded with data, it can be run. This usually involves taking the

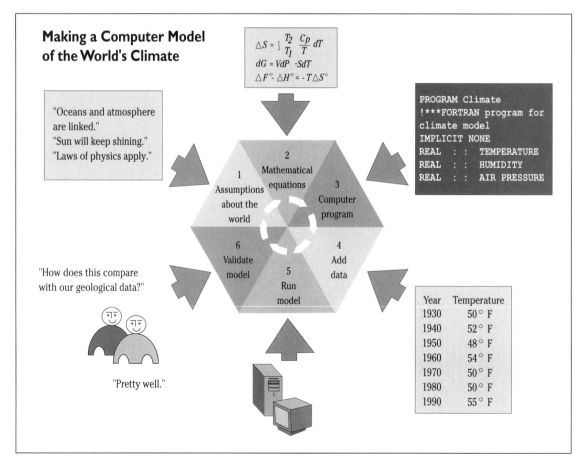

Making a Computer Model of the World's Climate

$$\Delta S = \int_{T_1}^{T_2} \frac{C_p}{T}\, dT$$
$$dG = VdP - SdT$$
$$\Delta F° - \Delta H° = -T\Delta S°$$

"Oceans and atmosphere are linked."
"Sun will keep shining."
"Laws of physics apply."

```
PROGRAM Climate
!***FORTRAN program for
climate model
IMPLICIT NONE
REAL   :   TEMPERATURE
REAL   :   HUMIDITY
REAL   :   AIR PRESSURE
```

1 Assumptions about the world
2 Mathematical equations
3 Computer program
4 Add data
5 Run model
6 Validate model

"How does this compare with our geological data?"

"Pretty well."

Year	Temperature
1930	50° F
1940	52° F
1950	48° F
1960	54° F
1970	50° F
1980	50° F
1990	55° F

There are six distinct stages involved in making a computer model. Once a model has been constructed and run, the predictions it makes are compared with data that is known to be correct (a process known as validating the model). If the model gives incorrect results, the assumptions are modified and the process is repeated until the model gives the correct results.

known state of the model at a particular moment and running time forward into the future to see what happens. In a climate model, scientists may start out with a set of measurements of temperature, atmospheric pressure, humidity, and wind speed across the globe up until the 1990s and attempt to find out how the climate changes throughout the decades of the 21st century. In this type of speculative modeling, no one can tell whether the model is right or wrong. For this reason, a model is typically tested against a known set of results to see whether it can make correct predictions. Thus, instead of being run into the future, a climate model might be run back into the past to see whether it can make predictions that correspond with real, historical data. This process is known as *validating* the model.

Historical Development

It comes as no great surprise to find that the development of computer modeling mirrors the development of computers. Indeed, one of the pioneers of modern digital (number-processing) computers, U.S. mathematician John von Neumann, was also a pioneer of computer modeling in the 1940s and 1950s; his work led to the establishment of the National Center for Atmospheric Research (NCAR) in Boulder, Colorado, in 1960. Controversy over the use of computer models probably dates to 1972, when a pessimistic report on the future of humanity, *The Limits to Growth*, was published by an influential group of academics and industrialists known as the Club of Rome. Based on a computer model of human resource use and population, the report painted a bleak picture of how life on Earth would collapse as the world's population continued to deplete its resources. Computer models that predicted a growing hole in the ozone layer and progressive warming of Earth's climate became controversial later in the 1970s and 1980s and drew attention both to the enormous number of potential applications of computer modeling and to their enormous limitations. Although the models were attempting to answer often politically uncomfortable questions, computer technology and scientific understanding were not sufficiently advanced to enable the models to provide definitive answers.

Despite these controversies, the 1980s and 1990s witnessed an explosive growth in computer modeling, due largely to supercomputers becoming both much more powerful and much more accessible. Greater computing power meant that more accurate and detailed models suddenly became feasible. Personal computer (PC) spreadsheet programs became widespread from the late 1970s onward, enabling small businesses to construct simple financial planning models. Workstations (large PCs with some of the power of a supercomputer) became popular during the late 1980s and 1990s, enabling architects, product designers, and scientists to construct complex models more cheaply.

Until the late 1980s, computer modeling was largely the domain of research scientists who knew how to feed vast quantities of numerical input into their supercomputers and interpret the vast quantities of numerical output generated in response. Then came a new type of modeling known as virtual reality, which was a method of creating three-dimensional, computer-generated models of the world in which people could walk around, touch virtual objects, and even interact with other virtual people. Virtual reality is expected to revolutionize modeling by creating models that people can interact with much more intuitively.

Uses, Effects, and Limitations

Computer modeling is an essential tool in many areas of science and technology, especially in climatology (studying the weather). Today's weather forecasters use supercomputers to run *general circulation models* (GCMs), which typically represent the entire globe using a three-dimensional grid of points specified by latitude, longitude, and altitude. The points are approximately 100 kilometers (km) apart in latitude and longitude and 2 km apart in altitude. This means that GCMs are unable to predict the behavior of storm regions (typically a few kilometers across), and this is one reason why the accuracy of weather forecasts is limited. Climatologists have used GCMs to investigate the problem of global warming by asking questions such as "What if the world's energy supplies were switched from oil to renewable sources such as wind power?" or "What if rapidly industrializing nations such as China and India consumed energy at the same rate in 20 years time as the United States consumes today?"

Product designers use computer models to test virtual prototypes (versions of their designs that exist only inside computers). Automobile manufacturers increasingly use computer models and virtual reality to test the crashworthiness of their vehicles. In 1995, engineers at German automaker BMW found that a total of 91 virtual test crashes were cheaper and more quickly executed than two crashes using real cars and crash-test dummies; and the computer models could be rerun instantly to test design modifications. Future models are predicted to be far superior to using prototypes and crash-test dummies, because virtual dummies will be able to contain detailed models of the human anatomy, enabling engineers to devise vehicle-body structures that protect people's vital organs more effectively.

Virtual crashes have also been used by accident researchers to understand how real tragedies have happened in the past. The California-based company Failure Analysis used computer modeling to investigate the notorious car accident that killed movie star James Dean in 1955. Using information about the vehicles involved, where they ended up, and the layout of the intersection where the crash occurred, engineers produced a computer model that enabled the accident to be rerun from a variety of perspectives, even from behind the wheel of Dean's car. This suggested that Dean's Porsche was traveling at a modest 55 mph and that the driver of the other car involved was prob-

Computer modeling is an important tool in many areas of science and industry, especially in studying weather patterns and climate change (Photo Researchers/Bill Bachman).

ably at fault. Failure Analysis has produced many other computer reconstructions, including a model of how President John F. Kennedy was assassinated that demonstrated that a single gunman could indeed have shot the president from the room where Lee Harvey Oswald was supposed to have acted.

Computer models have many advantages over physical models, but they are limited by the accuracy of the assumptions on which they are based, the quality of the data used, and the amount of computing power available. Thus modeling is a trade-off between increased accuracy and complexity, on the one hand, and delivering useful results in a reasonable amount of time, on the other: A perfect model of the world's weather is useless if it takes three days to compute tomorrow's forecast. Another limitation is that models are unable to predict unusual events beyond the experience of the modelers: Even if computer models had been around in 1912, they probably would not have predicted that the "unsinkable" ocean liner *Titanic* could have been downed by a combination of human error and a massive iceberg.

Issues and Debate

Computer models have made the world a very different place. Cars are safer, weather forecasts are more accurate, airplanes are more economical, and human beings generally have a better understanding of how the world works. Testing with models is safer and less controversial than testing in the real world: For example, trying out new drugs using computer models entails neither the controversial practices involved in testing drugs on animals nor the risks of testing them on humans. But models are models; by definition, they differ from the real world. Do the ways in which drugs work inside computers fully equate to how those drugs will work inside humans? All models are based on theories about the world, which may be incomplete. Very similar models can produce very different

results. Even tiny differences in the initial conditions of complex systems such as the weather can lead to enormous differences in how they turn out. It was this that led meteorologist Edward Lorenz to pose his now-famous question at a 1972 conference: "Does the flap of a butterfly's wings in Brazil set off a tornado in Texas?" The "chaotic" nature of complex systems such as the weather is fundamentally the reason why forecasts cannot extend much beyond 7 to 14 days in advance.

It is such variability that makes computer modeling so controversial, because government policies are often decided with the help of computer models. For example, environmentalists have used GCM results predicting global warming, the gradual warming of the Earth's climate as a result of a buildup of greenhouse gases such as carbon dioxide (CO_2), to argue that people in developed nations must drastically reduce their energy consumption. However, industrialists such as oil company leaders, who would be affected by such drastic changes, point to wide variations in the results among computer models and argue that more research is needed before models can be used to guide policy. Yet even if computer models cannot convincingly settle such big debates, they can at least stimulate research and focus attention on particular areas of concern.

But computer models raise a far bigger question: If humanity was equipped with models that could provide perfect answers to every question, would people want to know those answers? Solutions to world hunger or a cure for AIDS (acquired immune deficiency syndrome) are one thing, but what about computer predictions of when we will die, who we should marry, whether we should terminate pregnancies, how fast we should drive, or what career choices we should make? Psychologists believe that people need to experience a certain amount of risk and uncertainty just to feel alive. It is possible that this, rather than the ingenuity of modelers, the power of computers, or even the chaos of complex systems, may be the limiting factor in the development of computer models.

—Chris Woodford

RELATED TOPICS
Artificial Life, Supercomputers, Virtual Reality

BIBLIOGRAPHY AND FURTHER RESEARCH

BOOKS
Casti, John. *Reality Rules: Picturing the World in Mathematics*. New York: Wiley, 1994.
Casti, John. *Would-Be Worlds: How Simulation Is Changing the Frontiers of Science*. New York: Wiley, 1997.
Devlin, Kenneth. *Life by the Numbers*. New York: Wiley, 1998.
Devlin, Kenneth. *The Language of Mathematics: Making the Invisible Visible*. New York: W.H. Freeman, 1998.
Gleick, James. *Chaos: Making a New Science*. New York: Viking Penguin, 1987.
Jayaraman, Sundaresan. *Computer-Aided Problem Solving for Scientists and Engineers*. New York: McGraw-Hill, 1991.

Lorenz, Edward. *The Essence of Chaos*. Seattle, Wash.: University of Washington Press, 1993.

Press, William, Brian Flannery, Saul Teukolsky, and William Vettering. *Numerical Recipes in C: The Art of Scientific Computing*. Cambridge, Mass.: Cambridge University Press, 1988.

Schneider, Stephen. "The Science of Climate-Modeling and a Perspective on the Global Warming Debate," in *Global Warming: The Greenpeace Report*, Jeremy Leggett, ed. New York: Oxford University Press, 1990.

PERIODICALS

Gibbs, W. Wayt. "Taking Computers to Task." *Scientific American*, July 1997, 64.

Pournelle, Jerry. "Digital Models." *BYTE*, November 1995, 257.

Schneider, David. "Tectonics in a Sandbox." *Scientific American*, July 1998, 22.

Schneider, Stephen. "Climate Modeling." *Scientific American*, May 1987, 72.

Stewart, George. "Forecasting the Future." *BYTE*, June 1994, 237.

Stix, Gary. "A Calculus of Risk." *Scientific American*, May 1998, 70.

Thomke, Stefan, Michael Holzner, and Touraj Gholami. "The Crash in the Machine." *Scientific American*. March 1999, 72.

Vacca, John. "Faking It Then Making It." *BYTE*, November 1995, 66.

INTERNET RESOURCES

Great Buildings Online: Architecture, Design, History, Images, 3D Models, and More!
http://www.greatbuildings.com

"The Inside Story," by Corey Powell, *Scientific American* Online, September 9, 1996
http://www.sciam.com/exhibit/090996exhibit.html

International Research Institute for Climate Prediction
http://iri.ldeo.columbia.edu/

An Introduction to Computer Modeling: Lecture given by Dr. Kenneth W. Johnson, SCRI Research Scientist
http://www.scri.fsu.edu/~pettusn/mbasic.html

"Kaboom, Inc.: When Technology Goes Splat, Failure Analysis Sends in a Squad of Detective Engineers Wielding Heavy-Duty Computer Wizardry," by Caren Potter, *BYTE* Online
http://www.byte.com/art/9511/sec5/art1.htm

National Science Foundation article: "NSF Scientist's Computer Model Links Fire and the Atmosphere"
http://www.geo.nsf.gov/adgeo/press/pr9636.htm

Scientific American Presents: Feature Article: Bringing Life to Mars
http://www.sciam.com/1999/0399space/0399mckay.html

COMPUTER VIRUSES

A computer virus is a program that self-replicates and produces an undesired effect. Viruses spread among computers usually through the sharing of files, disks, and software. When an infected application is used, the virus copies itself into a computer and enacts its secondary and active component, the effects of which can range from a minor corrupting of files to complete loss of data on a hard drive.

Researchers experimented with self-reproducing systems beginning in the 1950s, but the first official computer virus did not appear until 1983. The initial virus to be found "in the wild" (circulating among the general population of computers) was the Brain virus, discovered in 1987. Since then, computer hackers have developed thousands of virus variations. Although the law has done little to inhibit virus writers, the best protection against viruses involves preventive measures and antivirus software.

Scientific and Technological Description

Computer viruses borrow their name from their biological counterparts. Both can infect the host without detection and have the ability to lie dormant for long periods before doing damage. In addition, both can spread without the knowledge or consent of the host. Finally, victims of both types of viruses must take specific measures to eliminate them. A computer virus is not an error or bug in legitimate software, but an intentionally programmed set of instructions designed to spread among computers, at times causing damage to data or programs (see figure). Computer viruses have two main components, the first of which is self-replication. When the user opens an infected application, the virus copies itself to the computer, and later, to other files. The virus then enacts its secondary component—anything the programmer designs it to do.

Three principal types of computer viruses exist: file infectors, boot-sector viruses, and macro viruses. *File infectors* are viruses that infect ordinary program files, usually .EXE or .COM files, which are the executable files that allow the MS-DOS operating system to carry out vital functions. They are often the first files to run when the computer is turned on. When the user runs an infected application—any type of program that involves the use of infected files—the virus first copies itself into the computer's memory and then allows the computer to return to the original program without the user noticing that a virus was enacted. From there, the virus can copy itself to other files once the user runs them.

Boot-sector viruses enter a computer if the user leaves an infected disk in the drive while booting up (starting) the computer. The computer reads the disk first and mistakes the virus for the operating system. The virus then loads itself into the computer's memory and can transfer its programming code to any disk the user subsequently places in the drive.

Macro viruses are the most recent virus variation to develop. They affect word processing and spreadsheet programs, which contain scripts, or macros. Scripts are instructions for specific sequences of actions, such as mathematical calculations or shortcuts within a program that are embedded within a document. The virus writer creates the virus so that the computer mistakes it for a legitimate script. The virus can then copy itself to other documents and subsequently erase or alter data.

Viruses spread through various means. One way is through the sharing of disks. Downloading programs from the Internet (especially from an unreliable source such as an unfamiliar Web page) can also infect a computer if those programs are then run. Yet another way to cause infection is by running pirated software (software that has been illegally copied rather than purchased), because there is no way of knowing how the software may have been tampered

How a Virus Infects a Computer

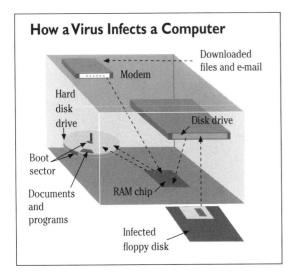

One way for a virus to invade a computer is via an infected floppy disk or downloaded file received via a modem. Once the disk is inserted, the virus can then copy itself to random-access memory (RAM), which temporarily stores programs and data. Any program in the hard drive, the computer's main storage site, may be damaged.

with. Viruses have spread over e-mail, but usually through attachments. The user activates the virus (usually a macro virus) upon opening an attachment. Some virus variations send infected attachments to people listed in a person's e-mail address list. Others then receive and open the attachments. Subsequently, their computers become infected. As the process repeats, the virus spreads rapidly. Recently, viruses have been developed that come directly through the e-mail message itself, rather than an attachment.

Historical Development

In the early 1950s, John von Neumann, a pioneer in modern computer science, first developed the concept of a self-reproducing automaton, a set of logical instructions that could replicate itself (see ARTIFICIAL LIFE). However, the first virus-like programs did not come into existence until 1962, when researchers at AT&T's Bell Labs in New Jersey developed a game called Darwin. The researchers created self-replicating programs that competed for memory space in a virtual arena. The most fit programs survived and reproduced. One programmer, Robert Morris, Sr., created a set of instructions that was able to pass on what it learned to its offspring. Consistently, his program dominated the game.

The first official computer virus did not appear until 1983, when it was created by Fred Cohen, a graduate student at the University of Southern California, who coined the term. Cohen realized that he could attach a self-replicating program to a host program. On a small network, he unleashed a virus that allowed him access to a user's name and files. Although he would eventually write his thesis on the new topic of computer viruses, the university administration, fearful of the potential consequences, prevented

Cohen from doing extensive research on the subject. Thus the advances in viruses in the following years were made primarily by computer hackers.

In 1986, two brothers who ran a computer store in Pakistan wrote the first virus that eventually appeared in the computer world at large, or "in the wild." The brothers, who harbored anti-U.S. sentiment, created a program targeted at U.S. customers who bought pirated versions of the MS-DOS software sold at their store. They distributed it on software sold to their American, but not Pakistani, customers. They also hoped to profit from the virus: When files crashed, the computer displayed a message telling the user to contact their store for help. Called the Brain virus, it was first found in 1987 at the University of Delaware and eventually erased hard drive data on more than 100,000 computers, 10,000 of them at George Washington University alone. (Universities are particularly susceptible to viruses because of the volume of people who use the same computers.) However, the Brain virus was a boot virus that operated only on a specific type of floppy disk. Because these disks are rarely used today, the virus is infrequently seen.

In November 1988, Robert Morris, Jr., a Cornell University student and son of the AT&T programmer, discovered a security flaw in a computer network at the Massachusetts Institute of Technology. To demonstrate the general lack of computer security at the time, he unleashed what researchers later called the Internet worm. A *worm* differs somewhat from a standard virus in being a self-contained program that propagates by itself rather than utilizing some form of human assistance, such as the exchange of floppy disks or the sharing and running of software. Whereas a virus requires an infected carrier, such as a file or disk, for transmission, a worm can be deposited on a network and wreak havoc all by itself. Although Morris's worm did not destroy any programs, data, or hardware, it caused more than 6000 Internet host machines to lose power, or crash. Morris lost track of the Internet worm, and it quickly shut down whole networks at universities and corporations, affecting millions of users.

Before this time, many people had not heard of computer viruses. Others believed them to be myths. By the end of 1988, however, viruses began increasing in prevalence, and the media took notice. Researchers identified 30 viruses in the wild by 1989, the same year that IBM released its first antivirus program. The Jerusalem virus then made its first appearance. This virus, which erases files on Friday the 13th, is still found today. The next widely publicized virus, Michelangelo, was discovered in 1991, and it was the first to threaten home computer use. The programmer of this virus designed it to delete files on March 6, the Italian painter's birthday. The media exaggerated the potential damage the virus would cause, with some reporters predicting that millions of computers would shut down. Although the virus affected thousands of computers, it did not cause the widespread devastation expected.

The Windows 95 operating system arrived in 1995, and with it came macro viruses, which were the first to infect both Macintosh- and IBM-compatible systems. At the end of 1996, Concept, the initial macro virus, was the most common virus in the world. As of 1999, between 20,000 and 40,000 computer viruses exist, although many are confined to laboratories for research purposes and have never infected an outside computer. Nonetheless, experts estimate that about six new viruses are created every day, and the total number of viruses doubles almost every year.

In 1999, viruses that spread over e-mail became prevalent. The most publicized of these, the Melissa virus, spread rapidly in April. This virus, a Microsoft Word 97/Word 2000 macro virus, affected users of the Microsoft Outlook e-mail system. It took e-mail addresses from the infected person's address book and sent the virus, in attachments, to the people at those e-mail addresses. As recipients opened the attachments, which appeared to come from people they knew, the virus spread rapidly. The Melissa virus caused computer shutdowns in many institutions, whose systems were overloaded by the proliferating e-mail. Investigators traced the original message that contained the virus to a posting on an Internet newsgroup. They eventually tracked down the exact phone line from which the virus was posted. Police subsequently arrested David L. Smith, a New Jersey computer consultant, charging him with wrongful access to computer systems, theft of computer service, conspiracy, and interrupting public communication. If convicted, the alleged virus writer faces up to four years in prison and $480,000 in fines.

Uses, Effects, and Limitations

Viruses have no beneficial effect. They enter a computer and operate without the user's consent. However, the severity of the damage they inflict ranges greatly. Some viruses merely duplicate themselves, a trivial effect. Others are pranks that may display messages on the screen such as "I feel good" or "Yankee Doodle Dandy." A different prank may cause the computer to make a clicking sound when the user types an "x." These viruses are actually the most common forms, and they do not interfere with the computer's ability to function. However, about 35 percent of all viruses are destructive. They can alter or delete executable program files or delete all hard-disk files. Many people consider a slow-acting virus to be the most damaging. This type of virus alters data over a long period, leaving no clue as to what was changed. This type of destruction is particularly devastating to a business or research institution that stores its numerical data on a hard drive. Financial records or years of research can be altered without the user's knowledge. Other malevolent viruses seek out passwords and spread them in the hope that others will use them for the wrong purposes. In the worst-case scenario, a virus could disrupt medical records, air traffic control systems, or other computing operations that directly affect people's safety.

Although legitimate programmers create viruses for research purposes, many virus writers create them for the sheer art of programming. Experts believe that virus writers are mostly males between 14 and 24 years of age who enjoy pushing the limits of technology and welcome the challenge of outsmarting antivirus programs. For some, the greatest thrill is to have a virus posted in the news worldwide. Some malicious virus writers wish to sabotage specific organizations. However, others are experimenting with programming and do not realize the extent of the damage they can unleash.

Issues and Debate

In 1986, Congress passed the Federal Computer Fraud and Abuse Act, a law that allows the prosecution of those who write damaging viruses. For many reasons, though, few have been convicted under the law. First, because there is little legal precedent in this area, prosecutors may shy away from spending the money needed to bring what may be a test case. Second, people rarely know who the virus writer was, and even if investigators could identify the person, collecting sufficient evidence to charge the person would prove difficult. In addition, computer network victims are frequently unwilling to report a virus infection for fear of publicizing their vulnerabilities. Finally, the perpetrator will rarely have enough money to compensate the vast number of victims involved. In the case of Robert Morris, Jr., reports estimated that the Internet Worm caused $100 million in direct and indirect damage. Morris had no means to compensate the affected. He was eventually tried, sentenced to community service, and given a $10,000 fine.

Experts estimate that about 35 of every 1000 computers will encounter a virus over the course of a month. Computer owners can significantly reduce these chances, however, by following a number of preventive procedures. Virus researchers consider the sharing of disks, CD-ROMs, and software to be risky behavior that conscientious computer operators should avoid when possible. Users should also avoid placing any disk or CD of unknown origin into their computer, especially before booting up the system. In addition, users should download software only from reliable Internet sources. Finally, computer owners should invest in antivirus software, to prevent and limit virus damage. These programs monitor the computer, checking to see if certain files have been modified. They also check programs periodically for suspicious changes, and they allow users to scan disks and attachments for viruses.

Because many virus writers alter existing viruses rather than creating entirely new ones, many viruses have similar signatures that antivirus software can easily identify. Furthermore, most viruses have an initial step of self-reproduction that is indicative of a virus, making even new variations readily recognizable. Nonetheless, users are safest if they regularly update their antivirus software. Often, software manufactures offer updates free over the Internet.

Although erasing and reinstalling affected programs and files is a fail-safe way to eliminate a virus, this is usually impractical. Antivirus programs can help repair some of the damage caused by infection, but virus researchers suggest that users should back up files regularly. To prevent important files from being damaged, computer users should print out the document in a format that can easily be scanned. In this manner, the information will not be lost in the event that backup files become infected.

Most experts believe that viruses will be an increasing problem in the future. Despite the recent efforts of investigators in tracking down the alleged writer of the Melissa virus, such persons are rarely identified. Because of the sheer volume of new viruses that appear every year, it is doubtful that authorities will discover a way to identify all virus writers. Because of this, victims of virus infections will probably never be compensated for losses incurred.

—Paul Candon

RELATED TOPICS
Artificial Life, CD-ROMs, Cryptography, Internet and World Wide Web

BIBLIOGRAPHY AND FURTHER RESEARCH

BOOKS
Fites, Philip, Peter Johnston, and Martin Kratz. T*he Computer Virus Crisis*, 2nd ed. New York: Van Nostrand Reinhold, 1992.
Haynes, Colin. *The Computer Virus Protection Handbook*. Alameda, Calif.: Sybex, 1990.
Levy, Stephen. *Artificial Life: The Quest for a New Creation*. New York: Random House, 1992.
Solomon, Alan, and Tim Kay. *Dr. Solomon's PC Antivirus Book*. Boston: Newtech, 1994.

PERIODICALS
Cannell, Michael. "Computer Viruses." *Science World*, October 19, 1997, 18.
Corbitt, Terry. "Data Files in Danger." *Accountancy*, January 1999, 59.
Kephart, Jeffery O., Gregory B. Sorkin, David M. Chess, and Steve R. White. "Fighting Computer Viruses: Biological Metaphors Offer Insight into Many Aspects of Computer Viruses and Can Inspire Defenses Against Them." *Scientific American*, November 1997, 88.
Magid, Lawrence J. "Personal Technology: Get Your Virus Vaccination." *Los Angeles Times*, October 12, 1998, 4.
Wayland, Hancock. "Understanding Computer Viruses." *American Agent & Broker*, September 1998, 14.

INTERNET RESOURCES
History of Computer Viruses
    http://antivirus.about.com/library/weekly/aa073198.htm?rf=dp&COB=home
Overview of the Virus Problem
    http://arachnid.qdeck.com/quarc/OverviewoftheVirusProblem.htm
What Is a Computer Virus?
    http://www.academic.marist.edu/papers/spitzerg/paper.htm
The WildList: List of the Computer Viruses in the Wild
    http://www.wildlist.org

CRYPTOGRAPHY

Cryptography is the science of writing secret messages. It involves the use of a mathematical transformation of text and a key that allows for encryption and decryption of a message. Three main cryptographic techniques exist: substitution, transposition, and fractionization. Modern cryptography uses a complex combination of these techniques to mask the original message.

Governments, militaries, and banks are the principal users of cryptography. However, what these institutions consider secure cryptography changes along with evolving technology. Currently, encryption programs exist that would take millions of years to crack. The U.S. government considers these strong encryption programs to be dangerous and therefore restricts their export. Many software manufacturers argue that these restrictions suppress their competitiveness on the international market, where foreign businesses may sell much stronger encryption software.

Scientific and Technological Description

Cryptography is the science of disguising messages, making them unreadable by all but their intended recipients. It is related to *cryptanalysis*, the science of breaking secret messages and codes. The term *cryptology* encompasses the study of both cryptography and cryptanalysis. In cryptography, the original message that is written in standard language is called *plaintext*. *Encryption* is the process that converts plaintext into a disguised message, or *ciphertext*. *Decryption* refers to the reverse: the conversion of ciphertext to plaintext. The encryption process involves the use of algorithms and keys. An *algorithm* dictates the particular mathematical transformation that scrambles the text or information. It can be a set of instructions or a computer program designed to make the text unreadable by all but the intended recipient. The *key* is the specific scrambling process that is used; it allows for the encryption and decryption of a message (see figure).

Three methods of encrypting text exist. The first is *substitution*, in which each letter of the alphabet is replaced by a different letter. Julius Caesar used such a method, called the Caesar Cipher, to communicate with his generals and other officials. He would shift the letters of the alphabet by three characters, such that an A would be replaced with a D, and a B would be represented by an E, and so on. For example:

Key:	abcdefghijklmnopqrstuvwxyz =
	defghijklmnopqrstuvwxyzabc
Plaintext:	I will invade Gaul tomorrow.
Ciphertext:	L zloo lqydgh jdxo wrpruurz.

Often, cryptography becomes more secure if one removes word divisions from the ciphertext. Customarily, cryptographers put ciphertext into blocks of five: lzloo lqydg hjdxo wrpru urz. In this simple example, the algorithm would be, "Shift letter by n characters in the alphabet." The key would consist of identifying the variable represented by n; in this case, "$n = 3$." The substitution technique is fairly easy for modern cryptanalysts to decipher. In all languages, certain letters occur more often than others do. In English, the letters E, T, A, I, O, N, R, and S are the most

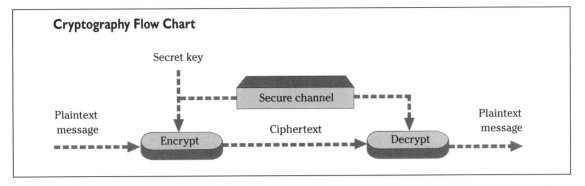

Cryptography Flow Chart

Secret key

Secure channel

Plaintext message

Encrypt

Ciphertext

Decrypt

Plaintext message

The secret key must be sent through a secure channel to the receiver. The sender encrypts a message with this key, changing it to cipher-text. The receiver uses the same key to decrypt the message, changing the ciphertext back to the plaintext message.

frequently occurring letters (easily remembered as "senori-ta"). Because substitution does not alter the frequency or pattern of letters, a count of the letters in the ciphertext allows a third party to decrypt a message with little trouble.

Other methods of encryption involve *transposition* and *fractionization*. When using transposition, cryptographers rearrange letters in the text instead of altering their identi-ties. The letters are transposed according to a predeter-mined pattern, or algorithm. Transposition can also be com-bined with substitution. In the example below, the letters have first been substituted, according to the same key as that used in the first example. They have then been trans-posed: The letters in each block of five have been reversed.

Key: abcdefghijklmnopqrstuvwxyz = defghijklmnopqrstuvwxyzabc (AND) reverse each block

Plaintext: I will invade Gaul tomorrow.

Ciphertext: oolzl gdyql oxdjh urprw zru.

Fractionization involves the use of more than one sym-bol or letter to represent a plaintext letter. Cryptographers then combine this method with transposition, thereby fur-ther disguising the plaintext. In the following example, two alphabets are used (an "a" can be be represented by either a "~" or a "d"), the ciphertext is divided into blocks of five, and each block is reversed. One can see how a combination of cryptography techniques results in ciphertext that appears increasingly unlike the plaintext. However, with the key, one can easily decrypt the message.

Key: abcdefghijklmnopqrstuvwxyz = defghijklmnopqrstuvwxyzabc (OR) ~!@#$%^&*)(-+?`_=\[]{};':" (AND) reverse each block

Plaintext: I will invade Gaul tomorrow.

Ciphertext: -o*;l #d}q* o{~jh \rp`w zru.

Using computers, modern cryptographers employ var-ious combinations of all cryptographic methods, creating extremely complex cryptographic systems. Rather than being simple variables, modern keys are long bits of infor-mation made up of binary units with a value of zero or one. A 56-bit key is commonly used in the United States. This has

2_{56} (or 72 quadrillion) possible values, making it exception-ally difficult to crack without a great amount of effort and computing power.

Historical Development

Cryptography may be as old as language itself. However, the documented history of cryptography begins in Egypt, where archaeologists have uncovered intentionally altered messages in tombs dating to 1900 B.C. Evidence suggests that ancient Greeks and Romans also used cryptographic systems. The Caesar Cipher is one documented example of this use. However, cryptography did not emerge as an art and science until the Middle Ages. Johannes Trithemius, a Benedictine abbot in Germany, wrote the first book on cryp-tography, *Polygraphiae*, in 1499. From the 16th to the 19th centuries, the commonly used cryptographic systems employed the substitution method in addition to a minimal code. This code disguised high-frequency letters, syllables, words, and phrases that appeared in a document. Because cryptanalysis at the time involved mainly frequency counts, code was intended to foil this type of analysis.

A major advance in cryptography came with the advent of the telegraph in 1843. Because telegraph lines were easy to tap and older styles of symbolic code could not be transmitted through Morse code, a new cryptographic system emerged, called the *polyalphabetic system*. Using this system, the substitution alphabet changed with every letter or after a certain number of letters. Cryptographers used slides or two interlocking disks to encrypt and decrypt text. Slides moved in a certain pattern, changing the alphabet that corresponded to the letter. Interlocking disks were of differ-ent sizes so that when one disk made a complete revolution, the second had not, thereby changing the corresponding ciphertext letters. Only those who knew the starting posi-tions or pattern (what we now call the algorithm) could break the code. Until World War I, many experts considered this type of cryptography unbreakable.

During World War I, a need for stronger cryptography techniques emerged. In 1917, the American inventor Edward Hebern developed a rotor machine that, through the use of a typewriter keyboard, could encrypt plaintext into ciphertext according to the location of the letter in the text and the sequence of the letters preceding it. The machine could produce an intricate cryptographic message. European inventors had independently created a similar device around the same time. The German version of the machine, called the *Enigma*, appeared in 1923. In the 1930s, the Germans produced a variation that they considered secure. However, in the 1920s, Polish authorities began to study Enigma, and they cracked its code in 1932. Before the onset of World War II, Polish cryptographers shared their information with French and British officials. Although the Germans changed their key frequently during the war, the British continued to crack the German code. Americans had similar successes breaking the cryptographic systems of the Japanese. With the close of World War II, 20th-century cryptography had evolved from an individual endeavor to a mechanized computational effort requiring teamwork.

As computer technology progressed in the following decades, so did cryptographic techniques. Although governments had been the main users of cryptography through history, industry and individuals became increasingly involved in the latter half of the 20th century as computers became a more common tool in commerce and at home. Today, the average computer user has access to algorithms that could take centuries to crack using the most powerful computers. The goal of cryptography has metamorphosed from merely disguising text to eliminating any association between the plaintext and ciphertext.

Uses, Effects, and Limitations

Governments historically have used cryptography to convey diplomatic messages, wartime communications, and scientific secrets. Military officials must communicate though encrypted signals, as enemy cryptanalysts will constantly attempt to monitor and decrypt messages that may affect life and death. After government and military users, banks are the largest users of cryptography, employing it to mask financial transactions. With the advent of the Internet and ELECTRONIC COMMERCE, the need for strong encryption in business has increased drastically.

A good encryption system should satisfy a number of criteria. Nineteenth-century Dutch linguist Auguste Kerckhoffs originally outlined these criteria, which are still applicable. First, a good cryptographic system should provide appropriate security. Because absolute security is rarely necessary, the ability to decode a message successfully should be delayed until after the message has lost its value. If Caesar's Cipher took a number of days to solve without the key, the invasion may have taken place by the time enemies deciphered the message, making the cryptanalysis worthless. Therefore, the strength of a system should be proportional to the amount of time the message needs to remain encrypted. Second, a system should be easy to use. Cryptographic methods that require complex implementation will either be avoided by users or used improperly, conveying incorrect messages or no message at all. Finally, the security of a system should rely on the specific key used, not the algorithm. Keys can be changed or kept secret more easily than can algorithms, which tend not to remain secret when used by many people.

Modern-day cryptanalysts chiefly attack an encryption system through brute-force searches. After uncovering the specific algorithm used, a cryptanalyst will use a computer to generate all possible variations of the key. Depending on the key length, the time a brute search takes can vary. An effective cryptographic system uses such a large key that a brute-force search will take too long to crack to be of any worth. A number of successful brute-force attacks have demonstrated the limitations of some modern cryptographic systems. In the summer of 1995, two French graduate students set out to crack the key to Netscape Navigator's 40-bit encryption code, commonly used to disguise messages, such as bank transactions and e-mail text, that are sent over the Internet. The students used 50 computers that were left in an empty office for the summer vacation. They set out on a brute-force attack, using the computers to generate all possible combinations of the key. The search took them slightly longer than a week.

Then in July 1998, the Electronic Frontier Foundation, an Internet civil rights advocacy group, cracked a message that used a 56-bit key using a homemade supercomputer. These examples demonstrate that the relative strength of an encryption algorithm changes as the available technology improves. Systems that were once unbreakable are no longer considered secure. Therefore, many banking systems now use 128-bit encryption. All analyses indicate that this type of encryption cannot be cracked for millions of years using a brute-force search.

Issues and Debate

The United States considers strong encryption algorithms to be munitions, similar to guns and warheads, because unfriendly nations or groups can use encryption to conceal their activities from U.S. law-enforcement agencies. The United States and other major nations have long regulated the export of high-technology tools such as SUPERCOMPUTERS because they can be used to help create nuclear, chemical, and biological weapons. Since encryption software has become available, the U.S. has tightly restricted its export as well. Until recently, the export of anything larger than a 40-bit encryption program was illegal. In 1996, the Clinton administration began to relax those restrictions, allowing the export of some 56-bit encryption programs. Within the United States itself, few restrictions exist for encryption programs.

Because 128-bit key encryption is widely available both domestically and internationally, and powerful enough that even the largest government computers cannot crack it, many law enforcement officials want to restrict the use of strong encryption software products. They believe that the advancing technology is helping criminals by restricting the government's ability to monitor and eavesdrop on what may be criminal communications. Some law-enforcement advocates claim that the government should limit domestic use of strong encryption as well as the export of such programs.

Before 1991, governments and large corporations were the primary users of encryption programs. However, as use of the INTERNET AND THE WORLD WIDE WEB began to expand in the 1990s, so did the use of encryption programs among smaller businesses and average computer users. Online customers are increasingly demanding assurances that their electronic transactions are private and secure, and in response, many businesses now require the use of strong encryption to remain competitive. Privacy advocates and businesses want to make online transactions as secure as possible, a desire that often conflicts with the concerns of law-enforcement officials. In addition, many U.S. software companies, businesses, and privacy advocates believe that the government should lift encryption export restrictions so that U.S. sellers of encryption technology may compete freely on the international market, where demand for the strongest available encryption products is high but U.S. firms may not offer them.

This issue has been tried in the courts a number of times. In 1999, a federal court overturned a lower court's decision and ruled that export controls are a restraint on speech and an attack on scientific expression, a finding advocates of strong encryption products applauded. The federal government had argued that the policy did not violate free-speech protections because encryption code is speech only when written on paper. Although the government lost this case, the issue may progress to the U.S. Supreme Court, whose justices could decide the ultimate fate of cryptography and its legal uses.

—*Paul Candon*

RELATED TOPICS
Electronic Commerce, Internet and World Wide Web, Supercomputers

BIBLIOGRAPHY AND FURTHER RESEARCH

BOOKS
Brassard, Gilles. *Modern Cryptology: A Tutorial*. New York: Springer-Verlag, 1988.
Kahn, David. *The Codebreakers: The Comprehensive History of Secret Communication from Ancient Times to the Internet*. New York: Scribner, 1996.
Konheim, Alan G. *Cryptography: A Primer*. New York: Wiley, 1981.
Newton, David E. *Encyclopedia of Cryptology*. Santa Barbara, Calif.: ABC-CLIO, 1997.
Smith, Richard E. *Internet Cryptography*. Reading, Mass.: Addison-Wesley, 1997.

PERIODICALS
Cravotta, Nicholas. "Encryption: More Than Just Complex Algorithms." *EDN* (*Electrical Design News*), March 18, 1999, 105.

McWilliam, Bruce. "Keeping Secrets: The New Rules for Cryptography." *Computing Canada*, October 5, 1998, 13.
Oriez, Charlie. "AITP—Encryption." *Information Executive*, April 1999, 4.
Peschel, Joe. "Understanding Encryption." *InfoWorld*, December 23/30, 1996, 51.
Schneier, Bruce. "The Crypto Bomb Is Ticking." *BYTE*, May 1998, 97.
Stark, Thom. "Encryption for a Small Planet." *BYTE*, March 1997, 111.
Wallich, Paul. "Cracking the U.S. Code." *Scientific American*, April 1997, 42.

INTERNET RESOURCES
Cryptography
    http://www.efa.org.au/Issues/Crypto/
Cryptography Timeline
    http://www.clark.net/pub/cme/html/timeline.html
RSA Laboratories' Frequently Asked Questions about Today's Cryptography
    http://www.rsa.com/rsalabs/faq/html/questions.html
Center for Democracy and Technology
    http://www.cdt.org/crypto/

DATA MINING

Data mining is a computer-aided technology used to search large databases and extract from them information and patterns that can be translated into useful applications. A data-mining analyst will attempt to discover trends within a data set without forming a prior hypothesis, a major fact that distinguishes data mining from the use of traditional statistics. The analyst searches for previously unknown relationships between specific factors such as customer demographics and purchasing trends. Based on historical and current patterns, businesses or organizations implement new strategies to attract new customers, better serve their current clients, or gain other competitive advantages.

Data mining evolved out of statistical procedures and problem-solving techniques that are rooted in the 1950s and 1960s. Widespread use of computers and a dramatic increase in computer power have made data mining a profitable process for many organizations to undertake. The information gained from data mining can be turned into advantages not only for businesses, but also for any organization that has created a database of relevant statistics. Although current data-mining software allows smaller organizations to perform data-mining operations on their own computer systems, a data-mining or statistics expert is often needed to guide the process. Otherwise, the procedure may lead to misleading, inaccurate, and unusable information.

Scientific and Technological Description

The data-mining process involves five steps. First, the business or organization must collect, store, and clean (i.e., eliminate errors and duplications in) its data. Often, corporations already maintain information on purchases, transactions, demographics, and related customer data without having collected the data specifically for the purpose of data mining. However, different persons or departments may have gathered the data, and therefore the data may be on different oper-

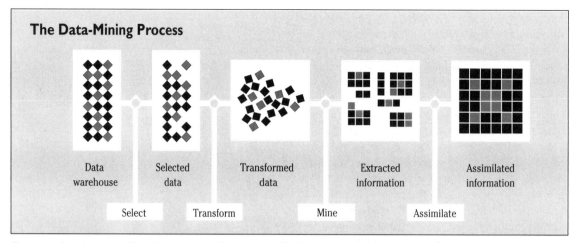

The Data-Mining Process

| Data warehouse | Selected data | Transformed data | Extracted information | Assimilated information |

Select   Transform   Mine   Assimilate

Data are collected and stored in a data warehouse. From this, specific data can be selected and mined. The information extracted from the data-mining process is then assimilated into relevant information. (Adapted with permission from Evangelos Simoudis, "Reality Check for Data Mining," IEEE Expert, © 1996 IEEE.)

ating systems or in different formats. An analyst must prepare the data so that they exist in a uniform format that is compatible with the specific data-mining software being used. For example, "male" and "female" might be converted to "1" and "2," missing values are filled in when possible, and all errors should be corrected. Upon inspecting their data, many companies find numerous duplicates in their information (e.g., they may find both a "Mr. A. Brown" and an "Anthony Brown"). In the data-cleaning process, these duplicates are eliminated. This step alone has saved some mail-order businesses thousands of dollars in unnecessary mailings. Clean, organized data are essential for successful data mining.

Data selection is the second step. The investigator should determine what question the data-mining process will address and what subset of data is appropriate to use for the particular query being posed. For example, a large department store may want to know when bed sales rise and fall over the course of a year and what additional items people most often buy when purchasing a bed.

The third step involves the actual mining of the data. Depending on the processing power and computer memory available, the analyst will decide on the number of factors that he or she can examine at the same time. The analyst will also determine which data-mining algorithms are appropriate to use in the investigation. (An algorithm is a step-by-step procedure or statistical analysis used to solve a problem.) The computer or mainframe system then analyzes the data. The resulting output arranges the data, showing correlations, rules, relationship graphs, or models that describe and predict data patterns.

In the fourth step, the investigator interprets and analyzes the results. Depending on the outcome, he or she can refine the questions being posed in order to uncover different or more precise results. The analyst may identify an unusual pattern or correlation and construct a rationale or hypothesis to explain it. One company discovered that in the fall of the year, sales of beds and chainsaws have a positive correlation. Although such an association can be due to chance or useless information, upon closer inspection, an investigator posed the following theory: In the fall, people are preparing to occupy new or refurbished winter cabins and therefore purchase new beds; these people also buy chainsaws to cut wood for the cabins.

In the final step, the company must apply its newfound knowledge, assimilate it with other relevant data, and create a program to utilize it. If the investigator believes that people are buying both beds and chainsaws to supply winter cabins, the store can place these items in close proximity to one another. In this way a customer who buys one item will be more inclined to purchase the other. In addition, the store could go even further by discounting one item when customers purchased the first. In this example, the store created a winter-cabin sales package, thereby increasing the revenue typically generated by these items.

Historical Development

Data mining did not develop out of a specific discovery, but evolved from statistical procedures and problem-solving techniques that have existed for decades. The roots of data mining can be traced to the mid-20th century, when researchers began to use early computers for electronic data processing. Investigators were hopeful that machines that were provided with enough data or observations could discover rules or trends that would be difficult for human experts to infer. In the 1960s, various researchers began to develop neural networks—problem-solving mechanisms (see ARTIFICIAL INTELLIGENCE). Although limited in their usefulness at the time, neural networks spawned further research into the fields of electronic data processing and problem solving.

Knowledge engineering was developed in the 1970s. With this technique, problem-solving computers were programmed with knowledge and rules and then were expected to describe data accordingly. However, at the time, computers were unable to analyze large quantities of data, and these systems consistently fell short of expectations. The 1980s witnessed a tremendous increase in available computing power and a decrease in the cost of storing and utilizing large databases. The combination of these factors caused a renewal of interest in machine-learning research and led to improvements in this and related fields. Financial services, banks, and direct marketing companies put the improved technology to use in analyzing customer behavior and the benefits and risks involved in past marketing campaigns. Nonetheless, much of the data analysis still had to be done by experts outside the corporation.

In 1995, the first International Conference on Knowledge Discovery in Databases (KDD) took place in Montreal, Canada. The KDD process (soon after known as data mining) described the entire procedure used to derive rules, patterns, and knowledge from large databases. Data-mining software is now available for smaller businesses and organizations. This has led to a great rise in the popularity and use of data-mining processes.

Uses, Effects, and Limitations

The information gathered from a data-mining operation can take five forms. Data-mining information is most commonly used to *classify* customers. With this procedure, the investigator analyzes an existing group of items or people, such as customers who cancel their credit cards. Data mining is used to uncover the characteristics of these customers and to develop a model that can help explain and predict this behavior. It can also determine what promotions and efforts in the past have been cost-effective in maintaining these customers and what efforts may be unnecessary.

Second, an analyst can discover in the data an *association*, two events that are likely to occur simultaneously. Customers who purchase corn chips may also buy soda 65 percent of the time; however, a store promotion may raise this combination buying to 85 percent. Third, the mined data can reveal a *sequence*, two or more events likely to occur over time (e.g., 45 percent of those who buy houses will also buy a new oven within one month). An investigator may also uncover *clustering*, specific groupings within the data. For example, people who buy at least one compact disk each month may be more likely to have multiple magazine subscriptions. The last type of information that can be gained from data mining is *forecasting*. Based on historical patterns in the data, analysts can estimate future sales, buying patterns, or other likely customer behaviors.

Armed with this information, businesses hope to better comprehend customer behavior and therefore gain a competitive edge in their marketing strategies. This may include creating specific in-store promotions or gearing certain advertising campaigns toward some customers and not others. For example, a large mail-order company may dis-

cover that people who have recently moved are most likely to buy tables, phones, and decorative items, while they are unlikely to buy expensive electronics. The company can then formulate a catalog geared specifically to this type of customer, thereby increasing the likelihood of a sale. Credit card companies may use data-mining information to determine which customers are credit risks or which transactions may be indicative of credit card theft.

Data mining has been used in nonconsumer arenas as well. The Washington State Department of Social and Health Services used data mining to find patterns in data that were originally collected regarding payment information in its foster care system. After mining the data collected regarding the children in long-term care, investigators found that these children usually arrive as infants. The foster care system then directed effort into getting infants adopted rather than placing them into foster care.

The National Basketball Association (NBA) also uses a data-mining system in dealing with its statistics. Every play of every game is recorded and entered into the NBA's Advanced Scout system. These statistics are posted and available to all NBA teams. Coaches can access these statistics, and with data mining they can look in the data for patterns that can influence coaching decisions. With this computerized system, coaches can spend less time sifting through statistics and more time on strategy.

Some organizations, especially larger corporations, collect data more quickly than they can be used. These businesses create data warehouses where information can be stored and utilized. Although the entirety of the data is input directly to a large mainframe computer, a portion of the data can reside on a computer network server. The server-based data can receive information from the mainframe monthly, weekly, or daily, depending on the organization's needs. These data can then be divided into *data marts*, smaller subdivisions containing more directed or specific information. In this way an investigator can more easily analyze data on subsets such as "frozen foods" or "household items" without having to search all the data on the mainframe. Nonetheless, setting up data warehouses and marts can be extremely expensive, and this part of the process, like all the other aspects, requires the input and direction of a data-mining expert.

Issues and Debate

As with all uses of statistics, data mining can be interpreted inaccurately and lead to misinformed policy decisions. The main downside of data mining is the ease with which one can infer a relationship that either is coincidental or not useful. Because one does not begin data mining with a hypothesis, explanations of results must be determined after the fact, often making it difficult to determine which relationships are valid. For instance, when mining a health care database, one may find that motorcycle accidents and tattoos are correlated. Yet this is unlikely to be translated into an effective policy. However, some supermarkets have found that both beer and diapers sell well on Friday nights. When placing them close to each other on the store floor, sales of both increased. (The explanation was that fathers who were sent out to buy diapers for the weekend purchased beer for themselves.) Determining which correlations are valid and useful can prove difficult and may lead to poor business decisions.

Some of the most difficult issues concerning data mining revolve around the complexity of the processes involved. Many companies can be crippled by the magnitude of data-mining projects. The actual mining of data is only one step, and not the first one, in a much larger process. A statistics or data-mining expert is usually needed to direct this process. With the new data-mining software now widely available, many people unfamiliar with statistical processes are tempted merely to enter data and perform the analyses themselves. However, attempting to mine a data set without proper instruction is likely to yield useless results. An expert will be more aware of which data subsets to analyze, which questions to address, and which algorithms to use. The process then may involve returning to previous steps and refining questions and analyses until useful data are produced. Often, analysts will use visualizations of the data, such as histograms (graphs that summarize data frequency) and scatter plots (diagrams that display patterns and groupings in data), to illustrate and summarize the data and to help direct queries. At times, analysts will integrate data mining with the use of traditional statistics to better address certain questions.

Some observers have also raised privacy concerns with respect to data mining. Citizens may object to the fact that pieces of information about them, such as their income level, home address, and pharmacy record, can be collected, stored, linked together, and used to help businesses market products to them. They may feel that their privacy is violated when companies gather and use information for marketing purposes without their explicit consent. Concerns have also arisen regarding the sale of consumer data, with some consumer groups complaining that information they willingly gave to one company at one time has been sold to another business without their knowledge or consent. Direct marketers counter that more advanced data mining will improve people's lives, as businesses, armed with ever-more-detailed customer information, will no longer waste people's time on solicitations to which they are not likely to respond. In the coming years, debate will no doubt continue as to whether government regulation is required to help balance businesses' need to collect consumer information with citizens' privacy concerns.

—Paul Candon

RELATED TOPICS

Artificial Intelligence, Computer Modeling, Intelligent Agents, Parallel Computing

BIBLIOGRAPHY AND FURTHER RESEARCH

BOOKS

International Business Machines Corporation. *Discovering Data Mining: From Concept to Implementation.* Upper Saddle River, N.J.: Prentice Hall, 1998.

Thuraisingham, Bhavani. *Data Mining: Technologies, Techniques, Tools, and Trends.* Boca Raton, Fla.: CRC Press, 1999.

PERIODICALS

Alexander, Steve. "Users Find Tangible Rewards Digging into Data Mines." *InfoWorld*, July 7, 1997, 61.

Brethenoux, Erick. "Buried Treasure: Some Dos and Don'ts of Data Mining." *Chief Information Officer*, October 1, 1996, 38.

Deck, Stewart. "Data Mining." *Computerworld*, March 29, 1999, 76.

Edelstein, Herb. "Mining Data Warehouses." *Information Week*, January 8, 1996, 48.

Hedberg, Sara Reese. "Searching for the Mother Lode: Tales of the First Data Miners." *IEEE Expert*, October 1996, 4.

Hedberg, Sara Reese. "The Data Gold Rush." *BYTE*, October 1995, 83.

Inmon, Bill. "Wherefore Warehouse?" *BYTE*, January 1998, 88NA1.

Krivda, Cheryl D. "Data-Mining Dynamite." *BYTE*, October 1995, 97.

Miller, Thomas W. "Done Right, Data Mining Discovers Diamonds." *Marketing News*, January 4, 1999, 17.

Simoudis, Evangelos. "Reality Check for Data Mining." *IEEE Expert*, October 1996, 26.

Small, Robert D. "Debunking Data-Mining Myths—Don't Let Contradictory Claims About Data Mining Keep You from Improving Your Business." *Information Week*, January 20, 1997, 55.

Watterson, Karen. "The Data Miner's Tools." *BYTE*, October 1995, 91.

INTERNET RESOURCES

Data Analysis for Forecasting, Fraud Detection, and Decision-Making at Los Alamos National Laboratory
http://www-xdiv.lanl.gov/XCM/research/genalg/members/hillol/datam_lanl.html

"Data Mining Digs In" (*Washington Post*)
http://www.washingtonpost.com/wp-srv/washtech/techcareers/strategy/jobsmarts/hth031598b.htm

How Data Mining Works
http://www.pilotsw.com/r_and_t/whtpaper/datamine/dmwork.htm

DEEP-SEA VEHICLES

The world's deep oceans are still relatively unexplored, but the past few decades have seen the rapid development of technology capable of revealing them. Submersibles (vehicles like submarines, but capable of withstanding much greater depths and specifically designed to carry out various tasks while underwater) have explored some of the world's deepest ocean trenches, revealed new scientific wonders, and helped in the rapid development of undersea industries.

As we enter the 21st century, both industry and science are using deep-sea vehicles more and more. The vehicles are helping oceanographers to learn more about the fragile environment of the ocean depths, but may be endangering that environment by encouraging its exploitation.

Scientific and Technological Description

The main problem faced by any vehicle attempting to explore the great ocean depths is the enormous pressure exerted by the sea above. Most undersea vessels are

The *Alvin*, a Deep-Sea Submersible

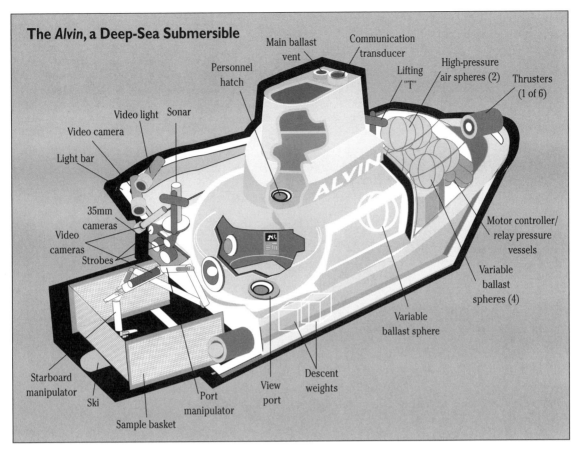

Main ballast vent
Communication transducer
Personnel hatch
Lifting "T"
High-pressure air spheres (2)
Thrusters (1 of 6)
Video light
Sonar
Video camera
Light bar
35mm cameras
Video cameras
Strobes
Motor controller/ relay pressure vessels
Variable ballast spheres (4)
Variable ballast sphere
Starboard manipulator
Ski
Port manipulator
View port
Descent weights
Sample basket

The Alvin, *owned and operated by the Woods Hole Oceanographic Institution, has been used in deep-sea research and retrieval tasks since the mid-1960s. The* Alvin *makes between 150 and 200 dives each year and has a depth capacity of 4,000 meters (14,764 feet). Its titanium personnel sphere allows three crew members to man the vessel. (Adapted with permission from an illustration by E. Paul Oberlander, © Woods Hole Oceanographic Institution.)*

designed to provide surface-like operating conditions for their occupants, so they are supplied with air at normal atmospheric pressure. However, this pressure is not capable of countering the water pressure around the vessel, giving deep-sea vehicles a natural tendency to implode. Only a strong outer shell (hull) and clever design can prevent this.

Shallow-diving submersibles can be designed in the same way as normal submarines: The strength of their steel hulls is enough to resist the water pressure down to depths of a few thousand feet. But for exploration at the greatest depths—typically between 6000 and 20,000 feet—the irregular shape of these submersibles is unsuitable; the pressure exerted on different parts of the vessel can cause the hull to tear and buckle. The deepest-diving submersibles, often called *deep submergence vehicles* (DSVs), borrow their design from traditional diving bells. The crew occupies a steel or titanium sphere, which is the only part of the vessel kept at normal atmospheric pressure. The sphere is surrounded by a superstructure that carries the propulsion system, air tanks, and a variety of other instruments, all of which are allowed to flood. The result is that the water pressure pushes in equally on all sides of the sphere, and because it has no weak corners or edges, it is almost impossible for it to implode.

The other major problem encountered by deep-sea vessels is one of buoyancy and depth control. The sea exerts an upward pressure on any object, equal to the weight of water the object displaces. This upward force counteracts the object's weight and makes the object effectively weigh less in water. Any submersible must be designed with neutral buoyancy: Its weight must exactly equal the weight of the water it displaces so that the submersible will hover underwater at any depth.

To adjust the submersible's depth, tanks on board are frequently filled with ballast (material heavier than water that increases the vessel's weight) or with liquids such as gasoline that are lighter than water. Replacing the heavy ballast with water decreases the vessel's weight, allowing it to rise to the surface, whereas pumping out the light gasoline and replacing it with heavier water causes the submersible to sink again. Submarines, which operate in shallower waters, have tanks filled with air (instead of gasoline) inside their hulls, but in a submersible, where the tanks are carried outside the pressurized hull, a liquid alternative, which prevents the tanks from crushing under pressure, must be used. Submersibles are distinct from submarines in both form and function. Whereas submarines are mainly used for purposes of transportation and warfare, submersibles are used for exploration. For that reason it is important that crew members be able to see out. Toughened glasses and clear plastics allow windows to be put into the walls of the crew cabin without compromising its pressure-resistant properties, but these have to follow the spherical contours of the rest of the vessel. A good example is the U.S. Navy submersible *Seacliff*, which has its cabin mounted at the front of the vessel and a large proportion of the sphere replaced with a toughened glass window.

To operate successfully, all submersibles have to carry a wide variety of other equipment. The precise equipment varies depending on the submersible's mission but typically may include lights, cameras, sonar, cutting equipment, and robot arms, all operated by the crew from within the sphere. Frequently, submersibles are used in conjunction with TELEOPERATED OCEAN VEHICLES (TOVs), robotic explorers tethered to the submersible and capable of exploring areas that would be inaccessible or too dangerous for the submersible itself.

Historical Development

Undersea vehicles have developed over many centuries. According to folklore, the first *submarines* were used in warfare during the time of Alexander the Great (ca. 330 B.C.). The Italian Leonardo da Vinci and English mathematician William Bourne both designed watertight craft capable of traveling underwater during the 16th century, but the first person known to have built a submarine was Cornelius Drebbel, a Dutch inventor at the court of England's King James I. Drebbel took members of the British aristocracy for trips under the Thames River in London on board a wooden-hulled, waterproofed, oar-powered vessel as early as the 1620s.

Throughout the 18th and 19th centuries, submarine technology developed, with the introduction of compressed air supplies, metal hulls, and diesel engine propulsion. By World War I, submarines had become formidable weapons of war. However, they still operated only in comparatively shallow waters, such as the English Channel, rather than the deepest ocean depths.

A simple alternative to a submarine is the *diving bell*, invented by British scientist Edmond Halley (best known for predicting the reappearance of a comet that was later named after him). The diving bell is a spherical craft with a base open to the sea. The bell is lowered into the sea, and air is pumped down to it from the surface. The pressure of this air prevents the water from filling the bell, allowing the bell's crew to remain dry and to view their surroundings through windows in the top of the bell. Although the diving bell has no means of propulsion, it provides simple access to the water and is still used to provide a base of operations during underwater construction projects.

In the early 1930s, American oceanographer Charles William Beebe recognized the advantages of the diving

Attempts to explore and recover items from the wreck of the Titanic *have been conducted by the* IFREMER Nautile, *a manned submersible able to operate at a maximum depth of approximately 20,000 feet. Nautile's exterior encases a titanium sphere that houses a crew of three in an area about seven feet in diameter (Corbis).*

bell's spherical shape and set out to design a vessel that would travel to much greater depths. Beebe's *bathysphere* was a sealed steel sphere, like a closed-off diving bell, with its own internal air supply. However, it still had to be lowered by cable from a support vessel on the surface above and had no means of propulsion. Nevertheless, Beebe and his engineer, Otis Barton, conducted dives in the bathysphere to depths of 3028 ft (923 meters).

The bathysphere was the direct ancestor of the first *bathyscaph* (self-propelled deep-sea craft), built by French scientist Auguste Piccard in 1948. Piccard's bathyscaphs inspired today's generation of deep-sea submersibles— smaller and more maneuverable craft with a wide range of instruments and equipment that enable them not just to travel to the ocean depths, but to work there. This vessel used a steel sphere, but Piccard's later vehicle, *Trieste*, launched in 1953 and purchased by the U.S. Navy, incorporated a new and stronger titanium design. In 1960, Piccard and naval lieutenant Don Walsh set a record for the world's deepest dive that still stands: a depth of nearly 35,800 ft. (almost 11 kilometers) in the Mariana Trench of the Pacific Ocean.

Uses, Effects, and Limitations

Deep-sea vehicles have a huge range of different uses. In terms of pure scientific research, the deep seas are still a largely unexplored frontier of our planet, and many important discoveries undoubtedly await us in the ocean depths. For instance, scientists have only recently become aware of the huge abundance of strange forms of life in the hostile conditions around geothermal vents—undersea volcanoes, belching superheated poisonous chemicals, which may have given rise to the first life on Earth. Submersibles can also map and study geological conditions on the seabed, revealing clues to the history of our planet's crust. Although the time a submersible can spend underwater is limited by its fuel and air supply, the vehicle can frequently deploy instrumentation or even autonomous TOVs on the seabed and retrieve them weeks or months later.

Most submersibles in operation today are used for commercial purposes, however, such as mineral and oil exploration and exploitation. Robot arms and cutting tools are used for underwater construction projects, cleaning and repair of oil pipelines, and drilling rock samples from the ocean floor for analysis on the surface. At present, most commercial exploitation of the seas is concentrated in relatively shallow waters around coastlines, and the design of commercial submersibles reflects this—they are not all built to withstand the tremendous pressures of midoceanic trenches.

Submersibles are also used in commercial salvage work and accident investigation. The discovery of the wrecked *Titanic* in 1985 would have been impossible without the Woods Hole Oceanographic Institute's DSV *Alvin*, and more recently, submersibles and TOVs have been used together to recover wreckage from crashed aircraft. Although most submersibles are designed to maintain normal atmospheric pressure within, some shallow-diving vehicles are equipped with a separate cabin whose atmosphere can be kept at higher pressures. This allows divers using high-pressure air supplies to enter and leave the submersible without having to worry about dangerous pressure changes that could otherwise cause the bends (the painful formation of nitrogen bubbles in the bloodstream as the result of pressure changes).

Although not strictly vehicles, *diving stations* are living quarters lowered to depths of about 100 m (300 ft.), where divers can live and work for long periods. The pressure and even content of the atmosphere within a station can be altered slowly to acclimatize the divers to the air mixture they will be breathing when diving.

Issues and Debate

The exploration of the deep seas has become a major field of study in the late 20th century, but, while scientific research is in itself unquestionably valuable, the exploitation of the seas and seabed that it encourages is the subject of fierce debate. At present, large-scale fishing and oil and gas drilling are the major areas of controversy, but an increased understanding of the ocean floor will undoubtedly lead to the development of new industries under the sea. Currently, diving stations and permanent scientific research colonies are the only settlements on the ocean floor, but some scientists have predicted that humans might one day colonize the sea beds.

One of the major achievements of deep-sea exploration so far has been to improve our knowledge of the effect that limited human exploitation is already having on the oceans. Submersibles and TOVs offer a unique way of studying the marine ecosystem without disturbing it. They reveal that the sea is a delicately balanced environment. Stocks of some fish have plummeted in the past few decades, partly as a direct result of overfishing, but also because of global warming and pollution disturbing natural ocean circulation patterns. Conversely, the information that deep-sea science has gathered about fish in their natural habitat has allowed aquaculture (farming of fish in pens at sea) to become a major contributor to the global fishing industry.

Ultimately, extracting resources from the seas must be done very carefully. Mining and drilling for rare metals and minerals may have few adverse effects on ocean environments, for example. But if major new oil reserves are found, some observers contend that that will provide an excuse for world leaders not to address the impending fossil fuel crisis—the fact that natural resources such as coal and oil are finite—and encourage further use of those fuels, damaging the climate still further. The burning of fossil fuels releases CO_2, one of the chief greenhouse gases that scientists believe are gradually warming the Earth's atmosphere.

The oceans are a valuable resource, but they are also an intrinsic part of the global environment. Studies indicate that so far their vast size and ability to recover from accidents such as oil spills may be cushioning many of the effects caused by human pollution. The risks involved in upsetting the seas still further could have far-reaching consequences. Deep-sea vehicles have an important role to play in the further industrialization of the oceans, but they also offer scientists a unique platform from which to observe the changing oceans and the effect we are having on them.

—*Giles Sparrow*

RELATED TOPICS
Robotics, Teleoperated Ocean Vehicles

BIBLIOGRAPHY AND FURTHER RESEARCH

BOOKS
Ferris Smith, P., ed. *Underwater Photography: Scientific and Engineering Applications.* New York: Van Nostrand Reinhold, 1984.
Society for Underwater Exploration. *Submersible Technology.* Boston: Graham & Trotman, 1986.
Society for Underwater Exploration. *Submersible Technology: Adapting to Change.* Boston: Graham & Trotman, 1988.

INTERNET RESOURCES
Society for Underwater Exploration
    http://www.underwaterdiscovery.org/
U.S. Submarines, Inc.
    http://ussubs.com
Woods Hole Oceanographic Institute
    http://www.whoi.edu/

DIGITAL LIBRARIES

A digital library is a collection of electronic books and reference works stored in computers and accessed over computer networks such as the Internet. Some of the world's greatest libraries, such as the New York Public Library, the British Library in London, and the Bibliothèque Nationale de France in Paris, are converting their archives to digital (computer-encoded) form. Numerous other libraries have been created purely to store digital materials.

Digital libraries offer many benefits over traditional libraries, including enormous savings in storage space and the ability to carry out instant electronic searches on vast quantities of archived material. But librarians are concerned that digital technology can actually be more expensive than paper, and that it may lead scholars to neglect entirely any materials not converted into digital form. Also, they are concerned about archiving the digitally stored materials. No one knows how long electronic storage media will last. Should newly stored works be stored only in electronic form? And will electronic media last as long as paper?

Scientific and Technological Description

Just as in any traditional library, storage and retrieval of information are key issues in the design of digital libraries. Books and journals are converted into electronic form using optical SCANNERS, which turn the printed pages into machine-readable form. This represents the words on a printed page as long strings of binary digits (zeros and ones) that can be stored and processed in computers. Electronic books and journals may be stored and displayed in a variety of different ways (see ELECTRONIC BOOKS and CD-ROMs). The scanned page images can either be stored as they are, which has the advantage that users of the digital documents will see pages that closely resemble the original documents; or the images can be converted into text files, which means that they can be edited into new documents or searched rapidly to find particular keywords.

Cross-references may be converted into hypertext links, which are automated connections between electronic documents that users can click on to go straight from one document to another. Just as a conventional cross-reference identifies a target document and tells readers how to get there, so a hypertext link contains simple computer coding that identifies the relevant file to which the reader should be sent. A hypertext link also involves the computer recording the exact location of the file from which the link is made, so the reader can return there later if necessary.

In a traditional library, details of books may be listed on a computer catalog, which takes the form of a dummy computer terminal (a combined keyboard and monitor with no built-in computing power) connected to a mainframe computer (a large, centralized computer system). Typically, all the libraries in a town or state are served by similar terminals connected to the same computer, which is sited in a central location, but the terminals can be accessed only by going to a library building. In a digital library, both the catalog and the books it refers to can be accessed either through a terminal in a library building or remotely using a computer network such as the Internet. This means that readers can access library materials using Internet-con-nected computers in their homes at any time of day or night. But it also means that library servers (the central computer systems that send information to a personal computer from a Web site) set up to serve a particular college or community may be slowed down by access requests from all over the world. So some digital libraries restrict access to their archives by requiring readers to enter a user name and password.

Retrieving information from digital libraries is simply a matter of finding the appropriate file and displaying it on a computer screen or electronic book reader. Most digital libraries allow their files to be searched automatically by keyword, similar to the way the Internet can be explored using a search engine (see INTERNET SEARCH ENGINES and PORTALS). But in scientific libraries, which may have millions of items covering a very similar subject area, this approach is far from ideal; a simple keyword search may yield thousands of likely hits. The most advanced digital libraries involve a team of human librarians who classify each item using a dozen or more keywords to identify main topic areas. A supercomputer is then used to find keywords that often occur together (such as *coronary* and *bypass* or *global warming* and *greenhouse effect*). This creates a collection of keywords closely related by meaning, known as a *semantic index*, which forms the basis of the catalog in the most advanced digital libraries. Instead of doing a keyword search, a researcher can use the semantic index to find items truly relevant to his or her search. Thus, in a medical library, a search on *coronary* will show up probable hits using words such as *bypass*, *heart attack*, *thrombosis*, and similar terms that the researcher might not think to look for.

Historical Development

The central idea behind digital libraries—storing information on computers—is rooted firmly in the 20th-century computer revolution. From the 1890s, when statistician Herman Hollerith used punched cards to automate processing of the U.S. census, to the 1960s, when American Airlines and other corporations pioneered the use of large mainframe computers in ELECTRONIC COMMERCE, information came to be thought of as something quite separate from the medium on which it was stored. Card files were replaced by computer files, paper was replaced by magnetic media such as tapes and computer disks, and it was only a matter of time before traditional books would be replaced by electronic books.

Digital libraries can trace their beginnings to a single incident in 1971. Personal computers had not been invented and computing jobs were carried out by time sharing on mainframe computers, with each user allocated a certain amount of time (often measured in equivalent money terms) on the computer. While working at the University of Illinois, Michael Hart found himself allocated $100,000,000 of computer time, largely because the university had more computing resources than it knew what to do with. He

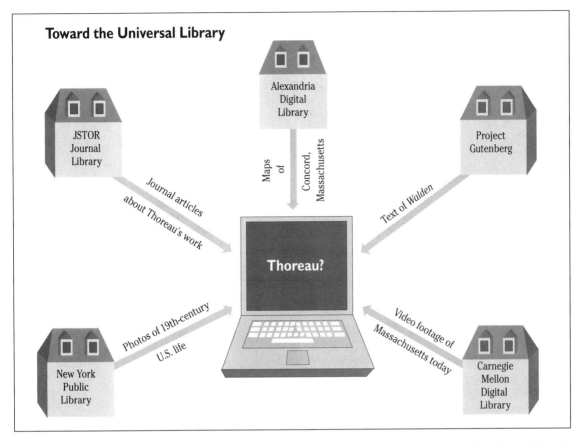

Toward the Universal Library

JSTOR Journal Library

Alexandria Digital Library

Project Gutenberg

Journal articles about Thoreau's work

Maps of Concord, Massachusetts

Text of *Walden*

Thoreau?

New York Public Library

Carnegie Mellon Digital Library

Photos of 19th-century U.S. life

Video footage of Massachusetts today

One strand of digital library research is looking at ways of linking different libraries together to form what has been described as a "universal library." Researchers envision that without leaving his or her desk a user could enter a simple query for information about the writer Henry David Thoreau, and a computer would automatically return details of relevant information in digital libraries around the country, or even around the world.

decided to create the world's first digital library, and typed in his first electronic "book": the U.S. Declaration of Independence. Thus was born Project Gutenberg, the best-known digital library, which now distributes at no cost a stock of about 2000 volumes across the Internet.

JSTOR, a digital library that specializes in storing academic journals, was conceived in 1992 at Denison University, Ohio, in a similarly accidental way. Governors of the university were discussing plans for an expensive new library building when William G. Bowen, president of the philanthropic Andrew Mellon Foundation and a trustee of Denison, noticed that the physical library space would be needed mostly for journals. He proposed a digital library for journals instead. With financial help from the Mellon Foundation, JSTOR became a not-for-profit organization at the University of Michigan in 1995.

Digital libraries really took off in 1994, when the U.S. Library of Congress announced the $60 million National Digital Library Program, which aimed to convert 200 collections of the library (some 5 million individual items) to digital form by the year 2000. Also in 1994, the U.S. National Science Foundation, the Advanced Research Projects Agency (ARPA), and the National Aeronautics and Space

Administration (NASA) announced a four-year project known as the Digital Libraries Initiative (DLI). This launched several large-scale partnerships between libraries, telecommunications corporations, publishers, and universities, including a digital video library at Carnegie Mellon, a digital library specializing in engineering sciences at the University of Illinois, and a digital library focused on earth and space sciences at the University of Michigan.

Uses, Effects, and Limitations

The best-known digital library, Project Gutenberg, stores approximately 2000 electronic volumes and distributes them over the Internet. The wide selection includes classic texts by authors such as Shakespeare and Henry David Thoreau, encyclopedias, dictionaries, and documents such as the U.S. Constitution. All of the books are available at no charge, but this is possible only because Project Gutenberg restricts itself to public-domain books (those whose copyright has expired, so there are no royalties to be paid to authors or publishers) and because volunteer editors prepare the texts for distribution (so there are no salaries to pay).

But the biggest advantages of Project Gutenberg are also its biggest drawbacks. Making books available over the

The Bibliothèque Nationale de France in Paris, one of the world's most important libraries, is converting its archives to digital form. Digital libraries may make it easier for scholars to access material from libraries all over the world. Shown here is the Bibliothèque Nationale's Reading Hall (Art Resource/Erich Lessing).

Internet in a form that can be read by any computer means that they must be stored as a type of least common denominator, unformatted text known as ASCII (American Standard Code for Information Interchange); this reduces quality and does not allow pictures to be stored. Under U.S. copyright laws, most books enter the public domain 50 to 100 years after the death of the author, so Project Gutenberg is restricted primarily to supplying classic volumes first published many years ago.

Project Gutenberg is a digital library without pictures, but other digital libraries specialize in storing mainly pictorial information. The Alexandria Digital Library (ADL), based at the University of California at Santa Barbara, is a collection of 7 million items that can be accessed over the Internet, including maps, satellite photographs, seismic (earthquake) data, and color photographs taken by the Space Shuttle.

What ADL has done for maps, JSTOR is doing for academic journals. JSTOR (Journal Storage Project) holds back issues of around 120 academic journals and is available to approximately 450 academic institutions around the world. Unlike a traditional library, JSTOR never closes; journals are available around the clock.

Digital libraries have many advantages over traditional libraries. Books and journals are becoming ever more expensive, so librarians are particularly interested in anything that will reduce their costs. Materials are generally easier to find in digital libraries than in traditional libraries, and space is no longer an issue. Once *brittle books* (a librarian's term for fragile documents) have been converted to electronic form, they can be examined by any number of people with no risk to their preservation; in theory, this means that a much greater number of scholars could have access to any book or journal. But the question of whether to convert fragile or important documents to electronic form is not always straightforward. Since 1497, Parliament in London has stored all the laws it makes on parchment rolls made from goatskin, at a cost of more than £28 ($45) per page, because this is the only material known to be capable of surviving for hundreds of years. Although the Parliament now stores all its documents on computer systems and makes some of them available over the Internet, it is still required to keep parchment copies, because no one knows exactly how long the electronic materials will survive.

Traditional libraries often specialize in a particular subject area (e.g., the Folger Shakespeare Library in Washington, D.C., is devoted purely to the works of William Shakespeare), so scholars often have to visit more than one library to find the information they require—for example, if someone were doing research on Shakespeare and his contemporaries. Traditional libraries can share materials using a system known as *interlibrary loan*, which allows a borrower in one town or state to access materials held by libraries elsewhere in the country. Some digital libraries are exploring a 21st-century version of this idea, connecting to other digital libraries to make what has been termed the *universal library* by digital library pioneer Raj Reddy of Carnegie Mellon University. The vision is of people being able to access the entire body of human knowledge using the Internet, without leaving their desks (see figure).

However, some of these much-touted benefits have turned out to be illusory. Far from saving money, digital libraries have often proved to be more expensive than their paper cousins, partly because of the cost of converting materials and partly because library users still expect both paper materials and their electronic equivalents to be available. Digital libraries that focus on providing easily searchable text often do so at the expense of losing pictures and photographs. And digital libraries that mix paper and electronic resources have noticed that scholars tend to ignore materials that are not available in electronic form—precisely the opposite effect to that intended.

Issues and Debate

Digital libraries are very popular with scholars, for whom they seem to offer only benefits: more convenient access to a wider range of information, almost instant keyword searches through vast archives, and access to old or obscure information that might otherwise be impossible to track down. They also offer important benefits for developing nations, where schools find it hard to afford up-to-date educational materials, and to blind people, for whom digital materials can be read out loud automatically using computerized speech synthesizers (see LANGUAGE RECOGNITION SOFTWARE).

Librarians have been more wary of the new technology, however. Trained as information specialists, they have found themselves wrestling with cutting-edge technology and with having to make sometimes arbitrary decisions about which parts of their archives to digitize within their limited budgets. A particular concern is what happens to a text once it has been converted to electronic form. With paper resources, it is easy to keep track of an original text, even if it is photocopied or notes are scribbled into its margins; but with electronic books and journals, which can easily be modified without trace, the concept of a definitive text becomes more problematic. If an important manuscript is converted to electronic form, how can anyone be sure that it has not been tampered with?

Another question is who will pay to convert paper resources into digital form. Libraries cannot afford to convert everything that has ever been published, but partial conversion can lead to users neglecting non-digital materials entirely. Book, journal, and newspaper publishers are increasingly making their own materials available in electronic form through CD-ROMs or Web sites, but they may charge users to access them. Copyright issues mean that free-access digital libraries such as Project Gutenberg may be restricted from expanding their archives to include modern material. Libraries may no longer be the definitive repositories they once were, and the "free information" ethic that has guided libraries since their inception may be changing.

The success of digital libraries will depend on many factors, among them to what extent libraries will have access to resources, whether publishers will have an incentive to make digital materials available, and whether scholars will be able to access the materials they need. In the first case, librarians must have access to all the resources their users expect them to provide; in the second, authors and publishers must be able to make money from the materials they produce; and in the third, scholars must be able to gain inexpensive, easy access to the materials they need without having to search in many different places. Traditional libraries have managed to satisfy these competing criteria since at least the time of the ancient Greeks, but whether digital libraries can continue that tradition remains to be seen.

—*Chris Woodford*

RELATED TOPICS
CD-ROMs, Electronic Books, Electronic Commerce, Internet and World Wide Web, Internet Search Engines and Portals, Language Recognition Software, Scanners, Supercomputers

BIBLIOGRAPHY AND FURTHER RESEARCH

BOOKS
Graubard, Stephen, and Paul LeClerc, eds. *Books, Bricks, and Bytes: Libraries in the Twenty-First Century*. Piscataway, N.J.: Transaction Publishers, 1998.

Heim, Michael. *The Metaphysics of Virtual Reality*. New York: Oxford University Press, 1993.
Lesk, Michael. *Practical Digital Libraries: Books, Bytes, and Bucks*. San Francisco: Morgan Kaufmann, 1997.
Romano, Frank. *Digital Media: Publishing Technologies for the 21st Century*. Torrance, Calif.: Micro Publishing Press, 1996.

PERIODICALS
Alper, Joseph. "Assembling the World's Biggest Library on Your Desktop." *Science*, September 18, 1998, 1784.
Bloom, Floyd. "Refining the On-Line Scholar's Tools." *Science*, January 26, 1996, 429.
"Digital Libraries." *Computer*, February 1999, 45 (special issue devoted to digital libraries).
Dutton, Gail. "Turning Electronic Pages." *Popular Science*, August 1998, 33.
Keller, Michael. "Libraries in the Digital Future." *Science*, September 4, 1998, 1461.
Lesk, Michael. "Going Digital." *Scientific American*, March 1997, 50.
Lynch, Clifford. "Searching the Internet." Scientific American, March 1997, 44.
Okerson, Ann. "Who Owns Digital Works?" *Scientific American*, July 1996, 64.
Rothenberg, Jeff. "Ensuring the Longevity of Digital Documents." *Scientific American*, January 1995, 24.
Stix, Gary. "The Speed of Write." *Scientific American*, December 1994, 72.

INTERNET RESOURCES
"Access and Archiving as a New Paradigm," by Margit A. E. Dementi, *Journal of Electronic Publishing*, March 1998, Volume 3, Issue 3
http://www.press.umich.edu/jep/03-03/dementi.html
Kent State University: Future of Print Media Symposium 1998
http://www.jmc.kent.edu/futureprint/
The Library of Congress National Digital Library Program
http://lcweb.loc.gov/homepage/digital.html
"Search No More: An Electronic Database of Academic Journals Known as JSTOR Saves Researchers Time and Librarians Stacks of Storage Space," *Philadelphia Inquirer*, April 15, 1999
http://www.phillynews.com/inquirer/99/Apr/15/tech.life/JSTOR15.htm
"Why Digitize?" by Abby Smith, Council on Library and Information Resources (CLIR) Report, February 1999
http://www.clir.org/pubs/reports/pub80-smith/pub80.html
Digital Library Sites
Alexandria Digital Library
http://www.alexandria.ucsb.edu
Bibliothèque Nationale de France: Making the Most of Modern Technology
http://www.bnf.fr/institution/anglais/plaogb.htm
British Library: Digital Library
http://www.bl.uk
Digital Library Project, University of California at Berkeley
http://elib.cs.berkeley.edu
Electric Library
http://www.elibrary.com
Internet Public Library
http://www.ipl.org
JSTOR (Journal Storage Project) Digital Library
http://www.jstor.org
Libweb—Library WWW Servers
http://sunsite.berkeley.edu/Libweb/
New York Public Library: Digital Library Collections
http://149.123.1.8/index.html
Project Gutenberg
http://www.promo.net/pg

DIGITAL VIDEO TECHNOLOGY

The digital video disk (DVD), also known as the digital versatile disk, is the latest stage in the development of optical storage technology, technology that stores digital information on a disk that can be read using a laser. The DVD might best be thought of as a superior compact disk (CD).

The DVD's vast storage capacity will remove the need to compromise on the amount of image and sound information that can be stored on today's more common forms of media. The DVD will bring cinema-quality pictures and sound directly into the home, allowing the viewer to experience a movie the way a director intended. Hailed by many as the all-encompassing replacement for audio CD, CD-ROMs, videotape, laser disk, and video game cartridges, the DVD is also likely to be developed for non-entertainment purposes. With enough storage space for seven entire sets of the Encyclopaedia Britannica on each disk, the DVD is likely to prove an ideal data storage medium for large businesses and government departments.

Scientific and Technological Description

Because they are the same size and are constructed from the same materials, the DVD and the CD appear identical to the naked eye, although the standard DVD can store almost seven times more information than the CD, 4.7 gigabytes (a gigabyte is equal to roughly 1000 billion bytes) to the CD's 650 megabytes (a megabyte is equal to roughly 1 million bytes). In both cases information is stored on the disk in the form of indentations that correspond to binary code. Binary code is a number system based on just two numbers, 0 and 1. During playback, this information is read by a laser in the disk player, and minute variations in the amount of laser light reflected off the disk are converted into video and audio signals (see figure).

Data Conversion Process

Sound is converted into digital information prior to storage on the disk by dividing the original sound waves vertically into tiny sections. The strength of the signal in each section is then measured and assigned an appropriate binary number. Similarly, film is divided into individual frames, which are in turn reduced to a fine grid of colors. (These can be thought of as resembling the tiny colored dots that make up a newspaper picture.) Each colored dot is then assigned a binary number that corresponds to its color.

Once all the necessary information has been converted in this way, it is stored in blocks of binary code called *files*. Each file is assigned a code that signifies file name and type, not unlike the files stored on a personal computer. Because the process of digitizing all this information produces truly enormous files (20 seconds of broadcast-quality video translates into approximately 600 megabytes of digital code), enough to fill an entire CD-ROM, it is necessary to compress the information before it is transferred to the DVD.

The compression system currently favored by DVD producers is called *MPEG-2* (named for its developers, the Motion Picture Expert Group). In simple terms, this system examines each frame of film in sequence and then stores only the changes that occur between each frame. On fixed shots this means that several seconds' worth of film (equivalent to hundreds of frames) can be stored as a single held image. Fast-action sequences obviously require more storage space. The MPEG system offers compression ratios of around 200:1, reducing files to just half of 1 percent of their original size.

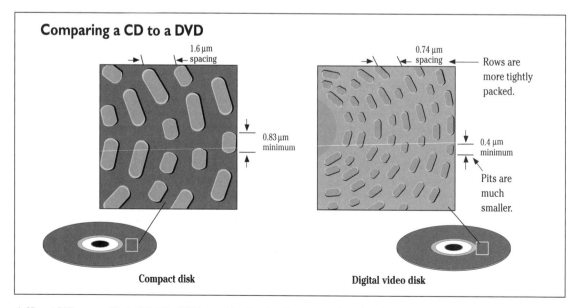

Comparing a CD to a DVD

1.6 μm spacing

0.74 μm spacing

Rows are more tightly packed.

0.83 μm minimum

0.4 μm minimum

Pits are much smaller.

Compact disk

Digital video disk

A CD and DVD appear identical, but the DVD is capable of storing almost seven times the information as the CD. That is because the pits on a DVD are much smaller and more tightly packed than those on a CD. In addition, a DVD may store four layers of information on the same disk (two on each side). (Adapted with permission from C-Cube Microsystems.)

Storage space can be increased further by creating dual-layer disks, which can hold almost twice as much information as a conventional DVD. The most common method for doing this is to place two layers of information on the disk, one of which is semitransparent. The first layer of information is read from the center to the outside of the disk by the laser before it makes a return trip (outside to center) to read the remainder of the data. This type of disk, known as a *reverse-spiral dual layer* (RSDL) *disk*, is the current standard. Earlier storage methods involved stamping information on both sides of the disk, just as with old-style vinyl records, but this method was abandoned after viewers complained about having to flip disks over.

Transfer to DVD

Once all the necessary audio, visual, and other information has been gathered and organized, the process of transferring it to the digital video disk can begin. The creation of a digital video disk begins with the mastering stage. Digital information is transferred to the light-sensitive base layer of the master disk by means of a series of laser pulses. (This operation must be carried out in an environment free of dust or vibration, as these can adversely affect the laser's focus.) After the master disk has been coded, it is immersed in an etching fluid, which reacts with those areas of the disk that have been exposed to laser light. This produces pits, or holes, in the surface of the master. Production copies, known as *stampers*, are created from the master using an electroplating technique, the same process a printer uses to create the printing plates from which books and magazines are produced. The stampers are then used quite literally to stamp the coded information onto blank disks, which are sealed in acrylic prior to final packaging.

After insertion into a DVD player, the surface of the mass-produced disk is read by a low-powered laser. The beam of laser light is focused onto the disk by a lens in the DVD player. This extremely fine beam of light is reflected by the disk onto a photodetector, which senses minute variations in the amount of light bouncing off the disk. These light signals are, in turn, translated into video and audio signals that can be understood by television or sound systems.

Historical Development

The DVD is the latest in a line of optical storage devices that began to appear in the early 1980s. Despite being used primarily as a means of viewing movies, the DVD has little in common with either film or videotape. The first true optical storage device was the compact disk. This digital disk was introduced in Japan in 1982 before being released in the U.S. and European markets in 1983. The CD revolutionized the music industry, as consumers eagerly embraced the much clearer sound the digital format delivered. CDs replaced the previous music industry standard, the analog-format vinyl record.

The next significant development in optical storage was the CD-ROM (compact disk read-only memory). Introduced in 1985, the CD-ROM can carry text, images, sound, animation, and video. Unfortunately, memory space on a standard CD-ROM is limited to approximately 650 megabytes and is therefore unsuitable for use as a digital medium for full-length feature films. The much larger but similar laser disk, introduced in 1985, was, however, capable of carrying the amount of information required to store a film. Yet the laser disk received only limited support from home entertainment enthusiasts, as most people were reluctant to pay the extremely high prices demanded for laser disks and disk players. Also, limited support from the major film studios led to a shortage of movies available in this format. The sheer size of the laser disk also meant that it was prone to twisting. This sometimes broke the acrylic seal on the disk and the resulting surface corrosion rendered it unplayable.

Philips Electronics introduced the world's first CD-I (compact disk–interactive) console in 1991. Intended as an all-around multimedia entertainment system, the CD-I was ultimately doomed to failure after being rejected by consumers and manufacturers alike. Lack of consumer demand, the product's failure to live up to its manufacturer's promises, and the fact there were very few CDs available combined to seal its fate. Philips appears, however, to be faring better with its release of the DVD, which was introduced in the United States in 1996 to an enthusiastic welcome. Although the DVD was launched in the rest of the world in 1997, DVD players and disks from the United States also reached the United Kingdom in 1996. Sony was quick to develop DVD players, and now Hitachi, Matsushita, Toshiba, and a host of other consumer electronics companies are manufacturing DVD players.

Uses, Effects, and Limitations

For the foreseeable future the primary use for the DVD will be home entertainment. The vast storage space available on a DVD means that both picture and sound quality are superior to those of the laser disk and far exceed those of ordinary video. Picture resolution on a DVD is roughly twice as sharp as that of a conventional videotape. On wide-screen movies the picture resolution is increased still further by use of a technique called *anamorphic imaging*. This involves recording the image at full frame and then squashing it evenly to the correct size for viewing on a television set. Slow motion and freeze-frame from a DVD incur none of the blurring normally associated with inferior mediums such as videotape.

In addition to carrying over two hours of very high quality digital video, the standard DVD can carry up to eight tracks of digital audio and 32 tracks of text (for subtitles). Common features also include instant rewind and fast-forwarding and search by title, track, or time code. The DVD also offers a previously unheard of degree of interactivity. By means of pull-down menus, the user can, when viewing specially produced films, opt for multiple story lines, multilingual translations, and up to nine different camera angles, allowing scenes to be viewed from a variety of perspectives.

Unfortunately, most home entertainment systems would require upgrading to take full advantage of the superior features offered by DVD. This may well prove expensive, especially when the high cost of the DVD player, starting between $200 and $300, is added to the equation. At present the number of titles available in the DVD format is limited, but the situation is improving on an almost daily basis. DVD films can cost as little as $15. That price may drop even further as long as there is no price fixing by the major film studios. On the other hand, the price may have been set artificially low to make the format appear more attractive, and its true market price may be higher. Although the DVD delivers a quality of picture and sound that is far superior to any comparable system, it currently has one major drawback: It is not re-recordable. Although no DVD manufacturer is prepared to put a date on it, all are claiming that the re-recordable DVD will be available within the next few years.

Issues and Debate

The principal concern about DVD among consumers is that it may very quickly become obsolete. There are obvious comparisons to be drawn between the DVD and CD-i. DVD enthusiasts are keen to point out, however, that the new format has the full support of the major film studios and most electronics manufacturers. In addition to this, the near studio-quality sound and pictures offered by DVD make it undeniably the best-quality format ever offered to the general public.

The major film studios are, however, seeking to protect their copyright by making it impossible to make tape copies of a movie from a DVD by means of complex encoding systems, called *copy protection systems*. Because the DVD format produces such high-quality sound and pictures, DVD-format films provide the perfect master from which to make top-quality video copies. Although there are four different copy protection systems in use at present, each covering different aspects of the copying process, the major film studios are still examining ways of improving copy protection for DVDs.

To maintain control over distribution, the major film studios have also insisted that manufacturers of DVD players produce machines that conform to a regional coding system. Each player is coded during manufacture with one of six regional codes, depending on where in the world it will be sold. DVD players produced for the North American markets will, for example, be designated code 1 machines. This means that they will, in theory at least, play only those disks that share the same code. A disk purchased in another market (Japan, Europe, South Africa, and the Middle East have been assigned code 2) will not be recognized by a code 1 machine.

The system was adopted because film studios do not want sales of DVD-format films to cut into sales from first-release films in theaters and later releases in video stores. If a movie is released in the United States, for example, but has not yet been released in the United Kingdom, film studios do not want people in the U.K. to be able to buy the DVD disk because they will be unlikely to see it in the theater and fewer people will buy copies of the film on videocassette. Several consumer groups have already complained that this system is unfair on the grounds that it reduces consumer choice by restricting access to DVDs. At present there are, however, several sites on the Internet where the technically minded can learn how to circumvent the regional coding system, with the understanding that if equipment is altered, manufacturers' warranties will no longer be valid.

—*Mike Flynn*

RELATED TOPICS
CD-ROMs, High–Definition Television, High–Fidelity Audio, Lasers

BIBLIOGRAPHY AND FURTHER RESEARCH

BOOKS
Cooper, Brian, Susan Schlacter, and John Watson, eds. *Multimedia: The Complete Guide*. London: Dorling Kindersley, 1996.
Taylor, Jim. *DVD Demystified*. New York: McGraw–Hill, 1997.

PERIODICALS
Barker, John. "DVD: Report from Strasbourg." *Inside Multimedia*, December 21, 1998, 10.
Barker, John. "DVD Video." *Inside Multimedia*, December 21, 1998, 11.
"Confused about DVD? You're Kidding!" (editorial). *Inside Multimedia*, December 7, 1998, 5.
"Content Protection for DVD" (editorial). *DVD Intelligence*, March 18, 1999, 6.
Poor, Alfred. "21st Century Storage." *PC Magazine*, January 21, 1997, 52.
Quain, John R. "DVD Players: Do Try This at Home." *PC Magazine*, January 21, 1997, 57.

INTERNET RESOURCES
The DVD Forum
    http://www.phillips-conferences.com/dvd.htm
Frequently Asked Questions About DVD
    http://www.dvduk.freeserve.co.uk/
PC Magazine Online (July 30, 1998), DVD special
    http://www.zdnet.com/pcmag/firstlooks/9807/f980729b.html
Regional Coding
    http://www.unik.no/~robert/hifi/dvd/world.html
Technical Information
    http://www.videodiscovery.com/vdyweb/dvd/dvdfaq.html
World's Largest Supplier of DVDs via the Internet
    http://www.reel.com/

DNA FINGERPRINTING

DNA fingerprinting (also called DNA typing or DNA profiling) is a method of identifying people by analyzing DNA (deoxyribonucleic acid), the genetic material in our cells. The technique serves primarily as a forensic tool, providing evidence for or against the guilt of an alleged criminal. In the late 1990s, DNA fingerprinting has been employed with increasing frequency in countries where the required technology and scientific expertise are available.

DNA fingerprinting represents an unprecedented and powerful way of protecting innocent suspects and of matching perpetrators to crime scene evidence. But the technique has also been the focus of some controversy. The science behind DNA fingerprinting is complex, and as a new player in the courtroom, it has yet to gain the public's firm trust.

Scientific and Technological Description

DNA can serve as an identification tool because each person's DNA is unique (except that of identical twins). Although some sections of human DNA are common across the population, other sections vary significantly from person to person. DNA-based identification takes advantage of this variability. Understanding DNA fingerprinting requires looking at the structure of DNA. DNA is packaged in long thin structures called *chromosomes*; each chromosome is made up of two strands of coiled DNA. The core of a chromosome consists of a ladder of chemical units called *bases*, of which there are four distinct kinds: adenine (A), cytosine (C), guanine (G), and thymine (T). The sequence of bases in a person's DNA represents a code instructing cells to carry out their functions. However, only a small fraction of our DNA actually contains these instructions. The vast majority of human DNA, approximately 95 percent, has no known function; it is called *junk DNA* or *noncoding DNA*.

Sprinkled throughout a person's noncoding DNA, there are stretches of DNA made up of repeating units. The units themselves contain a particular sequence of bases. For example, the sequence GAT may be repeated numerous times without intervening bases: just GATGATGAT-GATGATGATGAT. The number of repetitions varies greatly from person to person. One person may have 89 repetitions of the GAT unit at a certain location within her DNA, and another person may have only 31 GAT units at that location. At another location of repeated units, person A may have 42 repetitions (of a different sequence of bases) to person B's 109 repetitions.

RFLP Analysis

Repeat sequences of noncoding DNA are the basis of one type of DNA fingerprinting. *Restriction fragment length polymorphism* (RFLP) *analysis* begins with the isolation and purification of DNA from a biological sample such as blood or semen. Scientists use chemicals called *restriction enzymes* to cut the DNA into many small fragments and the technique of *gel electrophoresis* to separate the fragments by size. This results in a gel containing hundreds of DNA fragments, spread out according to size and invisible to the unaided human eye. DNA fingerprinting is concerned with only a handful of fragments, those that consist of repeat sequences. Therefore, scientists have devised a way to view those specific fragments on the gel. They have fashioned probes that attach themselves selectively to fragments consisting of particular repeat sequences. The probes are labeled with markers made of radioactive chemicals that glow when exposed on photographic film. The glowing markers make the repeat sequences visible on the gel.

A person's DNA fingerprint is made up of the pattern that appears on an exposed gel after the steps outlined above have been performed for several distinct repeat sequences in that person's DNA. More than five repeat sequences are probed in the creation of a typical DNA fingerprint. Because the lengths of people's repeat sequences vary greatly throughout a population, it is extremely unlikely that two people will have the same DNA fingerprint. The odds of two people having the same DNA fingerprint may be anywhere from one in several thousand to one in several million.

PCR Analysis

A second type of DNA fingerprinting takes advantage of the process *polymerase chain reaction* (PCR). Used when only a

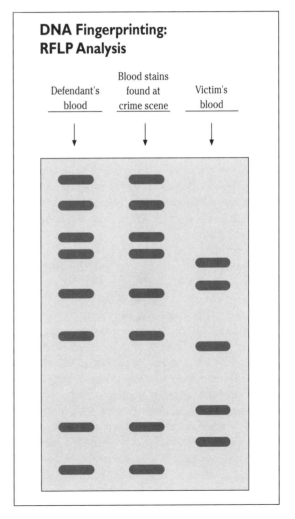

An RFLP analysis consisting of a gel exposed on photographic film. The defendant's blood matches the blood evidence found at the crime scene.

tiny amount of sample DNA is available (as little as one cell's worth), PCR can turn out billions of copies of a single molecule of DNA in an afternoon. PCR may be used to amplify (make billions of copies of) a specific chunk of DNA, such as a repeat sequence. (Other types of sequences, which also vary from person to person, are also looked at in PCR analysis. Such sequences are within coding DNA, which varies far less from person to person.) Armed with billions of copies each of several repeat sequences, scientists can complete an analysis similar to the one followed for RFLP, but the large number of copies allows them to complete the process far more quickly.

Historical Development

DNA fingerprinting gets its name from the true fingerprint, the kind our fingertips leave on surfaces. This type of fingerprint has been the prevailing form of biological identification used in criminal cases over the past century. The first description in the West of the uniqueness of human fingerprints occurred in 1880, when the journal *Nature* published letters by British scientists Henry Faulds and William James Herschel discussing the individuality and permanence of fingerprints. In 1924 the U.S. government established an original fingerprint file, which had grown by the 1990s to contain more than 90 million computerized entries. When a perpetrator leaves fingerprints behind, fingerprinting offers evidence that has long been established in the public eye as a reliable method of identification. Jury members can relate to fingerprints, because they can view and compare the evidence directly. DNA fingerprinting has yet to gain as sure a footing in courtrooms, partly because the evidence is taken from complicated laboratory procedures with which the average person has no experience.

The history of all DNA technologies begins with the 1953 discovery of the structure of the DNA molecule, by Francis Crick, James Watson, Rosalind Franklin, and Maurice Wilkins. Knowing DNA's structure paved the way for scientists to investigate the finer chemical makeup of DNA. In the early 1960s, research revealed the *genetic code*, the code (based on units of three bases) by which cells translate DNA into proteins. By the late 1970s, scientists were able to determine the sequence of bases along a given stretch of chromosome, a process dubbed *sequencing*. Sequencing caused a blossoming of DNA technology, because it allowed scientists to pin down the function—or lack of apparent function—of any given series of bases.

As scientists began to sequence sections of the human genome, the extent of the noncoding portion of our DNA (the portion that is not involved in the production of proteins) became apparent. In 1984, British geneticist Alec Jeffreys was studying noncoding DNA when he discovered that there are numerous stretches of repeated units within it. He also found that the length of these stretches varied

DNA fingerprinting is primarily used by the criminal justice system. During the O.J. Simpson double murder trial in 1995, in which Simpson was accused of killing his ex-wife, Nicole Brown Simpson, and her friend Ronald Goldman, blood evidence, and the manner in which it was collected, was of crucial importance. Above, Henry Lee, a forensic scientist and witness for the defense, testifies. Simpson was ultimately acquitted of both murders (AP/Reed Saxon).

from person to person. This finding was among the first that showed extensive variation in people's DNA. Together with advances in DNA analysis technology, these repeat sequences (and other variations identified subsequently) led to the development of RFLP analysis. The first court cases involving DNA fingerprinting took place in the late 1980s, and the technique began its journey toward becoming an acceptable forensic player.

Uses, Effects, and Limitations

Although scientists use DNA fingerprinting techniques in various fields (e.g., animal behavior, animal classification, agriculture, environmental chemistry, medicine), DNA fingerprinting's primary use is in the criminal justice system. DNA fingerprinting is valuable when an offender leaves a piece of biological evidence at the scene of a crime: blood, semen, saliva, hair, or skin or other tissue. Forensic scientists create a DNA fingerprint from DNA isolated from the crime scene tissue sample and compare it to DNA taken directly from a suspect.

If a suspect's DNA fingerprint does not match the DNA fingerprint of a given crime scene tissue sample, the suspect did not leave that tissue sample; this can be asserted definitively. DNA fingerprinting thus offers powerful protection for the innocent. Imagine an assault during which the victim scratches the perpetrator's skin. Authorities may be able to extract enough tissue from beneath the victim's fingernails to perform a DNA analysis. If a suspect's DNA fingerprint does not match that of the tissue sample, the suspect cannot be the person who was scratched by the victim.

A match between a suspect's DNA and that of a crime scene tissue sample is a little less straightforward, because of the possibility of a false positive. If scientists could sequence every base in two samples of DNA and compare them easily and quickly, there would be no question as to whether the two samples came from the same person. But sequencing every base is not currently possible; in fact, the first effort to sequence the entire set of human DNA (which contains 3 billion bases) has been under way for many years and involves the work of numerous institutions worldwide. The only practical way to use DNA as evidence is to analyze DNA fragments.

The question thus arises: Is a handful of DNA fragments enough to identify a person? That is, is it possible for two people to have the same DNA fingerprint? Scientists have calculated the odds of a false positive in DNA fingerprinting to be anywhere between one in several thousand and one in several million. That is, there may be as little chance as one in several million that two people could have the same DNA fingerprint. In most situations, such odds are considered too small to be significant, and a matching DNA fingerprint is interpreted as strong evidence that the suspect is guilty as charged.

DNA fingerprinting is limited by the possibility of human error. Samples may be mislabeled; and if stored improperly, DNA may begin to break down prior to analysis. Tissue samples may be contaminated by microorganisms or by tissues from other humans. If the DNA of a given sample mingles with foreign DNA, analysis of that sample will probably be flawed. However, forensic scientists take great care with the preparations and procedures associated with DNA fingerprinting, and most authorities agree that DNA analysis is not typically more prone to error than are other types of evidence. DNA fingerprinting is also hampered by being publicly viewed as untrustworthy, as discussed below.

Issues and Debate

When DNA fingerprinting serves as evidence in a criminal case, it has the power to decide a person's fate. To live up to its potential, it must be embraced as a trustworthy method of identification by the public. But because it is a new technology, which is not easy for a layperson to comprehend, many juries have been reluctant to accept DNA fingerprinting evidence. For people to trust a technology, they must perceive that the scientists who developed the technology agree on its interpretation. Some authorities believe that the appearance of a lack of scientific consensus has fueled public distrust of DNA fingerprinting. Two geneticists may well cite very different odds within the wide range noted earlier, and the difference between the low and high ends of such statistics can sound to a jury like a major distinction. Jurors are likely to be suspicious of the accuracy of a method that two scientists appear to interpret so differently. But authorities argue that the method itself is very

accurate; different quotes arise from what is taken into account when calculating the probability. Some DNA sequences might occur with more or less frequency within one racial subgroup, for example, and this may or may not be considered when a calculation is performed.

Another controversial issue surrounding DNA fingerprinting is its relation to national and statewide DNA databases. Some organizations have advocated the compilation of databases holding the DNA fingerprints of every citizen. Proponents argue that such a database would provide an efficient tool for identifying suspects when biological evidence is left at a crime scene. But many people are vehemently opposed to this idea, because they believe that DNA databases would violate individuals' privacy rights. The information contained in a person's DNA fingerprint may carry information related to genetic diseases and general health, information that people would not want to see fall into the hands of employers and insurers, for example, for fear of discrimination.

—Tamara Schuyler

RELATED TOPICS
Cloning, Gene Therapy, Genetic Engineering, Genetic Testing

BIBLIOGRAPHY AND FURTHER RESEARCH

BOOKS

Billings, Paul R. *DNA on Trial: Genetic Identification and Criminal Justice*. Plainview, N.Y.: Cold Spring Harbor Laboratory Press, 1992.

Coleman, Howard, and Eric Swenson. *DNA in the Courtroom: A Trial Watcher's Guide*. Seattle, Wash.: GeneLex Corp., 1994.

Inman, Keith and Nora Rudin. *An Introduction to Forensic DNA Analysis*. Boca Raton, Fla.: CRC Press, 1997.

Levy, Harlan. *And the Blood Cried Out: A Prosecutor's Spellbinding Account of the Power of DNA*. New York: Basic Books, 1996.

Robertson, J., ed. *DNA in Forensic Science: Theory, Techniques, and Applications*. New York: Ellis Horwood, 1990.

Sheindlin, Gerald. *Genetic Fingerprinting: The Law and Science of DNA*. Bethel, Conn.: Rutledge Books, 1996.

PERIODICALS

Buckley, William F., Jr. "Why Be Alarmed at Giving of One's DNA?" *Houston Chronicle*, March 4, 1999, 28.

Dawkins, Richard. "Arresting Evidence." *The Sciences*, November–December 1998, 20.

Edwards, Rob. "Fissile Fingerprints." *New Scientist*, August 19, 1995, 22.

Holden, Constance. "DNA Fingerprinting Comes of Age." *Science*, November 21, 1997, 1407.

Jayaraman, K. S. "DNA Fingerprinting Evidence Questioned." *Nature*, September 11, 1997, 109.

INTERNET RESOURCES

An Interview with DNA Forensics Authority Dr. Bruce Weir
http://www.gene.com/ae/AB/IWT/Interview_Weir.html

DNALC: About DNA Fingerprinting
http://vector.cshl.org/resources/aboutdnafingerprinting.html

Show 1305 DNA Fingerprinting
http://www.eecs.umich.edu/mathscience/funexperiments/agesubject/lessons/newton/dna.html#two

Use of DNA in Identification
http://esg-www.mit.edu:8001/esgbio/rdna/landerfinger.html

DOPPLER RADAR

Doppler radar is a type of radar (radio detection and ranging) system that uses the Doppler effect to help monitor and predict atmospheric conditions. The Doppler effect describes what happens when the frequency of light, sound, and radio waves increases as the source object is approaching a receiver, or decreases as the object moves away. Analyzing this shift in frequency allows meteorologists to determine the direction and speed of the precipitation encountered by microwaves emitted from a radar antenna. Doppler radar uses information from the Doppler effect, in addition to other conventional radar information, such as the location and intensity of precipitation, to gain an accurate portrait of the weather.

In 1985, the U.S. National Weather Service installed Doppler radar systems across the country. The current technology allows meteorologists to analyze the movement, speed, and intensity of precipitation as well as the movement and speed of wind and potentially deadly conditions such as tornadoes. Although Doppler radar is the most advanced tool for weather prediction, atmospheric conditions are inherently unpredictable. Meteorologists cannot predict deadly storms with absolute accuracy. Weather prediction is the most common use of Doppler radar, although it has other applications. Doppler radar has also been used to monitor the movement of satellites, meteors, and planets, to aid navigation on ships and aircraft, and to determine the speed of automobiles.

Scientific and Technological Description

In a Doppler radar system, a high-energy transmitter first produces a short burst, or pulse, of microwave energy. The transmitter sends this pulse to an antenna that focuses many pulses into narrow beams and emits them into the atmosphere. The antenna can send out as many as 1000 signals per second and slowly rotates the angle at which it emits the signals. At the speed of light, the microwave beams radiate outward until they make contact with any objects in the environment. These could include water droplets, ice, snow, planes, birds, or insects. When the microwaves hit an object, some of the waves scatter, and some are reflected back to the antenna, which also acts as a receiver of the reflected microwaves, called echoes (see figure).

Once the antenna receives the echoes, it sends the information to a computer that analyzes and interprets the varied aspects of the returned signal. First, the computer identifies the echoes that are the result of anything unrelated to weather. This process is called Doppler filtering. Birds, planes, and other objects in the atmosphere can be identified and rejected according to their size, speed, and manner of movement. After filtering the clutter, the computer can determine the distance, intensity, and movement of any precipitation that reflected the waves. The time lapse of the echo (the time it takes for the microwaves to return) indicates the distance of the precipitation. Because the waves travel at a known speed—the speed of light—the amount of time it takes for the signal to return determines the distance of the precipitation. In addition, the number of echoes is indicative of the intensity of the precipitation. A greater amount of rain or snow will reflect more waves to the antenna.

Finally, the frequency shift of the echoes will reveal the direction and the speed of the target's movement. If the precipitation is moving away from the antenna, the electromagnetic frequency

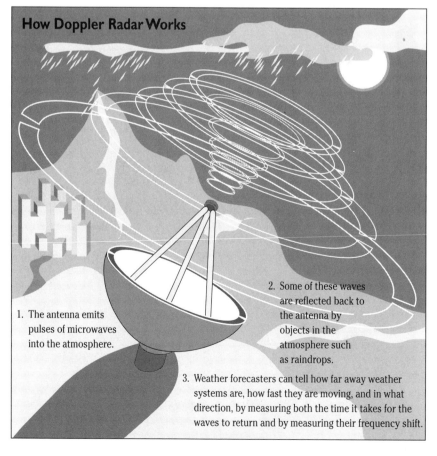

How Doppler Radar Works

1. The antenna emits pulses of microwaves into the atmosphere.

2. Some of these waves are reflected back to the antenna by objects in the atmosphere such as raindrops.

3. Weather forecasters can tell how far away weather systems are, how fast they are moving, and in what direction, by measuring both the time it takes for the waves to return and by measuring their frequency shift.

of the waves will be decreased when reflected. If the precipitation is approaching the antenna, the frequency of the waves will be greater upon return. This is due to the *Doppler effect*. A passing police car's siren is a good example of this phenomenon. As the police car approaches, the pitch of the siren increases. The greater the speed of the car, the more the pitch increases. This happens because the motion of the siren compresses the sound waves in front of it. The frequency shifts as the distance between the observer and the source becomes smaller. The rate at which this shift occurs is determined by the speed at which the distance is reduced. After the car passes, the pitch of the sound decreases because the force of the motion no longer compresses the sound waves. If both the car and observer are stationary, the frequency of the sound waves remains constant.

In the final step of the Doppler radar process, the computers display information about the location, intensity, and movement of precipitation, which is superimposed on a map. Blue, green, yellow, and red areas on the display represent the intensity of a storm. (Blue areas have the least precipitation, red the most.) Meteorologists then synthesize all this information to arrive at short-term weather predictions.

Historical Development

The Austrian physicist Christian Andreas Doppler (1803–1853) first described the Doppler effect, in 1842. He noticed that objects in space that were moving away from Earth had a blue color, while objects moving toward Earth had a red color. Doppler theorized that light and sound waves shifted their apparent frequency, relative to the observer, due to their velocity. This frequency shift, the Doppler effect, was not effectively integrated as a tool until the development of radar a century later.

In the 1870s, the Scottish physicist James Clerk Maxwell discovered the general rules of physics that govern electromagnetic fields, thereby laying the groundwork for the eventual discovery of radar. He theorized that light, sound, and radio waves would all be subject to the same physical laws. A decade later, the German physicist Heinrich Hertz showed that radio waves do in fact obey the same laws as sound and light. He showed that emitted radio waves will reflect from metallic objects.

In 1904, German engineer Christian Hülsmeyer developed and patented a radar-like device for obstacle detection and ship navigation using the principles demonstrated by Hertz. Hülsmeyer brought his invention to the attention of the German navy, which showed no interest. In fact, radar's potential did not arouse great interest until the 1930s, when long-range bombers were developed that were capable of carrying large and devastating weaponry. In 1935, Sir Robert Watson-Watt, a Scottish physicist, produced the first effective radar system. Throughout the remainder of this decade, many countries independently experimented with and advanced radar technology.

During World War II, radar was used as a warning system so that those on the ground could be notified of potential naval or air strikes. Aircraft and ships were also equipped with radar systems to aid navigation. With the radar systems of the time, antennas emitted radio signals that were reflected by incoming planes, ships, or other objects. Radar technicians could identify the location of an object by the reflected signals. However, technicians noticed that precipitation caused interference on their displays, making target objects undetectable. This observation, at the time a nuisance, led to the use of radar as a weather forecasting tool.

By the 1950s, scientists began to integrate the Doppler effect into their analyses of radar, calculating the frequency shift between initial and reflected signals to measure the movement of target objects. In the following decade, the U.S. National Weather Service started to experiment with Doppler radar for weather prediction. However, the technology at the time made Doppler radar impractical for widespread use. As digital technology advanced in subsequent years, so did the effectiveness of Doppler radar. In 1982, the National Weather Service installed a prototype of a Doppler weather system in Montgomery, Alabama. By 1985, Doppler technology had been refined and the National Weather Service began to install Doppler systems across the country. As the price of the technology dropped, many local television stations purchased their own Doppler radar systems for weather prediction purposes.

Currently, the National Weather Service uses a version of Doppler radar called Nexrad (for "next generation radar"). This is presently the most advanced form of weather forecasting radar, with the ability to detect air motion in a thunderstorm and even the movement of insect swarms. With this well-developed and detailed form of Doppler radar, the Weather Service is better able to detect and provide warnings of severe weather.

Uses, Effects, and Limitations

With Doppler radar, meteorologists can predict and monitor many aspects of weather. First, Doppler radar displays the location and type of precipitation that is occurring in the atmosphere. Different types of precipitation (e.g., rain, snow, or hail) will reflect different patterns of echoes to the radar source. Through analysis of the Doppler effect, the movement of precipitation and water droplets in clouds can be observed. Meteorologists can also analyze the structure and movement of wind. Because wind often carries in it dust, dirt, bugs, or water droplets, it can also be detected by Doppler radar.

With all this information, meteorologists are able not only to predict the weather more accurately, but also to better identify the severe atmospheric conditions that will affect lives. Tornadoes, hurricanes, and thunderstorms often cause property damage and cost many people their lives each year. Doppler radar allows meteorologists to witness the development of tornadoes before they touch down. Therefore, the public can be warned of potential tornadoes up to a half-hour before they are seen, giving people time to prepare. Tornadoes show up on Doppler radar as water droplets moving away (a decreasing frequency) and moving closer (an increasing frequency) in a tight range. Similar predictions can be made about hurricanes, again saving lives and property. In addition, because Doppler radar can monitor the intensity of precipitation, meteorologists can give flash-flood warnings well in advance of the danger.

Nexrad systems have the ability to monitor and measure rainfall for a distance of up to 460 kilometers (km). They can also ascertain frequency shifts up to a distance of 230 km from the source signal. Often, a number of antennas will monitor a given area or storm. Using more than one radar signal allows a storm to be analyzed from different angles, therefore allowing for a more detailed, three-dimensional representation of weather conditions. Furthermore, the digital processing of radar data allows even nonexpert personnel to readily interpret and comprehend Doppler radar information.

Many airports also have their own Doppler radar systems. These short-range Doppler systems are used to monitor dangerous weather conditions and phenomena that could affect incoming and outgoing flights. One of these phenomena is called a *downburst*, or *microburst*, which is a sudden, downward blast of wind that can bring an airplane to the ground. These are believed to be the cause of many weather-related plane crashes. The Doppler radar systems at airports look for indications of microbursts and other severe conditions, and planes can avoid these areas accordingly.

Although Doppler radar is most commonly used by meteorologists, it has other applications as well. Astronomers use variations of Doppler radar systems to analyze the movements of objects such as satellites, meteors, and planets. Planes and ships use similar systems to monitor their own environments and the movement of other craft. Law-enforcement officials use simple Doppler radar systems to determine the speed of passing cars. The same system is also used to analyze how fast a pitcher is throwing in baseball games.

Issues and Debate

When Doppler radar information indicates imminent severe weather, meteorologists can warn the public by means of television or radio, or through warning sirens. Many areas (especially those prone to tornadoes) have warning sirens that sound in advance of potentially dangerous conditions and can be heard up to 3 km away. However, the accuracy of Doppler radar systems used for these warnings is not absolute. Weather is inherently unpredictable, with small, undetectable factors in constant interplay that affect overall weather patterns. This is a limitation intrinsic to weather prediction that may never be overcome regardless of technological advances. In addition, because Doppler radar systems are so sensitive, what appears to be a tornado forming may not be. Therefore, some estimates indicate that 50 to 80 percent of tornado warnings are false alarms.

Nonetheless, Doppler radar technology has saved many lives. With advanced warnings, tornadoes that would otherwise kill hundreds of people now result in many fewer deaths, and often none. During the 1930s, tornadoes killed 1945 people in the United States. In the decade between 1986 and 1995, tornadoes caused 418 deaths. Yet some experts believe that a false confidence in the technology has emerged. Many people assume that a tornado warning is given in time to allow an automatic 10 to 20 minutes to flee, when in fact half of the warnings are given less than 10 minutes in advance of a storm. Furthermore, advanced Doppler tornado warnings have been interpreted by some as a signal to film impending storms with a home video camera. Many amateurs risk this dangerous behavior because of the price that television stations often pay for such films. Nonetheless, with the advancement of technology, Doppler radar will become even more refined and will enable meteorologists to predict ever more accurately the weather and potentially dangerous conditions.

—Paul Candon

RELATED TOPICS

Computer Modeling, Global Positioning System, Microwave Communication, Remote Sensing

BIBLIOGRAPHY AND FURTHER RESEARCH

BOOKS

Doviak, Richard J., and Dusan S. Zrnic. *Doppler Radar and Weather Observations.* San Diego, Calif.: Academic Press, 1984.

Skolnik, Merrill I., ed. *Radar Handbook.* New York: McGraw-Hill, 1990.

PERIODICALS

Condella, Vince. "Is It Coming or Going?" *Earth*, April 1998, 56.

Cook, William J. "Ahead of the Weather." *U.S. News & World Report*, April 29, 1996, 55.

Hendrick, Bill. "Storm Warning Systems Far from Perfect Radar Images: Improved Systems Can't Take Art Completely Out of Science." *Atlanta Journal and Constitution*, March 29, 1998, C6.

Nash, J. Madeleine. "Science: Unraveling the Mysteries of Twisters Armed with Powerful Computers, Doppler Radar, and Plenty of Nerve, Meteorologists Are Getting a Handle on Nature's Most Terrifying Storms." *Time*, May 20, 1996, 58.

INTERNET RESOURCES

Accu Weather
http://personal.accuweather.com/iwxpage/paws/dopplerfaq.htm

Doppler Radar
http://cimss.ssec.wisc.edu/oakfield/radar.htm

How Radar Works
http://www.tco.com/weather/how.htm

National Weather Project
http://www1.tor.ec.gc.ca/doppler/about/evolution_e.htm

DRUG DELIVERY SYSTEMS

Drug delivery concerns getting drugs inside the body and getting them to where they are needed to have their effect. Although conventional methods, such as swallowing pills, can be very effective, over the past 40 years new techniques for drug delivery have been developed. Skin patches deliver drugs steadily over a long period without patients even needing to think about taking their medication. Sustained-release implants do this too, but they also deliver drugs exactly where they are needed inside the body.

Other techniques, such as needleless injections, are very user friendly. Some delivery systems, particularly viruses and liposomes, are making it possible to use drugs that are difficult or even impossible to deliver by conventional methods. Modern techniques of drug delivery are often costly, but the benefits for clinicians and patients can be worth the extra expense.

Scientific and Technological Description

Most people are familiar with swallowing a pill or receiving an injection, two traditional routes of drug delivery. However, modern medicine is producing many drugs that are not effectively administered through these routes, and even old medicines can be more effective when administered by modern, sophisticated drug delivery systems. Several diverse types of drug delivery systems have been developed, the most common of which are discussed below.

Sustained (slow)-release capsules represent one of the first and still most common methods to control how fast and for how long a drug is delivered. Coated pellets containing the medication are packed into a capsule that dissolves quickly in the stomach, releasing the pellets, which then slowly release the drug.

Needleless injections are a significant recent improvement over injections by needles. A powdered drug can be forced through the skin at supersonic speed with a high-pressure jet of gas, such as helium. One commercially available device to achieve this is the PowderJect. The procedure causes no pain, although it leaves a slight but temporary reddening of the skin (see figure). Needleless injections can be used for any drug that can be dried into a powder or that can be attached to the minute powder particles. In fact, many drugs have a longer shelf life as a powder than in liquid form.

Gases and aerosols (small droplets or particles suspended in air) can be absorbed through the membranes of the lung through *inhalation* and can then pass quickly into the bloodstream. A drug can be made so that it has an effect in the lung only if it is attached to a particle that cannot cross the lung membranes. Steroids for the treatment of asthma are given in this way, allowing high doses to be administered that would cause severe side effects if the steroids could move through the rest of the body. *Dry powder aerosols* have been tested for the delivery of drugs that are either not suitable for use in liquid or air aerosols, or difficult to give in large enough doses. For example, insulin and human growth hormone have a longer shelf life when in a powder form than when dissolved in liquid, and it is difficult to deliver sufficiently high doses of these drugs using

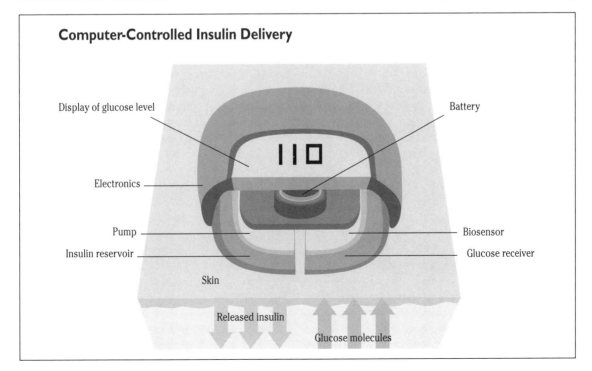

Computer-Controlled Insulin Delivery

Display of glucose level

Battery

Electronics

Pump

Biosensor

Insulin reservoir

Glucose receiver

Skin

Released insulin

Glucose molecules

A device currently being tested would automatically sense the level of glucose in a diabetic's bloodstream and release insulin as needed. (Adapted with permission from Jared Schneidman Design.)

liquid aerosols. Clinical trials for giving insulin as a dry powder aerosol began in 1998.

Since the introduction of slow-release capsules and tablets, more sophisticated methods have been developed to deliver drugs over an extended period. *Drug reservoirs* are materials that act like sponges to hold drugs. Using *sustained-delivery implants* in the body, a drug steadily leaches from the reservoir at a rate that depends on how large the "holes" in the spongelike reservoir are. As well as allowing drugs to be given to patients over long periods of time (for days or weeks), drug reservoirs can also deliver drugs locally, to the site where they are needed most.

Drug reservoir implants made from materials that are inert in the body—that is, materials that are neither affected by the body nor affect the body itself—have been developed since the 1970s. The first implants were approved for use in the United States in 1989 and 1990, for the treatment of endometriosis (with the drug Zoladex) and prostate cancer (with Lupron Depot). More recently, *biodegradable materials* have been used as reservoirs, so that the drug reservoir is digested by the body without the need for surgical removal. The first approved use of a biodegradable implant in the United States was in 1996, for a brain implant that gives sustained delivery of the drug carmustine directly to the site of a tumor.

Computer-controlled pumps have been used to administer insulin to diabetics. Users insert a small needle under the skin of their stomach, and the needle is connected to a small unit worn under the clothes or in a pocket. This unit contains a supply of insulin and a pump controlled by a microprocessor. The microprocessor is programmed by patients or doctors to control when and how much insulin is pumped into the body. For example, the device could give either a steady dose all day or a larger dose in the morning. Diabetics can also give themselves a large dose of the drug after meals, just as nondiabetics do naturally. The amount of insulin that diabetics need to take at any given time depends on how much sugar their blood contains. A device currently being tested links a microprocessor to a biosensor (see BIOSENSORS) that continually monitors the level of sugar in the blood. The biosensor then automatically triggers the release of insulin as the body requires it.

Skin patches have been developed that slowly release low levels of a drug. Nicotine (for people who are trying to quit smoking), nitroglycerine (for sufferers of angina), and estrogen (as hormone replacement therapy for menopausal women) have all been administered successfully in this way. For drugs that are not absorbed through the skin, *active patches* are being developed that use electricity from miniature batteries to force a drug through the skin. These can work because many molecules can be moved around if they are in an electric field. Researchers have also found that sound waves and lasers can be used to make small areas of skin "leaky" without causing pain. The leaky skin is able to

The PowderJect device blasts drugs through the skin at supersonic speed in a jet of high-pressure helium and is one of a new generation of drug-delivery systems (Courtesy of PowderJect Technologies, Inc.).

absorb a drug, provided by means of a skin patch, for several minutes or even hours before becoming sealed again.

Liposomes are small, water-filled "balls" of fatlike molecules. Every cell that makes up the human body can be thought of as a drop of water surrounded by a thin layer of fatlike molecules. In this sense liposomes are similar to small cells. However, unlike cells, which have complex interiors, liposomes just contain what their manufacturers want them to: for example, a drug. Drugs can either slowly leach from the liposomes, or the liposomes can fuse with cells in the body, so delivering the drug directly into those cells.

Some potential drug molecules cannot enter our cells, and this is particularly true for large molecules such as proteins and DNA. Other drugs do not dissolve in water; instead, they can dissolve in the fatlike substances from which liposomes are made. Liposomes were developed in the 1980s as delivery vehicles for these types of molecules. Researchers are currently experimenting with ways to target liposomes specifically to particular tissues in the body, so that only those tissues receive a dose of the drug. Lipsomes are already used to deliver daunorubicin and doxorubicin for the treatment of an AIDS (acquired immune deficiency syndrome)-related cancer, Kaposi's sarcoma, and liposome-based vaccines have also been licensed for use. Lipsomal amphotericin is used for the treatment of some fungal infections.

Viruses are commonly used as vaccines against other disease-causing viruses. Used in this way, viruses can be thought of as delivering a drug, the drug being a protein on the outside of the virus (the viral coat) that triggers an immune response. Today, researchers are experimenting with changing the virus coat proteins so that they trigger immunity to disease-causing organisms that are not viruses. Viruses deposit their genetic information (molecules of DNA or RNA) inside human cells, usually as part of a process that leads to illness. But RNA and DNA can be used as drugs in gene therapy (see GENE THERAPY), and one of the biggest problems in gene therapy is how to get the DNA or RNA into cells. One way to do this is with liposomes (see above), but another method is by using viruses. Adenovirus is possibly the most promising virus for this purpose, and it is being developed to deliver therapies for cystic fibrosis. Adenovirus can deliver genetic therapies to many different cells in the human body, although researchers would like to be able to target specific cell types as well.

Prodrugs are molecules that are not drugs in themselves but are converted to drugs by the body, often at or near their site of action. This property can be useful. For example, phenytoin (an anticonvulsant) is toxic when injected, but the prodrug fosphenytoin (cerebyx), which is converted to phenytoin once inside the body, is safe to inject. Between 15 and 20 percent of new drugs are in fact prodrugs.

Historical Development

In 1905, the German chemist and bacteriologist Paul Erlich (1854–1915) hypothesized *magic bullets*, drugs that would be very specific in the effects they have on the body and would not cause unwanted side effects. These drugs, in theory, would accumulate in the body where they were needed and would be found only at low levels in the rest of the body. Over the past 100 years, drug developers have put great efforts into producing drugs that do not have unwanted side effects, and in the past 50 years methods for getting drugs to where they are needed in the body have improved tremendously. Ehrlich's vision of magic bullets has yet to be fully realized, but modern drug delivery systems are bringing us closer to it.

Since the 1960s, companies have been formed to benefit commercially from drug delivery systems. One of the earliest such companies was the U.S.-based Alza, which developed sustained drug release systems. Pfizer was one of the first companies to use Alza's technology, which it used to deliver a heart drug, Procardia XL. By incorporating the sustained-release technology, Pfizer was able to cut the dose rate of this drug from three times per day to once per day, and as a result tripled their sales. Sustained-release drugs were the first major advance in drug delivery systems, proving both their clinical and commercial value. Another notable development came in 1969, when Glaxo introduced Ventolin inhalers, which revolutionized the treatment of asthma. Inhalers marked a major improvement in the lives of asthma sufferers and were sophisticated home-use systems. Since then inhalers have become commonplace for the treatment of diseases affecting the lungs.

Today, many pharmaceutical and drug delivery companies cooperate in strategic alliances, in recognition of their codependency, following the example set by Pfizer and Alza. Drug delivery systems generate around $22 billion in sales for pharmaceutical companies and delivery system companies each year, a figure that is expected to grow at a rate of 15 percent annually for the foreseeable future.

Uses, Effects, and Limitations

A number of factors determine how particular drugs are best administered. Three important factors are:

1. *Getting the drug to where it is needed.* All drugs must get to a particular place in the body in order to carry out their function. A drug to treat bone cancer needs to get into the bone tissue, for example. Different drugs move around the body in different ways, and the combination of how a drug is administered and how it travels through the body determines where the drug can get to. Some drugs, such as anticancer drugs, cause undesirable side effects that can be minimized by giving the drug in a way that keeps it localized (topical administration) and away from where it can cause harm. Other drugs, such as aspirin, are given systemically; that is, the aspirin becomes distributed around the body. A general aim of drug delivery systems is to get a large dose of the drug to where it is needed in the body and to reduce how much of the drug gets to places where it is not. For example, inhalers can deliver large doses of steroid hormones right into the lungs of asthma sufferers. In fact, the amounts of steroids given would cause bad side effects if they were given by injection into the blood. Similarly, carmustine-releasing brain implants deliver high doses of this toxic drug straight to the site of brain tumors. Liposomes and viruses should even make it possible to use new types of drug molecules that are difficult to get into the body effectively. A good example are the DNA drugs used in gene therapy.

2. *Dose control.* Some drugs work effectively when given in a large dose over a short period. Other drugs work better if given at a low dose that is maintained over a long period. Drug delivery systems have been designed that give the body just the right dose of a drug over the right period. A particular problem with taking tablets or having injections is that the body receives a large dose of a drug immediately, perhaps even a harmful dose. But then the levels of the drug fall until there is not sufficient drug in the body to have a beneficial affect. Sustained-release implants and insulin pumps, for example, are good at maintaining a constant and desirable amount of drugs in the body.

3. *User friendliness.* Many modern drug delivery systems are designed to make administering and receiving drugs as easy and painless as possible. Needleless inejections and skin patches are particularly user friendly. It is

also important to ensure compliance—getting patients to follow their medication instructions properly. By using skin patches and implants, patients don't have to think about taking their medication.

Issues and Debate

Patients benefit from new drug delivery technologies because they may have to take medication less often, with less pain or difficulty and with less chance of taking the wrong dose. This often helps clinicians, too. One-tenth of hospital admissions are due to patients falling ill through not taking their medications as they are supposed to, and half of all patients with serious chronic illness don't take medications as prescribed. Treating these "unnecessary" patients is expensive, so any drug delivery system that increases the chance that patients will take their medication properly can be potentially cost-effective. Although saving on the expense of unnecessary medications, however, these new technologies are often themselves quite expensive. It is possible that in the future only the financially better-off patients will have access to the most comfortable and convenient drug delivery systems, such as needleless injections.

Further advances may only worsen the stratification between those who are able to afford the latest drug delivery technologies and those who are not. Liposomal and viral drug delivery systems are expensive to develop and produce, but they are making it possible to use new types of molecules as drugs. Drug molecules such as DNA (for use in gene therapy), as well as the delivery systems for such drugs, will be costly to use. These costs must eventually be passed on either to patients themselves or to society as a whole.

—Matthew Day

RELATED TOPICS
Biosensors, Gene Therapy, Light-Activated Drugs, Rational Drug Design

BIBLIOGRAPHY AND FURTHER RESEARCH

BOOKS
Aldridge, Susan. *Magic Molecules: How Drugs Work.* Cambridge: Cambridge University Press, 1998.
Henry, John A., ed. *The New Guide to Medicines and Drugs.* London: Dorling Kindersley, 1997.
Julien, Robert M. A *Primer on Drug Action*, 7th ed. New York: W.H. Freeman, 1995.

PERIODICALS
Berressem, Paul. "The Birth of New Delivery Systems." *Chemistry in Britain*, February 1999, 29.
Hecht, Jeff. "Skin Dipping." *New Scientist*, June 20, 1998, 11.
Judge, Michael. "Sugaring the Pill." *New Scientist*, August 24, 1996, 24.
Kefalides, Paul. "New Methods for Drug Delivery." *Annals of Internal Medicine*, June 15, 1998, 1053.
Tonks, Alison. "More Precise Targeting of Drugs in Pipeline." *British Medical Journal*, February 7, 1998, 412.
Wilson, James M. "Adenoviruses as Gene-delivery Vehicles." *New England Journal of Medicine,* May 2, 1996, 1185.

INTERNET RESOURCES
Alkermes, Inc.
    http://www.alkermes.com/Pages/alkermes_corporate.html
Elan Pharmaceutical Technologies
    http://www.elan.ie/drugdel/
Information from North Carolina State University
    http://www5.bae.ncsu.edu/bae/research/blanchard/www/465/textbook/otherprojects/drugDeliver_97/
Site devoted to drug delivery
    http://members.tripod.com/~Chaubal/index.html

ELECTRONIC BOOKS

An electronic book (sometimes shortened to e-book) is a computerized book in which information stored in electronic form is read from a screen instead of from a printed page The simplest electronic books are scrolling text versions of printed books that can be downloaded from the Internet and read on a personal computer (PC); the most sophisticated are handheld devices that aim to combine the convenience of paper books with the power of computers.

Electronic books promise easy access to vast amounts of information at low cost, but they will need to be as simple to use, portable, and durable as their paper cousins if they are to achieve similar popularity.

Scientific and Technological Description

Electronic books work by storing text and pictures in an electronic form that can be displayed on a screen of some kind (see figure). This means that electronic book designers have two major problems to tackle: how to store information and how to display it. The most important component of electronic book technology is information storage. Some books are easier than others to store in electronic form. Novels, for example, use only a single font (typeface) and seldom have pictures. They can be stored using the *American Standard Code for Information Interchange* (ASCII), a way of representing the 256 most common characters (letters, numbers, and symbols) that virtually all computers can understand. Graphically designed documents that use mixed fonts and pictures cannot be stored using ASCII alone. One technique commonly used in CD-ROM books (electronic books on compact disks) stores text and pictures as *bitmaps* (images made up of thousands or even millions of dots), but this technique treats words as pictures and means that they cannot be searched.

Electronic book designers use more advanced methods of information storage to handle text and images equally well. Graphic software maker Adobe Systems' Acrobat software converts word-processed documents into a page-description language, which describes in detail how headings, fonts, hypertext links (electronic cross-references), and pictures should be treated. The resulting files, known as *portable display format* (PDF), can be read with special Acrobat reader software and transmitted easily across the Internet.

An Electronic Book

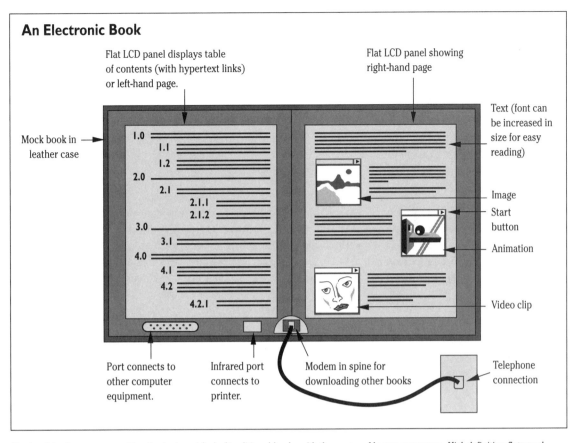

Flat LCD panel displays table of contents (with hypertext links) or left-hand page.

Flat LCD panel showing right-hand page

Mock book in leather case

Text (font can be increased in size for easy reading)

Image

Start button

Animation

Video clip

Port connects to other computer equipment.

Infrared port connects to printer.

Modem in spine for downloading other books

Telephone connection

Books of the future may combine the look and feel of traditional books with the power of laptop computers. High-definition flat-panel screens replace pages. A built-in modem allows new books to be downloaded from the Internet and an infrared port allows pages to be printed out just as they appear on the screen.

The most versatile (and most complex) method of document storage available to electronic publishers is known as *standard generalized markup language* (SGML). SGML is a set of highly structured rules for creating what are known as *markup languages*. The World Wide Web (WWW) document language HTML (Hypertext Markup Language) is an example of SGML in which users "mark up" Web-page text with bracketed symbols (tags) such as <P> to produce a paragraph and <H1> to produce a main heading. How those tags ultimately appear depends on how the screen on which they are displayed (or printer on which they are printed) is set up. Although this makes it difficult for authors to know exactly how their material will turn out, it has the advantage that a single SGML or HTML file can be used to produce output on a variety of screens and printers without modification. This explains how WWW pages can be displayed on many different computers set up in different ways. It also means that a paper book that has been coded in SGML can easily be turned into an electronic book by changing only how the tags are ultimately interpreted, without changing any of the original files.

Electronic book designers must also devise ways to display text electronically, yet as readably as a printed book. The display capabilities of most cathode-ray tube (television-style) monitors and liquid crystal or flat panel displays (LCDs) are crude compared to those of printed paper. However, in 1999, researchers at IBM announced that they had developed a screen capable of displaying 200 pixels (picture dots) per inch, which is four times greater than the resolution of the best available computer monitor and which the researchers claim rivals the readability of paper. Researchers at the Boston-based company E Ink are trying to solve this problem in a different way by creating what they call *electronic ink*. This consists of billions of microscopic capsules about 40 microns (micrometers; millionths of a meter) across embedded in a paper page, which turn white or black with an applied voltage. The pages are expected to look as sharp as paper, but refresh as quickly as a computer screen.

Historical Development

U.S. electrical engineer Vannevar Bush is widely regarded as the father of the electronic book. In 1945, Bush, who was a pioneer of the digital (number-processing) computer and worked on the Manhattan Project to build the first nuclear bomb, sketched out a hypothetical device that could help a person store and organize his or her entire memory and knowledge. Bush called this device the Memex and suggested that it should consist of a large desk on which were mounted display screens and a keyboard. It would be able to store more than 5000 pages of new information a day on microfilm, including books and magazines that could be purchased and loaded straight in. To this, users could add their own notes, photographs, and other memories. Most important, the information would be indexed for easy retrieval and would feature links between related items. Although the Memex was never developed into a working device, Bush's ideas are now clearly visible in electronic books, hypertext, and the World Wide Web.

Since the Memex, various attempts have been made to bridge the gap between computers and printed books. Project Gutenberg, which makes books available over the Internet, was begun in the 1970s. Markup languages, which led to SGML and HTML, were pioneered by IBM in the 1970s and 1980s with its GML and BookMaster products. Initially, these were used for creating the large amounts of documentation that accompanied IBM's products.

In 1983, music began to be distributed on compact disks (CDs), and the technique was later modified for storing computer data. Compact disk read-only memory (CD-ROM), as this new format was known, led to the production of multimedia electronic books that featured fully integrated text, sound, video, and animation. In 1990, Sony produced a hand-held electronic book reader based on CD-ROM technology known as the Data Discman (later the Bookman), but it was a commercial failure, due in part to its high cost and in part because publishers failed to produce enough CD books to accompany it. The year 1994 brought the first major governmental commitment to electronic books with the announcement of the Digital Libraries Initiative (DLI) funded by the U.S. National Science Foundation. This promoted a number of major initiatives to store information in electronic form and distribute it via networks such as the Internet.

Meanwhile, during the 1990s, the part of the Internet known as the World Wide Web (WWW) developed into a global electronic library accessible from any desktop equipped with a PC, an Internet connection, and browser software. The invention of HTML by Tim Berners-Lee at CERN particle physics laboratory in Switzerland in the early 1990s led to new ways of creating and distributing electronic information. The key to HTML's success is that it is an ASCII-based markup language. This means that HTML files can be understood by almost any type of computer, which can decide how to interpret and format the HTML tags according to its own capabilities and the user's preferences.

A new wave of hand-held electronic book readers appeared beginning in the fall of 1998. The SoftBook, RocketBook, and Everybook aim to resemble paper books

as closely as possible; they are effectively books whose pages have been replaced by flat panel displays.

Uses, Effects, and Limitations

There are currently several different types of e-books on the market. The SoftBook by SoftBook Press is a leatherbound, 1-inch-thick computer that sold for $299 to $599 in 1999. It is designed around an 8- by 6-inch screen that can display 256 shades of gray and stores text using a proprietary format based on HTML. The RocketBook by NuovoMedia uses a much smaller monochrome screen and cost $499 at its launch in fall 1998. Unlike SoftBook, it uses standard HTML to store books. Costing about $1500 at its launch, the Everybook Dedicated Reader from Everybook Inc. is a more sophisticated product, with two 13-inch full-color displays that can mimic the twin pages of a book. Unlike SoftBook and RocketBook, Everybook stores books in Adobe Acrobat format.

These three electronic hand-held book readers have similar features. They all have full-page displays to eliminate scrolling (a key problem with traditional computer screens), meaning that the reader gets a piece of text as a "page" and then turns to the next. They are all single-purpose computers: They don't double-up as palm-top personal computers or diaries. They contain various features to make electronic reading easier, such as keyword searches, built-in dictionaries, the ability to make notes or add bookmarks, and the ability to change font size. They also have a means of loading new books into memory: SoftBook and Everybook have built-in modems (devices for connecting computers to the telephone network) to enable new books to be downloaded from the Internet, whereas RocketBook connects to a Windows PC, which downloads the new books from the Internet and copies them across to the RocketBook reader.

The book readers also have similar limitations. They are expensive, currently beyond the budget of most students. They can be heavy (Everybook weighs about 3.5 pounds), although their manufacturers claim that one electronic book could take the place of many pounds of college textbooks. They are battery powered, which means that they can give between 4 and 30 hours of continuous reading before new batteries are needed, depending on the model. Although their screens eliminate scrolling, they are still based on flat panel technology (see FLAT-PANEL DISPLAYS), which remains far inferior to paper in sharpness and readability. Finally, although publishers such as Random House and Simon and Schuster and bookstores such as Barnes and Noble are working with the manufacturers of these devices, few electronic books are currently available.

This does not mean that books are unavailable in electronic form, however. Some traditional publishers, notably Dorling Kindersley, Encyclopaedia Britannica, and Oxford University Press, have produced best-selling electronic encyclopedias and dictionaries on CD-ROM. IBM has been distributing technical manuals on CD-ROM since 1990 using book reader software known as BookManager, which can run on most computers, from the largest mainframe to desktop PCs; each CD-ROM carries the information published in several hundred thick manuals and is updated quarterly to give customers the benefit of bug fixes and other technical improvements. Adobe Acrobat is used increasingly to distribute books, magazines, corporate and government reports, and brochures via the Internet.

For many people, the Internet is now a seamless electronic library; some Web sites are electronic libraries in themselves. The best known of these, Project Gutenberg, makes available approximately 1500 classic books whose copyrights have expired. However, they are supplied in simple ASCII format and have to be read in a long scroll on a PC screen.

Other electronic book projects include a system at the British Library in London, which contains scanned images of precious manuscripts such as the notebooks of Leonardo da Vinci. A touchscreen allows readers to turn the virtual pages, zoom in, and even reverse Leonardo's famous mirror writing to make it easier to read. (Scholars debate why the artist wrote in reverse; some say it was to protect the secrecy of his inventions.)

The scientific problem of making books electronic is not just a matter of designing booklike computers; making information easier to find and access is just as important. Internet search engines enable keyword searches of the vast World Wide Web in a few seconds, but they are not ideal: There is a skill to using them and they assume some knowledge of the subject in question (see INTERNET SEARCH ENGINES and PORTALS). Electronic books are prompting the development of better methods of indexing and searching for information. Good paper books rely on the skills of an indexer who can tease out important concepts in a way that makes sense to readers. Similarly, electronic books and DIGITAL LIBRARIES are beginning to use semantic indexing (based on a knowledge of what keywords mean and how different concepts are related to one another) and ARTIFICIAL INTELLIGENCE to enable fast and easy retrieval of information.

Issues and Debate

Compared to paper books, today's hand-held electronic book readers seem crude and expensive, but they offer many potential benefits. For example, electronic books can be updated instantly over the Internet (the Texas Board of Education is already looking at whether it would be cheaper to issue electronic book readers to students rather than trying to keep its stock of textbooks up to date); and if electronic books can be combined with speech synthesizers (devices that convert computer output to speech), blind and visually impaired people will have vastly improved access to written knowledge.

Electronic books raise questions not just for readers but also for publishers. Distributing material freely over the Internet raises copyright problems (which is why Project Gutenberg distributes texts that are not copyright protected). The makers of electronic book readers and their publishing partners have devised ways of charging users to download book files from online stores using encryption (coding) technology. The new delivery system has prompted questions. Will the money earned by publishers flow back to authors in royalty payments? As e-books evolve, which books will become available electronically? Who will pay to convert the entire corpus of literature into electronic form? If the answer is no one, does that mean that electronic publishing will be confined to best-sellers? Electronic books can be reproduced at almost zero cost, but a publisher must still pay the cost of researching, writing, and editing the information to begin with, which means that it will never make sense to distribute electronic books for free.

Many public libraries were set up by philanthropists such as U.S. industrialist Andrew Carnegie, who believed in the benefit to society of a free education. Some people believe that electronic books and libraries and a "pay as you go" Internet could signal the end of libraries and of the open-information spirit that people have long taken for granted. Others argue that it could mark the beginning of a new era of digital libraries that would give easier access to far more information and in which librarians would become "information architects" and "knowledge navigators." The World Wide Web demonstrates that vast amounts of useful information can be made freely available if everyone shares in the cost. But as the Web grows more sophisticated, more authoritative, and easier to use, the very concept of books (static collections of dated knowledge) may simply come to seem old-fashioned.

—Chris Woodford

RELATED TOPICS
CD-ROMs, Digital Libraries, Flat-Panel Displays, Internet and World Wide Web, Internet Search Engines and Portals

BIBLIOGRAPHY AND FURTHER RESEARCH

BOOKS
Bolter, Jay. *The Electronic Book*. Hillsdale, N.J.: Erlbaum, 1991.
Cunningham, Steve, and Judson Rosebush. *Electronic Publishing on CD-ROM: Authoring, Development, and Distribution*. Huntsville, Ala.: CD Info Company, 1996.
Heim, Michael. *The Metaphysics of Virtual Reality*. New York: Oxford University Press, 1993.
Multimedia: The Complete Guide. London: Dorling Kindersley, 1998.
Preece, Jenny. *Human–Computer Interaction*. Harlow, Essex, England: Addison-Wesley, 1994.
Romano, Frank. *Digital Media: Publishing Technologies for the 21st Century*. Torrance, Calif.: Micro Publishing Press, 1996.

PERIODICALS
Alper, Joseph. "Assembling the World's Biggest Library on Your Desktop." *Science*, September 18, 1998, 1784.
Bloom, Floyd. "Refining the On-Line Scholar's Tools." *Science*, January 26, 1996, 429.
Coover, Robert. "Hyperfiction: Novels for the Computer." *New York Times Book Review*, August 29, 1993, 2.

"Electronic Books on the Horizon" (editorial). *Science*, July 17, 1998, 335.

Gibbs, W. Wayt. "The Reinvention of Paper." *Scientific American*, September 1998, 22.

Keller, Michael. "Libraries in the Digital Future." *Science*, September 4, 1998, 1461.

Okerson, Ann. "Who Owns Digital Works?" *Scientific American*, July 1996, 64.

"Turning Electronic Pages" (editorial). *Popular Science*, August 1998, 33.

Vizard, Frank. "Electric Tales." *Popular Science*, June 1997, 97.

INTERNET RESOURCES

The Alexandria Digital Library at University of California
http://www.alexandria.ucsb.edu

"As We May Think," article outlining Memex device by Vannevar Bush, *Atlantic Monthly*, July 1945
http://www.ps.uni-sb.de/~duchier/pub/vbush/vbush.shtml

The British Library
http://www.bl.uk

Future of Print Media: Virtual Symposium on Digital Transformation of Printing and Publishing
http://www.jmc.kent.edu/futureprint/

An Introduction to SGML
http://www.pineapplesoft.com/reports/sgml/index.html

Project Gutenberg
http://www.promo.net/pg

"Some Electrifying Reading" by Chris Stamper, ABC News Online, July 16, 1998
http://abcnews.go.com/sections/tech/dailynews/eink980715.html

ELECTRONIC COMMERCE

Electronic commerce (usually known as e-commerce) means trading over the Internet; virtual stores, online banking, cyber malls, and digital cash are its hallmarks. E-commerce is not in itself a new technology, but a new method of transacting commercial business that involves several key Internet- and computer-related technologies, some of which have been developed specifically to facilitate e-commerce.

The Internet evolved from the computer networks established by the U.S. government and universities in the 1970s and 1980s. In the 1990s, the growth of the World Wide Web (WWW), the navigable graphics-rich portion of the Internet, has taken off. As it has, corporations have begun to see the Web as a new way to make money. Today, enterprises launched on the Internet have become instant multinational corporations, simply because the Internet is accessible worldwide, with much lower costs than those of traditional businesses. However, the drive toward e-commerce could turn traditional commerce upside down, potentially making retail stores obsolete and threatening many jobs.

Scientific and Technological Description

Traditionally, trade involves attracting customers to a store where they can inspect products and decide whether to buy them. If so, they engage in a transaction in which money is exchanged for products. The transaction may also involve the storekeeper updating an inventory of the products in stock and arranging delivery to the customer's address. With e-commerce, some or all of these processes happen over the Internet (see figure).

The basic requirement for e-commerce is a Web site that simulates a store or bank. The site's home page needs to pull in shoppers from cyberspace in the same way that a traditional storefront attracts shoppers from a sidewalk. Some sites are the virtual equivalent of department stores, with similar products grouped together (*virtual* means existing inside a computer instead of in physical space); others encourage shoppers to browse through the site, wandering around graphical representations of supermarkets and dropping purchases into virtual shopping carts. Computerized stores use software known as *catalog servers*, which convert product details from a database running on a server (a computer at the store's headquarters) into Web pages that appear on a browser (a software program such as Netscape Navigator or Internet Explorer) running on the customer's personal computer (PC).

E-commerce Web sites usually enable customers not just to browse through products, but also to buy them online. *Transaction processing* (TP), as this is known, is nothing new; computerized TP systems have been around since airlines pioneered automated ticketing systems in the 1960s, and most banks and stores had sophisticated TP systems long before the Internet came along. TP computerizes all aspects of the purchase transaction, including connecting store computers to bank computers, checking account balances, and transferring money from the customer's account to the store's. Banks, stores, and large corporations that already have substantial TP systems use pieces of software known as *gateways* to connect their existing systems to the Internet. Although some reprogramming is required, the systems process Internet transactions in essentially the same way that they process transactions from automated teller machines (ATMs) or checkouts in stores; in other words, transactions launched from the Internet are just another form of input. Web-only enterprises use less sophisticated TP systems, which are designed from scratch, and which may be built into their catalog server software. Either way, TP systems may process hundreds or thousands of Web transactions every minute.

Once storefronts and transaction processing have been converted to the Web, the final step is to link in inventory control and goods dispatch. Typically, this involves separate computer systems that monitor stock availability (and order new stock automatically from suppliers) and issue instructions to automated warehouses that can dispatch goods within minutes of an order being received. E-commerce has one other key feature that traditional commerce takes for granted: security. Buying goods over the Internet involves sending credit card and other sensitive information over public telephone and computer networks that are subject to attacks from criminals and *hackers* (people who break into computer systems). In practice, cryptography (coding) is

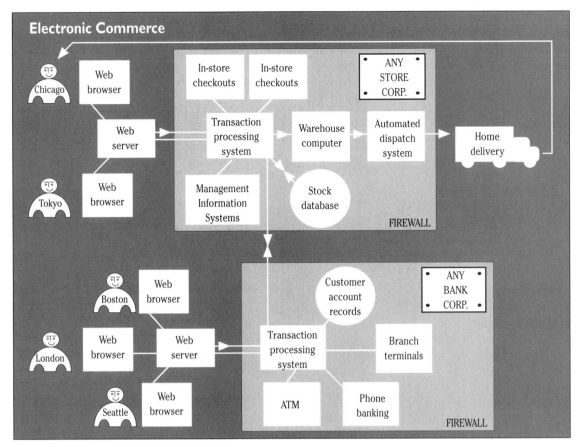

Electronic Commerce

A customer in Chicago orders goods using a Web browser on her home PC, which links to a Web server at the store's headquarters over the Internet. The Web server communicates with the store's main transaction processing system, which also handles purchases through in-store checkouts. The transaction processing system checks a stock database to make sure the goods are available, communicates with the customer's bank to ensure she has sufficient funds in her account, and finally sends dispatch details to an automated warehouse, which arranges for the goods to be delivered to the customer's door. Both the store and bank computer systems are protected from the Internet by "firewalls" that restrict customer access to a single, high-security route.

used to scramble any information that travels between the server (the store computer) and the browser (the customer's PC) (see CRYPTOGRAPHY). Many corporations using the Internet also have a security device known as a firewall, which is a virtual barrier between their sensitive internal computer systems and the public Internet that restricts customer access to a single high-security route.

Historical Development

Although e-commerce seems to have appeared overnight, it is only the latest development in the computerization of commerce, which dates back over 100 years. It was in 1887 that a statistician named Herman Hollerith working for the U.S. Census Bureau automated the process of tabulating large amounts of population data. Hollerith's contribution to history was to realize that the machine he had invented, which read its data from holes punched in cards, would have important commercial applications. In 1896, he set up his own company, and in 1924 that company changed its name to International Business Machines (IBM).

IBM dominated the computerization of commerce from the 1920s onward. From the 1960s, massive mainframe computers such as the IBM System/360 brought computerized transaction processing to airlines, banks, and stores throughout the world. Mainframe computers meant that large corporations could reduce costs and improve efficiency, but this digital revolution was largely invisible to ordinary consumers. That began to change in the late 1970s when the first desktop microcomputers (PCs) became available. By the early 1980s, even the smallest businesses were buying their own computers and running payroll, accounting, and financial planning software. Microcomputers also spurred the growth of proprietary networks and online services, such as CompuServe, America Online (AOL), and Prodigy, which pioneered shopping from home computers. The explosive growth of the World Wide Web from 1993 onward provided the next logical development: connecting all these PCs together so that interested parties could do business with one another. Internet-based enterprises were soon set up, including the

Internet Bookshop, the Amazon.com bookstore, and the Virtual Vineyards wine store.

Despite the immediate enthusiasm for electronic trading in the 1990s, e-commerce progress has been hampered by a lack of security for electronic transactions. During the mid-1990s, Internet browsers began to use cryptography, and a worldwide standard for Internet commerce known as Secure Electronic Transaction (SET) was proposed. In August 1997, Germany became the first nation to pass legislation enabling digital signatures (unique electronic codes originated by a person using a computer) to be accepted in place of a handwritten signature in electronic transactions. Concerns about online security continue to create a drag on e-commerce. In the United States, industry groups and privacy advocates have lobbied the Clinton administration to change its policy on encryption software exports, which the Commerce Department regulates. The federal government bars the export of scrambling software beyond a certain strength, which critics say hinders widespread use of the strongest available encryption software as well as hurting the U.S. industry's competitiveness abroad.

Between 1996 and 2000, the business consultancy firm Forrester Research, based in Cambridge, Massachusetts, forecast that Internet spending on consumer goods will have increased from approximately $500 million to about $6.6 billion. Economic analysts estimate that Internet-based retail transactions will be worth over $35 billion by 2005.

Uses, Effects, and Limitations

Electronic banking (e-banking) is one the many faces of e-commerce. In some cases, traditional banks such as Wells Fargo and Citibank in the United States and the Cooperative Bank in the United Kingdom have set up Web sites that offer an extra way for customers to manage their accounts, in addition to ATMs, telephone banking, and personal transactions at a branch. In other cases, banks, such as Security First National, are virtual banks that exist only on the Internet. Electronic banking implies replacing paper money with virtual money; DigiCash and Cybercash are two software systems that allow e-money to be uploaded and downloaded from a traditional account and spent in electronic transactions over the Internet. Software packages such as Intuit's Quicken enable customers to pay bills by writing checks on a home PC. The virtual checks are transmitted electronically to Intuit's headquarters in Chicago, then to a plant in Logan, Utah, where they are printed out and mailed in the normal way. Personal money manager programs such as Quicken also allow financial data, such as a person's latest credit card transaction, to be downloaded from credit card companies over the Internet.

Internet shopping is more of a departure from traditional business practices than electronic banking, because it involves a complete reinvention of how enterprises attract, relate to, and retain their customers. It is now possible to buy almost anything over the Internet, through Web

As electronic commerce expands, and as more consumers begin to do shopping online, demand for delivery services such as United Parcel Service (UPS) may expand rapidly (Courtesy of UPS).

sites set up by traditional retailers, through virtual stores such as Amazon.com that exist only on the Internet, or through one of the many Internet shopping malls such as iMALL, where retailers cluster together.

E-commerce has many benefits for companies as well as consumers. For a smaller investment than that required to set up a single store, a virtual enterprise can become an instant multinational, attracting customers from around the world. Virtual enterprises need few staff members and small premises, and this means low trading costs. For example, a single Internet banking transaction costs a bank about $0.13, compared to $0.54 cents for telephone banking, and $1.08 for personal banking at a branch. Although sophisticated Internet catalog servers can cost millions of dollars to set up, even the smallest store could set up a crude Web site in a few hours and with a very modest investment. A small-business owner could create a Web site for the cost of a few phone calls.

Big or small, e-commerce sites share another important benefit: the ability to make up-to-the-minute product or marketing information available to customers. Another upside for companies is the demographic profile of those customers: Web shoppers tend to be young, educated, and affluent. For example, market research company Student Monitor reports that 95 percent of college students use the Web; New York business consultancy eMarketer forecasts that 61 percent of U.S. teenagers will spend a total of $1.3 billion online by 2002. Over the past several years, consumers have benefited from a blossoming of online shopping opportunities, which have allowed them to grocery shop from home, send someone flowers without leaving their desk, and book airline and travel accommodations with greater ease.

But e-commerce also has limitations. The biggest of these is that most shoppers are still not connected to the Internet; even by 2002, little more than half of U.S. homes are expected to be online. Industry gurus also believe that people will not shop over the Internet until simple-to-use and cheap network computers (slimmed-down PCs designed specifically for accessing the Internet) become available. Yet even this may not be enough of an attraction to pull people online to shop. Where shopping can be an enjoyable family or social activity, e-commerce tends to be a solitary affair; the Internet has not yet found a way to attract customers in the way that traditional stores use the aroma of fresh-baked bread, ripe oranges, or leather car seats. Surprisingly, it can also take much longer to locate stores in cyberspace than in the physical world, and prices for goods (which include home delivery) are often higher.

Finally, ensuring the security of online commerce has been and continues to be a limitation. Many people avoid online transactions for fear that their personal information may be stolen and misused. Until systems are developed that assure people that they are not taking a risk by doing business online, security concerns will continue to rein in the growth of e-commerce.

Issues and Debate

E-commerce appears to offer enormous opportunities for enterprises that can quickly capitalize on it; Forrester Research estimates that the entire U.S. Internet economy will be worth $200 billion by the year 2000. But what about those businesses who don't or can't stake a claim in cyberspace? When manufacturers and wholesalers can sell direct to the public across the world through a single Web site, what future is there for traditional retail stores? When Internet banking transactions are a tenth of the cost of in-person branch banking, what future is there for bank personnel?

There are about 1900 enclosed malls and 42,000 shopping centers in the United States, but the Internet's biggest cyber mall, iMALL, already has over 1,600 stores and millions of customers. If Internet malls are so much cheaper than physical real estate, what will happen to city centers and retail jobs? Some people have suggested that the trend toward e-commerce could create an introverted society in which people cease to interact with other people except through the Internet; others argue that e-commerce will take the drudgery out of shopping and so give people more time to socialize.

Economists have worried less about the social impacts of e-commerce than about the way it challenges the competitiveness of traditional enterprise and the very way that business is transacted. Some economists doubt whether the huge sums of money currently being invested in online businesses will ever show a profitable return. Others question whether Web sites that mix free information content with commerce are selling goods and services efficiently. E-traders counter that "high-content" sites build customer relationships and will be more profitable in the long term.

Many practical issues need to be resolved before society begins to debate the wider questions associated with e-commerce. These include the difficulty of making digitally signed Internet transactions completely secure, but also legally valid and compatible worldwide; the problem of monitoring, regulating, and taxing domestic and international trade by companies that do not exist in any conventional sense; and the problem of making e-commerce available to the vast majority of humankind for whom the Internet is currently an irrelevance.

One of the most interesting issues is the extent to which global trade through the Internet will affect languages and cultures around the world. U.S.-based multinationals such as McDonald's, Coca-Cola, and Microsoft have sometimes been accused of exporting Western values and undermining the rich diversity of global culture. But if that is true, could e-commerce reduce the world to an "e-monoculture"? Or will struggling cultures find wider distribution, a bigger voice, and a new lease on life due to the global reach of cyberspace? E-commerce offers at least as many opportunities as it does threats.

—Chris Woodford

RELATED TOPICS
Cryptography, Intelligent Agents, Internet and World Wide Web, Internet Search Engines and Portals, Personal Digital Assistants

BIBLIOGRAPHY AND FURTHER RESEARCH

BOOKS
Cataudella, Joe, Ben Sawyer, and Dave Greely. *Creating Stores on the Web.* Berkeley, Calif.: Peachpit Press, 1998.
Dyson, Esther. *Release 2.1.* New York: Broadway Books, 1998.
Gates, Bill. *The Road Ahead.* New York: Viking/Penguin, 1996.
Henning, Kay. *The Digital Enterprise.* London: Random House, 1999.
Leebaert, Derek, ed. *The Future of the Electronic Marketplace.* Cambridge, Mass.: MIT Press, 1998.
Patel, Alpesh. *Trading Online: A Step-by-Step Guide to Cyber Profits.* London: Financial Times/Pitman, 1999.
Seybold, Patricia. *Customers.com: How to Create a Profitable Business Strategy for the Internet and Beyond.* New York: Times Business/Random House, 1998.

PERIODICALS
Articles referred to in *BYTE* are available online at http://www.byte.com.
Carroll, Mark. "Internet-Commerce Security." *BYTE*, May 1997, 25.
Ericson, Louis. "Shopping the Net." *Popular Science*, January 1996, 70.
Flohr, Udo. "Electric Money." *BYTE*, June 1996, 74.
Mauth, Rainier. "Digital Signatures to Power E-Commerce." *BYTE*, January 1998 (special insert), 55.
O'Malley, Chris. "Wired to the Bank." *Popular Science*, June 1996, 97.
Pountain, Dick. "The Component Enterprise." *BYTE*, May 1997, 93.
Seachrist, David. "Hanging Out an Internet Shingle." *BYTE*, April 1997, 136.
Singleton, Andrew. "The Virtual Storefront." *BYTE*, January 1995, 125.
Ward, Mark. "Knowing Who to Trust Is Crucial to Net Trade." *New Scientist*, August 29, 1998, 15.

INTERNET RESOURCES
Amazon.com
    http://www.amazon.com
E-Commerce Times
    http://www.ecommercetimes.com/
Electronic Commerce Resource Centers
    http://www.ecrc.ctc.com/

Empire Mall
   http://www.empirena.com
Geocities Shopping Center
   http://www.geocities.com/Shopping_Center/
iCatmall Shopping Mall
   http://www.icatmall/com
iMALL
   http://www.imall.com
Peapod Inc. Grocery Store
   http://www.peapod.com
Security First National Bank
   http://www.sfnb.com
Virtual Vineyards
   http://www.virtualvin.com
Wells Fargo
   http://www.wellsfargo.com
Yahoo! Store
   http://www.yahoo.com

FETAL TISSUE TRANSPLANTATION

Fetal tissue transplantation, the transfer of tissue from in vitro (test tube) embryos or aborted fetuses for treating medical conditions, have been conducted in the United States and other countries for several decades. During the late 1980s, the spotlight of publicity fell on the experimental use of fetal nervous tissue in treating Parkinson's disease (a disease of the nervous system characterized by muscle tremors and weakness). The influence of the strong antiabortion lobby in the United States has severely curtailed federal spending on fetal tissue transplantation. However, in 1998, two privately funded U.S. research teams published results showing that they had isolated embryonic human stem cells. These cells may offer the potential for developing a wide range of transplantable tissues. In response to these developments, United States policy on funding such research is currently being reviewed.

The therapeutic potential of fetal tissue transplants—whether as an end in themselves or as a steppingstone to other forms of medical treatment—appears great. However, as of 1999, the use of transplanted fetal tissue in treating medical conditions is still firmly in the experimental phase, and those who oppose abortion on ethical grounds appear likely to continue to object to fetal tissue use for medical purposes.

Scientific and Technological Description

Fetal tissue may be defined as comprising cells extracted from aborted fetuses or cells from human embryos derived from IN VITRO FERTILIZATION (IVF). This account excludes therapeutically useful cells obtained from the umbilical cord or placenta of a full-term baby (see TISSUE TRANSPLANTATION). Fetal tissues are of great interest to researchers and clinicians because they contain actively dividing cells that may have the potential to be implanted in the human body, where they can replace or compensate for dead or damaged tissue. Fetal tissue transplants as a treatment for Parkinson's disease were, until recently, the most publicized of such applications (see figure). However, other kinds of transplant, such as the use of fetal thymuses (the thymus is a gland that is part of the immune system) to treat newborn babies, have been carried out for several decades with much less public awareness. Since late 1998, scientific and media interest has focused on fetal stem cells, cells that are, as yet, undifferentiated (not specialized for a particular function). These cells have the potential to be cultured to differentiate into a variety of other types of tissues for possible transplantation.

Parkinson's disease is attributed to the degeneration of nerve cells in a specific region of the brain called the *substantia nigra* (meaning literally, "black substance"), part of the basal ganglia. The basal ganglia form part of a relay system that controls large-scale movements of the body, such as hand and arm movements, and walking. Disruption or damage to this region can produce recognizable signs and symptoms. In the case of Parkinson's disease, muscular tremors of the hands, arms, or legs may be evident at first, sometimes with accompanying rigidity, stiffness, or slowness of movement. In time, face or jaw tremor may develop and walking may develop into a characteristic shuffling gait that breaks into small, running steps as though the person were falling forward. The face may take on a masklike expression. Eventually, many years later, intellect may become impaired. Before that, everyday activities such as eating, washing, and dressing commonly become difficult. According to the Parkinson's Disease Foundation in the United States, at least one person in 200 will develop Parkinson's disease. Public figures who suffer from Parkinson's include former boxer Muhammad Ali, evangelist Billy Graham, and Attorney General Janet Reno.

In Parkinson's disease, more than 50 percent of dopamine-secreting nerve cells in the darkly pigmented substantia nigra area of the brain are lost or damaged, and dopamine secretion is reduced by 80 percent or more. Dopamine is a chemical messenger or *neurotransmitter*, a substance that carries signals from one nerve cell to another. Parkinson's is regarded as a dopamine-deficiency syndrome. However, other neurotransmitters, including acetylcholine, noradrenalin, and serotonin, are also affected in Parkinson's disease. Acetylcholine and dopamine seem to have a reciprocal relationship; an increase in one is associated with a decrease in the other. Drug treatment of Parkinson's disease centers on substances that intensify the action of dopamine (dopaminergic medication) or limit the action of acetylcholine (anticholinergic medication).

Many therapies, ranging from drug treatment to microsurgery and electrical implants, are available to alleviate the signs and symptoms of Parkinson's. However, these treatments cannot replace the absent or faulty nerve cells, nor can they stop the condition from progressing. Implanting fetal nervous tissue seeks to arrest progression of the disease and may initiate a partial or complete reversal of signs and symptoms. Success hinges on survival of the implanted cells and ideally, the establishment of nerve connections between fetal nerve cells and the recipient's tissue.

The treatment of Parkinson's disease is but one of several active areas of research in which fetal tissue transplants are being tested as potential treatments for medical conditions. Early positive results have been reported in using fetal tissue to treat two other neurodegenerative disorders: Huntington's disease (a condition characterized by dementia and involuntary, jerking movements) and Alzheimer's disease (a condition characterized by brain shrinkage, and the commonest cause of dementia in old age). Nonnervous fetal tissue is being tested on conditions such as diabetes, leukemia, and immune disorders. Since the mid-1990s, fetal nervous tissue has been used experimentally to treat spinal injuries. The treatment of macular degeneration, breakdown of the central region of the retina at the back of the eye, is one of the latest to employ fetal tissue. In this case, the eye tissue is obtained from second-trimester fetuses (at least 13 weeks old) that have been legally aborted for medical reasons.

Historical Development

Human fetal tissues from aborted fetuses have been used in transplant experiments in the United States for many years. In 1939, fetal pancreatic tissue was used to treat an adult with life-threatening diabetes. The experiment was unsuccessful. However, by the 1970s, many investigations had been performed using fetal thymus tissue to replace that missing in newborn babies. Babies suffering from DiGeorge syndrome, lack of a functioning thymus, usually benefited from this treatment.

Research on the implantation of nerve cells from the substantia nigra of aborted human fetuses began at the Karrlinska Institute in Stockholm, Sweden, in the late 1980s. This research drew upon animal studies, where fetal implants into adult animals showed encouraging results. In humans, transplants of dopamine-secreting tissue from patients' adrenal glands to their brains showed that the surgical procedures worked, in allowing the brain surgery and placement of tissue to be performed correctly, but the effect on alleviating Parkinson's disease symptoms was minimal. Better success was achieved using fetal nervous tissue injected into the basal ganglia during microsurgery. Positron emission tomography (PET) scans showed that the implanted tissue survived and that it

could develop a blood supply and establish new nerve connections at the implant site. However, the benefits, if any, were highly variable from person to person.

By the late 1990s, over 300 people in the United States and a similar number in Europe had received fetal tissue transplants to help alleviate the signs and symptoms of Parkinson's disease. In November 1998, two teams of researchers, one led by James Thomson of the University of Wisconsin and the other by John Gearhart at Johns Hopkins University, reported that they had isolated stem cells derived from human embryos. Those cells have the capacity to differentiate into one of many different tissue types given an appropriate physical, chemical, and biological environment. The two teams of researchers had succeeded in isolating such cells and could cultivate them in a state of perpetual infancy, ready to be triggered into differentiating into nervous, skeletal, or gut tissue given the appropriate signals. Among the challenges for the forthcoming decades will be learning how to "trigger" the cells to produce customized tissues that may be used in transplants.

The Wisconsin team obtained their tissue—very early cells—from several-day-old embryos donated for research by couples undergoing in vitro fertilization. The Johns Hopkins team isolated their cells from four- to nine-week-old embryos or fetuses obtained from medically induced abortions. In the latter case, the cells resembled germ cells,

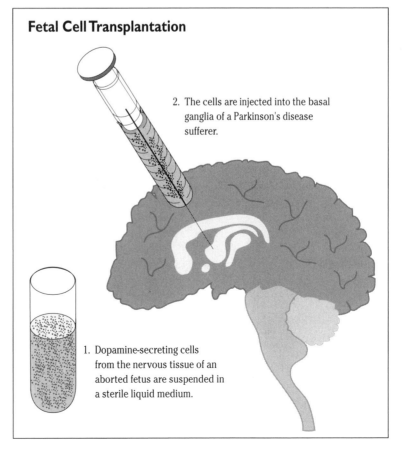

Fetal Cell Transplantation

2. The cells are injected into the basal ganglia of a Parkinson's disease sufferer.

1. Dopamine-secreting cells from the nervous tissue of an aborted fetus are suspended in a sterile liquid medium.

those that ultimately give rise to eggs or sperm. The research by these two teams was funded privately because of uncertainty about the federal stance on the funding of such research. The situation regarding the use of federal funding for such purposes is now being clarified.

Uses, Effects, and Limitations

Studies in Europe and North America since the early 1980s suggest that the effectiveness of fetal tissue transplantation in treating Parkinson's disease varies greatly from person to person. Some patients undergo significant remission of signs and symptoms. In some, progression of the disease is halted temporarily; in others, any effect is negligible or absent. Positive changes, where they occur, often relate to increasing the effectiveness of drug treatment or postponing the requirement for drug medication. Currently, fetal tissue transplants are not used in the treatment of Parkinson's disease to effect a cure. Rather, as with drug therapy, their use seeks to control signs and symptoms and so prevent or reduce disability. In time, old signs and symptoms may reemerge and new ones may appear. Typically, those who have undergone fetal tissue transplantation to treat Parkinson's continue drug treatment afterward but perhaps at reduced dosages.

Fetal tissue implantation as a procedure for treating Parkinson's remains in the experimental phase. As Marjan Jahanshahi and David Marsden point out in their 1998 book *Parkinson's Disease*, many questions remain unanswered. For example, will long-term suppression of a patient's immune system be necessary to prevent rejection of implanted fetal tissue? If so, for how long should immunosuppression be continued, and what potential risks are involved? Will the implanted tissue itself degenerate over time? Dopamine is not the only neurotransmitter system affected in Parkinson's disease. What about other neurotransmitters that are not replaced or are affected by the implanted tissue? Precisely where in the basal ganglia is fetal tissue best implanted? Clearly, much research remains to be completed before fetal tissue transplantation becomes established as a feasible treatment for Parkinson's disease. Research into the use of other types of transplanted fetal tissues for treating medical conditions is also at a very early stage.

Issues and Debate

A moratorium against federal funding of research using fetal tissue implants from elective abortions, which had been in place since 1988, was lifted by President Clinton in January 1993. Prior to this, government-funded research was active in Sweden, the United Kingdom, Australia, France, and several other countries, including Mexico and Cuba. Even though fetal tissue transplantation now has the U.S. stamp of approval, such work still poses ethical difficulties for scientists engaged in the research because of its association with the debate over abortion. Pro-life groups oppose the use of fetal tissue out of fear that it will encourage abortions, espe-

Pro-life groups oppose the use of fetal tissue in medical research, arguing that it will encourage abortions, which they believe to be morally wrong. Advocates of fetal-tissue research argue that the potential benefits of treating people with disabling or life-threatening conditions justify the use of fetal tissue that is obtained from abortions, which they argue will be performed anyway (AP/Richard Drew).

cially if fetal transplants are shown to be successful. They argue that a market for fetal tissue will be created, encouraging even more abortions than those already performed. They also argue that the use of fetal tissue for transplants will bring greater legitimacy to and wider acceptance of abortion, which they believe to be morally wrong.

In response, some members of the scientific community and many patient advocacy groups claim that the potential benefits of fetal tissue implantation in treating hundreds of thousands of people with disabling or life-threatening conditions outweigh moral reservations about the source of the material. In any case, they argue, this is tissue that would be disposed of anyway. Researchers, meanwhile, report some progress in treating a range of conditions using fetal tissue. Some bioethics experts, among them Arthur Caplan, author of *Am I My Brother's Keeper?* (1997), argue that there is little or no evidence that successful transplants of fetal tissue will encourage people to have abortions and so increase abortion rates. On the other hand, there have been a few highly publicized claims that some people will become pregnant and will abort in an attempt to create transplantable tissue, whether for personal or family use or even, perhaps, for sale. The latter possibility is regarded as unethical by medical establishments the world over.

Embryonic stem cells as a source of transplantable material have been the subject of great scientific debate since late 1998. In January 1999, the U.S. National Institutes of Health (NIH) announced that, contrary to what many scientists had feared, federal law does not bar funding for the burgeoning field of embryonic stem cell research. *Science* magazine reported that Harriet Rabb, general counsel of the U.S. Department of Health and Human Services, had announced a distinction between a human embryo or organism and stem cells. Stem cells are not organisms, or even precursor organisms, in her view. They cannot be implanted in a uterus to grow to become an embryo or fetus, so cannot be proscribed by current legislation that prohibits the federal funding of embryo research, she reasoned. As of spring 1999, the ethics of embryonic stem cell research was being reviewed by the National Bioethics Advisory Commission (NBAC) in Washington, D.C. Federal funding of embryonic stem cell research seems set to begin by the end of 1999.

Although requiring the use of fresh tissue in the short term, embryonic stem cell research could, in time, eliminate the need for fetal tissue in experimental transplants. Once perpetually dividing cell strains have been developed, the need for fresh tissue will be largely eliminated. Meanwhile, research is under way to genetically engineer human tissues that might, for example, produce desired neurotransmitters such as those depleted in Parkinson's disease.

—Trevor Day

RELATED TOPICS
Artificial Organs, Artificial Tissue, Cloning, Gene Therapy, In Vitro Fertilization, Organ Transplantation, Tissue Transplantation

BIBLIOGRAPHY AND FURTHER RESEARCH

BOOKS
Caplan, Arthur L. *Am I My Brother's Keeper?* Bloomington, Ind.: Indiana University Press, 1997.
Jahanshahi, Marjan, and C. David Marsden. *Parkinson's Disease.* London: Souvenir Press, 1998.
Kimball, Andrew. *The Human Body Shop*, 2nd ed. Washington, D.C.: Regnery Publishing, 1997.
Pearce, J.M.S. *Parkinson's Disease and Its Management.* Oxford: Oxford University Press, 1992.

PERIODICALS
Beardlsey, Tim. "Culturing Human Life." *Scientific American*, June 1998, 9.
Henkel, John. "Parkinson's Disease." *FDA Consumer*, July 1998, 13.
Miller, Linda J. and Floyd E. Bloom. "Publishing Controversial Research." *Science*, November 6, 1998, 1045.
Richardson, Sarah. "Forever Young." *Discover*, January 1999, 58.
Stocum, David L. "New Tissues from Old." *Science*, April 4, 1997, 15.

INTERNET RESOURCES
American Parkinson's Disease Association
    http://www.apdaparkinson.com
National Institute of Neurological Disorders and Stroke
    http://www.ninds.nih.gov
National Parkinson Foundation
    http://www.parkinson.org
Network of European CNS Transplantation and Restoration (NECTAR)
    http://www.bm.lu.se/~nectar/NECTARINDEX.html
Parkinson's Disease Weblinks
    http://mednav.com/zones/Illness/Parkinsons/index.html
Pluripotent Stem Cells: A Primer
    http://www.nih.gov/news/stemcell/primer.htm
A Pro-Life View of Fetal Tissue Research
    http://user.mc.net/dougp/stopftr.html

FIBER OPTICS

Fiber optics is a method of piping light along hair-thin glass or plastic cables to transmit information from place to place. Its many applications range from medical probes known as endoscopes to high-capacity telecommunication links.

Industry analysts have predicted that fiber optics will have as big an effect on computing and communications in the 21st century as the silicon chip had in the 20th century. But the world has an enormous investment in electronic equipment, which will be expensive to replace with optical technology.

Scientific and Technological Description

Water can be funneled down pipes; electricity can be sent down wires; and light beams can be channeled along strands of glass or plastic known as fiber-optic cables. Fiber optics relies on a property of light known as internal reflection: If a light beam traveling inside a glass rod strikes the edge at a shallow-enough angle (less than 42°), instead of passing outside as might be expected, it is reflected back inside as though the edge of the rod were a mirror; the rod acts as a *light pipe*.

Just as a copper cable can be used to send information in the form of electrical impulses, so a fiber-optic cable can be used to send information as pulses of light. A tiny laser or *light-emitting diode* (LED) sends digital information (data stored numerically, in binary code) down a fiber-optic cable. The light is detected at the other end by a *photodiode* (an electronic component that generates a small electric current when it detects light). Devices known as *repeaters* boost the light pulses every so often by converting them to electrical signals, amplifying them, then converting them back to light. Intervals therefore have to be left between the pulses so they can be identified, and this limits the rate at which information can be sent (the bandwidth). Another form of transmission, known as *wavelength-division multiplexing* (WDM), sends different *streams* of information along a single fiber at the same time, using a different wavelength of light for each stream. The streams (effectively, separate information channels) are combined into a single pulse using a mathematical technique known as multiplexing; this type of fiber optics increases the bandwidth of a cable at least 20-fold.

Fiber-optic cables consist of two concentric layers of high-grade glass surrounded by a protective plastic coating (see figure). The inner layer (the *core*) has a higher refractive index (it is a more dense material in which light travels more slowly) than the outer layer (the *cladding*). The cladding reflects the light more effectively back inside the cable than

Types of Fiber-Optic Cable

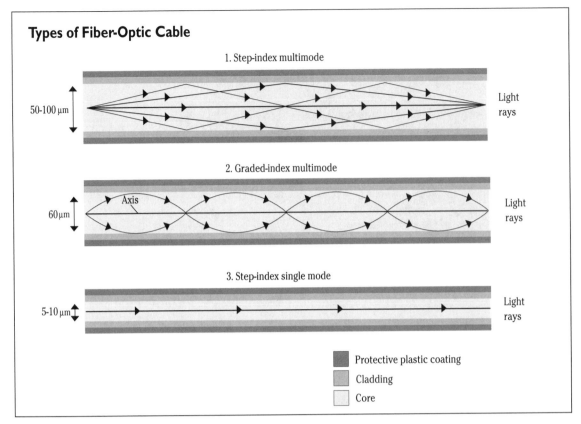

1. Step-index multimode

50-100 μm

Light rays

2. Graded-index multimode

Axis

60 μm

Light rays

3. Step-index single mode

5-10 μm

Light rays

■ Protective plastic coating
▨ Cladding
□ Core

There are three types of fiber-optic cable available for use in a range of different applications. Step-index multimode cable allows light rays to take a variety of different paths. This fiber is suitable only for transmitting information over short distances, such as in medical endoscopes. Graded-index multimode fiber forces light rays to travel in a curved path. This type of fiber is suitable for medium-distance transmissions, such as in local-area networks (LANs). Step-index single-mode fiber is very thin. All light rays take the same path, which makes this fiber suitable for long-distance telecommunications.

does air alone. The simplest type of fiber has a thick core 50 to 100 microns (micrometers; millionths of a meter) in diameter with a uniform refractive index; the refractive index of the cladding is a "step" lower. This fiber is known as *step-index multimode*, because light rays can travel along such a fiber in different paths, or *modes*. Some of these modes are straight lines along the axis of the cable: others bounce back and forth along the sides. The farther the pulse travels, the greater the difference between the distances traveled by different light rays; in other words, the components of the pulse are dispersed, the pulse is "blurred" in transmission, and the information is degraded. This means that step-index multimode fiber is suitable only for transmitting light over short distances: for example, in endoscopes.

In *graded-index multimode fiber*, the refractive index of the core (approximately 60 microns [μm] in diameter) decreases gradually between the axis (center of the core) and the cladding. This makes light rays bend back from the cladding toward the core. It also means that light travels more quickly near the cladding than near the axis, so in practice the components of a pulse tend to arrive at the same time no matter what path they follow during their

journey. These fibers are suitable for medium-range transmissions such as in local area networks (LANs), which connect computers over short distances.

Step-index single-mode fiber has a very thin core (typically, 5 to 10 μm in diameter) of constant refractive index. This means that virtually all light beams travel near the axis, which avoids the problem of dispersion. Single-mode fiber is used for cable TV and telephone networks, because it allows signals to be sent over 60 miles (100 kilometers [km]) without amplification (boosting).

Historical Development

The principle of fiber optics, internal reflection, was demonstrated in 1870 in London at the United Kingdom's oldest and most prestigious scientific institution, the Royal Society, by Irish physicist John Tyndall. He shone light into a pitcher of water and then released some of the water through a small hole. As the water poured out, the light lit up its curved path. Internal reflection was used to explain natural phenomena such as rainbows, in which droplets of rain bend different wavelengths (and therefore colors) of light by different amounts.

Crude devices using this principle, developed from about the 1920s onward, included plastic rods used to illuminate aircraft instrumentation panels. However, the principle of channeling light through thin strands of glass fiber was discovered only in 1955 by Narinder Kapany at Imperial College, London. In 1966, scientists at the Standard Telecommunications Laboratory in Essex, England, proposed using Kapany's fibers to transmit telecommunications signals over long distances.

A major breakthrough in fiber optics occurred in the late 1960s, when scientists at the Corning Glass Works in New York developed a method of manufacturing fibers so pure that light could travel around 16 miles (30 km) without attenuation (reduction in strength). Although both core and cladding were made from a high grade of glass known as fused silica, a small quantity of impurities was introduced to the core of the fiber to reduce its refractive index (a process known as *doping*).

In April 1977, the first fiber-optic telephone cable, spanning the 5.5 miles (9 km) between Long Beach and Artesia, California, was laid by the General Telephone Company. By the early 1980s, U.S. companies were laying fiber cables at the rate of about 4000 miles (6400 km) a day. Eight cables now cross the North Atlantic Ocean and six more cross the Pacific. In 1997, a major transoceanic cable, known as FLAG (fiber-optic link around the globe) was laid between London and Tokyo, taking in or branching off to other countries, including Spain, India, and China.

Physicists are constantly trying to find better ways of transmitting information using fiber optics. In the late 1980s, David Payne at the University of Southampton, England, and scientists at Bell Laboratories in the United States pioneered a method of doping using the mineral erbium and amplifying signals using tiny lasers. Known as an *erbium-doped fiber amplifier* (EDFA), this development led to all-optical *photonic networks* that were many times faster because they did away with the need for electronic repeaters.

Uses, Effects, and Limitations

Fiber optics has found a wide range of applications, from simple lighting effects to advanced telecommunications. Fiber-optic imaging builds up pictures using an array of cables, similar to the way that television pictures are built up from an array of dots, or an embroidered picture is made up from many individual stitches. This technique is used in the *endoscope*, a collection of fiber-optic cables with a tiny lens at one end and an eyepiece at the other, that is used as a medical probe. Endoscopes of various diameters are used to inspect different body parts (from the stomach and the colon to the bronchi of the lungs and inside the arteries), without the need for expensive, invasive, and traumatic surgery. Industrial endoscopes, known as *fiberscopes*, are used to examine inaccessible machine parts such as turbine blades in airplanes and critical components in nuclear reactors. Fiber-optic cables are also used in bar-code readers, pressure- and temperature-measuring devices, and liquid and gas sensors.

Telecommunications is by far the most significant application of fiber optics. A single strand of glass fiber could transmit up to 40 million telephone calls at the same time using WDM technology, which has a theoretical capacity of 200 trillion bits (binary digits) of information per second (over 200 times the current total traffic on the Internet). With a total population of nearly 2 billion people placing extreme demands on telephone and Internet connections, China's government has been one of the most enthusiastic advocates of fiber optics, and planned to have laid around 3.4 million kilometers of optical cable by 2000. Fiber already connects much of the world and forms the "backbones" (major interconnections) of the Internet in most countries.

There are many other ingenious applications of fiber optics. In Tel Aviv, Israel, in 1998, fiber-optic cables were used to simulate 3500 stars in the ceiling of a shopping center. Also in Israel, scientists at Ben Gurion University have proposed using fiber-optic cables to channel sunlight into cheap and simple medical tools that could be used in field hospitals in developing nations where lasers are unavailable. Swiss researchers have created an optical stylus, which uses a single optical fiber that is 1000 times lighter than a traditional diamond stylus and is used to play fragile vinyl records without damaging them.

The benefits of fiber optics center on the technology's ability to send vast amounts of information over long distances using relatively little power; high-frequency light can transmit information at a much higher rate than electricity or radio waves. Also, unlike electrical cables, glass fibers are not subject to electrical interference, are more secure against tapping (listening in to phone lines), and are much cheaper to manufacture (glass is made from silica, a constituent of sand; wire is made from expensive materials such as copper and gold). These advantages are expected to lead to the systematic replacement of wire networks by all-optical networks.

The biggest limitation of fiber optics is that until recently it has been capable only of transmitting information; to switch, amplify, or decode signals, signals had to be converted back into an electrical form. This means that although optical networks theoretically work at the speed of light, in practice their speed is restricted to the much slower speed of electronic switching gear. All-optical networks are expected to solve this problem. The huge capacity of optical fibers also poses a significant vulnerability and therefore weakness; in theory, damage to a single cable could result in the loss of service to an entire continent. Faults and bottlenecks in optical fibers are difficult to pinpoint for the same reason.

Issues and Debate

Fiber optics is fundamentally a benign technology that has already benefited society in many ways. Endoscopes have

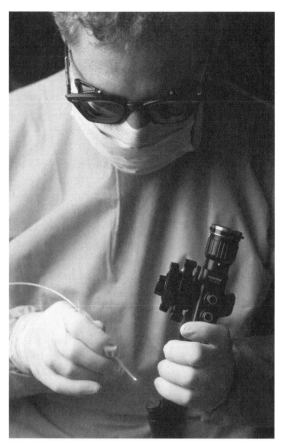

Fiber optics have applications in many fields, especially in telecommunications and medicine. Above, a doctor uses an endoscope, a instrument made of an array of fiber-optic cables with a tiny lens at one end and an eyepiece at the other. Endoscopes are small enough to explore body cavities, and their use has prevented invasive and unneeded surgery in many cases (Corbis/Leif Skoogfors).

improved the early detection of cancer. Fiberscopes have made it easier to check the safety of aircraft and nuclear plants, and optical fibers light up the Internet's "information superhighway." The most controversial aspect of fiber optics is the extent to which it threatens to render traditional "wire" technology obsolete. Telephones and computers, in particular, still rely substantially on copper-wire technology. Fiber optics will prompt not just the replacement of wire connections between appliances, but also the replacement of electrical and electronic components in those appliances. All-optical networks are predicted to lead to all-optical telephone systems and optical computers that can properly take advantage of speed-of-light processing.

Telecommunications corporations such as AT&T in the United States and British Telecom (BT) in the United Kingdom have a considerable investment in wired networks, which were constructed at enormous cost. Industry analysts believe that optical fibers will revolutionize the telecommunications industry by reducing the cost of bandwidth to trivial levels. But when the entire telecommunications needs of

a company can be supplied by a single optical fiber costing a few cents, how will telecommunications companies make money? If it costs tens or hundreds or billions of dollars to convert to all-optical technology, which provides today's service at almost zero cost, how can telecommunications companies afford to make such an investment?

The logical consequence of practically infinite bandwidth at practically zero cost is that it will be much cheaper to transmit information than to process or store it. This is likely to fuel the growth of pay-on-demand books, movies, and information services delivered over the Internet. Thanks to fiber optics, it will be far cheaper to use the Internet than to travel, transport goods, run traditional retail establishments, commute to work, or do many other activities in the traditional way. Fiber optics are predicted to transform society as much as steam engines and computers have.

—Chris Woodford

RELATED TOPICS

Electronic Commerce, Internet and World Wide Web, Internet Protocol Telephony

BIBLIOGRAPHY AND FURTHER RESEARCH

BOOKS

Booth, Kathryn, and Steve Hill. *The Essence of Optoelectronics*. London: Prentice Hall, 1998.

Davis, Christopher. *Electro-Optics: Fundamentals and Engineering*. New York: Cambridge University Press, 1996.

Dyson, Esther. *Release 2.1*. New York: Broadway Books, 1998.

Hecht, Eugene. *Optics*. Reading, Mass.: Addison-Wesley, 1998.

Jonscher, Charles. *Wired Life*. New York: Bantam Books, 1999.

PERIODICIALS

Chan, Vincent. "All-Optical Networks." *Scientific American*, September 1995, 56.

Gibbs, W. Wayt. "Bandwidth Unlimited." *Scientific American*, January 1997, 30.

Graham-Rowe, Duncan. "Guiding Lights." *New Scientist*, July 18, 1998, 20.

Hills, Alex. "Terrestrial Wireless Networks." *Scientific American*, April 1998, 74.

MacChesney, John. "Working Knowledge: Fiber Optics." *Scientific American*, August 1997, 76.

Marks, Paul. "Saved by the Light." *New Scientist*, February 27, 1999, 20.

Morrison, Philip, and Phylis Morrison. "Bandwidth Galore." *Scientific American*, July 1997, 81.

Stix, Gary. "Light Work." *Scientific American*, July 1996, 23.

Stix, Gary. "Nothing but Light." *Scientific American*, December 1998, 9.

Thomas, Jim. "Voices from the Past." *New Scientist*, September 13, 1997, 6.

INTERNET RESOURCES

Ciena Corporation
  http://www.ciena.com/revolution/metro/index.html
Corning Optical Fiber
  http://www.corning.com/maps/index.html
Fiber Optics Online
  http://www.fiberopticsonline.com
Lucent Technologies Optical Networking
  http://www.lucent.com/opticalnet/
Southampton University Optical Research Centre
  http://www.orc.soton.ac.uk/

FLAT-PANEL DISPLAYS

Flat-panel displays are displays that don't use the standard cathode-ray tube (CRT) found in most television sets and computer monitors. Instead of shooting a beam of electrons at a screen from a distance, flat-panel displays use arrays of electrodes, crystals, or vinyl polymers to create small points of light.

Flat-panel displays are used in everything from digital watches to cellular phone displays and laptop and other types of newer computer monitors. Future innovations with flat-panel displays aim to increase their physical flexibility and reduce their weight, cost, and power requirements. This may lead to higher-quality computer monitors with better screen resolution, as well as foldable digital books, which would be reusable and would store hundreds of books or newspapers.

Scientific and Technological Description

For the last 50 years the standard mode of home video display has been the television set, with its cathode-ray tube (CRT). This tube, used in television sets and computer monitors, sprays a beam of electrons onto a phosphorescent screen that lights up individual dots, called *phosphors* or *pixels*, into one of three colors: red, blue, or green. Mixing these colors forms an image in any possible color. The simplest flat-panel displays are those used in digital watches and cellular phones (see PCS PHONES). These liquid-crystal displays (LCDs) use magnetic fields to selectively darken certain areas of a display. The magnetic fields cause black particles to congregate in certain regions, forming letters, numbers, or other images. When the field is off, they dissolve away. The main problem with these displays is that they are rigid, monochromatic, and have low brightness and contrast.

Larger monochromatic flat-panel displays such as those found in older laptop computers aren't lit with an electron beam as are desktop cathode-ray computer monitors. They are, instead, lit with a series of small glowing *diodes*. Diodes are parts of electrical circuits and transistors that have a positive and a negative end. Each diode forms one pixel, a microscopic point of light. A layer of diodes is sandwiched between two rigid glass sheets that contain the circuitry (transistors) that provides the electric charges to light the diode. Circuitry in the computer controls where electric currents travel through the display; wherever an electric current gets to, a pixel lights up, shining white. The millions of pixels on the screen are lit (or not) to form a picture. The glass also contains a *polarizing filter* that allows light to travel through the monitor in only one direction, either perfectly horizontally or vertically. This eliminates scattering of the light and makes the display much sharper. A typical monitor of this style, found in older laptop computers, has a blue or purplish background and displays white letters.

Color flat-panel displays control the color of the pixel in two different ways. The first is to use three different-colored diodes and mix the colors red, green, and blue just as a television set does. A newer, still experimental method uses successive colored layers. A series of "windows" in these layers (red, green, and blue again) open and close to create a color filter over white pixels. Each pixel in a display is exactly the same size. Most monitors today contain pixels that are 0.28 millimeter or smaller. The smaller the pixel, the better the screen resolution. Some companies have also experimented with altering the shape of pixels on cathode-ray-tube screens from their typical circular form into elliptical shapes to cut down on "wasted" space between pixels. Flat-panel displays use rectangular pixels to waste as little space as possible.

Flat-panel displays make the best images when they are working at the resolution they were designed to work at, their *native* resolution. For example, a flat-panel display screen with a resolution of 800 x 600 pixels will work best at only that resolution. (Screen resolutions are measured in pixel counts.) The best screens today have resolutions of approximately 1500 x 1200. If the screen were to try to work at a lower resolution, say 640 x 480, it would have problems, since pixels would now overlap and perform a job similar to the one next to them. The decision of where overlap, called *antialiasing*, will occur and not occur, has to be made by the computer, slowing down the speed of other tasks. Computers without this feature tend to show text that appears blocky.

A variation on the flat-panel displays mentioned above that doesn't use a metal diode between the glass plates is *field-emission display* (see figure). Instead of a metal diode, there is an airtight vacuum. On the backing glass panel, tiny pyramids made of diamond (a network of carbon molecules) or carbon nanotubes (hollow tubes of carbon) stick out of the surface. These raised points, called microtips, serve as the conductors: After an electric charge accumulates in the point, it jumps across the vacuum to the screen, lighting up a pixel (phosphor) in a fashion similar to that of a cathode-ray-tube display. In this case, however, the distance to the display is only a few millimeters, whereas a CRT display electron stream must travel more than a foot.

Another type of flat-panel display being developed uses tiny balls, white on one side and black on the other, to produce images and text. Producing an image is done by flip-flopping the balls to show either white or black. The problem with this display is that it produces images that are still very grainy. To get good-quality text there must be a resolution of at least 300 dots per inch (dpi). Any laser printer can easily surpass this—most home models currently offer at least 600 dpi. Past 2580 dpi, the human eye cannot perceive dots. But to pack this many circuits and balls (600 x 600 = 360,000 balls per square inch) into a display is still a serious technical challenge.

The speed at which these windows, pixels, and balls open and close or turn on and off still cannot match the speed at which the image on a CRT can change. This is not much of a problem when working with static graphics or text, but it becomes a problem with more complicated content. Video games and movies, for example, show a distinct

Field-Emission Display

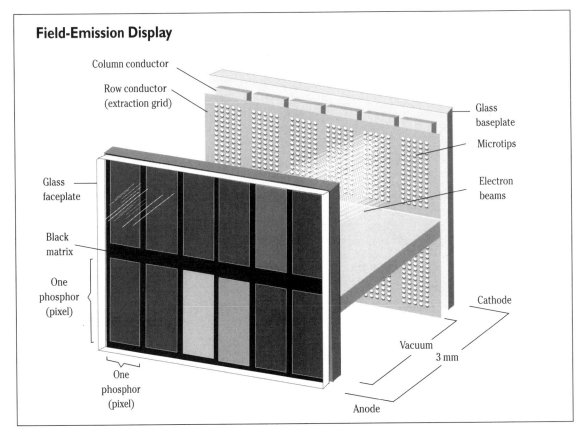

A field-emission display lights up colored phosphors by sending a charge across a distance of a few millimeters. Electric charges collect on microscopic tips (microtips) on the back panel of the display. After a large enough charge develops, the current crosses a vacuum to the front panel, lighting the closest phosphor. Mixing phosphor colors of red, green, and blue forms a color image. (Adapted from Motorola Flat Panel Display Web Site.)

lag when fast-moving images cross the screen and leave behind trailers or ghost images.

Historical Development

CRTs have been used for over 100 years. In the latter half of the 19th century, scientists used them to trace electric currents and to study the composition of atoms. This also led them to discover that the electron was a particle with mass. In the 1930s, CRTs were adapted for use in the first television broadcasts, which began in 1936. One of the first color television broadcasts was NBC's Tournament of Roses Parade on New Year's Day in 1954.

Liquid crystals were discovered in the middle of the 19th century when W. Heintz reported that the molecule stearin melted from a solid to a cloudy liquid to an opaque fluid and a clear liquid between the temperatures 52 and 62.5°C. In 1888, Friedrich Reinitzer discovered that cholesteryl benzoate had similar properties and coined the term *liquid crystal*. Research on liquid crystals continued into the 1940s. Strangely, they were almost forgotten in the years that followed. In 1958 the field was rejuvenated when Glenn Brown, an American chemist, published an article in *Chemical Reviews* that brought renewed interest to the field. The first working liquid-crystal displays were produced in

1971. The first displays made available to consumers were LCD watches produced by the Seiko Corporation in 1973. Since then LCDs have been progressively scaled up to produce modern flat-panel displays more than 1 foot in diameter.

The first attempt at a modern flat-panel display that used reversing balls was made by researchers at Xerox PARC Research Center in the 1970s, also a hotbed of computer networking research. The researchers attempted to create a rubberized sheet that held tiny balls, white on one side and black on the other. Electric charges would rotate the balls to produce black-and-white images. But the resultant images were very grainy and the idea was abandoned quickly. In the 1990s, this system was restarted by Joseph Jacobson of the Media Laboratory at MIT. Newer materials for the screen and tremendous increases in the speed of computer processors allowed his team of researchers to develop flexible, reusable flat-panel displays where the PARC researchers had failed. Instead of strewing the magnetized particles randomly over the screen, Jacobson encased each ball in its own transparent shell.

In 1999 Sony introduced the first television sets with flat screens. These sets still use large CRTs to create images but have a different mechanism for focusing electrons at the screen. These screens have brighter displays and can be seen from higher viewing angles than those of conventional-display sets, but such sets are still extremely heavy and power hungry.

Uses, Effects, and Limitations

Four limitations are common to all cathode-ray-tube displays. The first is distortion: Until recently all CRTs had to have a curve in their surface to accommodate the trace of the electron beam. The curve in a standard CRT always distorts the picture a small amount at the edges. The image also becomes difficult to view from the side. Flat-panel displays don't distort images, but they still must be viewed close to straight-on; otherwise, the display quickly fades. Second, increasing the size of a cathode-ray screen necessitates that the length of the electrons' path from the gun to the screen must be increased similarly. Third, the *refresh rate* of the screen has to be sufficiently high to avoid flickering or shadowy images that can strain and tire the eye. Finally, CRTs consume a great deal of power in accelerating electrons to the screen.

Flat-panel displays overcome the first two problems easily, since the image is not produced with an electron beam but with luminescent elements. These elements also require much less power than that required by an electron gun. The third problem, that of the refresh rate, is a different concern. A pixel in a flat-panel display remains on as long as it is activated by the circuits, unlike the high-speed flashing of a television, so flat-panel displays don't really have a refresh rate. But flat-panel display pixels can't switch on and off as quickly as cathode-ray-tube pixels, meaning that trailing and ghost images are still a problem.

The biggest problem facing engineers today is creating circuitry to control the display. Traditional copper wires aren't flexible enough to work in anything other than a rigid, fragile display; in addition, copper isn't transparent. Most flat-panel displays are made with circuit boards sandwiched between two thin glass plates. This makes the display completely rigid and easy to break. Many researchers are searching for ways to overcome this. Two promising candidates are transparent molecules that can conduct electricity: the metal indium tin oxide and certain types of vinyl polymers.

At the pixel level other researchers are looking at replacing metal diodes. One idea is to use different, flexible materials. One promising candidate is organic light-emitting POLYMERS. Polymers are long molecules made of a simple repeating unit. These polymers are a special type of plastic. Many companies are currently experimenting with these types of displays. These polymers can emit a broader variety of colors, need less power, and can switch on and off at faster speeds than can traditional inorganic diodes.

A flat-panel display will typically outlast any CRT monitor, since the amount of power being pushed through the screen elements is so much smaller. But on a typical flat-panel monitor with over 2 million pixels, it is not uncommon for some to get locked into a permanent on (a bright defect) or off (dark defect) state. Flat-panel displays have many advantages over cathode-ray-tube displays. A flat-panel display is a direct representation of an image: No distortion is introduced by the curved tube of a CRT. This allows graphic designers to have a better idea of what will appear in the finished version of their work. Flat-panel displays are also much lighter than CRTs. Since they have no lengthy electron tube, their thinness keeps their weight down.

Two characteristics for which flat-panel displays don't match cathode-ray-tube quality are brightness and contrast. Flat-panel displays still don't have the power to create very bright images, but they are certainly increasing quickly in quality. Nor do flat-panel displays handle subtle differences in brightness at the brightest and darkest scales, leading to problems in creating contrast in areas of differing brightness.

Issues and Debate

If flat-panel displays do become widespread and adopted by millions around the world, they could change the way we receive, handle, and store information. People could carry entire libraries with them. Access to any kind of information could become quick and easy through cables or radio transmissions. A society where everyone has this kind of instantaneous information access is hard to imagine. Flat-panel displays could also be a boon to the environment, since reusable plastic books could replace the thousands of trees that are cut down to produce newsprint and fine paper for newspapers, magazines, and books.

Because of their unique flexibility, organic light-emitting polymers, if developed successfully, may transform society. This quality is what many companies hope will lead to flexible, electronic, reusable, customizable books (see ELECTRONIC BOOKS). Modern portable laptop computers are still distinctly heavy, unwieldy, inflexible, and fragile. Many hope to produce a "book" composed of about 100 flexible pages, which will hold two small batteries in its spine. A device like this could store hundreds of books, newspapers, or other data. Most important, some say, the device would have the feel and touch of a book, which would lead people to accept and use it as such.

Displays made entirely of lightweight, flexible plastic could one day replace conventional television screens and laptop computer displays. Some observers are already dreaming of displays that can be rolled up and carted away. Flat-panel displays are used most often today in locations where space is at a premium, such as financial institutions and hospitals. They are slowly being introduced for home use as computer monitors but remain much more expensive than a traditional cathode-ray-tube monitor—currently at around double the price—although their price is dropping steadily. As manufacturers develop cheaper-to-produce models, however, flat-panel displays may become more prevalent, changing the look of the world we are used to seeing.

—Philip Downey

RELATED TOPICS

Digital Video Technology, Electronic Books, High-Definition Television, PCS Phones, Polymers

BIBLIOGRAPHY AND FURTHER RESEARCH

BOOKS

Miller, Richard K. and Terri C. Walker. *Flat Panel Displays.* Norcross, Ga.: Future Technology Surveys, 1989.

O'Mara, William C. *Liquid Crystal Flat Panel Displays.* Dordrecht, The Netherlands: Kluwer Academic Publishers, 1993.

PERIODICALS

Gross, Neil. "Flatter, Brighter—and Easy to Make?" *Business Week,* October 19, 1998, 120.

Normile, Dennis. "Field Emitters Finding Home in Electronics." *Science,* July 31, 1998, 632.

Platt, Charles. "Digital Ink." *Wired,* May 1997, 162.

Poor, Alfred. "LCD Monitors." *PC Magazine,* April 20, 1999, 127.

INTERNET RESOURCES

Display Technology
http://www.zurich.ibm.com/Projects/ST/display.htm

History of Television
http://www.dvb.org/dvb_articles/dvb_tv-history.htm

Motorola Flat-Panel Display Division Home Page
http://www.mot.com/ies/flatpanel/whoweare.html#history

Panel Displays
http://www.pctechguide.com/07panels.htm

PC Computing Online: LEPs vs. LCDs
http://www.zdnet.com/pccomp/stories/reviews/0,5672,391699,00.html

Surfaces and Displays
http://future.enterprisecomputing.hp.com/vision/2.30/foc_surfaces_content.html

FOOD IRRADIATION

Food irradiation is a method of destroying microbes and pests in food by exposing it to high-energy radiation. Food irradiation is quite effective at destroying the microbes responsible for many forms of food poisoning, and it is generally considered safe. But public distrust of the process, sparked in large part by its associations with radioactivity, has prevented food irradiation from being widely used.

Scientific and Technological Description

Food is irradiated in one of two ways: by exposure to radiation or by exposure to an electron beam. Both processes kill bacteria, insects, and other pathogens by creating a large number of free electrons in the food. The free electrons cause damage to the DNA (deoxyribonucleic acid) of the pathogens. DNA is the genetic material of cells. Free electrons damage DNA by reacting with it and with other molecules in food. Molecules are made up of atoms, which are in turn made up of negatively charged electrons, positively charged protons, and neutrons. Molecules in food usually have enough electrons to balance out their protons and so have a neutral charge and are not highly reactive. But free electrons will react with even a stable molecule, pulling it apart into ions (charged molecules). The

ions are also highly reactive and are often unstable. The result is many chemical reactions, with a number of new molecules being formed.

DNA is especially sensitive to free electrons because it is a very complicated molecule. If the DNA of a single-celled organism such as bacteria is damaged, the organism can be weakened or killed. DNA damage can kill a multicelled organism or render it sterile.

Radiation

Radiation creates free electrons by knocking the electrons off the stable molecules already in food. Two types of radiation are used to irradiate food: X-rays and gamma rays. X-rays produced by machine have been used in laboratories to irradiate food, but the machines are not practical for large-scale commercial use. Gamma rays are produced by radioactive material. Commercial gamma-ray irradiation uses radioactive isotopes of cobalt or cesium. (*Isotopes* are atoms that contain extra neutrons; some isotopes are very unstable.)

The most commonly used isotope in the United States is cobalt-60, but some irradiation plants use cesium-137. The use of cesium-137 is controversial because it is a nuclear waste product, and its use is viewed by some as a way of promoting nuclear power. In addition, cesium-137 is usually handled in the form of a salt, which means that it can dissolve in water. This makes cesium-137 less safe because most irradiation plants store radioactive material in deep pools of water when it is not being used, to prevent people from accidental exposure to the radiation. (Water will not become radioactive from absorbing gamma radiation, but it will become radioactive if radioactive material is dissolved in it.) Cobalt-60 does not dissolve in water and is not a waste product; it is made especially for irradiation by bombarding normal cobalt with neutrons in a nuclear reactor. Both cobalt-60 and cesium-137 degrade into nonradioactive material over time.

Food to be irradiated is wrapped in nonreactive plastic to prevent the food from becoming contaminated after treatment. Since people cannot survive the doses of radiation used to treat food, the irradiation process is highly automated. The food is placed on a conveyer belt that moves it into a room where the cobalt-60 or cesium-137 is held. Radiation comes off the material in all directions, so the belt snakes around it to expose the food evenly on all sides (see figure).

Electron Beams

Another way to irradiate food does not rely on radiation at all. Instead, an electron accelerator shoots a beam of free electrons directly into the food. The main drawback of this method is that electrons are much less penetrating than X-rays or gamma radiation. Electron accelerators can be used only on thin food passed individually under the beam, whereas commercial gamma-radiation plants can irradiate

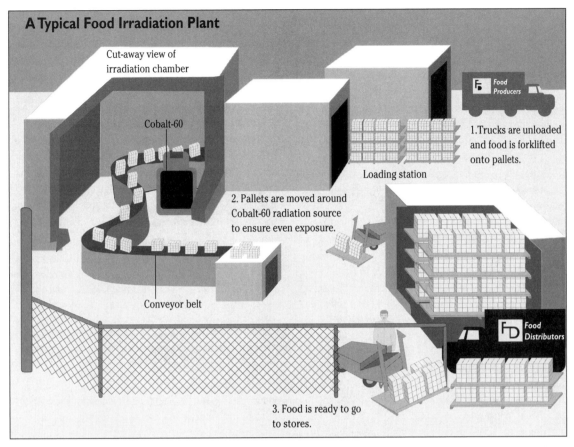

A Typical Food Irradiation Plant

Cut-away view of irradiation chamber

Cobalt-60

2. Pallets are moved around Cobalt-60 radiation source to ensure even exposure.

Conveyor belt

Loading station

Food Producers

1. Trucks are unloaded and food is forklifted onto pallets.

Food Distributors

3. Food is ready to go to stores.

Adapted with permission from Iowa State University Press.

large pallets of a greater variety of foods. Exposure to gamma radiation does not make food radioactive, but exposure to high-energy X-rays or electrons can. Consequently, the energy of X-rays and of electron-accelerator irradiators is tightly regulated. Although irradiated foods do not actually glow in the dark, they sometimes give off a faint luminescence that can be detected by equipment.

Historical Development

Ionizing radiation's potential for use in food safety was recognized early in the 20th century. German physicist Wilhelm Roentgen discovered X-rays in 1895 (see X-RAY IMAGING); the next year French physicist Antoine H. Becquerel discovered radioactivity. By the 1910s, scientists in several countries were proposing the use of X-rays or radioactive material to sterilize food, with the American Tobacco Company going so far as to commission an X-ray machine for use in irradiating insect-infested tobacco leaves. But the existing X-ray machines were unreliable for continuous commercial use, and radioactive materials were too rare to make their widespread use practical.

Interest in food irradiation was revitalized in the late 1940s. Two factors made irradiation practical: Electron accelerators were invented, and the race to build the first atomic bomb spurred the manufacture of radioactive mate-

rial. By the 1950s the U.S. Atomic Energy Commission had begun wide-scale research on food irradiation as part of President Eisenhower's Atoms for Peace program, an initiative designed to find peaceful uses for nuclear technology. Much of the research on food irradiation in the 1950s and 1960s was conducted by the U.S. Army.

Food irradiation gained attention in other countries as well, but questions about its safety prevented commercial development. A number of international organizations, including the United Nations' Food and Agriculture Organization (FAO), the International Atomic Energy Agency (IAEA), and the World Health Organization (WHO), coordinated studies of food irradiation during the 1960s and 1970s. In 1980 the Joint FOA/IAEA/WHO Expert Committee on the Wholesomeness of Irradiated Food announced that irradiation of food at levels below 10 kilogray was safe. (A *gray* is a unit of measurement that indicates how much radiation something has absorbed. One gray equals 1 joule of energy absorbed by 1 kilogram of material.) The announcement spurred many governments to legalize various forms of food irradiation.

In the United States, where low-level irradiation of wheat and potatoes had been legal since the 1960s, the Food and Drug Administration (FDA) responded by legalizing the irradiation of spices in 1983 and pork in 1985. The

FDA approved low-dose irradiation of all fresh foods and some dried food products in 1986, and higher-dose irradiation of chicken in 1990 and beef in 1997. The 1986 regulations also mandated that irradiated food be clearly labeled as such, although processed food that contains irradiated ingredients does not have to be labeled. But in the United States, as in many other countries, very few foods that can be legally irradiated commercially actually are. The most commonly irradiated foods are spices used in processed foods (which do not then have to be labeled as irradiated). Some independent food markets have sold irradiated and labeled meat and produce, but for most consumers irradiated food is essentially unavailable.

Uses, Effects, and Limitations

The effect of food irradiation depends on how much radiation the food receives. Irradiating food at low levels kills insects and prevents root vegetables, such as potatoes and onions, from sprouting during storage. Higher levels kill multicelled parasites such as tapeworm and the nematode that causes trichinosis (a disease caused by eating undercooked pork). Still higher levels kill most of the spoilage-causing bacteria in food, as well as most of the bacteria responsible for many types of food poisoning, including *Listeria monocytogenes, Escherichia coli,* and various species of *Salmonella* bacteria.

Very high levels of radiation will completely sterilize food. Although this may seem desirable, irradiating food has negative effects as well, and such high levels are not often used. High levels of irradiation can make food taste bad and destroy its appearance and texture. Many fruits cannot be irradiated at all because irradiation pits the skin and hampers the fruits' innate ability to heal wounds, making them more vulnerable to rot and mold. Some of these effects can be controlled by irradiating food with lower levels of radiation or at low temperatures. But both of these strategies make irradiation less deadly to pathogens. Radiation-sterilized food has been used by some hospitals for patients whose immune systems are severely damaged, but in general, irradiation, like pasteurization, only reduces the bacteria in food—it does not eliminate it.

Irradiation also destroys some of the nutritive value of food, especially reducing the amount of vitamin E, vitamin A, and some of the B vitamins found in food. The vitamin degradation often becomes more pronounced if the food is stored a long time. This degradation is a concern because, since irradiation destroys spoilage bacteria, irradiated food can often be stored much longer than can untreated food. In addition, cooking irradiated food appears to destroy more vitamins than would be expected by calculating the effect of cooking alone plus the effect of irradiation alone. Each process seems to make the other even more destructive, a phenomenon called the *synergy effect* (although, fortuitously, the synergy effect also holds true for pathogens). Again,

Food poisoning is a serious public-health problem that kills as many as 9,000 people in the United States every year. Above, an employee at a Jack in the Box fast food restaurant is shown using tongs to handle hamburger patties in 1995. The fast-food chain implemented new food-handling procedures after hamburger meat tainted with E. coli bacteria was traced to its restaurants in January 1993. In that outbreak, four children died and hundreds became ill. Advocates of food irradiation argue that the process could dramatically reduce food-borne illness (AP/Michael Poche).

vitamin losses can be minimized by using low levels of radiation or irradiating at lower temperatures.

Some bacteria and most viruses, molds, yeast, and fungi are quite resistant to radiation. For example, the bacteria that cause botulism make spores that can be destroyed only by unacceptably high levels of radiation. Botulism can be fatal (as can many other types of food poisoning) and is especially troubling because the bacteria that cause it do not make food go bad. Instead, spoilage bacteria, against which radiation is quite effective, cause the odor and appearance of food to become offensive. Usually, spoilage bacteria ruin food before the botulism bacteria can reproduce enough to be a problem, but some health experts worry that irradiation will wipe out the "warning" spoilage bacteria on food and thereby increase the number of botulism cases.

As the botulism example shows, irradiation alone does not guarantee the safety of food. Food that has gone bad before irradiation will continue to harbor bacterial toxins after irradiation. Even food that has been sterilized can be contaminated after irradiation by unsanitary handling.

Issues and Debate

Food irradiation has probably been studied more than any other form of food treatment or preparation. This fact alone indicates how deeply uncomfortable most people are with the idea of deliberately putting food near radioactive material for any reason. The fear many people have of radiation and nuclear technology is perhaps the single largest barrier to the widespread acceptance of food irradiation. But studies have generally shown food irradiation to be a safe way to kill disease-causing bacteria on food—and food poisoning is a serious public-health problem that kills as many as 9000 people in the United States every year. While opponents of irradiation strongly oppose what some call "nuking" food, proponents claim that it will save thousands of lives and billions of dollars in health care costs. Each side accuses the other of putting lives at risk. Not surprisingly, given the technology's historical link to the military, antinuclear activists have been some of the strongest opponents of food irradiation.

One area of disagreement is whether irradiation will form dangerous compounds in food. Irradiation of food does form what are called *radiolytic products*, some of which are carcinogenic. Most of these are identical to compounds found in untreated foods or formed by other food-processing methods, such as cooking, but a very small number appear to be unique to irradiation. Proponents and opponents disagree over the potential toxicity of these unique radiolytic products and whether, for example, the presence of small amounts of carcinogens in irradiated meat is more dangerous than the presence of small amounts of carcinogens in cooked meat.

Similar debates rage over the significance of the vitamin loss in irradiated foods and the likelihood that food irradiation will increase certain types of food poisoning. More fundamentally, proponents and opponents of food irradiation disagree over whether another layer of food processing is really necessary—whether the potential risks of irradiation are worth the potential gains of fewer cases of food poisoning. The potential gains are not guaranteed because many cases of food poisoning are caused by improper handling of food and unsanitary conditions on farms, in meat-processing centers, and in kitchens. Whether irradiation would be sure to prevent illness when a food handler uses the same knife on salad greens that was used on raw chicken, or when a chef undercooks a hamburger, is a question nobody can answer. But critics fear that widespread use of irradiation would lull producers and consumers into a false sense of security that will cause them to adopt more lax food preparation practices.

In the food-processing industry, a sort of catch-22 exists with regard to consumer demand for irradiated foods and product availability. Many industry members favor irradiation, especially irradiation of red meat and poultry, and want to see its wide implementation in food-preparation processes. Yet some firms are worried about being the first in the industry to begin irradiating meat, fearing that consumers will reject their products when browsing the supermarket aisles. Meanwhile, consumers will not be able to decide to buy irradiated foods if the products do not become available. A government- or industry-sponsored campaign to educate the public about irradiation's benefits and downsides could help bring a solution to the problem.

—*Mary Barr Sisson*

RELATED TOPICS
Nuclear Energy, X-ray Imaging

BIBLIOGRAPHY AND FURTHER RESEARCH

BOOKS

Diehl, J.F. *Safety of Irradiated Foods*, 2nd ed. New York: Marcel Dekker, 1995.

Johnston, D. E., and M. H. Stevenson, eds. *Food Irradiation and the Chemist*. Cambridge: Royal Society of Chemistry, 1990.

Murano, E.A., ed. *Food Irradiation: A Sourcebook*. Ames: Iowa State University Press, 1995.

Murray, David R. *Biology of Food Irradiation*. Taunton, Somerset, England: Research Studies Press, 1990.

Satin, Morton. *Food Irradiation: A Guidebook*. Lancaster, Pa.: Technomic Publishing Company, 1996.

PERIODICALS

Brody, Jane E. "No, the Food You Eat Will Not Be Radioactive." *New York Times,* October 12, 1994, C11.

Burros, Marian. "Eating Well." *New York Times*, August 26, 1992, C4.

Colby, Michael. "Food Irradiation: Magic Bullet or Threat?" *Baltimore Sun*, December 14, 1997, 1F.

Sugarman, Carole. "An End to Food Scares? Reconsidering Irradiation, with All Its Pros and Cons." *Washington Post*, November 12, 1997, E1.

INTERNET RESOURCES

Facts About Food Irradiation
    http://www.iaea.or.at/worldatom/inforesource/other/food/index.html
Food Irradiation: Solution or Threat?
    http://193.128.6.150/consumers//campaigns/irradiation/irrad.html
Position of the American Dietetic Association: Food Irradiation
    http://www.eatright.org/airradi.html
Radiation Pasteurization of Food
    http://www.cast-science.org/past_ip.htm
Foundation for Food Irradiation Education
    http://www.food-irradiation.com

GENE THERAPY

Gene therapy refers to the introduction of a gene or genes into a person's cells to treat disease. Gene therapy attempts to correct genetic diseases, such as muscular dystrophy and cystic fibrosis, and it is also under consideration for the treatment of acquired illnesses such as high blood pressure and cancer. Although gene therapy is believed to be theoretically feasible as a treatment for illness, it has not yet resulted in a long-term cure for anyone. Its foremost challenges involve finding an effective method of delivering genes into sufficient numbers of cells and ensuring that cells activate the corrective genes. Promising research results have led to numerous completed and planned clinical trials.

As have other technologies involving manipulation of genetic material, gene therapy has kindled debate. Modifying a person's genetic makeup—or that of a person's future children—has raised controversial ethical issues related to the prospects of genetic enhancement and alteration of the human gene pool.

Scientific and Technological Description

Gene therapy is concerned with human genetic material and how it causes disease when it malfunctions. The genetic

Virus-Mediated Gene Delivery

1. The healthy version of a human gene is inserted into the DNA of a virus.

2. The virus is injected into the patient's cells.

3. The virus inserts its DNA, carrying the human gene, into the cells' DNA.

4. If expressed, the human gene instructs the cells to manufacture healthy protein.

material in our cells is composed of DNA (deoxyribonucleic acid), the core of which is built of chemical units called *bases*, of which there are four: adenine, cytosine, guanine, and thymine. A gene consists of a distinct sequence of bases. A gene's job is to direct a cell to manufacture a particular protein; the sequence of bases within the gene represents a code instructing the cell how to build that protein. Proteins are the basic machinery of cells, performing many of the processes required for life. A gene may be in an active state of directing the synthesis of a protein, or it may be in an inactive state. When a gene is active, scientists say it is being *expressed*.

DNA typically exists in a healthy state, with genes made up of the correct sequence of bases for producing well-formed proteins. However, sometimes mistakes are made within DNA, resulting in genes with faulty sequences. These mistakes, called mutations, might be carried within a person's cells from birth, or they might be caused by internal or external factors after birth. When a gene has a sequence containing a mutation, it may produce a nonfunctioning protein, not enough of a protein, or too much of a protein. In addition, *mutations* that are not within genes themselves, but rather, within the sections of DNA that regulate the expression of a gene, can result in abnormal proteins. These abnormalities cause cells to malfunction, causing a genetic disease. Gene therapy attempts to correct genetic disease by providing a patient's cells with the correct version of a mutated gene. The technology requires delivering healthy genes to the patient's diseased cells and prompting the cells' machinery to express the genes delivered.

Gene Delivery

The most common way to introduce genes into a patient's cells is with a virus. Some viruses stay alive by entering a large number of cells and inserting their own DNA into the genetic material of the cells invaded. The viral DNA is then expressed along with the cells' DNA. Other viruses unload their genetic cargo into the *nucleus* of human cells (the cellular compartment that holds DNA) but do not splice their DNA into the cells' genetic material. Instead, the DNA of these viruses just sits in the nucleus, and the cells' machinery finds it and expresses it.

In gene therapy, a virus, which has been altered so that it cannot cause disease, is engineered to contain within its DNA the healthy copy of a human gene. Gene therapy can take advantage only of those genes that have been identified and cloned in the laboratory. A large number of human genes have been cloned and are readily available to genetic researchers for gene therapy applications. The engineered virus is delivered to the diseased cells of a patient. This may be achieved by removing a sample of the patient's cells, infecting them with the virus, and reinserting them into the patient. In other cases, the virus can be injected directly into the patient's tissue. If the viral DNA, which carries the therapeutic gene, enters the nucleus or squeezes into the cellular DNA as expected, it should be expressed and should yield the healthy version of the desired protein (see figure).

Gene Expression

Once the corrective gene is inside a patient's cells, it must be expressed. The expression of genes is controlled by a DNA sequence called a *promoter*, which is located in front of the gene it regulates. The promoter regulates expression in response to various stimuli, such as the presence of a particular chemical or the amount of the gene's protein required by the body at particular times or in particular circumstances. For gene therapy, it has been possible in some cases to include parts of a corrective gene's promoter along with the gene in the delivery vehicle. The promoter then naturally enables the gene to be expressed. In other cases, a promoter is included that responds to the presence of an external drug. The expression of a gene regulated by such a promoter can be controlled by giving the patient doses of the drug. When the drug is present, the promoter causes the gene to be expressed.

Germ-Line Gene Therapy

Among the newest branches of gene therapy is *germ-line gene therapy*, or germ-line genetic engineering. This is the application of gene therapy to the *germ line*, a collective term for the sex cells (egg cells and sperm cells) and the cells that give rise to them. In regular gene therapy, diseased cells (from the lung, blood, muscle, etc.) receive copies of a healthy gene, but the patient's sex cells are not involved. The patient's children, therefore, do not inherit the healthy gene. But in germ-line gene therapy, which is not yet possible in humans, the corrective gene would be delivered to a person's germ-line cells with the intention that the person's children would inherit the corrective gene. The recipients of germ-line gene therapy would not benefit from the treatment; it would be administered so that the future children of a person with a genetic disorder would not have the disease. Germ-line gene therapy would permanently alter the genetic makeup of all the recipient's descendents.

Historical Development

Gene therapy's history shares with other genetic technologies the series of discoveries that led to a molecular understanding of DNA and the process by which cells translate genes into proteins (see CLONING, GENETIC ENGINEERING, and GENETIC TESTING). In the 1950s, scientists discovered DNA's structure, and consequent investigations in the 1960s revealed the protein-building code that genes carry in their sequence of bases. Technological advances emerged from these findings, and by the late 1970s, geneticists could determine the precise sequence of bases in a given gene. A rigorous exploration of the genetic basis of disease followed, and research yielded an understanding of the genetic mutations that directly cause certain diseases.

Techniques for isolating and replicating genes and for transferring genes into vehicles such as viruses began in the late 1960s and developed quickly during the following decade. During the 1980s, scientists performed a myriad of gene therapy experiments on animals, attempting to correct genetic disease with the introduction of healthy genes to animal cells. The experiments had mixed results, but some were promising enough to warrant the first application of gene therapy to humans.

In 1990, the U.S. National Institutes of Health (NIH) sponsored an experiment to perform gene therapy on two young girls, Ashanthi DeSilva and Cindy Cutshall, who had the

genetic disease adenosine deaminase (ADA) deficiency. ADA deficiency is caused by a malfunctioning gene that when present in its normal version produces an enzyme that is crucial for immune system function. ADA-deficiency patients do not produce sufficient amounts of this enzyme; this condition means that even mild infections seriously threaten their health. Most children with ADA deficiency die young.

NIH scientists W. French Anderson and R. Michael Blaese led the team of researchers that administered the girls' gene therapy. The researchers removed large samples of T cells (cells that play important roles in the body's immune response) from the patients. They exposed the cells to inactivated viruses that carried the healthy ADA gene. Once the cells had been infected by the viruses, the researchers injected the T cells back into the patients. The procedure was repeated for two years. The T cells expressed the corrective ADA gene at low levels, and the therapy improved the girls' immune function. The experiment showed that gene therapy can be performed in a safe manner and have beneficial results. However, the therapy did not represent a clear cure in either case, and the girls have continued to receive various gene therapy and non–gene therapy treatments for the disease.

Numerous clinical trials of gene therapy followed the 1990 experiment. By 1995, more than 100 trials had been initiated, but none were as successful as media coverage had led the public to expect. In 1996, the NIH overhauled the approval process for gene therapy trials, making it more strict so that approved trials would have a greater chance of success. Research to solve the technology's shortcomings continued. In early 1999, more than 200 gene therapy trials were under way, some yielding encouraging results.

Uses, Effects, and Limitations

Gene therapy clinical trials have attempted to correct problems associated with numerous diseases and disorders. The earliest-treated examples include ADA deficiency, muscular dystrophy, cystic fibrosis, sickle-cell anemia, and familial hypercholesterolemia; these diseases are associated with simple and well-understood genetic mutations, making them reasonable targets for gene therapy. As the technology has developed, illnesses such as high blood pressure, heart disease, Parkinson's disease, hemophilia, various cancers, AIDS (acquired immune deficiency syndrome), and general aging processes have been the focus of gene therapy endeavors. Many trials have confirmed that gene therapy can be applied without harm to the patient, and some trials have added healthful years to patients' lives.

In the case of most genetic diseases, prevention is not possible. Mutations are present in a person's cells from birth, and the disease may be apparent immediately or arise later in life. Traditional treatments for genetic diseases focus on synthesizing drugs that offer what a given genetic mutation prevents a patient's body from being able to provide. For many genetic diseases, there is no known drug treatment or no completely effective drug treatment. For others, the treatments must be applied continuously and are prohibitively expensive. In theory, gene therapy offers the potential to treat the disease at its root. Optimally, gene therapy would deliver corrective genes in such a way as to ensure that patients' cells take up the gene as a permanent part of their genetic makeup. This would represent a definitive cure.

But the complexity of genetics has so far thwarted attempts to use gene therapy technology to cure genetic disease. A number of factors have prevented definitive success. First, a gene delivery vehicle that consistently functions effectively has yet to be identified. Far fewer cells than necessary now take up the engineered viruses or liposomes in a typical application of gene therapy. The source of this problem has been difficult to characterize, but improved gene delivery vehicles are increasing viral takeup. Following gene delivery, the correct expression of therapeutic genes is not assured. The natural expression of genes is regulated in a complex manner. Many applications of gene therapy have been plagued by low levels of gene expression, again for unidentified reasons. But it is also possible for a therapeutic gene to be overexpressed, yielding a harmful overdose of protein. Many researchers focus on solving these gene expression problems.

The primary risk associated with virus-delivery applications of gene therapy is the possibility that a corrective gene will disrupt normal function of other genes in a patient's cells. Viruses insert their genetic material into the infected cell's DNA at random locations. The viral DNA might end up spliced into the middle of one of the cell's genes, preventing that gene from functioning properly. Identifying a way to make viruses insert their genetic material at a specific location within a cell's genome would be a major step forward for gene therapy.

When the foregoing limitations are solved, gene therapy will be constrained by our understanding of genetic diseases. Gene therapy can be used only for those diseases and conditions whose associated genes have been identified and cloned; many such genes are readily available, but plenty are not. Some geneticists expect, however, that the swift rate of genetic discoveries will reveal the genetic components of all genetically related disorders within a decade.

Issues and Debate

That gene therapy has only been a qualified success has given rise to debates about whether it is worth pursuing. An enormous amount of public money has been spent on gene therapy research since the mid-1980s, but the technology has produced no cures. This makes some observers wonder if successful gene therapy is technologically impossible. Many scientists believe that this attitude is ill-guided. They argue that such skepticism emerged only because early public reports about gene therapy oversold the technology's potential and underestimated the time it would take to mature. They assert

that gene therapy is feasible and that it may actually reduce overall medical costs once it has been put into wide practice. Currently, considerable resources are spent on drug and other treatments to alleviate symptoms of diseases that could be overcome permanently by effective gene therapy.

Another issue surrounding gene therapy concerns unknown consequences. Since it is still a new technology, scientists do not know what the long-term effects of continual gene therapy treatments might be. Some disease-causing mutations actually give the person with the disease increased protection against other illnesses. For example, the faulty gene that causes sickle-cell anemia (a disease that affects red blood cells) helps combat malaria, a deadly infectious disease. Although gene therapy might cure sickle-cell anemia, it would remove the protection against malaria. There may be similar but unknown benefits that disease-related mutations confer which gene therapy would erase. But those who believe this issue is insignificant argue that many targets of gene therapy are lethal diseases and that the possibility of adverse consequences is not a good reason to ignore gene therapy as a possible way to save a patient's life.

Much of the controversy stirred up by gene therapy is related to genetic modifications of the germ line. Germ-line gene therapy would constitute making permanent changes in the makeup of human genetic material. Some critics assert that it would be wrong to bend the course of human evolution by tampering with the genes of entire families and their descendents. Others believe that unborn children have the right to inherit unmodified genetic material. Proponents respond that these concerns are inappropriate when the issue is preventing the birth of a child who carries a debilitating disease.

Some critics of germ-line gene therapy agree that the technique should be allowable for the prevention of disease but are worried that it will lead to genetic enhancement. If we are able to insert genes that counteract diseases, they ask, what is to stop us from inserting genes for height, strength, eye color, or nose size? Although the genetic elements of such traits as personality and talent are complex and poorly understood, it might be possible someday to shape these types of characteristics in unborn children. Some people believe that these enhancements would be wholly unethical. Others point out that the side effects of tampering with germ-line genetics are unknown and potentially harmful.

—*Tamara Schuyler*

RELATED TOPICS
Cloning, Genetic Engineering, Genetic Testing

BIBLIOGRAPHY AND FURTHER RESEARCH
BOOKS
Clark, William R. *The New Healers: The Promise and Problems of Molecular Medicine in the Twenty-First Century.* New York: Oxford University Press, 1997.

Frank-Kamenetskii, Maxim D. (trans. Lev Liapin). *Unraveling DNA: The Most Important Molecule of Life*. Reading, Mass.: Addison-Wesley, 1997.

Gonick, Larry, and Mark Wheelis. *The Cartoon Guide to Genetics*. New York: Harper Perennial, 1991.

Jenkins, Morton. *Teach Yourself Genetics*. Chicago: NTC/Contemporary Publishing, 1998.

Kitcher, Philip. *The Lives to Come*. New York: Touchstone, 1997.

Rifkin, Jeremy. *The Biotech Century*. New York: Tarcher/Putnam, 1998.

PERIODICALS

Jaroff, Leon. "Fixing the Genes." *Time*, January 11, 1999, 68.

Kmiec, Eric B. "Gene Therapy." *American Scientist*, May/June 1999, 240.

Moulton, Gwen. "Panel Finds In Utero Gene Therapy Proposal Is Premature." *Journal of the National Cancer Institute*, March 3, 1999, 407.

Wade, Nicholas. "Gene Therapy Passes Important Test, in Monkeys." *New York Times*, February 23, 1999, 1.

Wheeler, David L. "Prospect of Fetal-Gene Therapy Stimulates High Hopes and Deep Fears." *Chronicle of Higher Education*, January 22, 1999, A13.

INTERNET RESOURCES

CancerNet from the National Cancer Institute
http://www.oncolink.upenn.edu/pdq_html/6/engl/600718.html

Gene Therapy
http://www.ultranet.com/~jkimball/BiologyPages/G/GeneTherapy.html

Gene Therapy Links
http://www.genetixpharm.com/genes.htm

Gene Therapy Sites and Links
http://www.mc.vanderbilt.edu/gcrc/gene/inttext.htm

Questions and Answers About Gene Therapy
http://imsdd.meb.uni-bonn.de/cancernet/600718.html

Superhumans
http://www.newscientist.com/nsplus/insight/clone/superhu.html

GENETIC ENGINEERING

Genetic engineering refers to technologies that move genes from one species to another to produce organisms with unique combinations of traits. Scientists create genetically engineered bacteria, plants, and animals for use in agriculture, pharmaceuticals, synthetic fabrics, toxic-waste cleanup, and pure research. Genetic engineering generates plants that are resistant to pests and bacteria that serve as microscale drug factories.

A fiery public debate about the ethics and risks of manipulating genes has arisen. While advocates of genetic engineering see the technology as just an enhanced—and highly beneficial—version of selective breeding practices, some critics view alterations of genetic material as unethical and unsafe. The long-term consequences of genetic engineering are unknown, and some opponents argue that the technology may leave a disastrous biological legacy.

Scientific and Technological Description

The goal of genetic engineering is to create organisms that differ from existing organisms by just one or a few traits. This is achieved by adding genes to or deleting genes from an organism's normal genetic material. Genes are the basic unit of hereditary information; they carry the code responsible for producing organisms' physical traits. Genes are short strands of DNA (deoxyribonucleic acid). They are made up of various combinations of four chemicals, called *bases*, strung together along a backbone of repeating sugars and phosphates. A gene's particular sequence of bases represents a code that instructs the cell to make a protein. Proteins, each encoded by a gene, perform most of the structural and functional jobs in the body. By this chain of events, genes determine organisms' physical makeup.

Through genetic engineering, scientists are able to add genes to an organism's DNA, thus adding one or more traits. For instance, bacteria do not normally secrete the hormone insulin. But geneticists can add the human gene that codes for insulin to the DNA of certain bacteria. The engineered bacteria produce insulin, which can be isolated and injected into humans as a treatment for diabetes. (People with diabetes require injections of insulin because their bodies do not produce enough of it.) Some of the laboratory techniques used in genetic engineering are described below.

Polymerase Chain Reaction

To work effectively with a gene, researchers need many copies of it. The most common method of creating multiple copies of a stretch of DNA is the polymerase chain reaction (PCR; see figure). PCR requires the following ingredients: the piece of DNA that is targeted for copying, excess individual building blocks of DNA, enzymes that activate chemical reactions among the building blocks, and stretches of DNA that serve as primers upon which new DNA can be built. The PCR procedure involves several steps of heating and cooling, during which the targeted stretch of DNA is unraveled and copied. Each new strand, a replica of the original, is then also copied, and the process repeats. Billions of copies of the original DNA can be created in a day.

Recombinant DNA

Recombinant DNA techniques are methods of delivering foreign genes into cells. The techniques employ biological entities called *plasmids*. Plasmids are tiny circles of DNA that reside in bacterial cells; they can be removed from bacteria and engineered to contain a foreign gene. The engineering process involves chemicals called *restriction enzymes*, which cut the plasmid at specific locations. The foreign gene is inserted into the cut in the plasmid, both ends of the foreign gene joining to the open ends of the plasmid. Engineered plasmids are injected into the target bacteria, and these bacteria activate the genes in the plasmids' DNA. The engineered bacteria then produce the protein coded for by the foreign gene. A second vehicle for gene delivery in recombinant DNA technology is the *virus*, a tiny infectious particle. Viruses, like plasmids, can be engineered such that their DNA contains foreign genes. Engineered viruses infect the cells of bacteria, plants, or animals, and the recipient cells activate the viral genes along with their own.

The Polymerase Chain Reaction (PCR)

After the starting materials have been assembled, the double-stranded DNA (one strand shown as grey, the other black) is heated to separate the strands (1). Then the DNA is cooled, causing primers to adhere to each strand (2). With primers in place, enzymes coax DNA building blocks to join in creating a new strand of DNA next to each original strand (3). The cycle repeats, doubling the number of strands with each repetition.

Microinjection

For microinjection, a fine-tipped glass needle is used to inject foreign genes directly into recipient cells—typically, animal and bacterial cells, because their outer membranes are easier to puncture than plant cell membranes. Once the foreign DNA is inside the recipient cell, it is activated and the foreign protein is produced.

Electroporation and Chemical Poration

Another method of introducing foreign genes into cells is by dotting the cell membrane with holes; the genes enter the cells passively through the holes. The holes are created by either electric shock (electroporation) or chemicals (chemical poration).

Bioballistics

Various projectile methods are also used to introduce foreign DNA into a cell. Microscopic metal slivers are coated with DNA containing the foreign gene, and the metal slivers are shot into recipient cells. In one ballistic technique, a shotgun with a slightly shorter barrel is used to launch the slivers. The gun is used on cultures of cells only, not living organisms. A thick guard plate catches the shell cartridge but allows the slivers to enter the cell. For plant cells, bioballistics is more effective than direct injection methods.

Historical Development

The Austrian monk Gregor Mendel is widely regarded as the founder of modern genetics. In the 1850s and 1860s, he performed extensive experiments with pea plants, analyzing patterns of trait inheritance. His results represented the first documented evidence that discrete units of hereditary information are passed intact from parent to offspring. The discrete units of heredity that Mendel proposed are now called *genes*. Reports of Mendel's experiments, uncovered in 1900, marked the beginning of genetics research. During the first five decades of the century, gradual progress led to an understanding that the molecule dubbed DNA was the seat of genetic information. In 1953, the discovery of DNA's structure touched off a revolution in genetics. Rapid advances followed; by 1970, scientists had described the processes by which DNA replicates and by which genes code for the production of proteins.

Biologist Paul Berg was the first to practice recombinant DNA technology, using restriction enzymes in 1972 to cut DNA into pieces and the enzyme DNA ligase to join pieces of DNA in new combinations. The following year, Stanley Cohen and Herbert Boyer created the first recombinant DNA organism. They altered the DNA of a plasmid (circular strand of DNA) using the techniques Berg had introduced and inserted the engineered plasmid into a bacterial cell. The cell activated the plasmid's DNA.

The development of the *polymerase chain reaction* (PCR) was among the most significant breakthroughs in the history of genetics-related technologies. Biochemist Kary B. Mullis conceived of the idea in 1983 while contemplating the fact that large amounts of DNA are necessary for effective DNA manipulation (tiny fragments, such as those that might be found at crime scenes, are too small to allow analysis). By 1985 he had refined the PCR technique, which has been used since to make millions of copies of a gene within hours. PCR made the copying of genes vastly more efficient

Scientists have begun to modify the genetic material of cows, sheep, and goats so that the animals secrete therapeutic human proteins in their milk; these substances can be isolated and may someday be used as drugs for humans. Researchers hope that herds of animals may one day be able to produce such drugs on a large scale (Photo Researchers/Margot Granitsas).

than previous methods, which involved inserting genes into bacteria, allowing the bacteria to multiply, then isolating the DNA from the resulting colony.

Uses, Effects, and Limitations

Agriculture is the most common area of application for genetically engineered organisms. Numerous genetically modified food crops, including tomatoes, squash, corn, and soybeans, are sold in supermarkets and used in processed foods. The genetic makeup of these plants is altered to achieve various results, such as resistance to disease, improved nutritional value or appearance, increased transportability, and ability to grow in harsh environments. Crops grown for nonfood uses are also targets of genetic engineering. Cotton has been engineered to be resistant to pests such as the debilitating cotton bollworm. Research is also focused on growing genetically engineered crops that could serve as sources for fabrics, detergent ingredients, substitute fuels, and medicines.

Bacteria, which are typically the easiest organisms to engineer, exist in genetically modified form for use in several industries. Some bacteria have been engineered to repel pests; they are applied directly to crops as pesticides. Bacteria have also been genetically modified to produce substances useful for food processing; for example, bacteria-produced rennet (a substance used to curdle milk) is used in cheese manufacturing. Certain genetically engineered bacteria produce hormones for animals. The most widely recognized example is bovine growth hormone, a substance produced by engineered bacteria and given to cows to increase milk production. Bacteria are also used in nonagricultural applications. Some engineered bacteria produce medicines, such as insulin. Others produce ingredients for synthetic fabrics. Environmental engineers have also modified bacte-

ria to break down particular toxic wastes; this application is part of a technology called bioremediation.

Genetically engineered amphibians and small mammals play a role in basic genetics research, helping scientists learn more about genes, cells, physiology, development, and disease. In research facilities, if not yet in commercial industries, genetically engineered goats and sheep secrete hormones and other substances into their milk; these substances can be isolated and may someday be used as drugs for humans. Scientists are investigating the possibility of genetically engineering pigs so that their organs are compatible with human blood and can be transplanted into humans without being rejected (see ORGAN TRANSPLANTATION).

Genetic engineering is associated with several benefits over traditional methods of manipulating organisms' traits. One advantage of genetic engineering is that it allows the introduction of genes across species boundaries. Traditional breeding relies on natural reproduction, and organisms can be mated only to others of their own species. Therefore, with traditional methods, there is no way to get a gene from one species into an animal of another species. Furthermore, in the case of bacteria, breeding is not an option in traditional methods, because most bacteria reproduce asexually; a bacterial cell simply duplicates itself to give a genetically identical offspring.

Genetic engineering also allows precise genetic modifications. Traditional breeders rely partly on chance when mating two animals or cross-pollinating two strains of plant. The mixing of genes between parent organisms is random, and the offspring may or may not exhibit the prized attributes of the parents. The technology is limited, however, by scientists' understanding of genetics and the pace of scientific progress. Genetic processes are complex, and the outcome of genetic engineering cannot always be foreseen. Genetic modifications are sometimes accompanied by undesirable changes. For instance, efforts to create leaner pigs have yielded pigs with arthritis, low fertility, and weak immune systems.

Issues and Debate

Numerous aspects of genetic engineering are currently discussed among scientific communities and members of the public. The debates center around risks and ethical concerns, mostly related to the technology as it is applied to agriculture. As with any new technology, the long-term effects of genetic engineering are unknown. Many opponents believe that performing genetic modifications will have dire consequences for ecosystems. They cite evidence of potential dangers to human health and the environment.

One danger to human health is an allergic reaction. Some people have severe allergic reactions to certain plants. In some cases, the protein responsible for the allergy has not been identified. Genetic engineering might move a gene that codes for an allergy-causing protein from one plant to another. In 1996, researchers showed that one company's plans to transplant a gene from the brazil nut into soybean plants could have been disastrous, because the gene codes for a protein to which some people are allergic. Advocates of genetic engineering claim that allergy problems can be solved by labeling genetically altered food with the source of transplanted genes. But critics argue that it would be unfair to expect people with allergies to read the ingredients of every food they buy. They worry that restaurants or other food vendors might use the engineered products and that their patrons would have no way of knowing.

Several other health risks are posed by genetic engineering that involve instances of organisms adapting or reacting to internal changes. Genetically modified plants might begin producing toxins, for example, or lose natural resistance to fungi that persist through food processing and threaten the health of consumers. Genetic engineering also threatens environmental health, say opponents of the technology. Traits conferred by genetic engineering may make a plant able to spread and thrive in places where it would have unwanted effects. It might choke out native plants and upset the natural balance of the ecosystem. Also, instances of genes jumping from genetically engineered plants to surrounding plants through natural cross-pollination have been documented. The results of foreign genes being transmitted to other plants could lead to a cascade of uncontrolled genetic changes in the environment, yielding an unknown outcome.

Another environmental concern is a potential acceleration of pesticide resistance in disease-causing organisms. Some plants are engineered to repel pests (e.g., bacteria, fungi, and worms) by producing a toxin. The plant produces the toxin throughout its life cycle, exposing the pest to the toxin continuously. In time, the pest population will probably develop a strong immunity to the toxin through processes of natural selection, and the plant will be more susceptible to infection than it was prior to the application of genetic engineering.

Alongside doubts about the safety of genetic engineering, some opponents have voiced concerns about the ethics of the technology. One set of issues emerges from religious views. Some people feel that altering the genetic makeup of organisms is wrong, because humans should not endeavor to change God's creation. Although not everyone shares this belief, many observers note that it is important to respect the wishes of those who do not want to consume genetically engineered organisms. They propose that genetically modified food should be labeled. However, it would be immensely difficult to keep track of food as it makes its way through various manufacturing processes. Moreover, food manufacturers might be reluctant to label their products as "genetically engineered," because such a description might sound like a warning that the food is not safe.

Other opponents of agricultural genetic engineering have raised questions about the ethics of increasingly turning food production over to the corporate chemical industry. Many small farmers believe that their businesses have

already been undermined by high-production, low-cost farming that relies on chemical pesticides. The technology of genetic engineering, some fear, may increase the dependence of farmers on chemical manufacturers. Some corporations that manufacture genetically engineered plant seeds also engineer their seeds to be infertile, so that farmers must buy new seeds every year. Such practices have kindled widespread debate.

Advocates of types of nonchemical farming, such as crop rotation, claim that the perceived need for genetically engineered food products could be mitigated by changing the face of agriculture. They argue that crop rotation should be considered an alternative to pesticides and genetically engineered pest resistance, because it controls pests without causing pollution and without the risks of genetic engineering. But documentation of the efficacy of crop rotation is lacking, and many people remain unconvinced of its potential.

Proponents of genetic engineering, particularly as applied to agriculture, respond to critics by arguing that the technology brings benefits that have not been offered by other means. They assert that genetic engineering will reduce pesticide use, increase the nutritional value of food, and allow food to be grown in the harsh environments of some nonindustrialized countries where hunger is a devastating problem. These goals have not yet been met, and the world is waiting to see whether genetic engineering will live up to its promise.

—Tamara Schuyler

RELATED TOPICS

Bioremediation and Phytoremediation, Cloning, DNA Fingerprinting, Gene Therapy, Genetic Testing, Organ Transplantation

BIBLIOGRAPHY AND FURTHER RESEARCH

BOOKS

Aldridge, Susan. *The Thread of Life: The Story of Genes and Genetic Engineering.* New York: Cambridge University Press, 1996.

Frank-Kamenetskii, Maxim D. (trans. Lev Liapin). *Unraveling DNA: The Most Important Molecule of Life.* Reading, Mass.: Addison-Wesley, 1997.

Gonick, Larry, and Mark Wheelis. *The Cartoon Guide to Genetics.* New York: Harper Perennial, 1991.

Grace, Eric S. *Biotechnology Unzipped: Promises and Realities.* Washington, D.C.: Joseph Henry Press, 1997.

Holland, Alan, and Andrew Johnson, eds. *Animal Biotechnology and Ethics.* New York: Chapman & Hall, 1998.

Jenkins, Morton. *Teach Yourself Genetics.* Chicago: NTC/Contemporary Publishing, 1998.

Rifkin, Jeremy. *The Biotech Century.* New York: Tarcher/Putnam, 1998.

PERIODICALS

Abate, Tom. "Grass May Be Tough, but It's No Match for a Powerful Shotgun." *San Francisco Chronicle,* March 22, 1999, B1.

Kennell, David. "The Risks of Genetic Engineering." *St. Louis Post-Dispatch,* March 17, 1999, B7.

Milius, Susan. "Change One Gene, Plants Get Healthier." *Science News,* June 13, 1998, 378.

Quick, Rebecca. "Bacteria and DuPont Brew New Polyester." *Wall Street Journal,* April 2, 1999, B1.

Sale, Kirkpatrick. "Monsanto: Playing God." *The Nation*, March 8, 1999, 14.

Travis, John. "Scientists Harvest Antibodies from Plants." *Science News*, December 5, 1998, 359.

INTERNET RESOURCES

Genetic Engineering and Its Dangers
    http://online.sfsu.edu/~rone/gedanger.htm
Genetic Engineering Home Page (Church of Scotland)
    http://dspace.dial.pipex.com/srtscot/geneng0.shtml
Genetic Engineering News
    http://www.genengnews.com/
What Is Genetic Engineering? (Union of Concerned Scientists)
    http://www.ucsusa.org/agriculture/gen.whatis.html

GENETIC TESTING

Genetic testing refers to a variety of laboratory techniques that analyze genetic material and detect abnormalities in humans, including unborn children. The presence of a genetic abnormality might confirm or reveal that a person has a particular disease or indicate an increased risk for developing a disease. Its presence can also signal that the person's children are likely to inherit the problematic genetic element.

Genetic testing offers many potential benefits. But it has come under increased public scrutiny because it raises new and complex questions related to ethics, economics, health, and law.

Scientific and Technological Description

Genetic testing is concerned with the analysis of DNA (deoxyribonucleic acid), the genetic material within our cells. DNA is packaged in long, thin structures called chromosomes; each chromosome contains two strands of tightly coiled DNA. The core of the DNA coil, or *helix*, consists of a series of chemical units called *bases*, of which there are four distinct kinds: adenine (A), cytosine (C), guanine (G), and thymine (T).

DNA contains *genes*, which are sequences of bases that when activated, or "turned on," direct cells to manufacture proteins. Proteins (also called *gene products*) perform various biochemical and mechanical jobs within our bodies. The particular sequence of bases in a given gene represents a genetic code that tells the cell which protein to synthesize. If that code is disrupted, problems occur because the cell produces none of the protein, too much or too little of the protein, or an altered form of the protein. These problems can be directly harmful or they can launch a cascade of harmful effects by influencing the activation of other genes or interrupting crucial biochemical pathways, the multiple-step reactions among molecules and larger compounds that result in properly functioning cellular components.

One potential disruption of the genetic code is a mutated gene, in which the sequence of bases has been altered. Such alterations include deletions, insertions, inversions, duplications, translocations, and replacements of bases (see figure). Tay-Sachs disease, which causes degeneration of the nervous system and is fatal by age 4, is caused by the insertion of four extraneous bases within a gene whose protein is involved in the maintenance of the nervous system.

Genetic testing employs a variety of techniques. Physicians can test for the number and shape of chromosomes, for missing or mutated genes, and for defective gene products. Chromosome testing involves *karyotyping*, in which chromosomes are dyed and viewed. Other techniques involve isolating, manipulating, and analyzing DNA (usually from a person's hair, tissue, saliva, or blood), identifying particular genes, and determining the exact sequence of bases in certain regions of DNA.

Several types of genetic tests exist; they utilize various combinations of the laboratory techniques described above.

1. *Carrier identification* shows whether an asymptomatic person carries within his or her DNA a genetic problem that might be passed on to a child. It is possible to carry (and consequently, pass on) one copy of an abnormal gene without exhibiting signs of the associated disease.

2. *Prenatal diagnosis* indicates whether a fetus carries problematic genetic material. It is done through amniocentesis, the testing of fluid that surrounds the fetus, or chorionic villi sampling, the testing of placental fetal tissue.

3. *Newborn screening* determines the genetic health of newborns. Tests for diseases such as *phenylketonuria* (PKU), which causes mental deficiency, and congenital hypothyroidism, which causes extreme fatigue, are routinely performed on newborns in many countries because effective treatments for those conditions exist.

4. *Late-onset-disorder* testing shows whether a person carries genetic material that will or might cause the development of a problem, such as Huntington's disease, heart disease, or certain cancers, later in life.

Historical Development

The history of genetic testing is rooted in the seminal discovery of the double-helical structure of DNA in 1953 by James Watson, Francis Crick, Maurice Wilkins, and Rosalind Franklin. From that discovery, researchers unveiled further details concerning DNA, such as how it replicates and directs protein synthesis, which in turn yielded various advances in DNA technologies in the 1960s, some of which were crucial to the emergence of genetic testing. For instance, by comparing the characteristics of DNA from a normal person to those of a person with a genetic disease, researchers could determine the general location of abnormal genes.

Further refinement occurred with the introduction of DNA sequencing. In 1977, Walter Gilbert and Frederick Sanger invented a way to determine the exact order of bases along a region of DNA, a practice now called *sequencing*. Early sequencing required a month of experiments to produce a stretch of 100 bases, but improvements have made it possible to determine thousands of bases in a sin-

Genetic Mutations

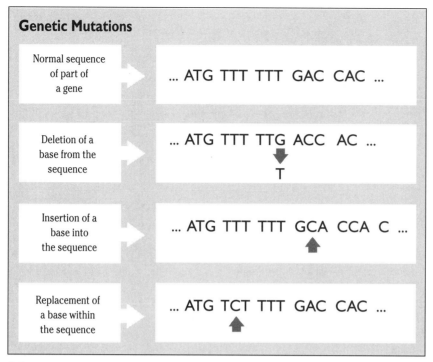

Normal sequence of part of a gene	... ATG TTT TTT GAC CAC ...
Deletion of a base from the sequence	... ATG TTT TTG ACC AC ...
Insertion of a base into the sequence	... ATG TTT TTT GCA CCA C ...
Replacement of a base within the sequence	... ATG TCT TTT GAC CAC ...

Deletions, insertions, and replacements of bases within the sequence of a gene can cause genetic disease by shifting or altering the reading frame of the gene. An incorrect reading frame leads to faulty protein production.

gle day. Sequencing allowed the precise identification of genetic abnormalities and showed which proteins those genes encode or fail to encode, revelations that have led to treatments for some diseases.

In the 1980s, "knockout" experiments in organisms such as mice led to the discovery of human disease–related genes. In a classic knockout experiment, the DNA of a mouse's sex cells is directly altered by mutation-inducing radioactivity or chemicals. That mouse's progeny will carry the mutated DNA and perhaps exhibit abnormalities caused by the mutations, which can be traced back to alterations in particular genes via DNA technology. Cancer-related mutations in human genes have been determined using such knockout techniques in laboratory animals.

In recent decades the techniques used in genetic testing have been refined. The sequencing of genes associated with diseases allowed the manufacture of markers that join to particular genes. In some genetic tests, a marker is added to an isolated sample of a person's DNA; if the DNA contains the gene in question, the marker will join to the gene and thus indicate its presence.

Uses, Effects, and Limitations

The four types of genetic testing are distinguished by a number of factors, including the circumstances under which they are performed, the results they achieve, and problems that limit their usefulness. *Carrier identification tests* are generally performed when there is a family history

of genetic disease. Genetic tests show whether parents carry faulty genetic material, and they give an accurate analysis of the chance that parents will pass a disease to their children. Couples can use this information to decide whether to conceive a child or to prepare themselves to treat a diseased child. Unfortunately, carrier identification does not offer a perfect forecast of which genetic material a child will inherit. Each parent carries two copies of every gene and passes only one copy to a child. Genetic testing might show that a parent carries one abnormal copy of a gene and one normal copy. The child has an equal chance of inheriting each copy.

Prenatal diagnostic tests are recommended when there is a family history of disease, when the mother has previously had two or more miscarriages, and when the mother is over age 34. A history of miscarriages indicates that the developing fetuses might have had severe genetic abnormalities and thus that another fetus might have the same problem. Women over age 34 are at higher risk of conceiving a child with Down's syndrome, a disease caused by an extra chromosome and resulting in physical deformities and mental deficiency. Prenatal diagnosis tests generally check for abnormalities that would severely disrupt or shorten the life of the child. The results provide parents with information that allows them to terminate the pregnancy or to be prepared (as with carrier identification) to treat the disorder upon birth. Prenatal diagnosis is limited by the dangers of amniocentesis and chorionic villi sampling. In the former, a needle is inserted into the membrane-bound sack of fluid surrounding the fetus, and there is a 1 percent chance of causing maternal bleeding or fetal death. The risks of chorionic villi sampling, in which an inserted tube sucks out a sample of placental tissue, are still unknown. Prenatal diagnosis is therefore recommended only in cases of suspected abnormality.

Newborn screening for some diseases is performed on most infants; for other diseases it is performed only on those who exhibit signs of a genetic problem or who have a sibling with a genetic problem. In most cases, test results indicate the appropriate treatment or preventive measures, which can be administered immediately. PKU illustrates one remarkable success of newborn genetic screening. PKU is a genetic disease caused by the accumulation of excess levels of the amino acid phenylalanine, resulting in severe mental retardation. But when children with PKU are reared from infancy on a special diet (low in phenylalanine, high in tyrosine), they develop almost normally.

One potential problem related to the testing of newborns and children is that it may uncover a genetic problem for which no cure exists. An example is Huntington's disease, which sets in between age 30 and 50 and causes neural deterioration and extreme physical pain. Families with a history of Huntington's might decide to test a newborn or child for the disease. The life of that child could be drastically altered by the knowledge that Huntington's awaits.

Late-onset-disorder testing, which analyzes a person's risk of developing a genetic-related disease later in life, involves factors that complicate many of its applications and interpretations. On the one hand, there are unambiguous cases. Genetic testing for several uncommon inherited cancers has been carried out for the past 10 years; these cancers are almost certain to develop when the abnormal genetic material is present, and thus regular surveillance is prescribed or tissues are removed as a preventive measure. On the other hand, some late-onset disorders are not as straightforward. Recently developed tests to determine a person's genetic predisposition for developing cancers of the breast, ovaries, and colon are problematic. A positive test result in these cases gives only a probability that the person will develop the disorder, because other factors influence the development of the disease. Some people with the faulty genetic material do not develop the disorder at all. The practical interpretation of what is meant by "increased risk" varies from person to person.

Issues and Debate

Genetic testing has generated complex and important concerns. The medical profession faces a variety of unanswered questions surrounding new testing techniques. Some genetic tests, such as newborn screening for PKU, are rarely controversial. However, more recently developed tests that indicate predisposition for late-onset cancer and heart disease have some physicians concerned. Who should be tested? Many practicing physicians do not have an advanced knowledge of genetics. They might worry about lawsuits if they do not prescribe genetic tests for everyone, a prospect that would increase medical bills and could lead to the widespread use of tests that have not been adequately checked for reliability.

Some physicians, researchers, and members of the public are concerned that regular genetic testing will result in job and insurance discrimination. Given access to people's genetic information, employers might choose not to hire someone with a genetic sensitivity to workplace toxins

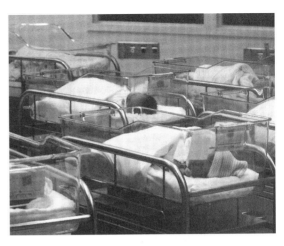

In many countries, newborns' genetic material is routinely screened to detect diseases, such as phenylketonuria (PKU), for which effective treatments exist (Photo Researchers/Larry Mulvehill).

or someone who is likely to develop Huntington's disease within 10 years. Similarly, insurance companies might raise the cost of health insurance for those with a genetic predisposition for disease, or refuse them insurance altogether.

These issues revolve around legal and ethical questions of privacy. Who has the right to a person's genetic test results? Perhaps relatives, who might carry similar genetic problems, should be notified of positive results. Many people worry that genetic information will be freely available someday—perhaps as part of a person's permanent medical file—to employers, insurance providers, the government, or researchers.

Economic concerns must also be addressed. Who is going to pay for genetic tests, genetic counseling, research, and the collection of statistical information? Will genetic testing be a public service, and if so, who will regulate the service and outline quality standards? Perhaps, instead, genetic testing will be provided only by private, for-profit companies. In that case, will only the affluent (and only those in countries where the latest genetic research exists) have access to genetic testing and its benefits?

In contemplating the future of genetic testing, some observers are concerned about the prospect of shaping populations through genetic manipulation. Few people are opposed to the use of genetic tests to prevent and treat diseases. However, it may be possible someday to screen fetuses for non-disease-related characteristics, such as physical and behavioral traits. The ethics associated with this issue sometimes lead to highly charged debates. These concerns are currently under scrutiny by government committees in several countries. In general, recommendations from these panels include testing only when disease interventions are possible and barring insurance providers and employers from practicing genetic discrimination.

Alongside the controversial issues, the science behind genetic testing is progressing rapidly. It is likely that even more diseases will be prevented, diagnosed, and treated with the help of genetic testing. Genetic testing could reveal information that enables the modification of biochemical pathways or the tailoring of traditional therapies, such as pharmaceuticals, to a person's particular genetic problem.

—*Tamara Schuyler*

RELATED TOPICS
Cloning, DNA Fingerprinting, Gene Therapy, Genetic Engineering

BIBLIOGRAPHY AND FURTHER RESEARCH

BOOKS
Annas, George J., and Sherman Elias, eds. *Gene Mapping: Using Law and Ethics as Guides.* New York: Oxford University Press, 1992.
Hubbard, Ruth, and Elijah Wald. *Exploding the Gene Myth.* Boston: Beacon, 1993.
Kitcher, Philip. *The Lives to Come.* New York: Touchstone, 1996.
Nelkin, Dorothy, and Laurence Tancredi. *Dangerous Diagnostics: The Social Power of Biological Information.* New York: Basic Books, 1989.

PERIODICALS
"Bad Luck Insurance: Confidentiality in the Use of Personal Genetic Information" (editorial). *Nature,* September 20, 1994, p. 214.
Harper, Peter S. "Should We Test Children for 'Adult' Genetic Diseases?" *Lancet,* May 19, 1990, 1205.
Holtzman, Neil A. "Predictive Genetic Testing: From Basic Research to Clinical Practice." *Science,* October 24, 1997, 608.
Kodish, Eric. "Genetic Testing for Cancer Risk: How to Reconcile the Conflicts." *Journal of the American Medical Association,* January 21, 1998, 179.
Nowak, Rachel. "Genetic Testing Set for Takeoff." *Science,* July 22, 1994, 464.
Ponder, Bruce. "Genetic Testing for Cancer Risk." *Science,* November 7, 1997, 1050.
Raeburn, Sandy. "The Big, the Small and the Average." *New Scientist,* May 3, 1997, 50.
Stephenson, Joan. "As Discoveries Unfold, a New Urgency to Bring Genetic Literacy to Physicians." *Journal of the American Medical Association,* October 15, 1997, 1225.
Vines, Gail. "Gene Tests: The Parents' Dilemma." *New Scientist,* November 12, 1994, p. 40.

INTERNET RESOURCES
Benefits, Distress and Uncertainty Related to Genetic Testing
http://www.cybertowers.com/selfhelp/articles/chronic/gentest.html
Genetic Testing
http://www.lbl.gov/Education/ELSI/Frames/genetic-testing-f.html
Introduction to Genetics and Genetic Testing
http://www.kidshealth.org/parent/healthy/genetics.html
OncoLink@ASCO: Presidential Symposium on Genetic Testing
http://nisc8a.upenn.edu/conference/asoc96/sat/weber.html

GEOTHERMAL ENERGY

Geothermal energy is electrical or heat energy obtained from the Earth. The energy is created by taking hot water or steam from the Earth and using it to generate electrical energy in turbines or as direct heating with hot water in homes and buildings.

Geothermal energy is used in various places around the world, mainly where there are natural hot springs or thermal vents near Earth's surface. Electrical energy has been generated in substantial amounts from geothermal sources since the 1960s. Future innovations aim to increase the efficiency of lower-quality geothermal energy sources.

Scientific and Technological Description

The Earth is composed of three different layers: the *crust*, the *mantle*, and the *core*. We live on the crust, which varies from 30 to 130 kilometers (km) in depth. The mantle extends about 2900 km downward; depending on its temperature it can be solid rock or semi-liquid magma. The solid inner core and liquid outer core, with a combined diameter around 5000 km, comprise Earth's center. At certain points on the Earth's surface where the crust is thin, hot magma from the mantle can bubble up to the surface as volcanic lava. In other regions, we can tell when magma is not too far from the surface, because the heated water in hot springs or geysers causes them to give off telltale signs of water vapor. This water travels up through channels and cracks in semiporous rocks and escapes into the atmosphere.

In such places, this heat energy can be employed to produce geothermal energy, which can provide heat and electricity (see figure). Heat can be produced by piping hot water directly into radiators in homes and factories, or underneath roads to keep them permanently free of snow and ice. Electric energy can be produced by controlling the path of the hot water and steam and directing it to turbine generators.

There are three related ways in which electrical energy can be generated. The first is to use hot steam directly from the Earth. This hot water vapor, called *dry steam* because there are no water droplets in it, can be shunted directly to an electric turbine. It is a simple system where the pressure of the steam escaping from the Earth spins the wheels of a turbine. Unfortunately, there are fewer than 10 places on Earth where dry steam reservoirs exist.

Therefore, the second and more commonly used method is to take hot water and use it to drive turbines. When water is heated under the Earth's surface, it is under tremendous pressure and can remain in its liquid form well above 100°C, its normal boiling point. Water temperatures at Cerro Prieto, Mexico, and Wairakei, New Zealand, have been measured at over 300°C. When this superheated water is brought to the surface and released under atmospheric pressure, it instantly "flashes" into *wet steam*, a mixture of steam and liquid water. After the wet steam is separated from the water, and small rocks and dissolved minerals have been removed, it can be used to drive a turbine.

Finally, when there is no water, but hot subsurface rocks remain easily accessible, a third method is used to create energy. Two separate pipe systems are used to produce wet steam. The first pipe drives cold water underground under pressure where it percolates through the hot impermeable rocks and is heated. The wet steam produced

Electricity Generation from Geothermal Sources

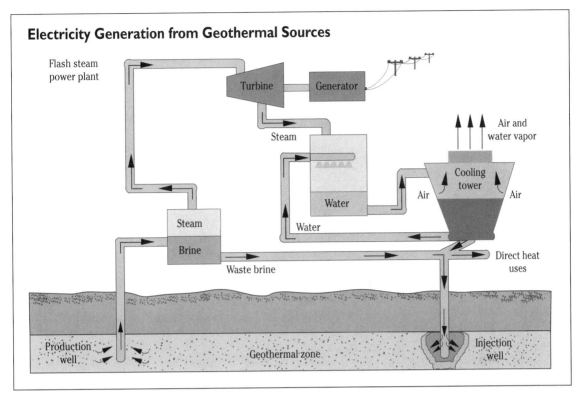

Electricity is generated using underground heat by pumping superheated water or steam from underground. Superheated water evaporates into steam once it reaches low atmospheric surface pressures. After being filtered to remove dissolved minerals and small rocks, this steam passes through a turbine. The pressure of the steam spins the turbine, and electricity is generated. The leftover water, called waste brine, can be used for heating systems. (Adapted with permission from U.S. Department of Energy.)

can escape only by traveling back up the other pipe, where it is trapped and used in the turbine.

Historical Development

The first recorded use of naturally heated water from the Earth was by the Romans, who created many resorts at hot springs for leisure purposes throughout Italy and Europe, ranging all the way to Bath, England. Many similar resorts exist worldwide today. The first people to use geothermal energy for heating were the Vikings. After landing at Iceland around the year 1000, they quickly noticed that the land there, which is solidified volcanic lava, was always warm. (Reykjavik, Iceland's capital, means "smoking bay.") Early Icelanders used the hot springs to wash clothes and themselves. They baked bread by placing their dough underground in a small oven. The first person to use geothermal energy to heat a house was Snorri Sturluson (1178–1241), a famous Icelandic historian and scholar. In addition to using the hot spring near his house to create a warm bath, he was the first to use a system of pipes to channel the water to heat his house. Unfortunately, his idea did not catch on for another 700 years.

By the 1930s Iceland was suffocating under a pall of black smoke produced by burning fuel oil. Engineers resurrected the Sturluson idea and were able to drill down thou-

sands of feet below Reykjavik and tap the energy of the pressurized hot water. Today, almost every home and building in Iceland is heated by geothermal systems. Once the hot water has made its first pass through home heating systems, it can be used in showers, baths, pools, and hot springs. The water is also used to support a vast system of greenhouses. Oil and gasoline are now used rarely and the air is much cleaner. Two U.S. cities that use similar heating systems are Boise, Idaho, and Klamath Falls, Oregon. The U.S. remains the world leader in the use of this technology to produce electricity

The first geothermal energy facility to produce electricity from dry steam was Larderello, a mountainous region south of Florence, Italy. In 1904, engineers there tapped a *fumarole* (a vent that continuously emits dry steam) and directed its steam to a small turbine. They were able to run five light bulbs simultaneously. The plant was expanded, destroyed during World War II, and rebuilt. Today the electricity produced there is used to run Italy's trains.

In 1847 the American explorer William Bell Elliott discovered steam pouring out of a canyon north of San Francisco near Cobb Mountain, an extinct volcano. He thought he had found the gates of hell, since the rising steam and sulfurous stench reminded him of the biblical depiction. He named the site The Geysers—a misnomer, since geysers spout at intervals, whereas this canyon continuously emitted steam and is properly called a fumarole. For many years The Geysers was used as a hot springs resort. In the early 1920s engineers were first able to generate electricity there on a small scale. Large-scale attempts failed, since not enough was known about geology to predict the best spot to dig. Most of the steam vents were too powerful to control (one vent dug in the 1950s continues to blow unchecked today) and minerals dissolved in the water quickly wore out and rusted the equipment of the time.

In the 1960s these problems were finally overcome by the Pacific Gas and Electric Co. That company's innovation was to spin the steam to centrifuge out any rocks and debris before it hit the turbine. Since then The Geysers geothermal field has grown to become the largest and most productive geothermal energy facility in the world, supplying a large percentage of the San Francisco Bay area's power. The government buildings in Santa Monica began receiving all their power from geothermal energy in 1998.

Uses, Effects, and Limitations

The two most important uses of geothermal energy are in heat and electricity generation. Heat can be obtained by piping the heated water and steam through metal pipes to wherever it is needed. Unfortunately, distance is a severe limiting factor, since the water cools rapidly once it reaches the Earth's surface. Once heat is transformed into electricity, however, it can be sent along power lines for many kilometers. Electricity can also be converted back into heat once it reaches its destination.

Finding geothermal energy sources can be either extremely simple or quite complex. Simple sources include locations near volcanoes, geysers, or fumaroles. Geologists also have an array of other techniques for finding hot spots near the Earth's surface. They can measure the Earth's temperature by drilling small sample wells and inserting thermometers on long wires that can travel miles down into the Earth. On average, the underground temperature increases about 17°C for every kilometer descended. An increase in this rate may indicate a hot spot. Geologists may also measure the Earth's electrical resistance, which is affected by water and temperature, and use *seismography*, which measures vibrations in the Earth; the data from these measurements can be used to create maps of the subsurface. Another way is to use pictures taken with infrared-sensitive cameras mounted on airplanes or satellites. Small differences in surface temperature can easily be detected this way. Also, radar can be used to determine underground faults and hollows that may indicate hot spots (see Remote Sensing).

The development of geothermal energy sources can be limited by high startup costs—initiating a geothermal well

Iceland has a long history of utilizing geothermal energy, the Earth's natural heat source (Photo Researchers/Bernhard Edmaiser).

can be very expensive. Drilling a steam well is more costly than drilling oil or natural gas wells, since special heat-resistant equipment must be used. But on the positive side, once a productive well has been tapped, it produces electrical power more cheaply than any other system. If a tapped well does in fact contain heated water, but not under enough pressure or at sufficient temperature to produce steam, it can still be used as a productive source of energy. The solution is to use the heat of the hot water to heat a liquid with a lower boiling point. One commonly used liquid is isobutane, which has a boiling point of -11.7°C. By passing the hot water alongside a pipe containing isobutane, the isobutane absorbs heat from the water and is converted into a hot gas that can drive the turbine. Another advantage gained from using this system is that isobutane will always be clean, unlike the briny water obtained from the ground. This system is used in Iceland, where the water is not quite hot enough to produce steam.

Issues and Debate

One problem with geothermal energy is its environmental impact. Wherever there are hot springs or fumaroles, the land surrounding them is often very beautiful, and no matter how hard designers may try, a power plant is difficult to integrate into any environment. Like coal, or nuclear plants, electricity-generating geothermal energy plants may require large cooling towers to vent steam. About 80 percent of steam obtained from a geothermal source is vented straight into the atmosphere. However, it is mostly water, unlike the smoke produced by a coal-burning plant, which contains sulfur, nitrogen, and carbon dioxide.

Various minerals can be dissolved in the water and steam. These include boron, which is toxic to plants, ammonia, which harms fish, and mercury, which harms people. There can also be smaller concentrations of other poisonous metals dissolved in the water. Once this water has been used for power generation, it is illegal to dispose of by pouring it into the nearest river or lake. The best solution to date seems to be to reinject the water back into the Earth, where it will be recirculated and reheated in the geothermal system.

Another issue concerns the physical effects that geothermal energy plants have on the areas in which they are located. Once a well has been tapped and the steam begins to rise, noise pollution and unpleasant odors may become a problem. As the steam rises, the ground may rumble and shake in the nearby vicinity, and the smell of rotten eggs may permeate the air as hydrogen sulfide (at concentrations that are relatively nontoxic) is released from the ground.

As the world seeks alternatives to fossil fuels such as oil and coal to use as energy sources, geothermal energy will continue to be investigated along with wind, solar, biomass energy, and hydroelectric energy. No one is sure how long geothermal energy sources will last. On average, the Earth's heat will continue to be available for tens of millions of years, but individual locations may become tapped out much sooner. Consequently, certain locations may have to undergo "rest periods" while the Earth's subsurface recharges itself with fresh magma.

—Philip Downey

RELATED TOPICS
Biomass Energy, Hydroelectric Power, Nuclear Energy, Remote Sensing Wind Energy

BIBLIOGRAPHY AND FURTHER RESEARCH

BOOKS
Jacobs, Linda. *Letting Off Steam: The Story of Geothermal Energy.* Minneapolis, Minn.: Carolrhoda Books, 1990.
Kiefer, Irene. *Underground Furnaces.* New York: William Morrow, 1976.
Lauber, Patricia. *Tapping Earth's Heat.* Champaign, Ill.: Garrard Publishing Company, 1978.
Yates, Madeleine. *Earth Power.* Nashville, Tenn.: Abingdon, 1980.

PERIODICALS
"Fighting Pollution and Cleaning Up Too." *Business Week,* January 19, 1998, 90.
Graham-Rowe, Duncan. "Dipping Down into the Earth's Core." *Guardian,* May 12, 1998, 8.
Morris, Jim. "The Brimstone Battles." *Houston Chronicle,* November 9, 1997, 4.
Sykes, Lisa. "The Power to Choose." *New Scientist,* September 6, 1997, 18.

INTERNET RESOURCES
Geothermal Education Office
    http://geothermal.marin.org/escrap.html
Sandia Labs' Geothermal Research Department
    http://www.sandia.gov/geothermal/
U.S. Department of Energy's Energy Efficiency and Renewable Energy Home Page
    http://www.eren.doe.gov/state_energy/technology_content.cfm?techid=5
U.S. Department of Energy's Geothermal Energy Technical Site
    http://geothermal.id.doe.gov/
U.S. Department of Energy's Geothermal Technologies Program
    http://www.eren.doe.gov/geothermal/techno.html

GLOBAL POSITIONING SYSTEM

The global positioning system (GPS) is a system of satellites in orbit high above the Earth, whose signals can be used by computers to generate accurate navigational information in latitude, longitude, and even altitude. GPS has revolutionized navigation in the last part of the 20th century. It is routinely used on board ships and aircraft, and automobile and hand-held GPS units are becoming increasingly common.

Scientific and Technological Description

GPS works by measuring a receiver's position relative to several satellites in orbit around the Earth. The position is measured in terms of the time delay taken for a radio signal to travel from each satellite to the receiver. Using computer technology, GPS generates a constantly updated "map" of all the satellite positions, allowing the user to know their exact position on or above the Earth.

The GPS system is based on a "constellation" of 24 satellites that orbit the Earth at an altitude of 17,700 kilometers (11,000 miles), so that they circle the Earth exactly twice each day. These Navstar satellites (developed by the U.S. Navy) are each 5 meters (m; 16 feet) long and carry a highly accurate atomic clock, an onboard computer, and a radio transmitter that broadcasts signals down to the Earth at two frequencies, 1575 and 1228 megahertz (MHz). These *carrier waves* are modulated like a frequency modulation (FM) radio signal in order to carry a wide variety of digital information, some of which forms a unique signature for the specific satellite. Power is provided to each satellite by solar panels that fold out once the satellite has been placed in orbit. The satellites are arranged so that at any point on Earth, between five and eight of them are above the horizon at any one time.

The other main element of GPS is the receiver unit. This can range from a hand-held computer the size of a mobile phone to a more complex unit linked to an autopilot system on an aircraft or ship. However, the important components of each receiver unit remain the same: a radio receiver, a clock, and a computer. A typical civilian GPS unit scans the 1575-MHz frequency until it detects a specific satellite. It then compares the time signal received from the satellite with the time according to the receiver's own clock. Because the speed of radio waves, like that of light, is limited to 300,000 kilometers per second (km/s; 186,000 miles per second), there will be a small but measurable time delay as the radio waves travel from satellite to receiver.

The receiver unit now consults a built-in "almanac" that gives accurate information on the orbit of each satellite. It is therefore able to calculate the satellite's exact position at the instant that it broadcast the signal. The time delay in the signal's arrival allows the distance between satellite and receiver

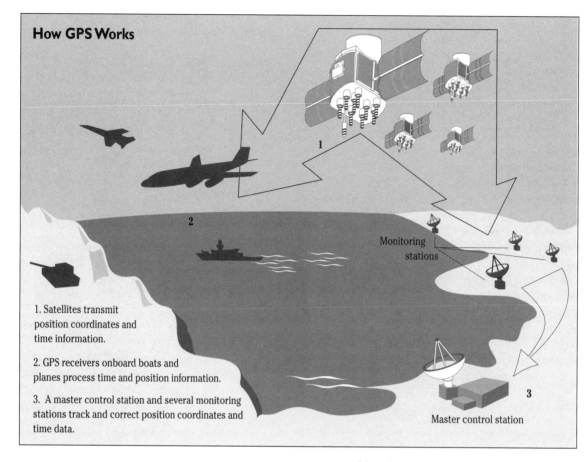

How GPS Works

Monitoring
stations

1. Satellites transmit
position coordinates and
time information.

2. GPS receivers onboard boats and
planes process time and position information.

3. A master control station and several monitoring
stations track and correct position coordinates and
time data.

Master control station

Adapted with permission from Salamander Books, Ltd.

to be found. This will be a minimum distance directly below the satellite, and circles can be drawn spreading out from this point corresponding to longer and longer time delays. With a single reading, the receiver can do no more than work out which of these circles it lies on, but because the satellites are themselves moving at speeds of about 3.5 km/s, it can do this with a high degree of accuracy. The unit then repeats this procedure for another available GPS signal, finding another circle on which the receiver must lie. The intersection of these two circles places the user at one of two possible points. A reading from a third satellite fixes the receiver's position at one of these points and enables the receiver to calculate its longitude and latitude. In practice, unless the receiver is at sea level or an independent altimeter is available, a fourth satellite is used, allowing the receiver to calculate its own altitude.

Historical Development

The principle of GPS predates the Space Age. During World War II, the allies developed LORAN (long-range navigation), a radio navigation system that measured an aircraft's position (but not altitude) by the difference in time signals received from two ground-based antennas. LORAN entered civilian use after the war and became the ancestor of an entire generation of instrument-based navigation systems.

LORAN systems were widely used for navigation on boats as well as aircraft.

Meanwhile, the first satellites were launched in the late 1950s. Sputnik 1, launched by the Soviet Union in 1957, carried a radio transmitter that sent a simple radio pulse back to Earth. Even in these early days, it was soon realized that if the satellite's orbit was accurately known, a ground-based receiver could calculate its own position. This could be done by measuring the *Doppler shift* in the satellite's signal. Doppler shift is the compression or extension of any wave's wavelength and frequency caused when the wave source and observer are moving toward or away from each other. This is the same phenomenon that changes the pitch of a siren when a police car passes the listener: As the car approaches, the pitch sounds higher, but as it speeds away, the pitch sounds lower (see DOPPLER RADAR). By applying the same principle to a satellite radio signal, the satellite's speed of approach or retreat can be found.

This principle was first put into practical use in the U.S. military's Transit satellite system, which became operational in 1962. It was intended for use by nuclear submarines and other ships, allowing them to find their position to within one-tenth of a mile and was soon made available for civilian navigators. At the same time, in the early

1960s, the Soviet Union was launching its own Tsyklon system, also based on the Doppler effect.

However, Doppler shift could provide only limited accuracy, and the principle of GPS was proposed by the U.S. Navy in 1973 as a logical next step in satellite-based navigation. The first Navstar launch took place in 1978, but it took 14 years, until 1992, for the full constellation of the current 24 functioning satellites to reach orbit. Over this period, the emphasis of the project had changed from the purely military, and GPS was also being made available to civilian users, subject to some limitations (see below). The system was declared to have reached fully operational status by the U.S. Air Force on July 17, 1994.

Uses, Effects, and Limitations

The most important application of GPS is, of course, in navigation. GPS is fully integrated into the world's air traffic and sea-lane control systems, and allows a navigator to find his or her position within 100 m horizontally and 156 m vertically. In addition, the Doppler shift of the GPS signals can be used to calculate a ship or aircraft's velocity and direction. Expensive, complex GPS units can be linked into autopilot systems on planes and boats, and smaller hand-held receivers can be used by individual travelers, expeditions, and rescue teams in hostile territory.

However, the civilian version of GPS (called the standard positioning system [SPS]) does not operate at the limits of the system's accuracy. A more accurate version, the *precise positioning system* (PPS), is capable of locating a receiver much more accurately, but its use is restricted to the military. This is done by allowing civilians access to only part of GPS's complex error-correcting techniques. The key to GPS's accuracy lies in the information carried by the radio signals that each satellite broadcasts. In orbit around the Earth, the Navstar satellites are subject to constant minute variations in their path, while the clocks in both satellites and receiver units are liable to drift. For the system to work properly, they have to be perfectly synchronized.

These problems are overcome by constant monitoring of the GPS satellites by a network of ground stations around the world, often called GPS's *control segment*. The information that these tracking stations gather is processed at Schriever Air Force Base, Colorado, and used to produce constantly updated almanacs for the satellite orbits, which are then relayed to the satellites themselves. From here, they are beamed down to the receiver units as part of the 1575-MHz radio signal. Time corrections to the satellite and receiver clocks are also relayed to users in this way, so that, as well as regular time signals, the GPS signals contain an entire library of information for the receiver, encrypted (sent in encoded form) in every 12.5 minutes' worth of signal.

Although the constantly updated information from the satellites allows the SPS system to operate on quite a small scale, the PPS system is accessible only to users with the equipment to receive the second, 1228 MHz, GPS signal. This signal is encrypted for security, and once decoded, enables the user to find a second time signal from the same satellite. A second time signal is useful because it enables the receiver unit to measure the amount of distortion and time lag introduced as the radio waves travel through the ionosphere (an electrically charged region of the upper atmosphere). The journey through the ionosphere always produces a small lag in the signal, and the SPS wavelength carries a "weather forecast" of conditions in the ionosphere that allows this lag to be estimated.

But the two waves used by PPS allow far greater accuracy. Because the ionosphere affects different frequencies by different amounts, the time lag between the two frequencies can be used to calculate the characteristics of the ionosphere at that moment, and thus to work out the time lag effects on each separate signal. The result is a system that can calculate position to within 22 m horizontally and 28 m vertically.

Even PPS is not the absolute limit of accuracy that GPS can provide, however. Surveyors and others equipped with suitable equipment can use GPS signals to find positions with accuracies within centimeters. This is done by comparing the phase (point in the wave cycle) of signals arriving at two receivers simultaneously. Signals arriving at both receivers are either recorded for later comparison or are compared instantly by a radio link between the receivers. Because the signal wavelength and frequency are being modulated constantly in order to carry GPS's timing and other information, it is possible to identify the same segment of signal as it arrives at each receiver and compare them.

The surveying receivers are studying the phase changes from just one satellite, so they do not gather the timing information that would enable them to locate themselves by traditional GPS. However, if one of the two receivers is set up at a position that is already accurately known, the time difference between a particular phase change arriving at each

GPS devices are primarily used for navigation on boats and on planes, but hikers and other individual travelers also use small hand-held GPS receivers to keep track of their location (Arizona Board of Tourism).

receiver shows their relative distances from the satellite and allows the moving receiver's position to be calculated. This method can be accurate to a fraction of a wavelength (one wavelength is approximately 19 centimeters for the SPS signal), but one important limitation is that the two receivers must be reasonably close (within about 30 km) to avoid signal variations caused by fluctuations in the ionosphere. As a general rule, the farther apart the receivers are, the longer the signals must be compared to get an accurate fix.

Issues and Debate

GPS is an interesting example of a military technology that has been made available for civilian use. However, unlike satellite surveillance, GPS is not really a "big brother" technology—GPS receivers are passive instruments that cannot reveal a user's location. However, GPS is still controlled by the U.S. military (while its nearest equivalent, Glonass, is controlled by Russia), and this means that it can be manipulated for political ends. A 1982 report recognized that GPS could be used by terrorists or enemy forces against U.S. interests (e.g., to accurately target the locations of U.S. military forces); that report led to the introduction of the dual SPS/PPS system. Civilian GPS was effectively "crippled" in order to limit its resolution to around 100 m, and PPS information was encrypted to prevent its use by unauthorized receivers.

Furthermore, the SPS system has a built-in "option to degrade." This means that in the interests of U.S. national security, the information provided in the SPS signal can be deliberately distorted, lowering the available resolution or even disabling the system completely for users without access to PPS. The system is only as good as the information the control centers supply to it.

In everyday life, GPS has been a technology with revolutionary benefits. It has simplified and improved the accuracy of navigation systems at sea and in the air, and undoubtedly prevented many accidents. As a personal navigational aid, it has brought accurate location finding to the individual user for the first time, and helped save many lives in inhospitable terrain. However, as with many other modern technologies, the greatest danger GPS poses could come from our overreliance on it. Fail-safe traditional systems should always be maintained. If they are not, an accidental failure of GPS could plunge the world's airports and shipping lanes into chaos.

—*Giles Sparrow*

RELATED TOPICS
Doppler Radar, Satellite Technology

BIBLIOGRAPHY AND FURTHER RESEARCH

BOOKS
French, Gregory T. *Understanding the GPS: An Introduction to the Global Positioning System.* Santa Fe, N.Mex.: OnWord Press, 1997.
Grubbs, Bruce. *Using GPS: Finding Your Way with the Global Positioning System.* Helena, Mont.: Falcon Publishing, 1999.
Hofmann-Wellenhof, B. *Global Positioning System: Theory and Practice,* 4th ed. New York: Springer-Verlag, 1997.

PERIODICALS
GPS World

INTERNET RESOURCES
"GPS Overview," by Peter H. Dana, University of Texas
http://wwwhost.cc.utexas.edu/ftp/pub/grg/gcraft/notes/gps/gps.html
Navstar GPS Program Office
http://www.laafb.af.mil/SMC/CZ/homepage/

HIGH-DEFINITION TELEVISION

High-definition television is a standard of television production and transmission in which viewers receive cinema-quality images and compact-disk-quality sound. What distinguishes HDTV from standard television is the number of lines it uses to form its pictures, its quality of sound, and the technologies it uses to make those improvements. True HDTV generally is thought to offer at least twice the number of lines of resolution as standard television, although formats with fewer lines still are considered high definition.

HDTV was conceived in the 1970s as a way to enhance standard television transmission. Since then, systems have been incorporated in Europe and Japan based on analog-signal HDTV, but these systems are likely to be replaced by digital-based HDTV, which industry observers see as the next television standard. Television stations in the United States have been required by the federal government to stop all analog-based transmissions by 2006, when they will send only digital signals. Broadcasters will need to decide how to use their allotted channel space by choosing either to send a single HDTV signal or to split a signal into several subchannels of lesser resolution.

Scientific and Technological Description

In standard television transmission, a television camera converts light into electronic signals that are combined with radio waves and transmitted as analog signals in bandwidths. When tuned to the correct bandwidth, a standard television set converts the signals, creates an image in two sweeps using an electron gun that scans every other line, and then fills in the gaps; the process is called *interlacing*. The resulting image, refreshed 60 times a second, offers 525 lines of resolution and a boxy proportion of width to height (or *aspect ratio*) of 4:3.

HDTV features far more lines of resolution than those of a standard television receiver, and true HDTV offers at least twice the number of horizontal and vertical lines. HDTV also features a wider, more movielike aspect ratio of 16:9 and high-fidelity audio. Current models of high-definition television sets offer images in one of two formats. The images are either interlaced as on a standard television set, but created by at least 1800 lines of resolution instead of 525 lines, or they are formed in one sweep at a resolution of 720 lines in a *progressive format*. Although many industry observers refer to the 720-line progressive format as HDTV,

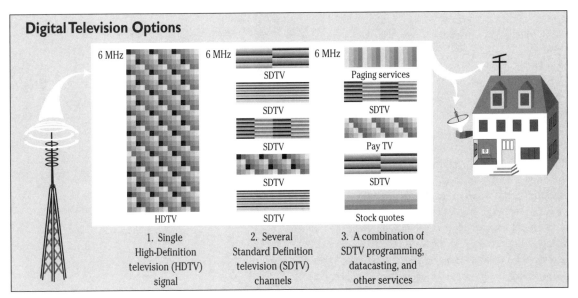

Digital Television Options

6 MHz

HDTV

6 MHz

SDTV

SDTV

SDTV

SDTV

SDTV

6 MHz

Paging services

SDTV

Pay TV

SDTV

Stock quotes

1. Single High-Definition television (HDTV) signal

2. Several Standard Definition television (SDTV) channels

3. A combination of SDTV programming, datacasting, and other services

Broadcasters will have many options as to how they will use the channel space they have been allotted for digital television. They may broadcast a single HDTV signal, which will take up the entire 6 MHz of the channel's bandwidth, or they may use the same bandwidth to broadcast a combination of SDTV and other services, such as stock quotes.

some contend that true HDTV must consist of at least 1050 lines. The highest definition conceived for HDTV using known technologies is 1800 lines in a progressive format, but this format is still in development.

Current HDTV technology can be separated into two groups: analog- and digital-based HDTV. *Analog-based HDTV* uses the same radio waves as standard television, but signals are compressed to carry more lines of resolution. Digital sound accompanies the waves, as does digital information that tells the television system how to recreate the picture. *Digital-based HDTV* is considered the successor to analog-based HDTV. Digital HDTV converts information into binary code, by dividing images and sound into small components and assigning a numerical code to those components, just as a computer does with data. The digitally recorded and broadcast signals are less vulnerable to interference than are analog signals. Because they receive information in the same way that computers do, digital-based television may also perform functions similar to those of a computer, such as sending and receiving e-mail.

Digital HDTV also allows broadcasters to use the bandwidths available to them in new ways as a result of a data compressor called *MPEG-2* (named for its developers, the Motion Picture Expert Group). Standard broadcast channels are not large enough to carry all the information contained in a digital HDTV broadcast. In fact, they are 50 times smaller than the bandwidth needed. To reduce the amount of data to be transmitted, MPEG-2 uses a video encoder that squeezes images by deleting redundancies within and between scenes and estimating the motion of objects. Pictures are analyzed and identical pixels are removed so that only unique features are sent. For example, the encoder would not compress hundreds of identical pixels to make a

blue ocean when it can recreate the ocean using only a few. The sound is encoded as HIGH-FIDELITY AUDIO.

An instrument called a *transport multiplexer* then combines the video and audio files and sends the information to a *vestigial sideband modulator*, a device that turns the data into a radio-frequency wave. The wave is amplified and broadcast as a channel. Exactly how and at what resolution the image appears on the screen depends on the format the broadcaster has chosen: whether progressively scanned or interlaced, and at how many lines per screen.

The Advanced Television Systems Committee, a group of broadcast specialists who set new technical standards for television and consult with the Federal Communications Commission (FCC), has accepted 18 formats as standard for digital television, and the three major networks have chosen to use two of the 18. CBS and NBC have decided to transmit signals at a resolution of 1080 lines in an interlaced format (1080I), and ABC has opted for 720 lines in a progressive format (720P).

Historical Development

HDTV resulted from a series of experiments conducted in the early 1970s in Japan by the state-owned company NHK. The Japanese researchers found that widening the aspect ratio to a format like that of a movie, as well as doubling the number of lines that made up the vertical and horizontal resolution, made images more realistic. Based on these results and emerging technologies, Japanese companies began building prototypes of HDTV and developed a standard called *MUSE* (multiple sub-nyquist encoding).

The Japanese version of HDTV was analog-based, just as traditional television was, but the NHK researchers had figured out how to compress the analog signals and send them via satellite directly to people's homes. This method allowed more information to be sent and higher-resolution pictures to be delivered. The drawback was that consumers would have to buy new and expensive high-definition television sets to take full advantage of the high-definition picture, or cheaper converters that cut off the sides of the wider HDTV images.

By 1986, NHK researchers had proposed specifications for an international standard of television transmission and production so that a program made in one country could easily be transmitted to another. U.S. broadcasters were eager initially, but European broadcasters balked. In an ongoing effort to integrate European markets, the European Commission, whose members were charged with initiating policies for the European Community (now the European Union), decided to develop its own HDTV system based on satellite-to-home transmissions. Sixty companies and research groups joined together to produce a transmission standard called *MAC*, (multiplexed analog components). Like the Japanese MUSE, the new European standard was analog-based.

U.S. politicians and corporations remained interested in HDTV, but the push to switch from analog lost momentum after the European decision to create MAC. U.S. broadcasters also became wary of the inevitable lead that Japanese companies would have in producing and selling technologies for HDTV. Then, in 1990, a U.S. company discovered a way to send images using digital rather than analog technology, and U.S. broadcasters considered leapfrogging over the analog systems entirely and developing a digital-based standard.

An FCC advisory committee called for researchers to propose new standards for the U.S. television industry. By 1992, the committee had narrowed its search to five prototypes, but the committee members agreed that none seemed clearly superior. They called for the groups that submitted the prototypes to work together. The resulting collective, formed in 1993, became known as the Grand Alliance. It was comprised of Zenith Electronic Corp., General Instrument Corp., AT&T Corp., the Massachusetts Institute of Technology, and a consortium of Thomson Consumer Electronics, Philips Consumer Electronics, and the David Sarnoff Research Center. In 1996, the federal government approved a standard for HDTV that allowed both progressive and interlaced formats, to encourage flexibility in the development of new technologies.

The FCC concluded its proceedings on digital television in April 1997. In an effort to make the transition to digital television more appealing to consumers, commission members encouraged broadcasters to package a range of digital products, not only HDTV. The commission suggested that services such as data transfer, subscription video, and interactive materials be offered. Under a congressional order, the FCC allocated one extra channel for digital transmission to each of the nation's 1500 broadcasters free of charge. The channel could be used to transmit either HDTV or an

A Philips high-definition television—a TV designed to receive and display HDTV programs—was on display at the International Consumer Electronics Show in Las Vegas in January 1999 (AP/Eric Draper).

enhanced version of standard television (SDTV) that leaves room in the bandwidth, or airspace, for other digital services.

The FCC and the nation's broadcasters have agreed on a timetable to bring digital television into the marketplace over several years, and some stations already have begun transmitting a digital signal. In the fall of 1998, 42 television stations in the top 10 U.S. markets began sending high-definition digitized versions of their usual analog-based programs, but only consumers with high-definition or standard television sets with special converters received the signals. Broadcasters may decide what form of digital television they will transmit, and whether or not they will transmit HDTV programming, as they make the conversion to digital television.

The FCC has mandated that by 2003, all TV stations must be converted to digital technologies, whether they choose true HDTV or a lesser format. In 2006, broadcasters are scheduled to stop sending analog signals, meaning that only consumers with sets capable of picking up digital signals will be able to watch free television. The FCC did allow a loophole for consumers and broadcasters; analog signals will be permitted to continue if fewer than 85 percent of viewers in a television market cannot receive digital signals.

Uses, Effects, and Limitations

HDTV was developed primarily as a way to improve the quality of television images and sound, but its uses have evolved to include options such as multicasting (using one channel to broadcast several different types of digital information), datacasting (broadcasting data, such as stock quotes), and computer interfacing. A single high-definition digital signal can take up the entire bandwidth that is allotted to one channel, or the signal can be split into a number of smaller subchannels. This ability has created a potential for multiple uses of high-definition digital television sets in conjunction with computer technology.

Television programs in the United States are transmitted at a bandwidth of 6 megahertz (MHz), and current technology allows for 19.4 million bits of information to be transmitted per second (19.4 Mbps). With digital transmissions, broadcasters can allocate those bits in many ways. Fast-moving programs, such as sports, use more bits; news programs, movies, and sitcoms use fewer. Depending on the television schedule, broadcasters can send a single channel of HDTV, which takes up the entire 6 MHz of bandwidth, or multiple programs at lower resolutions that take up less bandwidth space. The ability to split the channel means that broadcasters may offer, and viewers may choose, up to six shows at any one time on one station.

Because digital television sets use the same binary language as computers, they can be interfaced with computer systems. In fact, when the FCC's Advisory Committee on Advanced Television Services asked industry groups for new television standards, it specified that the digital transmission be able to integrate with known computer technologies. Leftover space in the broadcasting bandwidth can be reserved for datacasting, e-mail, paging, Internet downloading, and other options.

A transition to HDTV would also lead to new and expensive production equipment. Most traditional stations use television cameras that convert light into analog waves, which are videotaped or filmed. The analog tape is converted into a digital format so that it can be edited; then it is reconverted into analog waves for transmission. All-digital television requires digital cameras and updated editing equipment. Even sets will need to be changed because the wider aspect ratio of HDTV means that a wider picture is transmitted. Sets built to be shown on a boxy screen might bring props or stagehands into view. Experts estimate that each television station will spend between $8 million and $12 million to buy new transmission systems, cameras, and studio equipment.

Viewers will need to make their own purchases and adjustments to accommodate HDTV. Consumers may receive both analog and digital signals by buying a high-definition television set with a built-in receiver that converts the digital signal, or they may buy a less-expensive converter that attaches to their regular analog set. The set-top approach is attractive to consumers who anticipate upgrading their systems without replacing the entire set.

Issues and Debate

HDTV has generated much excitement among broadcasters and technophiles, but it has also created heated debate about its programming, cost, and usefulness. No technology can improve the content of what it carries. Multicasting means an exponential increase in the potential number of programs on the air, yet money to pay for the programs' production remains limited. Critics have wondered if HDTV's price tag is worth an increase in programming in which quality suffers because of the quantity of shows being produced.

Furthermore, some consumer groups have asserted that broadcasters should be required to dedicate time on subchannels to educational and children's programming in exchange for having received free use of channels for digital transmission. Usually, a television station pays fees for the use of channels, but Congress gave broadcasters the channels, a grant with an estimated value of $70 billion. Critics have said that broadcasters should therefore be required to serve the public with public interest programming.

To address these issues and others, the Advisory Committee on Public Interest Obligations of Digital Television Broadcasters, headed by Vice President Gore and also known as the Gore Commission, was convened. The commission's purpose was to determine what requirements should be imposed on broadcasters in exchange for the free channels. In 1998, commission members were widely criticized for recommending that broadcasters commit voluntarily to such programming. In the same report, the commission members asked that stations voluntarily devote five minutes each night to candidate-centered discussions during the last 30 days of national electoral campaigns. Some had hoped to make free air time available to politicians as a means of campaign-finance reform.

The controversy over HDTV goes beyond programs to include their transmission. Cable companies provide television programs to two-thirds of U.S. homes, yet they have hesitated to embrace HDTV. The FCC requires the companies to transmit local broadcasters' analog signals, and cable company representatives have complained that bandwidths are not large enough for them to carry their own as well as local programming via both analog and digital signals. The FCC may eventually require cable companies to carry the signals.

Digital signals are useless if viewers cannot afford the television sets to receive them, and perhaps what has been seen as the biggest obstacle to HDTV for consumers is its prohibitive price. In 1993, analog high-definition television sets sold in Japan for $30,000. In the United States, current digital models cost between $5000 and $10,000. Industry watchers predict that those prices will drop, but consumers are used to purchasing television sets for an average of $300. Some may find HDTV's clearer picture worth the price when watching a high-action sports game or a nature show, but others may feel that the extra money is wasted on soap operas, sitcoms, and cartoons. Consumers may also resist purchasing set-top converters for their analog sets, feeling that there is little point in upgrading sets that will soon be obsolete.

—Christina Roache

RELATED TOPICS
Flat-Panel Displays, High-Fidelity Audio, Web TV

BIBLIOGRAPHY AND FURTHER RESEARCH

BOOKS
Brinkley, Joel. *Defining Vision: The Battle for the Future of Television.* New York: Harcourt Brace & Company, 1997.

Casabianca, Lou. *The New TV: A Comprehensive Survey of High Definition Television*. Westport, Conn.: Mackler, 1992.

Dupagne, Michel, and Peter B. Seel, eds. *High Definition Television: A Global Perspective*. Ames, Iowa: Iowa State University Press, 1998.

Prentiss, Stan. *HDTV: High Definition Television*. Blue Ridge Summit, Pa.: TAB Books, 1990.

PERIODICALS AND REPORTS

Aucoin, Don. "Panel Urges Free Air Time." *The Boston Globe*, November 10, 1998, C8.

Booth, Stephen A. "Digital TV Turns On." *Popular Science*, April 1998, 76.

Booth, Stephen A. "HDTV from Space." *Popular Science*, April 1998, 41.

Corcoran, Elizabeth. "Picture Perfect: Europe Has Bet Heavily on HDTV." *Scientific American*, February 1992, 94.

Day, Rebecca. "Digital TV Is Finally Here." *Popular Mechanics*, April 1999, 86.

Faber, Scott. "An End to Snow (HDTV)." *Discover*, January 1994, 97.

Fleishmann, Mark. "All That HDTV Jazz." *Popular Science,* January 1993, 44.

Fleishmann, Mark. "And Unveiling Digital HDTV." *Popular Science,* June 1992, 69.

"High Definition Television." Hearings Before the Subcommittee on Telecommunications and Finance of the Committee on Energy and Commerce, House of Representatives, One Hundredth Congress, October 8, 1987, June 23, and September 7, 1988. Serial No. 100-188. Washington, D.C.: U.S. Government Printing Office, 1989.

Kirschner, Suzanne Kantra. "HDTV Transmissions." *Popular Science,* April 1998, 97.

Norman, Colin. "HDTV: The Technology du Jour." *Science,* May 19, 1989, 761.

INTERNET RESOURCES

"The Great HDTV Swindle"
http://www.wired.com/wired/archive/5.021/netizen.html?pg=2:topic=
HDTV Newsletter
http://www.web-star.com/hdtv/hdtvnews1.html
"HDTV: What's Wrong with This Picture?"
http://www.wired.com/wired/archive/1.01/negroponte.html
"Science and the Citizen: Digital Dilemma: March 1998"
http://www.sciam.com/1998/0398issue/0398infocus.html

HIGH-FIDELITY AUDIO

High-fidelity audio refers to many techniques of sound reproduction for home stereo systems and movie theaters. These include noise reduction systems and audio formats for broadcasting, such as THX and Dolby Surround Sound. Audio engineers are always striving to create perfect sound reproduction. They are plagued by several recurring problems: How is the sound stored? How is it played back? How can extraneous noise be eliminated? How can the spatial effects of sound be reproduced through a set of speakers? Although engineers have come up with many solutions, they have never achieved perfection.

Today, the predominant home audio standard is known as 5.1 (see figure). Audio and video recordings, and home and theater equipment, have all been optimized to play recordings in this format, which accurately replicates the movie theater experience. This format uses a collection of six speakers to reproduce sound. Future formats will increase recording qual-ity by making more precise recordings that better reproduce the setting in which they were made.

Scientific and Technological Description

There are many stages involved in making high-fidelity audio recordings. *High-fidelity audio* is a term that encompasses many different sound recording and playback technologies. At each stage of recording and reproduction, many different pieces of equipment must work together to produce good-quality sound. High-fidelity audio systems attempt to recreate the exact sound of the original performance and replicate the acoustics of that setting.

The technology needed to transmit sound electrically through wires is quite simple. A *microphone* consists of a mouthpiece that collects sound and a diaphragm that vibrates when hit by sound waves, connected to a magnet wrapped with wire. Vibrations in the diaphragm are transmitted to the magnet, causing oscillations in its field. These oscillations affect the electrical current passing through a wire wrapped around the magnet. The changes in current are a direct representation of the sound. At the other end of a wire is a *speaker,* which is essentially a microphone in reverse and can emit sound. At this point, one has the makings of a simple telephone, which works on the same principle.

In between the microphone and speaker can be the storage medium. (Without a storage medium, one can use an amplifier to create a public-address system or miles of wires to create a telephone network.) This medium can be anything from the grooves of a vinyl record, to the magnetic fields of a cassette tape, to the microscopic pits of a compact disk (see CD-ROMS; DIGITAL VIDEO TECHNOLOGY) or the data of a computer hard drive.

In 1965, Ray Dolby, a U.S. physicist, formed a company to deal with the problem of reducing the *noise* in audio signals. Noise, in this sense, does not refer to background noises such as clapping and coughing, but to the hisses and pops that come from random electrical fluctuations in wires, microphones, and storage media. Any high-fidelity sound system should eliminate all these noises, since they weren't present in the original performance. The measure of the amount of noise is called the *signal-to-noise* (S/N) *ratio:* A high ratio indicates less noise. Noise is present whether or not other sounds are being recorded; therefore, if the frequency (or frequencies) of the noise can be identified during a silent passage, it can be removed from the entire recording by electronic filters. The problem facing sound engineers is to do this without removing matching frequencies in the recorded sounds, which would adversely affect the sound quality. Dolby's first system of noise-reducing filters for home audio cassettes was released in 1970. Since then the Dolby company and others have turned out a variety of noise reduction systems that have been used by the music and movie industries.

When a movie is made, up to 100 separate audio channels can be used to record voices, dialogue, car chases, background noise, and sound effects. These scores of chan-

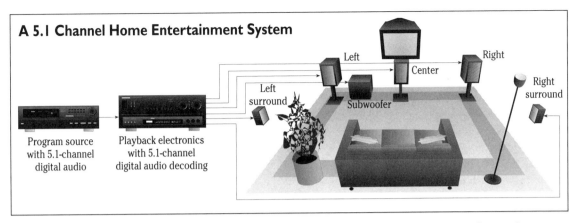

A 5.1 Channel Home Entertainment System

Program source with 5.1-channel digital audio

Playback electronics with 5.1-channel digital audio decoding

Left

Center

Right

Left surround

Subwoofer

Right surround

This diagram shows the recommended speaker placement and setup for a 5.1-channel home entertainment system. Four speakers are placed near the front of the room, representing the left, right, and center channels, as well as the subwoofer; two surround speakers are, placed to the left and right to complement the front speakers. All the speakers are controlled by an amplifier with circuits especially designed to sort out the six different speakers. (Original illustration courtesy of Dolby Laboratories Inc.)

nels all have to be "mixed down" to four or so channels for the film that will be used in theaters. Compressing music and sound in this fashion always leads to loss of sound quality and increased noise. Compounding the problem is the very narrow space on the film where the soundtrack must be placed. No matter how hard sound engineers may try, some details are lost when a soundtrack is replayed at high volume in a large theater. The resulting sounds have a flat feel. The four channels of a film are divided into the two front channels, which contain most of the dialogue and music and are broadcast from speakers at the front of the theater, and the back channels, which contain secondary sounds such as background noises and also serve to create three-dimensional effects, such as a car or airplane passing from right to left on the screen.

Historical Development

Sound reproduction began in 1876 when Alexander Graham Bell invented the telephone. His invention converted sound waves into electrical signals that could be sent through wires and converted back into sound at the other end. He patented his invention later that same year. Thomas Edison set about improving it, which he did by changing the microphone. In 1877, Edison invented what he called his personal favorite of his many inventions, the *phonograph*. He recorded sounds by connecting the microphone to a needle. When sound was collected, the needle vibrated in time with the sound waves and scratched tinfoil that was wrapped around a cylinder. Afterward, the needle would accurately reproduce sound when it followed the tracks in the foil. Since then, recorded music has been a constant of 20th-century life, and the development of high-fidelity audio technologies has been the focus of much effort.

In 1887, the German-born American inventor Emile Berliner replaced the cylinder with a flat platter that eventually led to the vinyl record, a format that is still used

today. The vinyl record survived for over 100 years before being replaced in consumer popularity in the 1980s by the compact disk (CD), a digital storage medium (see CD-ROMS). Digital recording converts sounds into binary code (a series of 1s and 0s), instead of "representing" it in the grooves of a record or the magnetic fields of tape (analog recordings). Other formats, such as magnetic tape, in the form of eight-track, reel-to-reel, cassette, and minicassette players, have all had their popular moments in this century. Some companies have introduced other digital formats to compete with the compact disk, including DAT (digital audiotape), made by various companies; Sony's Minidisc format; and MP3 audio file format, used by computers, which has quickly grown in popularity since being introduced in 1998.

Although records were the standard format for home audio for most of this century, movie theaters faced a problem: The audio track had to be placed onto the filmstrip itself to allow perfect coordination with the video. Getting four audio tracks placed beside the video was a challenge. Four (or more) were required: two for the left and right channels of the front speakers, and two more for the side and rear speakers. The solution was to place a magnetic strip similar to a cassette tape on the edge of the film. In the 1970s, when movie studios wanted to improve this method, there was no room for more audio channels. Dolby Laboratories provided a solution by switching from magnetic tape to an optical system. Instead of consisting of magnetic signals on the edge of a film, an optical soundtrack is placed in between the sprocket holes that guide the film through the projector. This innovation marked the introduction of stereo sound to movie theaters in 1975. One film could now hold an analog and an optical soundtrack, and could be shown with either kind of projector. Dolby went on to introduce digital soundtracks for films. This film standard, called *Dolby Stereo Digital*, was introduced in 1992 in the movie *Batman Returns* and has been used since in more than 1000 movies.

High-fidelity audio has taken great strides forward in the last two decades, thanks to the advent of home video and the introduction of digital technology. When movies became available for home viewing in the early 1980s, consumers wanted sound quality as good as a theater's. Television soundtracks had always been of fairly low quality and it had never been a problem before, but now consumers demanded something better. The two audio channels of the typical television system proved inadequate for reproducing the carefully crafted sounds of a movie. Extra channels and speakers had to be added to home systems. The two companies that led the revolution were Ray Dolby's Dolby Laboratories and George Lucas's Lucasfilm THX. Their primary achievements were in applying digital technologies to process sound and sort it through six speakers, which when properly arranged would accurately reproduce the sounds heard in a movie theater.

Dolby Laboratories introduced *Dolby Surround,* an attempt to replicate the four channels of a movie theater, for home systems in 1982. This was a four-channel system composed of four speakers: two placed at the front of the room beside the television set, and two placed beside or behind viewers. The Dolby Surround system made it possible for videocassettes to be produced with four channels instead of two. High-fidelity took another step forward in the 1980s when George Lucas, the creative force behind the *Star Wars* movies, decided that he wanted better-quality sound in theaters and at home. He formed a company, Lucasfilm THX, and developed the THX sound system. This system is a set of standards for recording and playing back sound. The generic name for a home system with this setup is *5.1,* which includes five speakers plus one bassbin, or subwoofer, that reproduces only low frequencies. The two main speakers and the bassbin are placed at the front of the room; the other three are placed behind or beside viewers to enhance certain sounds, such as a passing car or a person speaking from the left or right of the screen. Variations of this system are scaled up for movie theaters, which can contain more than 20 speakers. THX sound systems, for home or theater, must earn a certification from the company before being advertised as such. Dolby and Lucasfilm THX have agreed on the 5.1 standard, which is now used in both homes and theaters.

Uses, Effects, and Limitations

The range of sounds that can be detected by the human ear is usually stated as 20 to 20,000 hertz (Hz; Hertz is a measure of *frequency*, the number of waves per second). The average human being actually hears frequencies between approximately 35 and 14,000 Hz. Any good sound system should try to capture and reproduce all of these frequencies. Faithful reproduction depends on the quality of the microphone, the recording setting, the storage medium for the sound data, and the amplifiers and speakers that eventually generate the final sound.

All recordings are limited by how fast the equipment used records, transmits, and generates signals. All digital recordings have a *sampling rate,* which is the speed at which sound is split and recorded into separate *bins*. Compact disks are recorded at 44.1 kilohertz (kHz), which means that 44,100 recordings are being made each second. (The sampling rate is not related to the sound frequencies mentioned above, even though they use the same units.) At this rate one can capture any audible frequency. Regardless of the current level of accuracy, increasing the 44.1-kHz sampling rate should lead to even better sound recordings. Although no industry-wide standard has yet been agreed on, future recording formats will push the sampling rate to 96 kHz. Two currently vying for supremacy are DVD-Audio, an offshoot of the digital video disk format (DVD; see DIGITAL VIDEO TECHNOLOGY), and SACD (super audio compact disk), both of which will have a sampling rate of 96 kHz.

One problem that sound recordings and speakers may never overcome is the problem of reproducing the three-dimensional aspects of recording. Consider attending a concert by a symphony orchestra of 100 members spread across a stage: How can a pair of home speakers attempt to preserve and reproduce the three-dimensional experience? Sound engineers have many ways of compensating in the recording process, and can accurately reproduce the left and right aspects of a recording, but no one has been able to make a recording or design a set of speakers that can accurately reproduce the depth or height of a performance.

Issues and Debate

The development of high-fidelity audio technologies has involved aesthetic, technical, and economic concerns. One debate that will never end is the ongoing controversy over the superiority of analog or digital sound. Both sides have their adherents. Supporters of analog sound claim that digital sound gives a cold, impersonal flavor to recordings. They say this because in digital sound the sampling rate splits "continuous" sounds into evenly spaced groupings or bins, whereas analog recording takes place continuously and is therefore more exact. Digital recordings are always recordings in *steps*. Analog supporters maintain that the steps make those recordings sound cold, and that analog recordings give more warmth to the sound. Digital aficionados claim that a high-enough digital sampling rate can overcome anything that can be detected by the human ear.

Digital recordings are also longer lasting, an advantage that ultimately translates into an economic benefit. One problem with analog recordings is that they eventually wear out, because they depend on physical contact between the sound storage medium and the sensor. One example is the record player needle, which wears out both itself and any record it plays. Cassette tapes wear out with repeated playing as they pass over tape heads. Compact disks (CDs) bypass this problem by using reflected light instead of physical contact. The laser in a CD player shines a beam of light onto the CD surface; the light reflects off microscopic pits on the CD surface, and the pattern of the pits is converted into sound. CDs don't wear out with repeated playing, although a scratch on the plastic coating can stop the player from reading the pits.

Now that compact disks can be recorded at home, and music files can easily be traded in MP3 format over the Internet, many in the music industry are worried that these technologies may put them out of business. If consumers can now make perfect copies of recordings at home, there will be little need for them to buy their own personal copies. The music industry would prefer a method of buying and selling audio files that can be downloaded and played only by the person who has paid for them. To promote this idea, the Recording Industry Association of America (RIAA) has introduced the Secure Digital Music Initiative, a consortium of computer and music companies that plans to settle on and produce a high-quality secure digital format. The group has been slow to advance a feasible system, however, and some companies have broken away and announced plans to distribute music over the Internet in MP3 format. That format, which sounds almost as good as a CD and is easy to use, may well become consumers' preferred form of music. Several companies, such as Diamond Multimedia, have introduced portable Walkman-style players capable of storing and playing one hour of MP3 files. Other companies may soon join in supporting the MP3 format.

—*Philip Downey*

RELATED TOPICS
CD-ROMs, Digital Video Technology

BIBLIOGRAPHY AND FURTHER RESEARCH
BOOKS
Borwick, John, ed. *Sound Recording Practice*. Oxford: Oxford University Press, 1994.
Zaza, Tony. *Audio Design*. Upper Saddle River, N.J.: Prentice Hall, 1991.
PERIODICALS
Brockhouse, Gordon. "DVD-Audio vs. SACD." *Stereo Review*, January 1999, 26.
Ranada, David. "Educational Truancy." *Stereo Review*, January 1999, 36.
"What to Buy Now." *Video*, October 1998, 30.
INTERNET RESOURCES
Lucasfilm THX Ltd.
    http://www.thx.com
Dolby Laboratories, Inc.
    http://www.dolby.com
Home Theater Systems
    http://www.dtsonline.com/consumer/index.html

HIGH-SPEED AND MAGLEV RAILWAYS

High-speed trains are passenger trains that use fairly conventional technology to achieve average speeds of around 180 miles per hour (mph). Maglev (magnetic levitation) train technology exploits the power of magnets to raise trains above their tracks and propel them along at speeds of up to 271 (mph). Both designs are fairly recent attempts at vastly improving the speed with which large numbers of passengers can be moved around a rail network, a process that began in the early 1960s after it was assumed that the limits of conventional train technology

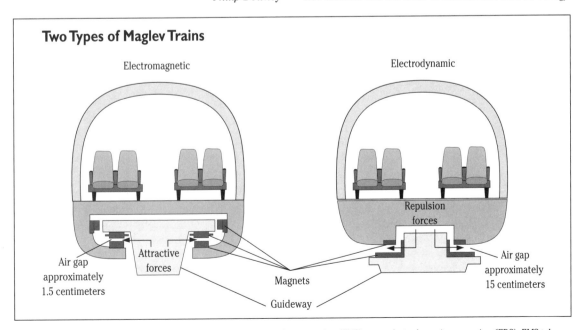

Two Types of Maglev Trains

Electromagnetic

Electrodynamic

Repulsion forces

Attractive forces

Air gap approximately 1.5 centimeters

Air gap approximately 15 centimeters

Magnets

Guideway

Maglev trains work in one of two ways—by relying on electromagnetic suspension (EMS) or on electrodynamic suspension (EDS). EMS takes advantage of the attractive force between magnets on the underside of the train and on the underside of the guideway. EDS uses the repulsive force of superconducting magnets to force the train to hover above the guideway. (Adapted with permission from George Retseck.)

had been reached. To this end various other research projects were undertaken, which resulted in vehicles capable of speeds far in excess of anything that had come before.

Unfortunately, however, these new technologies, impressive as they were, turned out to be too expensive or unreliable to be worthy of serious consideration. The research emphasis was moved away from pure high-speed travel and toward developing a fast but cost-effective railway system. To this end, the Maglev principle has largely been abandoned in favor of more conventional high-speed trains, such as France's Train à Grande Vitesse (TGV) and Japan's bullet trains, which grew out of existing technologies.

Scientific and Technological Description

Conventional trains are powered by electric motors that receive between 1500 and 3000 volts of power from wires above the track or rails on the ground. This electrical power is converted into mechanical power by a standard electric motor and used to drive the train. In this respect, the conventional train closely resembles its high-speed cousin. But high-speed trains must also incorporate crucial weight-saving measures, reinforced track, and additional raw horsepower to achieve their performance speeds of 180 (mph) and higher.

One type of high-speed train is the TGV (Train à Grande Vitesse), developed by the French during the early 1970s. The TGV is not simply a high-speed train, but rather, an integrated upgrade of existing technologies, which encompasses changes in train, track, and even signaling design. Aside from its eye-catching aerodynamic design, which brought it relatively minor improvements in speed, the most obvious feature of the TGV is the way in which its axles are distributed. Rather than having two or more axles, TGV carriages have a single axle at the rear, just like a trailer, and so must be supported at the other end by being semipermanently attached to the carriage in front. By designing the train this way, it was possible to gain enormous savings in weight, which are reflected in the speeds the trains can achieve.

The TGV has a locomotive, or power unit, at each end of the train. These draw power from electric cables suspended well above the track. The forward power unit receives electrical power directly from the overhead cable by means of a device called a *pantograph*. The rear power unit draws power in the same way, but this is then sent on by means of a cable running along the top of the carriages to the forward power unit, where it can be used as and when it is needed. This ensures that there is a continuous supply of power to the forward drive locomotive.

Because these trains regularly travel at speeds in excess of 170 (mph), conventional trackside signals are too difficult to read, so signaling is incorporated into the driver's cab. Signaling information is fed along the track as an electrical signal and picked up by antennas placed under the train. The signaling information is then processed by an onboard computer and displayed on a

A Magnetic Levitation (Maglev) train, operated by Japan Air Lines, is shown above. Maglev trains use the attractive and repulsive properties of magnets to propel train carriages along guideways (Photo Researchers/Japan Air Lines).

monitor in the cab. Aside from these modifications to existing technology, there is little to distinguish the TGV trains, and the similar Japanese bullet trains, from those of more traditional design.

Maglev train designs employ an entirely different technology from that of standard high-speed trains to achieve a similar but almost frictionless high-speed ride. By exploiting the repellent force of magnets of the same polarity, it is possible to build a vehicle that will levitate to a position just above the rail. By fitting the underside of a train with high-power magnets and placing magnetic coils of the same polarity into the track or guideway, it is possible to generate sufficient repellent magnetic force to raise the train. This type of maglev train relies on electrodynamic suspension—the repulsive force of magnets. Another type relies on electromagnetic suspension—taking advantage of the attractive force between magnets (see figure). These trains are capable of moving at speeds of up to 271 (mph). The magnets can also be manipulated to function as an efficient braking system.

Historical Development

The 1960s saw the first attempts at building new, superfast passenger trains capable of transporting people between cities at remarkable speeds. Experimental evidence seemed to indicate that the maximum possible speed achievable using conventional technology was unlikely to be much in excess of 200 (mph), so engineers began to look at other ways of building high-speed trains. Since then various attempts have been made at achieving high-speed rail travel. The first of these was the *turbo train,* developed by the United Aircraft Corporation, which was fitted with aircraft-style gas-turbine jet engines. These engines worked by drawing air in from the surrounding atmosphere and compressing it before feeding it into a combustion chamber, where it was mixed with fuel and ignited. The resulting jet of hot gases was then used to drive the train.

Because the train received its power from two engines incorporated into passenger cars at each end of the train, the turbo train, unlike conventional designs, did not have a separate locomotive. In a further attempt to improve the power-to-weight ratio of the vehicle, the turbo train was constructed almost entirely of aluminum. A special pendulous suspension system was also adopted to take full advantage of the additional speed offered by the gas turbine jet engines. By suspending all the carriages of the train above their common center of gravity and allowing them to swing freely within a purpose-built frame, it became possible for the entire train to bank on bends, achieving cornering speeds between 30 and 40 percent higher than those of conventional trains.

During the 1970s, this train design was used for a time between Boston and New York and also on the Montreal–Toronto line in Canada. Rapidly rising fuel prices during the oil crisis of 1973, however, soon caused the turbo train to be abandoned due to excessive operating costs. The true legacy of the turbo train lies in its suspension system, which is still in use on high-speed trains today.

While turbo trains were being tested in the United States, the French were working on the Aerotrain. Up to this point, high-speed trains had been based on the conventional flanged wheel design, in which the locomotive and carriages are all fitted with wheels that have internal flanges that sit on the inside of the rail, ensuring that the train remains on the track. The Aerotrain, however, floated like a hovercraft on a cushion of air that lay between it and a central guiding rail, which replaced the more conventional rail track. Air was pumped under the train using fans and trapped until sufficient pressure built up and the only way for the air to escape was by pushing its way out from under the vehicle. This raised the train off the track. As long as there was a continuous stream of air being pumped under the vehicle, the train would remain sitting about 4 inches above the central rail. Surprisingly little force was needed to propel the Aerotrain forward. By fitting it with a fan jet for lateral propulsion, it became possible to achieve a top speed of 235 (mph). Unfortunately, like the turbo train, the project was canceled in 1974 in the face of rising fuel costs and protests from environmental groups concerned about the amount of noise generated by the train's power unit. Engineers began to look at alternatives to the fuel-intensive air-cushion train.

Experiments in the early 1970s had shown that magnetic levitation could be used to raise a train off its track and soon efforts at alternative high-speed train building concentrated on turning the Maglev principle into a practical passenger train. Both Germany and Japan invested heavily in this technology. By 1979, the Japanese had built a small experimental Maglev train that achieved a speed of 321 (mph) in tests. (The full-sized, much heavier version of this train was later to reach a top speed of 250 (mph). The Germans had rather less success, although by 1990 they had built a full-sized Maglev train, which reached 271 (mph) in tests. Economic factors such as the cost

per mile of Maglev track and the incompatibility of the Maglev system with the existing European rail network eventually caused the Germans to abandon the Maglev idea in favor of fairly conventional technology.

Today, the standard for high-speed rail travel is that set by the legendary Japanese bullet trains and the similar French TGV system. The very first bullet train went into service in 1964, proving to the world that Japan had taken the lead in high-speed rail technology. The first train ran along a route from Tokyo to Osaka, the first of several lines linking Tokyo to the other major cities of Japan by means of purpose-built high-speed lines. Europe was slow to catch up, but by 1975 the French had begun work on the TGV. This was a systematic upgrade of existing technologies aimed at producing a rail network with average top speeds of around 180 miles per hour, the standard set by the Japanese. The first TGV service began in 1981, on a line between St. Florentin and Lyon.

Uses, Effects, and Limitations

In Europe and Japan, countries where mass rail transit is a crucial transportation alternative, high-speed trains move millions of people efficiently to and from their destinations each year. At present, the Japanese are planing a further upgrade of the entire bullet train system, which consists of seven routes linking Tokyo to major cities across the country. Japan's bullet trains carry an estimated 275 million passengers a year. Meanwhile, the French continue to expand the TGV network. This currently has three principal routes, which link Paris to Valence, Le Mans, and Calais, although four more internal routes are planned in addition to two international links. These will connect Lyon to Torino in Italy and Montpellier to Barcelona in Spain. Both international routes are due to begin operations in 2005.

The quest for improvements in the speed of travel was the driving force behind the development of ever-faster forms of rail travel in the 1960s. The development of such technologies as the turbo train or Maglev railways was justified on these grounds, and as experimental projects, they undoubtedly proved their worth. Unfortunately, however, it has not been possible to translate these experimental successes into a practical railways system. A Maglev system would have to run alongside existing routes or replace them altogether. This would lead to an enormous outlay of capital on new track, stations, and rolling stock, an expense that no government or private body has so far proved willing to bear. Any transport system that is run for profit must inevitably achieve a balance between quality and operating costs. It is for this reason that rail networks across the world operate on the most cost-efficient rather than technologically advanced basis. Although the Japanese continue to develop Maglev trains, there are at present no regular full-scale, intercity passenger services operating under this system anywhere in the world.

Improvements in existing technology, such as that represented by the TGV system, have proven to be a far more practical proposition, not least because these improvements can be incorporated into the existing rail system at a cost far below that of more exotic technologies. For this reason, if for no other, the future of high-speed rail travel is likely to be one of gradual improvements in existing technologies rather than a complete replacement of current systems.

Issues and Debate

Concerns about the noise generated by high-speed trains have led to the formation of a number of protest groups, which have lobbied railway companies and politicians in an effort to ensure that those who live near high-speed routes are treated fairly. When a number of different possible routes were proposed for a high-speed rail link between the Channel Tunnel—the undersea tunnel between England and France—homeowners whose properties lay along the proposed routes suddenly found that the value of their homes dropped dramatically.

The issue of safety is also a matter for concern. Although there has yet to be a single fatality on any of the three French TGV routes, a train crash at speeds in excess of 180 (mph) will inevitably lead to the deaths of many passengers. For this reason safety remains a matter of paramount concern among railway operators.

Although many governments appear willing to invest in new railway technology, there has been an increasing tendency toward partnerships between government and private companies, with costs and any subsequent profits being shared by both parties. The decision to build or not depends largely on the circumstances and even on the geography of various countries. Aircraft technology is a more practical means of getting around a vast continent such as North America. Consequently, its rail network is in decline, with passengers preferring to fly or even drive to their destinations. By contrast, in Japan, some 30 percent of all passenger transportation is by rail, due partly to the willingness of the Japanese government to invest heavily in what is now seen as the world's leading rail network.

—*Mike Flynn*

RELATED TOPICS
Alternative Automotive Fuel Technologies, Smart Highways, Smart Trains

BIBLIOGRAPHY AND FURTHER RESEARCH

BOOKS
Bourne, Russell. *Americans on the Move : A History of Waterways, Railways, and Highways.* Golden, Colo.: Fulcrum Publishing, 1998.
Hosokawa, Bill. *Old Man Thunder: Father of the Bullet Train.* Englewood, Colo.: Sogo Way, 1997.
Nice, David C. *Amtrak: The History and Politics of a National Railroad.* Boulder, Colo.: Lynne Rienner Publishers, 1998.
Vranich, Joseph. *Supertrains: Solutions to America's Transportation Gridlock.* New York: St. Martin's Press, 1993.

PERIODICALS
Anderson, Alun. "Trains, Lifts and Automobiles." New Scientist, October 2, 1993, 37.

Fox, Barry. "Going Up...and Sideways." New Scientist, October 14, 1995, 21.
Hadfield, Peter. "Japan on Track for Second Magnetic Train." New Scientist, September 2, 1995, 18.
Kleiner, Kurt. "Star Wars Express." New Scientist, April 6, 1996, 32–35.
O'Neill, Bill. "Beating the Bullet Train." New Scientist, October 2, 1993, 36.

INTERNET RESOURCES
High-Speed Ground Transportation Association
    http://www.hsgt.org/
International Union of Railways
    http://www.uic.asso.fr/
Japan Railway and Transport Review
    http://www.ejrcf.or.jp/html_rt03/rt03_c00.html
Japanese Bullet Trains (Byun Byun Shikansen)
    http://www.teleway.ne.jp/~dolittle/byunbyun/index.htm
TGV Pages
    http://mercurio.iet.unipi.it/tgv/tgvindex.html

HOLOGRAPHY

Holography is the science and manufacture of a device (the hologram) that records how energy reflects off an object. Holograms can be created with many types of energy, including sound waves and X-rays, but the holograms most people are familiar with are made with visible light. When exposed to light, a hologram of, for example, a coin, "replays" how light reflected off the actual coin. To the human observer, a three-dimensional coin—accurate in all visible detail except (usually) color—mysteriously appears.

Scientific and Technological Description

A hologram is in some respects like a photograph. A holographer exposes holographic film to light, and the light is recorded on the film. Then the film is developed using specialized chemicals, and a hologram is the result. The process of creating a hologram is referred to as *holography;* the film itself is a *hologram.* (Although holograms that record forms of energy other than light do exist, they are experimental and are not manufactured on the same scale as light holograms. This article focuses on light holograms, but the basic principles are the same for all holograms.) A hologram and a photograph are quite different, however. A photograph records only a two-dimensional image; a hologram records a three-dimensional image. That is because a hologram records more information than a photograph does about how light reflects off an object. Whereas a photograph records how intensely light is reflected off an object, a hologram records both the intensity of the reflected light and in what direction the light was reflected off the various surfaces of the object. All this information is contained in what is called the *interference pattern* of the light.

Like all electromagnetic radiation, light travels in waves. When one wave hits another wave, the two waves interfere with each other: They can cancel each other out, combine to make a stronger wave, or create a number of

Reflected-Beam Transmission Holograms

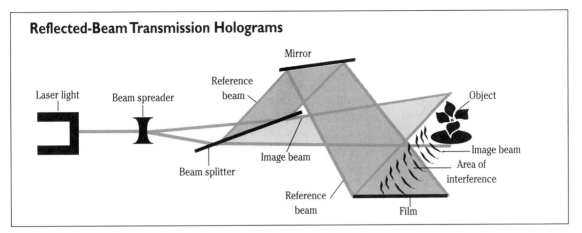

Adapted from Graham Saxby, Practical Holography, *Prentice Hall International, 1987.*

combinations in between. Light waves are constantly interfering with each other, yet this interference is usually too chaotic to make a visible pattern because light waves come in different lengths and usually travel *out of phase,* meaning that the crests and troughs of two light waves of the same length do not usually line up. But coherent light waves—waves of the same length and traveling in phase—can interfere with each other in a way that makes a distinct pattern. Holographic film can record this pattern. A holographer must thus produce coherent light. Coherent light needs to be produced in dark working conditions by a machine that makes light that is all the same frequency and is in phase.

Early holographers used special lamps and extensive filters, and their holograms were still not very clear. But modern-day holographers have a device at their disposal that creates coherent light—the laser (see LASERS). The holographer must use only one laser—otherwise the light waves would not be in phase. But a single beam of laser light does not interfere with itself without some help. So the holographer uses a device called a *beam splitter,* which is a piece of glass coated very thinly with reflective material. The coat is thin enough that a beam splitter reflects only some of the beam's light and lets the rest pass through. This splits the laser beam into two beams (see figure). One of the beams, called the *reference beam,* is pointed at the film. The second beam, called the *image beam,* is pointed at the object, which has been placed near the film. The image beam reflects off the object and then interferes with the reference beam. This creates a distinct interference pattern. The interference pattern is unique: If the image beam is reflected off a coin, it will create a different interference pattern than if it is reflected off a flower, for example.

The interference pattern is a microscopic mix of light spots and dark spots. Like photographic film, the holographic film is light sensitive, so the chemicals in it react only to the light spots. (Some holographic films are made of the same chemicals as photographic film, but photographic film cannot be used to make holograms because the grains of chem-

icals in the film are larger than the spots of the interference pattern.) The exact reaction depends on the type of film and developers used, but the chemicals hit by light usually either are removed or become transparent during developing. After it is developed, the holographic film acts like an incredibly complicated light filter. When light passes through it, the hologram screens out some light and bends some light to recreate the three-dimensional image of the object recorded.

Historical Development

Holography was first proposed in 1948 by the British scientist Dennis Gabor. Gabor was seeking a way to improve electron microscopy (the electron microscope has allowed scientists to view large molecules and cellular structures; see MICROSCOPY), and he theorized that recording the interference patterns of X-rays could create three-dimensional images of atoms. Gabor was, to put it mildly, ahead of his time. To demonstrate that holography was possible, he shined highly filtered light through an etched transparency and recorded the result on modified photographic film. When he developed the film and shone the same filtered light on it, the result was two images, one in relatively good focus and one badly out of focus. Neither was of very good quality, and they could not be seen unless the same light that was used to make the holograms was shone on it. In addition, the viewer saw the two images superimposed on each other, making it impossible to get a clear look at either image.

At the time, holography looked rather unimpressive (by 1971 it would look much more impressive, and Gabor would receive the Nobel Prize in Physics), so it was essentially ignored until the early- and mid-1960s, when several holography breakthroughs occurred on the heels of the creation of the first laser in 1960. Emmett Leith and Juris Upatnieks of the University of Michigan first discovered that laser light allowed the creation of holograms of much higher quality than did filtered light. In addition, they developed a recording technique that eliminated Gabor's two-image problem. Also in the early 1960s, Soviet scientist Y. N. Denisyuk created holograms with images that were viewable with regular light rather than the filtered or laser light that was used to make the image. Other scientists would later continue to improve holograms, making them brighter, crisper, and more easily viewable under normal lighting conditions.

Since light reflects off a holographic image the same way it reflects off the original object, it is possible to make a hologram of a hologram. But reproducing holograms on a large scale was a tedious business until the mid-1980s, when scientists perfected a method of embossing simple holograms. In this process the initial hologram is made on film that develops pits on its surface when exposed to light. The developed film is then used as a mold for a metal die. The resulting die has a pattern of tiny points that correspond with the exposed film's pits. The die can then be stamped onto unexposed film countless times, creating the same hologram over and over again. This method greatly reduced the cost of making holograms on a commercial scale.

Uses, Effects, and Limitations

Because they are produced using specialized equipment but cost relatively little to make, mass-produced holograms are well suited for use as markers of authenticity. These days, holograms can be found in many wallets: on credit cards, debit cards, security passes, and identification documents. Mass-produced holograms have also been used to make holographic jewelry, art, and novelty items available to consumers at a low cost. The holograms on credit cards and inexpensive jewelry are a type of hologram called *reflection holograms*. When viewed, the light comes in through the surface of the hologram and is reflected back to the viewer.

Another type of hologram that has been popular among holographic artists is the *transmission hologram*. This hologram is lit from behind, and the image appears to be suspended in air in front of the hologram. The image can be extremely large and quite striking, while the hologram itself sits inconspicuously behind or beneath it. Transmission holograms have some useful properties: light reflects off the image just like it would an actual object, yet the hologram itself is just a thin, extremely lightweight piece of film. Transmission holograms can be used to bend or focus light much as a lens does, and they can be placed into equipment without taking up the amount of space a glass or plastic lens would. These holograms, called *holographic optical elements,* have many uses, including forming part of the machinery of compact disk players (see CD-ROMS).

Any hologram of an object contains a tremendous amount of visual data about that object and stores that information in an extremely small space. Some countries have made holograms of treasured works of sculpture to create as complete a record of those objects as possible. The data-storage potential of holograms has excited researchers for years, but because of technical and cost problems, it remains an experimental use. Despite the large amount of visual data they record, holograms made in the ordinary way do not record color truly. The reason for this is that the light shone on an object to create a hologram is all the same wavelength, which means it is all the same color. Holographers can create full-color holograms by essentially making the hologram three times with three different-colored lasers. Since some holographic films work better with some colors of light than with others, full-color holography often requires that more than one type of holographic film be sandwiched together.

Holograms also do not record motion well—a seeming drawback that has actually turned into an advantage. Attempting to holograph a moving object will just create a "hole" in the resulting hologram because the light reflected off the object is too scattered from the movement to make a distinct interference pattern. But if the object is moving only a tiny bit—a movement far too small to be visible to the naked eye—a very useful thing happens. If the object is holographed both before and after the movement, and the two holographic images are projected on top of each other, dark rings appear on the image. The rings, which resemble a moiré pattern (a pattern that gives a washed or watery appearance) on fabric, are the result of light from the two images interfering with each other. The moiré effect has been a boon to engineers because it allows them to see microscopic alterations in the shape of objects caused by heat or by stress. The alterations can indicate weak spots in, for example, an airplane engine working under normal conditions, or an artificial hip joint.

Holography has very recently come full circle and been used for the purpose for which it was invented—to image atoms. In the late 1990s, almost 50 years after Gabor's original proposal, researchers were able to create three-dimensional holograms of atoms using X-rays. Earlier methods for determining three-dimensional atomic structures gave scientists an idea of their overall structure, but X-ray holograms can be used to see small local deviations from this structure.

Issues and Debate

When most people think of holograms, they probably don't think of the real holograms in their life—the stickers on their credit cards, for example. They are more likely to think of the full-colored moving holograms of the *Star Wars* movies (created by special effects) or the even more technologically wondrous solid, sentient holographic characters of the *Star Trek* television series (which are actual human beings). Holography's "gee-whiz" factor, fostered in part by the large public displays of transmission holograms that were popular in the 1970s, has contributed to an inaccurate impression of the technology in the public mind. Most people don't know what holograms are actually used for, they don't know how widely holograms are used, and they think holograms are good for things they do only with great difficulty, such as recording color and movement.

As a result, public debate surrounding the real-life application of holograms is almost nonexistent. The characters on *Star Trek* routinely debate the ethics and morality of their holographic technology, but debates surrounding the use of real holograms tend to be financial and the general public is usually not involved. In part this is because holograms are often used to replace such well-accepted existing technologies as glass lenses or photographs, and their foreseeable uses involve replacing such uncontroversial technologies as the microfilms of books and periodicals found in libraries. Some holographers appreciate the lack of public scrutiny, but others complain that their work is unacknowledged and unappreciated.

Perhaps the only use of holograms that has been hotly and publicly debated is the question of holograms as an art form. When public displays of holographic art became popular in the 1970s, some critics complained that the pieces were just reproductions of existing objects and should not be considered art. Partly in response to this criticism, a number of holographic artists created abstract and surrealist works. Although holographic artworks are now widely sold, holographic artists are still scrambling for legitimacy in the art world. Their relative lack of success became apparent in 1992 when the 15-year-old Museum of Holography in New York City shut down and its contents were purchased by the Massachusetts Institute of Technology.

Holography is heavily reliant on mathematics and optical science, which many people find daunting. As a result, many people consider holography to be something scientists can create but normal people cannot. The truth is that simple holograms can be made at home by people with no elaborate training and no special licenses. A person who wants to make a hologram needs a dark room in which to make the holograms, a table that can hold tools and objects securely, a photographic-type dark room in which to develop the holograms, and a laser. Cheap, low-energy lasers (such as laser pointers) are readily available to the public, and mirrors, beam splitters, and holographic film are commercially available through companies that supply laboratories and professional holographers. An average person can now create a holograph at home—one more example of the broader social movement that has made technology more accessible to the layperson.

—*Mary Barr Sisson*

RELATED TOPICS
CD-ROMs, Lasers, Microscopy

BIBLIOGRAPHY AND FURTHER RESEARCH

BOOKS
Abramson, Nils. *The Making and Evaluation of Holograms.* London: Academic Press, 1981.
Harihara, P. *Optical Holography: Principles, Techniques, and Applications,* 2nd ed. Cambridge: Cambridge University Press, 1996.
Saxby, Graham. *Manual of Practical Holography.* Oxford: Focal Press, 1991.

Unterseher, Fred, et al. *Holography Handbook: Making Holograms the Easy Way.* Berkeley, Calif.: Ross Books, 1992.

PERIODICALS
Anderson, Dana Z. "High Gains for Polymer Dynamic Holography." *Science,* July 25, 1997, 530.
Beardsley, Tim. "The Dope on Holography." *Scientific American,* September 1998, 41.
Fadley, Charles S., and Patrick M. Len. "Holography with X-rays." *Nature,* March 7, 1996, 27.
Nadis, Steve. "Two Versions of Holography Vie to Show Atoms in 3D." *Science,* May 3, 1996, 650.
Shen, Xiao A., et al. "Time-Domain Holographic Digital Memory." *Science,* October 3, 1997, 96.

INTERNET RESOURCES
Holo.com—Information on Holography, Lasers, and Related Topics
    http://www.holo.com
Hologram Production Site
    http://www.3dimagery.com/
HoloNet—Information on Holography, with Emphasis on Art and Design
    http://www.holonet.khm.de/
Internet Webseum of Holography
    http://www.holoworld.com/
Practical Holography
    http://www.holo.com/holo/book/book1.html

HYDROELECTRIC POWER

Hydroelectric power takes advantage of the energy of falling water by pushing it against turbines that produce electricity. When hydroelectric power made its modern debut in 1882, it was perceived as an ever-ready source of energy for small-scale lighting projects. As the electricity grid grew across the United States and in other countries in the early to mid-20th century, hydroelectric power contrasted starkly with cheaper but polluting fossil fuels. The dams that many hydroelectric plants built created reservoirs in which people could sail or swim, and helped control flooding.

But over the past 20 years, scientists and others have questioned the effects of hydroelectric power since it has become evident that reservoirs create several significant environmental problems and that the alteration of waterways can irrevocably change surrounding environs. Even as some new hydroelectric projects are being built, plans for new plants are being rigorously assessed in terms of their likely environmental impact.

Scientific and Technological Description

Hydroelectric plants are situated in or near waterways and most consist of six parts: the dam, penstocks, turbines, generators, transformers, and transmission lines. The most common types of hydroelectric plants, called *conventional plants,* direct the natural current of the water through their structures and out one side. They come in two principal designs. *Storage plants,* such as the Hoover Dam, create large reservoirs at the facility so that the turbines have a ready source of energy should water levels outside the plant drop. This constant supply of water ensures that electricity production continues uninterrupted. In contrast, the second type of conventional plant, *run-of-river plants,* do not

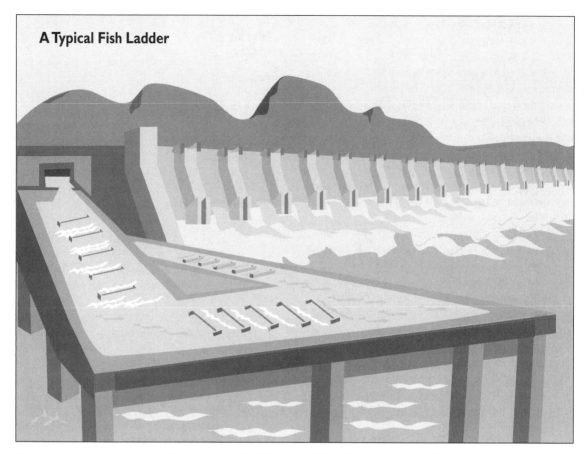

A Typical Fish Ladder

A serious concern about hydroelectric dams is that they block the paths of fish as they attempt to reach their spawning grounds, which may endanger fish populations. To enable fish to get past dams, engineers have designed "fish ladders" like the one shown here.

form reservoirs but instead allow the natural current of the water to run through the facility. Because these plants do not create reservoirs, they are perceived as more environmentally friendly than storage plants, but they also are more vulnerable to weather and seasonal changes, and their electrical production is less stable as a result.

Most common hydroelectric plant designs incorporate dams to capture the water, although not all dams create electricity. Once the dam is built, a reservoir forms upstream, and water moves from the reservoir to the powerhouse via penstocks. These steel pipes, or concrete-lined tunnels, traffic the water either around or through the dam, and valves inside the penstocks control the amount and speed of water that will reach the turbines.

Generally, in any hydraulic turbine, the water flows inward, then down into an area called the *runner*. The runner is made of buckets or blades, and the water turns the buckets or blades. They, in turn, rotate the drive shaft of the turbines. Generators connected to the turbines rotate to create electricity, and transformers convert the electricity into usable voltages. The electricity is carried away from the plant by transmission wires. The amount of electricity a hydroelectric plant produces depends on the volume of water that flows through the turbines and the hydraulic head, which is the height from the water surface to the turbines. Higher heads help produce more electricity because the energy of the water increases as it plunges from higher to lower areas.

There are several other categories of hydroelectric plants. *Pumped storage facilities,* for example, reuse water instead of discharging it back into the waterway. In this design, an upper and lower reservoir exchange water in a self-contained stream. Related plants are *wave-powered electricity generators* and *tidal power generators,* but neither has taken hold in a competitive market. Small hydroelectric plants have emerged as a renewable energy option for areas that do not have major waterway resources or do not connect to an electricity grid. As their name suggests, these plants do not produce as much electricity as the larger facilities, but they do support small communities and have fewer negative environmental impacts.

Historical Development

Human beings have harnessed the energy of flowing water to do mechanical work for thousands of years, but it wasn't until the late 19th century that the mechanical work of water mills was connected to electrical production. The first hydroelectric plant was built in Appleton, Wisconsin, in 1882. By 1889, 200 U.S. electric companies were using hydroelectric power. The development of the electric motor in the late 19th century drove the demand for more electricity, and utilities responded by increasing their dependence on hydropower. By early in the 20th century, more than 40 percent of the country's electricity supply came from hydroelectric plants.

With the growing interest in hydroelectric power came governmental scrutiny. In 1920, the Federal Water Power Act created the first Federal Power Commission. The commission was the forerunner of the current Federal Energy Regulatory Commission (FERC), which licenses hydroelectric plants among its other duties. To date, the FERC has licensed nearly 1700 hydroelectric projects that involve 2300 dams.

The 1930s saw several key turning points in hydropower and dam development. Plant designs became standardized, and for the first time, statistical methods were used to predict floods, rates of soil erosion, and other factors that affect the stability of dams and plant facilities. These technological advances, along with a growing need for electricity and flood control, sparked the "big dam" era in the United States, beginning with the largest public project seen on American land at that time—the Hoover Dam. Completed in 1936, the Hoover Dam spans the border of Nevada and Arizona on the Colorado River and towers 726 feet high. The next massive dam built was the 551-foot Grand Coulee Dam on the Columbia River in Washington in 1942. These dams were followed by six more giant plants, along with smaller ones, over the next 30 years.

Hydroelectric plant construction continued to expand in the United States following World War II, but the energy source faced fierce competition. The use of petrochemicals and fossil fuels outpaced development of hydroelectric power, but when oil prices skyrocketed from $3 per barrel in 1973 to $40 a barrel in 1981, policymakers rethought energy programs. During the price hikes, the U.S. Congress passed the Public Utility Regulatory Policies Act (PURPA) in 1978. This act required utilities to purchase electricity from qualified independent power producers, and small-scale hydroelectric plants benefited from an expanded market.

By 1986, however, environmentalists had raised enough concerns about the impact of large dams on waterways and wildlife habitats that Congress amended the Federal Power Act to give equal consideration to nonproduction factors, such as fish migration in rivers that were blocked by dams. Meanwhile, other countries, such as Canada, Brazil, Russia, and China, have developed their hydroelectric resources extensively—so much so that Canada leads the world in hydroelectric power output and Brazil shares with Paraguay the world's largest hydroelectric power plant, the Itaipú Binacional project on the Paraná River.

Uses, Effects, and Limitations

Hydroelectric power provides about one-third of the world's electrical output. In the United States, the energy source supplies about 12 percent of the electrical generating capacity of the country, a dramatic decrease from the 1930s, when it supplied almost 40 percent. In Canada, hydroelectric plants account for more than 60 percent of the country's total electrical output. Other countries, such as Russia, Brazil, and Norway, use hydroelectric plants extensively, but no country is planning a hydroelectric power scheme on the scale that China currently is planning. China intends to build the world's largest dam project along the Yangtze River. The project, named Three Gorges, calls for a dam measuring 1.24 miles long and 328 feet high. Its reservoir will flow for 372 miles upstream of the dam. The dam is expected to have a capacity of 18,000 megawatts of electricity. Hydroelectric power plants and dams in general can be used to redirect rivers to irrigate land or prevent flooding. Recreationalists can take advantage of the reservoirs that form behind dams for fishing, swimming, and sailing.

Any new construction of a hydroelectric plant can be made difficult by siting restraints. The best sites often are unsuitable because they involve rivers that surge down unruly terrain where transmission lines are difficult to mount. Other factors, such as human settlements, animal habitats, and protected parklands, affect siting decisions. To better identify appropriate spots in the United States, researchers working at the request of the Department of Energy have developed a computer model called hydropower evaluation software (see COMPUTER MODELING). The software integrates a range of environmental considerations for any given location. As a result of using this software, the department has identified 5677 sites with an undeveloped capacity of 30,000 megawatts of electricity. Whether these sites will be developed as hydroelectric plants remains to be seen.

Issues and Debate

For an energy source that was once considered to be wholly environmentally friendly, hydroelectric power in recent years

The Hoover Dam, and the artificial lake it created, Lake Mead, are located on the Colorado River between Arizona (right) and Nevada. The building of the Hoover Dam, beginning in 1931 and ending in 1936, marked the beginning of the "big dam" era in the United States (Photo Researchers/Joe Munroe).

has produced a plethora of concerns that have come to the forefront of environmental and legal discussions. When engineers choose a site for a hydroelectric plant, they must flood the land to create a reservoir, unless the plant is a run-of-river design. These floods now cover 372,000 square miles worldwide and have forced hundreds of thousands of people to resettle. The Three Gorges project in China, for example, will resettle more than 1 million people. The floods also destroy wildlife habitats and vegetation, although efforts are made to create animal reserves and to reintroduce species.

Once a plant is built, it can greatly affect its environs. Reservoirs can slow currents so much that water on the surface warms appreciably more than underneath, which causes a stratification effect. The colder water carries less oxygen to the plants and animals below, and the decreased oxygen can change their ecosystems drastically. Reservoirs also trap sediments upstream of the dam, which can affect a hydroelectric plant's efficiency as well as the plant and animal life around it. The Sanmenxia Dam on the Yellow River in China, for example, had to be rebuilt four years after it opened in 1960 because the reservoir had filled with silt. If too many sediments remain trapped behind a dam, they become a ready food source for organisms that could overpopulate the reservoir and deplete its oxygen. Meanwhile, the water that flows through the dam carries fewer sediments and moves more quickly than it would normally. The advanced speed erodes the banks of waterways faster than a naturally running system would, while the lack of sediments downstream can deprive organisms that feed on them of nutrition.

Another serious consideration is that fish may find their migratory paths barred by dams. Engineers have tried to address this issue by creating fish passages or "ladders" that direct the animals alongside or through the dams. Some fish, however, go through the turbines, which batter and sometimes kill the fish. As a result, the Department of Energy is working with research groups to design an advanced hydropower turbine systems program that, among other things, offers better means to direct fish safely through dams.

Reservoirs also may undermine what many consider the most attractive feature of hydroelectric power—that it is nonpolluting. Recent scientific studies implicate reservoirs in the creation of greenhouse gases. The trees and vegetation that become submerged under a reservoir decompose and can release carbon dioxide and methane. Researchers who have studied the Balbina reservoir on the River Uatumã in Brazil, for example, have estimated that the reservoir emitted an estimated 10 million tons of carbon dioxide and 150,000 tons of methane during 1988. They maintain that the reservoir over the past eight years has produced 16 times as potent a greenhouse effect as that of fossil fuel power plants. A similar study in Canada found that flooded forests and peat bogs, areas traditionally used for reservoir sites, produced a considerable amount of greenhouse gas. Both of these studies have been criticized for the way in which the gas amounts were calculated, and it remains unclear how significantly dam reservoirs add to the greenhouse effect.

Hydroelectric plants continue to be built, particularly in developing nations, and research into environmentally friendly designs progresses. Yet the future of existing small plants remains uncertain. In July 1999, the Edwards Dam on the Kennebec River in Maine became the first dam in the country's history to be destroyed by government order, and observers wonder if there will be more.

—*Christina Roache*

RELATED TOPICS
Computer Modeling, Nuclear Energy, Wave and Ocean Energy, Wind Energy

BIBLIOGRAPHY AND FURTHER RESEARCH

BOOKS
Doherty, Craig A. *Hoover Dam*. Woodbridge, Conn.: Blackbirch Press, 1995.
Goldsmith, Edward, and Nicholas Hildyard. *The Social and Environmental Effects of Large Dams*. San Francisco: Sierra Club Books, 1984.
McGuigan, Dermot. *Harnessing Water Power for Home Energy*. Charlotte, Vt.: Garden Way Publishing, 1978.
Payne, Sherry Neuwirth. *Wind and Water Energy*. Milwaukee, Wis.: Raintree Publishers, 1983.
Pisani, Donald J. *To Reclaim a Divided West: Water, Law, and Public Policy, 1848–1902*. Albuquerque, N. Mex.: University of New Mexico Press, 1992.
Pitzer, Paul C. *Grand Coulee: Harnessing a Dream*. Pullman, Wash.: Washington State University Press, 1994.
Quing, Dai, compiler; John G. Thibodeau and Philip B. Williams, eds. *The River Dragon Has Come! The Three Gorges Dam and the Fate of China's Yangtze River*. Armonk, N.Y.: M.E. Sharpe, 1998.
Stevens, Joseph E. *Hoover Dam: An American Adventure*. Norman, Okla.: University of Oklahoma Press, 1988.
United States Bureau of Reclamation. *Design of Small Dams*, 2nd ed. Washington, D.C.: U.S. Printing Office, 1973.

PERIODICALS
Baker, James. "Letting the Rivers Flow (Banning of Future Hydroelectric Dams in the Pacific Northwest)." Sierra, July–August 1988, 21–24.
Devine, Robert. "The Trouble with Dams." *Atlantic Monthly*, August 1995, 64–70.
DiChristina, Mariette. "The Hoover Dam." *Popular Science,* April 1998, 100.
Hathorn, Clay. "It's a Dam Shame (the Loss of Salmon Populations in Hydroelectric Dams)." *Wildlife Conservation,* July–August 1992, 54–59.
Hecht. Jeff. "High Time for Compact Tidal Power?" *New Scientist,* February 11, 1995, 20.
Lee, Kai. "The Columbia River Basin: Experimenting with Sustainability." *Environment,* July–August 1988, 6.
Pearce, Fred. "The Biggest Dam in the World." *New Scientist,* January 28, 1995, 25–29.
Pearce, Fred. "Trouble Bubbles for Hydropower (a Hydroelectric Reservoir Can Be More Polluting Than a Coal-Fired Power Station)." *New Scientist,* May 4, 1996, 28–31.
Svitil, Kathy. "The Coming Himalayan Catastrophe." *Discover,* July 1996, 80–84.
Topping, Audrey. "Ecological Roulette: Damming the Yantzge." *Foreign Affairs,* September–October 1995, 132–146.

Truchon, Myriam. "Hydro-Quebec at James Bay." *Ecologist,* November–December 1994, 239–40.

Warwick, Hugh. "Come Hell and High Water." *New Scientist,* March 30, 1996, 39–42.

Webb, Jeremy. "Anchors Aweigh for Wave-power Pioneers." *New Scientist,* July 29, 1995.

INTERNET RESOURCES

Foundation for Water and Energy Education site
http://www.fwee.org

U.S. Army Corps of Engineers
http://www.usace.army.mil

U.S. Bureau of Reclamation
http://www.usbr.gov

U.S. Department of Energy
http://www.eren.doe.gov/RE/hydropower.html.

U.S. Geological Survey site
http://wwwga.usgs.gov/edu/wuhy.html

INTELLIGENT AGENTS

Intelligent agents are software programs that act autonomously to perform tasks on behalf of a user. Common applications of intelligent agents include information searches and e-mail filtering. Intelligent agents may also possess the following capabilities: the ability to monitor and adapt to user behavior patterns; reasoning abilities to make decisions on behalf of the user; and communication skills to collaborate with other agents on complex tasks.

During the past two decades, intelligent agents have evolved from simple single-task software programs to adaptable, humanlike "personal assistants" that can predict a user's needs and act accordingly. Some of the most popular intelligent agents are Internet "bots" (short for robots), which efficiently complete such time-consuming tasks as conducting job searches on the Web.

Although widely favored by many people, intelligent agents have also attracted a few opponents, who argue that some agents—those that monitor people's Internet browsing patterns, for example—invade people's privacy. Additionally, some argue that the agents are not yet competent enough to perform critical tasks, which further opens the debate as to whether agents can and will replace people in the workplace.

Scientific and Technological Description

An intelligent agent's purpose is to facilitate computing by performing repetitive, routine, and/or labor-intensive tasks, such as searching for and delivering topic-specific news from the Internet. Conventional software can also perform repetitive tasks for a user. However, intelligent agents act *autonomously,* without user input and guidance. Conventional software operates through direct manipulation, in which the user gives the program directions based on options presented on the screen. Intelligent agents don't require this type of constant feedback; instead, they draw

How an Intelligent Agent 'Learns'

...likes to receive travel offers...
...likes to receive the latest sports scores...
...likes to check stocks on line...

...is only interested in air travel from New York to Boston...
...is only interested in the baseball scores of the New York Yankees and Boston Red Sox...
...is only interested in the performance of certain stocks...

1. The agent begins with a small body of knowledge about the user's preferences.

2. By monitoring user behavior, the agent learns more specifically what the user's preferences are.

3. As it learns more about the user, the intelligent agent is able to retrieve and deliver only the information that the user really wants, allowing the user to spend less time searching for information or sifting through unwanted e-mail.

on knowledge programmed into them as well as their observations of user preferences and behavior.

Intelligent agents can work in different ways. In creating agent software, many developers use a *knowledge-based approach,* in which they give the agent a model of behavior based on the agent's intended application and some assumed information about the user. This approach has several limitations. The developer must endow the agent with a large amount of knowledge and that knowledge cannot be updated or altered. In addition, if the agent appears to be too sophisticated and independent, users may find it hard to trust the agent, since they have little control over the agent's proceedings. For instance, a user might employ an intelligent agent to filter out e-mail and to delete spam (unwanted) e-mail automatically. However, knowledge-based agents may not gather user input as time goes on as to what constitutes spam e-mail. If the agent simply works under its own definition, independent of user input, it may delete valuable communications. Thus, the user would find it difficult to trust the agent and to employ it.

Another approach is to have the agent "learn" behavior from repeated interactions with the user. This form of agent may begin with a small background of knowledge and accumulate more as it monitors user behavior. It may also adapt its behavior based on its observations of user behavior. This type of agent requires repetitive-task regularities on which to draw its conclusions. Using the example of the e-mail filter agent, the agent would observe what the user classified as spam or junk e-mail. It would look for patterns in the unwanted e-mail. For instance, it might observe that the user deleted all e-mail with phrases such as "lose weight" and "earn money" in the subject line, or that the user always deleted e-mail that came from certain senders. Then the agent could learn to identify all incoming e-mail with those words in the subject line or from those senders as spam and remove the messages automatically.

Intelligent agents are also proactive in identifying and responding to opportunities for action; to accomplish this, they use a form of ARTIFICIAL INTELLIGENCE, which is the simulation of human intelligence processes, including learning and reasoning. Ultimately, intelligent agents do not require the user's stimulus, although they do engage in a cooperative process in which the user and agent can initiate communication, monitor activities, and complete tasks.

Historical Development

The genesis of intelligent agents began in research performed in the 1940s and 1950s. In the study of *cybernetics,* researchers aimed to mathematically unify studies of control and communication in animals and machines. Studies in cybernetics were most widespread at the Massachusetts Institute of Technology (MIT), which continues to maintain a strong program today in the field of intelligent agents.

In the 1950s and 1960s, cybernetics gave way to a new discipline, artificial intelligence. Researchers such as John von Neumann, Alan Turing, and Claude Shannon from MIT studied artificial intelligence in the pursuit of developing individual artificial systems that could think rationally. Beginning in the 1980s, scientists began to use past research in artificial intelligence to construct intelligent agents that could perform tasks on behalf of users. Many developers have since become interested in creating multi-agent systems that can deal with the coordination of intelligent behavior among a collection of agents.

Research on intelligent agents began to pick up speed in the early 1990s. An October 1992 symposium on agents, held at MIT, attracted a surprising 1,000 participants. As the World Wide Web was developed in the early 1990s and blossomed into wider use, researchers began to investigate agents that could gather and deliver information from the Web.

Today, intelligent agents are growing in acceptance and use, as developers create ways to make the agents affect a user's progress and performance in direct ways. In addition, research on earlier intelligent agents has spawned several types of agents, including *softbots* (software agents), *knowbots* (agents that seek and return information), and other autonomous computer processes. Scores of companies are now offering different types of agents.

Uses, Effects, and Limitations

Intelligent agents have a range of uses, which include performing tasks on behalf of a user, training or teaching users, collaborating on a project with other agents, and monitoring events and procedures that the user performs to better understand his or her behavior.

Task-Based Agents

Most agents perform some type of task for users. Many of these agents operate on the Internet, where they are commonly referred to as *bots.* Popular bots include search-and-retrieval agents that cull through the vast information on the Web and deliver information on specific topics. For instance, search agents can research news topics or search for specific types of job listings on the Internet.

Another popular task-based agent is the e-mail agent. E-mail agents observe a user's patterns of prioritizing, filtering, deleting, and sorting e-mail and then perform these actions on behalf of the user. From observing patterns of use, the agents develop e-mail filtering rules. For instance, an agent may notice that the user always places high priority on e-mail messages from a specific co-worker or boss. The agent can then examine the "From:" field of every e-mail to assign it a high priority if necessary. Or, the agent could learn to recognize spam mail (unwanted or junk e-mail) and delete it automatically. E-mail agents can also inspect the "CC:" and "Subject" boxes as well as the body of the message to determine how to handle each message. If the agent is ever unsure, it can query the user for direction. The user can set confidence levels in the agent's capabilities. A lower confidence level means that the agent would query the user more often before moving or deleting a message. The agent gradually gains competence over time through observing the user's interactions and acquiring examples of behavior.

Another potential use for a task-based agent would be a customer service agent that fields popular questions from customers. Agents are designed to understand natural-language queries from users, so a customer service agent could receive incoming questions and inquiries, search its database of potential solutions, and present the remedy to the customer. Service companies estimate that 80 percent of customer questions are repeat questions, so intelligent agents could handle these types of queries. When confronted with a question the agent can't handle, it could then defer the query to a human service representative. Customer service agents can work over the Internet and on voice calls, using text-to-speech technology (see LANGUAGE RECOGNITION SOFTWARE).

Other Types of Agents

Another type of intelligent agent is an entertainment selection agent. This type of agent can make entertainment (book, movie, or theater) recommendations based on user preferences. It doesn't work like typical search-and-retrieve agents, however. Typical information search agents would prompt the user for entertainment preferences and then examine Web sites for compatible information to present to the user. By comparison, an entertainment selection agent works in collaboration with other intelligent agents. For example, agent A will "learn" a user's preferences, store this information, and then search the Internet for other agents with the same set of user preferences. If agent A discovers that agent B has similar preferences, such as liking science fiction, and has recommendations on additional entertainment options, agent A will present this information to its user. After meeting agent B, agent A might discover that there is a new movie that appeals to people with science fiction preferences. It would then present the movie as a recommendation to its user. These types of agents appear on entertainment Web sites and Internet guides to American U.S. cities.

One of the limitations of this collaboration system is that it requires a large base of agents that can compare themselves to one another. However, developers have dealt with this drawback by creating "virtual users," each of which represent a particular taste, such as science fiction films or classic rock music.

Internet marketers also use intelligent agents to monitor the usage patterns of Internet browsers and potential customers. One application of this technology is to tailor banner advertisements on the Web to the individual user. For instance, an Internet shopping Web site could track the navigating and purchase patterns of a particular customer, who is identified by his or her log-on user name. The agent could then customize the advertisement selection to this particular user. Thus, two users could browse the same site simultaneously and never view the same banner advertisement.

Issues and Debate

Intelligent agents used for marketing purposes have provoked complaints among individuals concerned about Internet security and privacy. Opponents argue that intelligent agents are helping companies to "know" too much about the customers they monitor. Since many Web sites require users to "register" before they may access the site, the site operators have collected basic personal information about the user in addition to records of that person's browsing and purchasing patterns. Proponents for intelligent agents acknowledge that many people have this privacy concern. However, they argue that in large part, the companies that monitor their customers respect the customer and do not use the information in any way that might hurt the customer. They believe that intelligent agents allow companies to serve their customers better, by alerting them to the types of products they have been interested in previously, for example.

Another issue regarding intelligent agents is their competence. Some maintain that intelligent agents cannot be trusted to perform competently important tasks, such as choosing which e-mail to delete in a user's absence. Equipping agents with competent decision-making capabilities is the software engineer's greatest challenge.

Most analysts predict that intelligent agents will eventually possess the competence to perform some of the tasks currently completed by people. Some have speculated that there may be a small displacement of the workforce as efficient agents substitute for human employees. For instance, an agent could perform a human auditor's duties by conducting periodic audits to predict bankruptcy or detect fraud. Other analysts believe that agents will be used as aides to people presently employed rather than as substitutions for them.

—Christine Doane

RELATED TOPICS

Artificial Intelligence, Internet and World Wide Web, Internet Search Engines and Portals, Language Recognition Software, Robotics

BIBLIOGRAPHY AND FURTHER RESEARCH

BOOKS

Boden, Margaret. *Artificial Intelligence and Natural Man.* London: MIT Press, 1987.

Boden, Margaret, ed. *The Philosophy of Artificial Intelligence.* New York: Oxford University Press, 1990.

Williams, Joseph. *Bots and Other Internet Beasties.* Indianapolis, Ind.: Sams.net Publishing, 1996.

PERIODICALS AND INTERNET RESOURCES

"Agents Track the Once and Future Consumer." *New York Times,* June 12, 1996.

"Bots Answer Customer Questions." *Interactive Week,* January 19, 1999.

General Computing and Technology Information
    http://www.zdnet.com
Maes, Pattie. "Agents That Reduce Work and Information Overload."
    Powergrid Journal, version 1.01
    http://electriciti.com/1.01/index.html
MIT Media Lab
    http://www.media.mit.edu
Technology and Other Definitions
    http://www.whatis.com

INTERNATIONAL SPACE STATION

The International Space Station (ISS) is a permanent research facility that orbits Earth at an altitude of approximately 410 kilometers (km) and is designed to accommodate long-term human habitation in space. When fully assembled and operational in the year 2004, this "outpost in space" will have a total mass of approximately 460,000 kilograms (460 tons), require over 45 flights of the U.S. Space Shuttle and other launch vehicles to construct, and cost in excess of $50 billion.

Sponsors of the International Space Station present it as a shining example of peaceful cooperation in space. The ISS establishes a permanent laboratory in a realm where gravity, temperature, and pressure can be manipulated to achieve numerous scientific and engineering pursuits, they point out. Detractors of the project are not as enthusiastic. They see the facility as a volatile, politically driven undertaking that will produce very little "real" scientific return for the enormous expense involved.

Scientific and Technological Description

The International Space Station is a major international laboratory facility in orbit around Earth. Its development and operation draws upon the resources and scientific expertise of 16 nations and represents the largest cooperative aerospace project ever undertaken. In all, about 460 tons of structures, modules, equipment, and supplies will be placed in low Earth orbit (LEO) to create the space station by the year 2004. The U.S. Space Shuttle and expendable Russian Proton and Soyuz rockets will conduct more than 45 missions to launch and assemble the more than 100 elements that will comprise the completed ISS.

On-orbit assembly of the ISS began in late 1998 and is scheduled to continue until 2004. Upon completion, the station will contain laboratory modules supplied by the United States, Japan, the nations of the European Space Agency, and Russia, along with a versatile robot arm supplied by Canada. A crew of

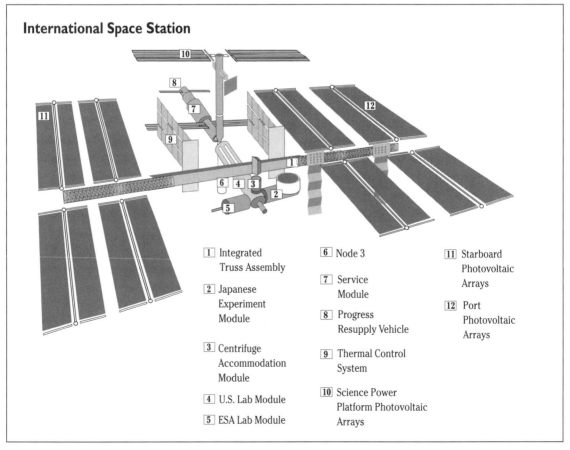

International Space Station

1 Integrated Truss Assembly

2 Japanese Experiment Module

3 Centrifuge Accommodation Module

4 U.S. Lab Module

5 ESA Lab Module

6 Node 3

7 Service Module

8 Progress Resupply Vehicle

9 Thermal Control System

10 Science Power Platform Photovoltaic Arrays

11 Starboard Photovoltaic Arrays

12 Port Photovoltaic Arrays

Adapted from an original illustration courtesy of NASA.

A digital artist's vision of the completed International Space Station passing over Florida. The ISS is scheduled to be assembled by 2003 (Courtesy of NASA).

seven will live and work in shirt-sleeve comfort for extended periods (typically, up to one year) in a total pressurized volume equivalent to that of two modern jumbo jets. The completed ISS will measure 109 meters (m) across (end-to-end width) and about 88 m long, essentially the dimensions of a football field with end zones and spacious sidelines. It will comprise almost an acre (about 4000 square meters) of solar panels and will provide 110 kilowatts (kw) of electrical power.

The ISS orbits Earth at an altitude of approximately 410 km with an inclination to the equator of 51.6°. This particular orbit allows the station to be reached by launch vehicles belonging to all of its international partners: Italy, Belgium, the Netherlands, Denmark, Norway, France, Spain, Germany, Sweden, Switzerland, and the United Kingdom. The orbit of the station takes it over almost 85 percent of Earth's surface and about 95 percent of the world's population.

The United States is responsible for developing and operating the major elements and systems aboard the station. These include three connecting modules (or nodes), a laboratory module (called Destiny); truss segments (for structural support); four solar arrays (for electric power generation); a habitation module (the crew's living quarters); three mating adapters (to connect various modules); an unpressurized logistics carrier (for supplies and equipment storage), and a centrifuge module (in which to conduct experiments at different levels of simulated gravity). The various support systems being developed by the United States include thermal control, life support, guidance, navigation and control, data handling, power systems, communications and tracking, ground operations facilities, and launch-site processing facilities.

The international partners—Canada, Japan, nations of the European Space Agency, Russia, and Brazil—are also contributing key elements to the ISS. Canada is providing the mobile servicing system (MSS), a 17-m-long robotic arm that will perform assembly and maintenance tasks on the station. Europe is building a pressurized laboratory (called Columbus) and three multipurpose logistics modules. Japan is constructing a laboratory (called Kibo, or "hope") with an attached space-exposure research facility. Russia is providing a service module (called Zvezda, or "star"), which will serve as early station living quarters, research modules, a science power platform that can provide about 20 kw of electrical power, logistics transport vehicles, and the *Soyuz* spacecraft for crew transfer and return. Brazil is providing a pallet for attached payloads (called *EXPRESS*).

Historical Development

At the start of the 20th century, a Russian, Konstantin Tsiolkovsky, became the first person to describe all the essential technical components needed for a permanent human presence in space. Throughout the first half of the 20th century, the concept of a space station continued to evolve. In his book *The Rocket to Interplanetary Space* (1923), the German physicist Hermann Oberth suggested that a space station could be used as a scientific research platform, an astronomical observatory, or an Earth-monitoring facility. After World War II, the German-American rocketeer, Wernher von Braun, assisted by the space artist Chesley Bonestell and the entertainment visionary Walt Disney, helped popularize the concept of a wheel-shaped space station in the United States.

Since its creation in 1958, the U.S. National Aeronautics and Space Administration (NASA) has been the technical forum for the U.S. space station debate. In 1960 a Manned Space Station Symposium was convened in Los Angeles. Experts from every part of the fledging aerospace industry gathered and agreed that a space station was a logical goal, but then proceeded to disagree on exactly what the space station was, where it should be located, or even how it should be built. This station "applications" debate continues to the present day.

For over four decades NASA has asked the scientific, engineering, and business communities over and over again: What do you want the space station to be? As responses have flowed in, NASA has developed a variety of space station concepts to help satisfy these projected requirements. The International Space Station is the latest in an evolving series of station concepts.

On May 14, 1973, NASA launched Skylab, the first American space station. This large facility was placed in orbit in one piece by a giant Saturn V booster rocket from the *Apollo* Moon landing program. Three teams of astronauts visited Skylab from 1973 through 1974 and demonstrated that people could function in microgravity ("weightlessness," a condition where the effects of gravity are virtually absent) for up to 12 weeks and, with proper exercise, could return to Earth with no ill effects. However, Skylab itself was not designed for a permanent presence in space. For example, the station was not equipped to maintain its own orbit, a design deficiency that eventually caused its fiery demise on July 11, 1979, as it reentered the atmosphere over the Indian Ocean and portions of western Australia.

While the United States was focusing on the *Apollo* Moon landing program, the former Soviet Union embarked on an ambitious space station program. As early as 1962, Russian aerospace engineers suggested that a space station could be made up of modules that were launched separately and then brought together in orbit. The world's first space station, Salyut-1, was launched on April 19, 1971, by a Proton booster. From 1971 until the present, the Russians have flown three generations of space stations, the latest of which is Mir ("peace"). Since 1986, the Mir space station has been a major part of the Russian space program. But now, because of Russia's active participation in the ISS program and severe budget constraints, Mir will probably be decommissioned sometime in the year 2000.

In his January 1984 State of the Union address, former president Ronald Reagan called for a space station program that would include participation by U.S. allies. By 1985, Japan, Canada, and the European Space Agency had each signed a bilateral memorandum of understanding with the United States for participation in this project. Reagan named the station Freedom in 1988. Never strongly embraced by either the U.S. Congress or the American people (as the previous Apollo program had been), Freedom's design experienced modifications and "downsizing" with each annual budget cycle as Congress clamored for reductions in cost. In 1993, the Clinton administration called for Freedom to be redesigned once again to reduce costs and to include more international involvement, especially that of Russia, in the wake of the breakup of the Soviet Union in 1989. The Clinton White House selected a design option called *Space Station Alpha*, a configuration that would use about 75 percent of the hardware designs intended for Freedom. Starting in 1994, after the Russians agreed to participate, the station became known as the International Space Station.

Phase I of the ISS officially began in 1995 and involved more than two years of continuous stays by a total of seven U.S. astronauts on the Mir space station and nine Shuttle–Mir docking missions. This phase ended in June 1998. Phases II and III involve the on-orbit assembly of the station's components: In phase II the core elements are assembled and in phase III the various scientific modules.

In late November 1998 a Russian Proton rocket placed the NASA-owned, Russian-built Zarya ("sunrise") control module into a perfect parking orbit, a stable orbit into which a spacecraft is placed to await the arrival of another spacecraft for rendezvous and docking operations. A few days later in early December, the Space Shuttle Endeavour (STS-88 mission) carried the U.S.-built Unity connecting module into orbit for rendezvous with Zarya. Astronauts Jerry Ross and James Newman then performed three space walks to complete the initial assembly of the ISS.

The Russian service module (Zvezda) is scheduled to be launched by a Proton rocket from the Baikonur Cosmodrome in Kazahkstan in November 1999. Once in orbit, preprogrammed commands onboard Zvezda will fully activated its systems. It will then become the passive vehicle for rendezvous and docking with the orbiting Unity–Zarya assembly. In March 2000 an international crew will begin living aboard the ISS, starting a permanent human presence on the station. When they arrive, the ISS will consist of just three modules: Zarya, Unity, and Zvezda. Zvezda will serve as the living quarters and onboard control center for the early station assembly operations. In 2004 assembly will be complete and the ISS will be fully operational.

Uses, Effects, and Limitations

According to NASA administrator Daniel S. Goldin, "the space station is being built to see how people can live and work safely and efficiently in space." Working for extended periods, scientists can study healthy humans in a unique environment of prolonged weightlessness. ISS-conducted life-sciences research could provide new insights into the possible prevention and treatment of diseases, including those that affect heart, lung, brain, and kidney function, as well as cardiovascular disease, osteoporosis (bone loss), and hormonal disorders. The ISS will also accommodate unique, sustained microgravity experiments involving materials and metal alloys, fluids, and combustion science. Station advocates claim that such research programs could unmask interesting properties of materials and lead to better manufacturing processes and products on Earth.

The ISS is also a platform for observing Earth and space. Such observations could lead to a better understanding of our home planet and the universe in which we live. However, the ISS has limitations. For example, despite its clear view of the heavens and Earth below, the station is simply too jittery to accommodate high-precision measurements. Consequently, formal remote sensing programs involving Earth monitoring or astronomy are not being included in the ISS's mission (see REMOTE SENSING; SPACE-BASED REMOTE SENSING TOOLS). Although many interesting long-duration plant and animal development experiments can be conducted only in the sustained microgravity environment found on the station, it is presently unclear exactly how useful such experiments will be. Besides, NASA has learned from previous orbiting platform experience that the care of animals in orbit is a very messy, labor-intensive activity that could place undue or unnecessary stress on the crew. Monitoring human beings in a confined, synthetic habitat continuously exposed to microgravity could introduce some fortuitous medical discoveries. However, the bulk of such discoveries may be relevant only to other long-duration space missions, such as a human mission to Mars. Such missions are neither funded nor officially planned.

Issues and Debate

The most hotly debated issue concerning the ISS is whether its huge expense is worth its potential scientific and technical return. Even the strongest station advocates currently admit that although interesting science can be performed in areas of biotechnology and materials processing, the promise of rewards from this line of research really can't justify the cost of the station. Rather, they say, the station must be viewed as an inspiring human endeavor that promotes international cooperation and establishes a permanent human presence in space.

Opponents view the station as a politically driven "space adventure" that is draining scarce resources away from more important fields of research both in space and on Earth. A wide variety of other technical, political, and social issues are also tied to the ISS. The station is a very large, complex engineering project that is being incrementally assembled "untested" on orbit. Will it work after all the parts are in place? Nobody can say for sure despite countless modeling and simulation experiences on Earth. The real test comes only as the pieces are put together in the hostile space environment. Among many technical questions is this very human one: How will an international crew get along? Issues such as food selection, the use of alcohol, and even bathing frequency will definitely stress a small heterogeneous community of isolated humans in ways as yet unimagined. Will such stress lead to fights or unapproved work stoppages? Again, no one can say for sure and the success of a multibillion-dollar program could hinge on such issues. Finally, assembling this large complex in orbit will require many hours of "space walking" or extravehicular activity (EVA) by astronauts and cosmonauts. EVA is a very hazardous activity and a deadly accident could occur, potentially bringing the ISS program to an emotional halt.

—*Joseph A. Angelo, Jr.*

RELATED TOPICS

Remote Sensing, Reusable Launch Vehicles, Satellite Technology, Space-Based Materials Processing, Space-Based Remote Sensing Tools, Space Shuttle

BIBLIOGRAPHY AND FURTHER RESEARCH

BOOKS

Angelo, Joseph A., Jr. *The Dictionary of Space Technology*. New York: Facts On File, 1999.

Cole, Michael D. *International Space Station: A Space Mission*. Springfield, N.J.: Enslow Publishers, 1999.

Logsdon, John M. *Together in Orbit: The Origins of International Participation in the Space Station*. NASA History Office, Monographs in Aerospace History 11. Washington, D.C.: U.S. Government Printing Office, November 1998.

Neal, Valerie, Cathleen S. Lewis, and Frank H. Winter. *Spaceflight: A Smithsonian Guide*. Toronto, Ontario: Macmillan, 1995.

PERIODICALS

Beardsley, Tim. "Science in the Sky." *Scientific American*, June 1996, 64.

Boyer, William, Leonard David, and Theresa Foley. "Special Report: Building the International Space Station." *Final Frontier*, December 1994–January 1995, 20.

Guterl, Fred. "Castle in the Air." *Discover*, May 1997, 90.

Kistler, Walter P. "Humanity's Future in Space." *Futurist*, January 1999, 43.

INTERNET RESOURCES

Boeing Web Site for Space Station
http://www.boeing.com/defense-space/space/spacestation/

NASA Home Page
http://www.nasa.gov/

NASA Kennedy Space Center Web Site for Space Station Activities
http://www.ksc.nasa.gov/station/

NASA Web Site for Human Aspects of Spaceflight Related to Space Station
http://spaceflight.nasa.gov/station/

National Academy of Sciences
http://www.nas.edu/cets/aseb/coss

Principal NASA Web Site for Space Station Activities
http://station.nasa.gov/station/

INTERNET AND WORLD WIDE WEB

The Internet is often described as a "network of networks." It is made up of a huge number of computers that have been linked together across continents to enable the exchange of information, ideas, and more recently, products. Using an ordinary computer and a modem that links it, via telephone lines, to other computers, it is possible to tap into this vast resource via the World Wide Web and enter what has come to be known as cyberspace.

Hailed by some as the future of civilization and by others as a destructive technology that will inevitably divide society into distinct classes—those with Internet know-how and computer skills, and those without—the Internet looks set to change the way we live our lives, be it for better or for worse.

Scientific and Technological Description

In much the same way as one might subscribe to the services of a telephone company to get access to all of the world's telephone networks, ordinary users of the Internet must enlist the services of an *Internet service provider* (ISP) to gain access to the Internet. These individual ISPs are linked to even larger ISPs, which are responsible for maintaining the fiber-optic links that connect entire regions, or even countries, together (see FIBER OPTICS). These links are in turn connected across the world via further fiber-optic lines, undersea cables, or satellite connections. By using a computer linked to an ISP via a modem and a telephone line, the user can, in theory at least, connect to any other computer user in the world, provided that user is connected to the Internet.

Information on the Internet is usually accessed using an information retrieval service, the most popular of these being the *World Wide Web* (WWW). Based on a computer language, or protocol, called *HyperText,* the World Wide Web allows users to access Web sites or pages that are related to each other by means of *hyperlinks,* or *links.* Links are electronic connections that appear underlined on screen and, when clicked on with a mouse, connect one document to another across the Internet.

Web pages are written using the *HyperText Markup Language* (HTML) and assigned a unique on-line address, called a *uniform resource locator* (URL). (The URL is commonly known as the *Web address.*) To view a Web site, the user usually begins by typing the unique on-line address of the site into a *Web browser,* software that enables users to navigate the Web. (Netscape Navigator is one of the most commonly used browsers.) This unique on-line address will contain all the information the browser requires to make a connection to the site.

Once a URL has been entered by the user, the browser makes a link to the appropriate Web site and sends a GET command. On receiving the GET command, the Web site server, a kind of computerized electronic middle man that acts as a host for the site, retrieves the file that corresponds to the Web address and sends it back across the Internet to the browser as a stream of data. Special pieces of code in the data file, called HTML tags, carry information about the formatting of the text and indicating which parts of the text are hyperlinks. These are read by the browser, which uses them to reconstruct the Web page on the user's computer screen. From here the user can follow hyperlinks carried by the HTML tags to other Web pages

should he or she so wish. In this way it is possible to surf the internet, moving seamlessly from one Web site or page to another.

Historical Development

The origins of the Internet lie in a U.S. Defense Department program called the *Advanced Research Projects Agency Network* (ARPANet). Established in 1969 to provide a secure communications network capable of withstanding substantial damage in the event of war, the network was modeled on a spider's web. This allowed communications to be rerouted should any one section of the Web cease to operate. In practice, the network was used as a tool for communication between organizations engaged in defense-related research.

Other academic groups engaged in nonmilitary research that used the network on occasion realized the many obvious advantages of such a set up—chiefly, the ability to share information with colleagues across the nation. In 1984, the National Science Foundation adopted much of the existing ARPANet technology to establish a similar, parallel network known as NSFNet. The National Science Foundation Network originally enabled U.S. universities to share the resources of five regional supercomputing centers, but this was expanded when NSFNet established a "network of networks" capable of handling data at a rate of 45 million bits per second.

The cost of connecting to the Net came down during the early 1990s, as more ISPs started offering Internet access and competition among them drove the price of access down. It soon became cheap enough for ordinary civilian users to get online via a home computer. As the amount of information available on the Internet grew throughout the 1980s, it became necessary to set up information retrieval services capable of searching for and indexing information on the Net. Early examples of these include Gopher and Archie, which might be thought of as primitive browsers. These retrieval systems used text protocols to search for and retrieve information from remote locations over computer networks. But most laypeople, who didn't understand computer programming languages or commands, did not use these tools and therefore did not access information on the Internet.

Development of the World Wide Web, the primary Internet information retrieval service, began in 1989. British scientist Tim Berners-Lee and others at CERN, the Geneva, Switzerland–based international scientific organization, invented a computer language called the *HyperText Transfer Protocol* (HTTP). This protocol standardized communication across the Net, allowed the transmission of formatted text and graphics (and later multimedia features such as sound and video), and

How a Web Page is Accessed

Server machine running a Web server

1. Your browser connects to the server and requests a page.

Your machine running a Web browser

2. The server sends back the HTML page requested.

enabled users to surf effortlessly between Web sites. The World Wide Web gained broad acceptance after September 1993, with the development of the first browsing software, a product called Mosaic. This browser, which had been developed by American Marc Andreessen at the National Center for Supercomputing Applications at the University of Illinois, enabled users to surf the Net using the same point-and-click techniques that were by then in common use on home computers. Andreessen then went on to establish the Netscape Communications Corporation, which, in December 1994, released Netscape Navigator. This became the world's primary Web browser, although Microsoft's Internet Explorer browser became quite popular in the late 1990s as well. To date, there are an estimated 200 million people using the Internet on a regular basis, although this figure is in flux and bound to continue to rise in the near future.

Uses, Effects, and Limitations

In most respects, the Internet presents a whole new world of opportunities for those who have access to it. Without the need to leave his or her home, the individual user is able to find just about any kind of information, be it the boiling point of nitrogen, the availability of flights to Baltimore, or the likelihood of the Earth being hit by an asteroid in the next hundred or so years. Messages and documents can be sent anywhere in the world in an instant and numerous virtual and online versions of galleries, museums, and libraries (see DIGITAL LIBRARIES), which in the real world are spread across continents, can be visited in a single afternoon.

The Internet is also proving to be an invaluable and indispensable business tool. Daily correspondence, customer support, and product marketing across the Internet have made it as essential in the world of business as the telephone (see ELECTRONIC COMMERCE), and few self-respecting companies are without at least an Internet address and more often than not a comprehensive Web site.

Advocates of the Internet have been keen to stress the educational opportunities it presents. It has been claimed, for instance, that access to the Internet is like having a million consultants on hand to offer advice and information on every subject under the Sun. In reality, however, the Internet is primarily a source of entertainment for most of its users. Bulletin boards, chat groups, and networked games account for much of the traffic on the Net, although inquiries about pornographic sites remain as the most popular request, accounting for nearly 80 percent of all Web searches, according to the *Guinness Book of Records* (2000).

In many respects, the opportunities offered by the Internet are limited only by the imaginations of those who use it. It is a world within itself, and the nature of this world, which is sometimes called *cyberspace,* is determined by those who inhabit it. Although access to the Net is unlikely ever to be universal, the tumbling costs of getting online, combined with a reduction in real terms of the cost of com-

As the ability to use computers and the Internet becomes more important in the job market, it becomes increasingly important that all children learn computer and Internet skills in school. Observers fear that continued growth of the Internet could mark a new class division in American society: those with and without Internet know-how (Photo Researchers/Will & Deni McIntyre).

puters, means that the Internet will eventually be an integral part of everyday life for billions of people. Online shopping and banking are already on offer to those customers who have Internet access, and it is conceivable that other services, such as those offered by the medical or legal professions, will be carried out across the Net as a matter of course within a generation (see TELEMEDICINE).

Issues and Debate

The rapid and unchecked growth of the Internet has proved to be a source of worry for many. Of primary concern to some is the amount of sexually explicit adult material available on the Net. Parents who might not be as computer literate as their children have understandably expressed worries that they have little control over what their offspring might be viewing when free from parental supervision. Yet on the other side of this debate are civil-liberties groups, which are concerned about censorship and the violation of free-speech rights on the Web.

Similar concerns have also been expressed about the almost casual way in which copyright laws are being flouted on the Net. Extracts from books, samples of music, and photographs appear regularly at sites all over the Net—yet rarely, if ever, with the permission of the author, performer, or artist responsible for the original work. By its very global and sprawling nature, however, the Internet is extremely difficult to police, and efforts to impose some control over its content have so far met with charges of censorship and limited success.

Another cause for concern is the safety and privacy of financial transactions carried out across the Net. As more companies go online, the opportunities for criminals to gain possession of sensitive information such as bank account details and credit card numbers have grown accordingly, creating a whole new area of criminal activity known as *cybercrime.*

Perhaps most worrying of all is the growing feeling that certain groups of people are being excluded from the brave new world that is the Internet. Although in most respects the Net is open to all, those who lack the necessary computer skills and equipment will inevitably be excluded. In most cases this tends to be the poor, the elderly, and those who lack education, especially those who simply are not computer literate.

As the number of users has increased there has been a noticeable slowing down of Internet traffic. During the busiest periods, Internet searches become frustratingly slow and users are sometimes temporarily refused access to the Net by oversubscribed ISPs. Some regular users have taken to limiting their Net searches to off-peak periods, and it is now common practice in Europe to carry out the day's Net activity in the morning, before America "wakes up." Because so much Internet traffic is deemed to be nonessential, the possibility of limiting access by means of a ranking system, or even of creating a separate Net for academic and business users, has been proposed. Many users consider these solutions to be unacceptable, as they appear to run counter to the spirit of the Net, dividing what has been hailed as the first truly democratic global community.

—*Mike Flynn*

RELATED TOPICS
Digital Libraries, Electronic Books, Electronic Commerce, Fiber Optics, Internet Protocol Telephony, Internet Search Engines and Portals, Telemedicine, Web TV

BIBLIOGRAPHY AND FURTHER RESEARCH

BOOKS
Gralla, Preston. *How the Internet Works.* Indianapolis, Ind.: Que Education and Training, 1998.
Kehoe, Brendan P. *Zen and the Art of the Internet.* Upper Saddle River, N.J.: Prentice Hall, 1993.
Levine, John R., Carol Baroudi, and Margaret Levine Young. *The Internet for Dummies,* 6th ed. Foster City, Calif.: IDG Books Worldwide, 1999.
Moschovitis, Christos J.P., et al. *History of the Internet: A Chronology, 1843 to the Present.* Santa Barbara, Calif.: ABC-CLIO, Inc., 1999.
Shapiro, Andrew L., and Richard C. Leone. *The Control Revolution: How the Internet Is Putting Individuals in Charge and Changing the World We Know.* New York: Public Affairs, 1999.

PERIODICALS
Browning, John. "Cyberview: The Networking Computer." *Scientific American,* August 1996, 17–18.
Browning, John. "The Internet Is Learning to Censor Itself." *Scientific American,* September 1996, 38.
Lusted, Hugh S., and R. Benjamin Knapp. "Controlling Computers with Neural Signals." *Scientific American,* October 1996, 58–63.
Okerson, Ann. "Who Owns Digital Works?" *Scientific American,* July 1996, 80–84.
Wallich, Paul. "Wire Pirates." *Scientific American,* March 1994, 72–80.

INTERNET RESOURCES
CNET: How Does the Net Work?
    http://www.cnet.com/Content/Features/Techno/Networks/ss02.html
David Mayr's The History of the Net
    http://members.magnet.at/dmayr/history.htm
Entering the World Wide Web: A Guide to Cyberspace
    http://uu-gna.mit.edu:8001/uu-gna/text/internet/notes/tutorial-www/index.html
"How Web Servers and the Internet Work," by Marshall Brain
    http://www.howstuffworks.com/web-server.htm
World Wide Web Frequently Asked Questions
    http://uu-gna.mit.edu:8001/uu-gna/text/internet/notes/tutorial-wwwfaq/index.html

INTERNET PROTOCOL TELEPHONY

Internet protocol telephony refers to communication services, including voice calls, faxes, voice messaging, and teleconferencing that are transmitted over Internet protocol (IP)-based networks rather than the standard public switched telephone network (PSTN). The largest public IP network is the Internet. However, IP telephony is also commonly deployed on private networks such as corporate local area networks (LANs). In IP telephony, the analog signal (i.e., voice) from the caller is broken into discrete samples, digitized, inserted into data packets, and transmitted over the network. On the receiver's end, software reassembles the packets and converts the digital data back into the analog format required by the receiving telephone.

Although IP telephony began as a novel way to make inexpensive long-distance phone calls between personal computers (PCs), the extensibility and flexibility of this technology has spurred the major telecommunications companies to develop long-term strategies for merging their existing voice and data networks into a single IP network. Industry analysts are predicting aggressive migration of long-distance traffic from public phone lines to IP telephony over the next few years.

Scientific and Technological Description

To better understand IP telephony, it is helpful to review the basics of traditional telephony. The existing *public switched telephone network* (PSTN) is built on a network designed solely for voice, as opposed to data, transmission. This network offers a single basic service: two-way voice service with short end-to-end delays and a guarantee that once a call is accepted, it will run to completion. This service guarantee is achieved by setting up a dedicated circuit between callers. When the telephone system was first developed, this circuit was actually a dedicated electrical circuit. Telephone switching stations now convert the analog signal from a caller's telephone into digital voice samples. These samples are forwarded across high-speed links (*backbones*) to the receiver's local switching station, where the samples are converted back into analog signals that the receiver's telephone can understand. Although modern telephone systems digitize voice for transmission over long distances, they still reserve a dedicated circuit from endpoint to endpoint for the duration of a call—data traveling from the sender to the receiver always travels over the same path. The PSTN is therefore known as a *circuit-switched network*.

In contrast, an IP data network transmits all data in discrete packets (data networks are often referred to as *packet-switched networks*). A *packet* is a data structure consisting of the content, or *payload,* and meta information about the data called the *header.* The header contains information such as the address of the sender, the address of the receiver, and the length and type of payload data. The self-describing nature of a packet allows much greater flexibility. For example, whereas circuit-switched data flows over a dedicated connection between endpoints, packetized data can be transmitted over multiple network paths (a *virtual circuit*) before arriving at the reassembly and resequencing location. Since a data network does not reserve fixed resources for a call, a data network can better share its total bandwidth (the data-carrying capacity of the network) among all users. Data can also be routed around congestion or broken network segments. These factors make data networks more scalable (able to handle increasing traffic volume), less expensive, and easier to maintain.

Because of the popularity of the Internet, and because so many Internet users connect to their service providers over dial-up telephone lines, many public telecommunications networks now carry significantly more IP data traffic than voice traffic. These networks, optimized for voice traffic, are poorly equipped to handle increasing data traffic volumes

and increasing call length of Internet traffic. Furthermore, customers continue to demand new communication features (such as caller ID, call waiting, and call forwarding) and more integrated voice and data services (such as videoconferencing). Since voice-only networks are difficult to adapt to new applications, IP data networks are becoming more important as a means of carrying both voice and data in the public telecommunications infrastructure of the future.

Historical Development

IP telephony is a young technology. The first company to demonstrate the possibility of voice communications over the Internet was VocalTec Inc. In February 1995, the company released its Internet Phone software, which allowed PC users to communicate with each other by speaking into their computers' microphones and listening over their computers' speakers. This PC-to-PC telephony worked only if the sender and receiver were running the same software. The software compresses voice signals and translates it into IP packets for transmission over the Internet. Since nearly all Internet users pay a flat rate for unlimited usage, this technology opened up the possibility of nearly free long-distance calls. The only disadvantages were poor voice quality and the fact that the system did not integrate with standard telephones. Because IP networks do not guarantee quality of service or a consistent transmission rate, and because packets in an IP network can

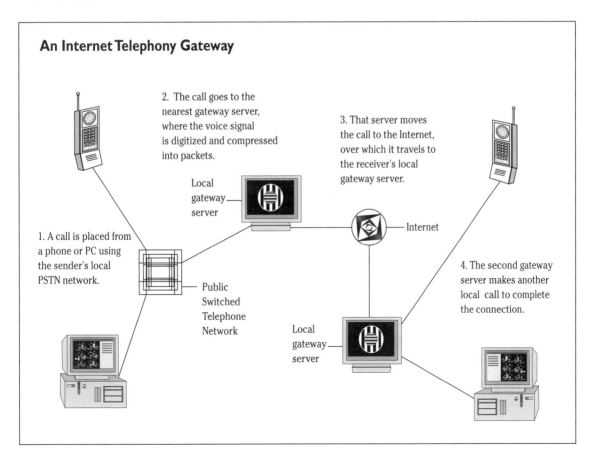

An Internet Telephony Gateway

2. The call goes to the nearest gateway server, where the voice signal is digitized and compressed into packets.

3. That server moves the call to the Internet, over which it travels to the receiver's local gateway server.

Local gateway server

Internet

1. A call is placed from a phone or PC using the sender's local PSTN network.

Public Switched Telephone Network

4. The second gateway server makes another local call to complete the connection.

Local gateway server

be lost or received out of order, early IP telephony conversations could be very choppy, with significant echo and delay, which made normal conversation difficult.

Despite these technical problems, within a year after VocalTec introduced Internet Phone, IP telephony had captured the world's attention. Even the early implementations provided a practical alternative to international telephone calling, based on cost alone. More efficient data compression techniques began to improve voice quality to the point where conversations became more natural. Currently, dozens of companies have introduced products to commercialize the technology, and every major telecommunications company has launched research to better understand this huge potential threat to their marketshare.

In March 1996, VocalTec and another company, Dialogic, announced that they were working together on the first IP telephony *gateway* (see figure). Gateways, which act as a bridge between an IP data network and a traditional PSTN, are the key to bringing IP telephony into the mainstream, since they offer the advantages of IP telephony to the cheapest, most abundant, and familiar endpoint in the world: the standard telephone. Gateways also solve the addressing problem in IP telephony. To send something to a remote network PC user, you must know their IP address. In contrast, gateway products allow callers to specify remote users by their telephone number. Calls go over the local PSTN network to the nearest gateway server, which digitizes the voice signal and compresses it into IP packets. Then those packets are moved onto the Internet for transport to a gateway at the receiving end. Gateways improve the computer-to-computer IP telephony picture by additionally allowing computer-to-phone, phone-to-computer, and phone-to-phone calls. Gateways also understand the low-level signaling system used by the PSTN (Signaling System 7, or SS7). This ability allows them to tap into existing "intelligent network" services such as local number portability, cellular phone authentication, and roaming and toll-free (1-800) service.

Uses, Effects, and Limitations

Although IP telephony is progressing rapidly, it still has some problems with reliability and sound quality, due primarily to limitations both in Internet bandwidth and current data compression technology. In light of these concerns, companies are finding it is easier to deploy IP telephony on their privately owned intranets (local IP-based LANs). Since intranets generally have more predictable bandwidth available than the public Internet, they can support full-duplex (two-way), real-time voice communications. IP telephony within an intranet enables a company to save on long-distance bills between corporate sites. Users can make inter-site calls via gateway servers attached to the LAN. In this configuration no desktop telephony software or Internet access is required.

Still, the ultimate goal of IP telephony is to allow reliable, high-quality voice service over public data networks such as the Internet or a future hybrid of the existing telecommunications network. The industry is currently addressing bandwidth limitations by upgrading the Internet backbone to *asynchronous transfer mode* (ATM). ATM, a switching technology designed to handle voice, data, and video traffic, allows software applications to specify their desired quality of service. Another emerging technology, *resource reservation protocol* (RSVP), is gaining popularity as another means of guaranteeing quality of service on IP networks. As this protocol is standardized and implemented in network routers, the quality of voice communications on the Internet will continue to improve.

The Internet industry is also attempting to solve problems of network reliability, sound quality, and vendor interoperability by adopting IP telephony standards. These standards are focusing on three core areas: the audio codec format (algorithms used to compress audio data), transport protocols (higher-level protocols above the Internet protocol, which specify how telephony data are transmitted), and directory services (the IP-telephony equivalent of white pages and yellow pages). Until recently, lack of standards had been a constant problem with IP telephony: Because products from different vendors did not work together, a company trying to build an IP telephony solution was locked into using products from a single vendor.

Issues and Debate

Technological debates revolving around the issue of establishing appropriate transfer standards are at the forefront of discussions about IP telephony. In May 1996, the International Telecommunications Union (ITU) ratified the H.323 specification. This document, based on the existing *real-time protocol* (RTP), defines how voice, data, and video traffic will be transported over IP-based local area networks. Early market IP telephony products, most of which used proprietary technologies to implement their services, are now being enhanced to support this emerging standard. Unfortunately, H.323 is extremely complex and some of its technical details are subject to interpretation, which has led to incompatibilities between products that claim to be H.323 compliant. Furthermore, some vendors create extensions to H.323 in order to give their products a market advantage. To some extent this defeats the purpose of standardization, since it limits vendor interoperability.

Even as H.323 continues to evolve and gain acceptance, a new proposal called *session initiation protocol* (SIP) has emerged with the potential to challenge its dominance. SIP, invented at Columbia University, is a much simpler text-based protocol modeled after the *hypertext transfer protocol* (HTTP), the foundation of the World Wide Web (see INTERNET AND WORLD WIDE WEB). Currently, SIP is being studied by the Internet Engineering Task Force (an international organization concerned with the evolution of the Internet) and is far from becoming a standard. As in many technology standards battles, H.323 may have established itself as the de facto standard simply because it was first to market.

Besides the debate over standards, government regulation of IP telephony is highly controversial. Telephone service has traditionally been heavily regulated. In many countries, governments or their agents retain monopolies for providing telephone service. IP telephony has stirred fears among traditional service carriers throughout the globe, many of whom are reacting by seeking regulatory protection from the new technology. In the United States, the America's Carriers Telecommunications Association (ACTA), a coalition of smaller long-distance carriers, filed a petition with the Federal Communications Commission (FCC) in March 1996 stating that "ACTA submits that the providers of [IP telephony] software are telecommunications carriers and, as such, should be subject to FCC regulation like all telecommunications carriers." At the root of the petition is the matter of local access charges. Whereas traditional long-distance telecommunications providers such as Sprint and AT&T must pay a surcharge to local service providers (in most cases, the "Baby Bells"), the FCC has thus far exempted Internet service providers (ISPs) from these charges. The U.S. Voice on the Net (VON) Coalition is one of several organizations devoted to educating legislators about the benefits of IP telephony and to minimizing the impact of future regulation.

In some countries, governments have taken radical action against IP telephony. In November 1998, the Czech Telecommunications Office banned the Paegas Internet Call product, citing that under Czech law the government has a complete monopoly in offering long-distance telephone service until the year 2001. Some governments, including those of India and Pakistan, have indicated they will regulate not only phone-to-phone IP telephony but also PC-based telephony. Other countries are still formulating their laws regarding this new technology.

While governments deliberate the course they will take, the worldwide telecommunications industry doesn't have any uncertainty about the ultimate outcome of IP telephony. Industry analysis firm International Data Corp. predicts that worldwide voice-over-IP use will increase to 41.2 billion minutes in 2000, up from 16.75 billion minutes in 1999. Market-research firm Jupiter Communications forecasts that 4.38 percent of all long-distance traffic will take place over IP by 2003; *TeleTimes,* an industry magazine, projects the same growth by 2001. Even PSTN carriers agree that voice and other types of data in the carrier arena are moving to IP. The question is whether Internet service providers or the established telephone companies will be the ones to provide the services that consumers demand.

—*Kevin Manley*

RELATED TOPICS
Fiber Optics, Internet and World Wide Web, Internet Search Engines and Portals, Web TV

BIBLIOGRAPHY AND FURTHER RESEARCH

BOOKS
Briere, Daniel D. *Internet Telephony for Dummies.* Indianapolis, Ind.: IDG Books, 1997.
Goncalves, Marcus. *Voice over IP Networks.* New York: McGraw-Hill, 1998.
Held, Gilbert. *Voice over Data Networks: Covering IP and Frame Relay.* New York: McGraw-Hill, 1998.
Kirk, Cheryl. *The Internet Phone Connection* (with CD-ROM). New York: McGraw-Hill, 1996.
Minoli, Daniel, and Emma Minoli. *Delivering Voice over IP Networks.* New York: Wiley, 1998.

INTERNET RESOURCES
"Study: IP Telephony Worth $8B To ISPs" (*Interactive Week* article)
  http://www.zdnet.com/intweek/daily/980305j.html
U.S. Telephone Association Home Page
  http://www.usta.org/index.html
Voice on the Net Coalition Home Page
  http://www.von.org/

INTERNET SEARCH ENGINES AND PORTALS

An Internet search engine is a system similar to the computerized card catalog at a library. It presents an interface to a database of indexed locations on the World Wide Web. The vast size of the Web, its lack of an organized structure, and the fact that the information it contains is constantly changing have made search engines a common starting point for users' travels on the Web.

Many search engines have begun taking advantage of this fact by transforming themselves into all-purpose Web gateways known as portals. Portal sites offer not only searching but also additional services such as e-mail, online chat, and shopping, all designed to keep visitors "stuck" to the site. Other large sites are attempting to cast themselves as portals by beginning to offer media services (news services, stock quotes, and video) or branded content such as a syndicated column, or a comic strip, and using search engine services provided by a third party. Although not every search engine site is a portal, every portal offers some level of search capability. Search engines and portal sites have adopted new business models on the Web and will continue to transform its character.

Scientific and Technological Description

It is important to clearly define the term *search engine,* since it is often misapplied to directory services or hybrid search services offered on the Web. Using a search engine, a user can enter a query ranging from a single keyword to a complex pattern of words. The search engine's software program matches the request against its database of resources, estimates the relevancy of the matches, and displays the resulting list back to the user in the form of hyperlinks. The user can then click through the links to visit the site of interest.

A *directory service* is a hierarchical index, similar to the Yellow Pages, that is compiled by human researchers, rather than a software program. Directories usually yield fewer but higher-quality results to user queries than search engines because human researchers manually evaluate the relevancy and quality of each site they list. The dominant directory service of the Web is currently Yahoo! Another site, the Mining Company, offers a directory compiled by a network of over 500 human "guides."

In contrast, every true search engine uses software programs to search and index Web sites automatically. Since these programs run continuously and follow every link on a page, they tend to find information beyond what is listed in directories. Queries to search engines usually yield more results at the expense of lower relevancy, since

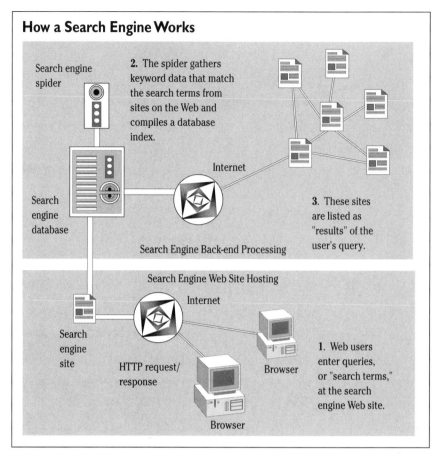

How a Search Engine Works

Search engine spider

2. The spider gathers keyword data that match the search terms from sites on the Web and compiles a database index.

Search engine database

Internet

Search Engine Back-end Processing

3. These sites are listed as "results" of the user's query.

Search Engine Web Site Hosting

Internet

Search engine site

HTTP request/response

Browser

1. Web users enter queries, or "search terms," at the search engine Web site.

Browser

they rely on simple text matching to rank (prioritize) "hits" (documents that contain those words). Search engines normally revisit their listed pages periodically to keep their indexes up to date. Among the currently dominant search engines are Excite, Lycos, Infoseek, AltaVista, and HotBot.

Search engine companies create their indexes using computer programs called *spiders,* which automatically visit Web sites and index the contents of the pages they find. Spiders examine the words in a document and store the frequency of each word, along with a reference to the document's URL, in a database. After processing a page, a spider follows each hyperlink it finds in the document, repeating the process until it runs out of links. A top-level or *root page* at a particular site may lead to an expanding tree of lower-level pages. When a spider has indexed all the pages in the tree, it has effectively indexed the entire site. At that point it moves on to the next site in its list and starts the process anew.

Most search engine companies operate popular Web sites where users can enter keywords into a form. A program on the company's Web server then searches its database and returns the URLs of the documents that match. The hits are usually ranked in order of relevance, with the highest-scoring results displayed near the top of the list. Some search engine companies, most notably Inktomi, do not run their own searching services. Instead, they contract with other Web publishers to provide them with search ser-

vices. Some search engine companies, such as Infoseek, both host their own search site and outsource search services to third parties.

Nearly all modern search engines run spiders on clusters of machines that are networked together. The spider on each machine is given a region of the Web to search. In some cases the data the spiders return are consolidated in a huge central database. Other companies use a distributed database that resides on several computers, with protocols that enable the individual machines to coordinate a search through the complete set of data.

Hybrid search services offer some combination of directory and search. In some cases the human-compiled directory is keyword searchable. In other cases, the directory is offered to supplement the database. Most directory and search engine companies are now finding that they must offer hybrid services to keep up with each other in the race for dedicated users. Portal sites are a natural extension of these hybrid search services, often aggregating content and services from other companies in a central location. Most of the major search engine sites are now considered portals, including Yahoo!, Excite, and Lycos.

Historical Development

The infrastructure of the early Internet, before the development of the World Wide Web, allowed people to access plain text-based documents (without hyperlinks, images, or formatting information) using what were called *gopher servers*. The first search engines targeted at gopher documents, called Veronica and Jughead, were developed in the early 1990s. These were text-based programs without graphical user interfaces that indexed documents and allowed document retrieval through the use of text-matching expressions.

What we now think of as modern search engines got their start in 1993 when six Stanford University undergraduates created the Architext system, the precursor to the Excite search site launched in October 1995. Around the same time, two Stanford Ph.D. students named David Filo and Jerry Yang began compiling a database of interesting Web sites they called Yahoo! The popularity of their searchable directory grew at an astounding rate, mirroring the general growth of the Web. Yahoo! is now the most visited site on the Web. Other notable search engines also came out of university research. Webcrawler was developed in early 1994 at the University of Washington, where it became so popular that its traffic overwhelmed the university network. It was purchased by America Online (AOL), which later sold it to Excite. Distributed computing technologies (see PARALLEL COMPUTING) have aided in the development of Web searching by contributing to the increasing capacity and performance of Web indexing systems.

Changes to HTML standards (standards for HyperText Markup Language, the language used to create Web-page text through coding with bracketed symbols, or tags, that designate how text will appear) have also helped improve automated searching. Early spiders simply stripped all HTML tags out of documents before examining them for keywords. An addition to the HTML standard around 1995 introduced attributes that produce META tags. These attributes, *keywords* and *description,* allow those who maintain Web sites to specify the keywords and summary information for their documents separately from the body of the document. Every search engine uses a different scheme for rating how relevant a document is to a user's search—most rank a document as more relevant if a search term appears among the META tag keywords in addition to the document text.

Uses, Effects, and Limitations

Information on the Web is growing and changing more quickly than it can be searched. In April 1998, a study reported in *Science* magazine estimated that there were over 320 million pages on the Web. Forrester Research of Cambridge, Massachusetts, estimates that there will be 1 billion pages available on the Web by the year 2000. A major measure of quality among the various search engines is the breadth of their precomputed databases. As of February 1, 1999, Search Engine Watch, a Web site dedicated to search engine issues, reported the following statistics for the major search engines. Note that even the largest index reaches less than half of the estimated pages on the Web.

Search Engine	Millions of Pages Indexed
Alta Vista	150
Northern Light	125
Inktomi	110
Excite	55
Lycos	50
Infoseek	45
Webcrawler	2

In general, a higher number of indexed pages improves the chances that a query will find relevant results. However, another important consideration is the method the search engine uses to categorize pages and judge relevancy. A frequent complaint among search engine users is that entering a common query can result in hundreds of thousands of matches. Since few users are willing to wade through so many possibilities, the overall number of indexed pages may not be as important as the methods used to return hits that are truly relevant.

Search engines and portal sites have found ample commercial uses and applications, as the possibilities offered by Web searching have been quickly exploited. Internet companies realized early on that the major factor in calculating their success would be the amount of overall traffic to a site—the raw number of page requests received from users' browsers. Web publishers think of traffic as inventory they can sell to advertisers. Every page view represents an opportunity to display a banner advertisement or deliver a marketing message. Web sites that display banners get paid by advertisers on the basis of the number of banners they display. They also receive a premium for each user who "clicks through" the banner to view the advertiser's site.

Although advertising alone was once considered a feasible revenue source for search engine sites, the industry standard unit pricing for ad banners, CPMs (cost per thousand banner views, or impressions), has declined steadily. Furthermore, a flaw in the banner advertising model is that when a user clicks through an ad, he or she is taken away from the site that hosted the ad and brought instead to the advertiser's site, possibly never to return. The portal strategy that developed out of these realizations centers around the notion of *stickiness,* that is, keeping users "stuck" to the portal site by offering such a wide range of services that they never feel the need to leave. To this end, the portals continue to add features that entice users to return.

The general-purpose nature of the first generation of search engines has created a market opportunity for specialty, or expert, search engines targeted to a single subject area. For example, devSearch searches only sites devoted to software development. Inomics provides a search engine tailored to economists. These sites are usually run by people who have a good knowledge of the content and who can guide their spiders to high-quality sites. What the sites lack in traffic they make up for in their ability to offer a more focused and better characterized user demographic. This allows finely targeted advertising, so these sites can command higher prices for their ad banner impressions and data about their members.

Users confounded by the fact that each search engine on the Web tends to produce different results for a given search can take advantage of a new trend called *metasearching*. Previously, to do an exhaustive search a user had to submit searches to several engines and compare the results manually. *Metacrawlers* solve this problem by providing a single location where a query can be dispatched to many search engines simultaneously. The metacrawler submits the query to each engine, reads the results, collates them by relevancy, and reformats the final output for consistency. Two popular metacrawlers are Dogpile and Mamma (see *Bibliography and Further Research,* below).

Search engine technology is fundamentally limited by the lack of a general structure among Web documents. Because there is no standard way of declaring the meaning of the text within a document, spiders by and large must rely on simple textual data to compile their indexes. This means that results can fail because of misspellings, use of related words that don't happen to appear in a document, or use of words that are spelled the same as words with different meanings. Some search engines also have trouble with "stemming"—for example, if a user enters the keyword golf, the engine must decide whether to also search for *golfing* or *golfer*.

Research into conceptual matching is attempting to solve these problems. Excite is currently the best-known commercial search engine that attempts to perform concept-based matching. It uses predefined rules that associate groups of related words with distinct concepts. By examining the frequency with which these words occur, the engine can make general inferences about the conceptual content of a document. For example, a search on *heart* and *lung* may turn up results regarding health and disease, whereas a search on *heart* and *flower* turns up results about Valentine's Day.

A final problem with search engines is the difficulty in programmatically judging the quality of the pages they index. Every search engine site wants to deliver links to high-quality content, but without human review, this is difficult to guarantee. This problem alone justifies the continued existence of human-compiled directories.

Issues and Debate

The way in which search engines are developed and applied over the course of the next few years will help to shape the course of the Web and how it disseminates information. The introduction of META tag attributes to aid search engine spidering has seen abuses similar to those associated with the plague of e-mail *spam* (unwanted e-mail solicitations, the electronic version of junk mail). Keyword spamming, or *spamdexing,* is the abuse of the META tag system by parties who misrepresent a Web page by including keywords in its META tag so that the page will be included in a list of search results. They may do this by repeating keywords, using keywords that are not actually pertinent to document content, or using another party's trademarks in keywords. The intent is to trick spiders into giving a document a higher ranking for certain keywords or to steal business from a competitor by using their trademarked terms. Many spiders now recognize spamdexing and will either ignore pages that use it or will blacklist the entire site from future listings.

Search engines have also been held accountable for the result links they serve. Lycos came under scrutiny in March 1999 for creating a searchable index of audio files known as *MP3 files*. The Recording Industry Association of America (RIAA), knowing that thousands of these files contain pirated (illegally copied) copyrighted material, sought to prevent Lycos from serving links to these illegal files. In one case, the legality of spidering a Web site at all was questioned. In early 1998, the *London Sunday Times* objected to the specialty search engine News Index indexing the *Times'* articles because the way News Index served links to the *Times* allowed users to bypass the *Times'* registration page. This case raised issues such as whether the *Times* might eventually lose customers as people found that they could access the same information through the News Index outlet; the paper might also lose potential income if it one day charged for their online delivery. Furthermore, the *Times* expressed concern that the indexing and abstracting of its content by an outside source was an infringement of its copyright. Although the parties settled amicably, similar cases will undoubtedly arise in the future.

—Kevin Manley

RELATED TOPICS

Intelligent Agents, Internet and World Wide Web, Parallel Computing, Web TV

BIBLIOGRAPHY AND FURTHER RESEARCH

BOOKS

Lind, Robin. *Webpointers Essential Search Engine Websites.* Manakin-Sabot, Va.: Hope Springs Press, 1999.

Szuprowicz, Bohdan O. *Search Engine Technologies for the World Wide Web and Intranets.* Charleston, S.C.: Computer Technology Research Corporation, 1997.

Testerman, Joshua O., et al. *Web Advertising and Marketing.* Rocklin, Calif.: Prima Publishing, 1998.

INTERNET RESOURCES

Internet and World Wide Web History

Learn the Net
   http://www.learnthenet.com

Selected Search Engine Sites

Alta Vista
   http://www.altavista.digital.com

DevSearch
   http://www.devsearch.com

Excite
   http://www.excite.com

Hotbot
   http://www.hotbot.com

Infoseek
   http://www.infoseek.com
Inomics
   http://www.inomics.com
Lycos
   http://www.lycos.com
WebCrawler
   http://www.webcrawler.com
Metasearch Engines
Dogpile
   http://www.dogpile.com
Mamma
   http://www.mamma.com
Search Engine Information
Search Engine Alliance
   http://www.searchenginealliance.com
Search Engine Watch
   http://www.searchenginewatch.com
Directories
The Mining Company
   http://www.miningco.com
Yahoo
   http://www.yahoo.com
Portal Information
Portal Hub
   http://www.portalhub.com
Selected Portals
Yahoo, Excite, Lycos mentioned above
America Online
   http://www.aol.com
Go Network
   http://www.go.com
Microsoft Network
   http://www.msn.com
Netscape Netcenter
   http://www.netcenter.com

IN VITRO FERTILIZATION

In vitro fertilization (IVF) is a technologically advanced method of treatment for infertility. It involves surgically removing an egg (oocyte) from a woman's ovary and fertilizing it outside her body. In vitro, meaning "in glass," refers to the glass Petri dish that is used for mixing the egg and sperm.

In many respects, IVF remains a method of last resort. It is highly expensive, and both physically and emotionally demanding for the couple involved. The technique remains highly controversial for ethical, legal, and medical reasons.

Scientific and Technological Description

In vitro fertilization refers to fertilization of the human egg cell outside the body. The notion of a "test-tube baby" (a child conceived by IVF), as popularized by the press, is really a misnomer. In IVF, fertilization takes place in a glass dish rather than a test tube and the young embryo is normally implanted in the mother's uterus (womb) at an early stage (see figure). IVF is one of a range of treatment options for *infertility* (an inability to conceive), or more correctly, for *subfertility* (a reduced capacity to conceive). Biologically, successful conception in humans is quite problematic. Data

compiled by organizations such as the American Society for Reproductive Medicine and the U.K. National Fertility Association suggest that among couples of reproductive age who wish to have children, between one in six and one in ten (between 10 and 16 percent) have problems conceiving.

In the United States and the United Kingdom, approximately one-third of infertility cases are attributed to factors that affect men, about one-third to female factors, and about one-third to problems that affect both partners or cannot be explained. Among men, absence of sperm or poor-quality sperm are the biggest factors. In women, failure to ovulate (release eggs) is the greatest factor. In the United States, 85 to 90 percent of infertility cases are treated using drug therapy, artificial insemination, or conventional surgery to repair reproductive organs. Less than 5 percent of infertility treatments involve IVF. IVF is used in cases where the woman has blocked, damaged, or missing Fallopian tubes (the tubes that carry eggs from the ovaries to the uterus), where men have poor-quality sperm or low sperm counts, or where infertility is otherwise unexplained. It offers hope to couples who would otherwise have no chance of parenting a biologically related child.

At its simplest, IVF involves the removal of ripe eggs from a woman's ovary; the collection and preparation of sperm to fertilize the eggs; creation of an environment outside the body where sperm and eggs are mixed together, fertilization occurs, and subsequent early stages in the growth of the embryo take place; and finally, the placement of viable embryos in the uterus of the mother-to-be. The main stages in IVF are as follows:

1. *Oocyte (egg) collection and preparation.* Ripe eggs are required. In some cases, the physician and patient opt to use the woman's natural fertility cycle. The physician closely monitors the woman's natural cycle by testing blood and urine for estrogens, examining the cervical mucus, and using ultrasound to visualize those ovarian follicles that are developing mature eggs. The physician is able to predict the time when the patient is about to ovulate (release one or more eggs). The egg (usually one) is collected just prior to ovulation. More often, the physician controls the normal reproductive cycle using superovulatory drugs or hormones so that the woman produces several mature oocytes in one cycle. By using several ripe eggs rather than just one, the chances of producing and implanting more than one viable embryo may be greatly increased. Whether the fertility cycle occurs more-or-less naturally or is primed to produce several mature oocytes in one cycle, the hormone human chorionic gonadotrophin is administered to trigger final egg maturation, and the eggs are obtained just prior to their release from the follicle. Oocyte collection is usually carried out under sedation or under general anesthetic using a needle inserted through the wall of the vagina or the bladder and guided by ultrasound imaging. The eggs are incubated under strictly controlled conditions prior to use.

2. *Sperm collection and preparation.* On the morning that the woman's eggs are collected, the male partner produces a semen sample by masturbating into a sterile tube. The semen sample is then washed in a liquid culture medium and spun in a centrifuge to separate the sperm from the seminal fluid.

3. *Insemination and fertilization.* A sample containing about 100,000 sperm per milliliter is added to the medium containing each oocyte. The sperm and oocyte are left in the incubator for up to 24 hours, during which time fertilization, if it is to occur, takes place. Fertilized oocytes (embryos) are graded according to appearance, and the best of these (in the United States, typically three or four) are selected for use. Any other embryos may be frozen for subsequent implantation, donated to other infertile couples, or given up for research.

4. *Embryo transfer.* Viable embryos are placed in the woman's uterus via a catheter that is inserted through the cervix (the entrance of the uterus). One of the greatest uncertainties with IVF is timing the transfer so that the embryo is ripe for implantation and the lining of the uterus is ready to receive it. Even naturally conceived embryos commonly fail to implant because of an apparent mismatch between embryo maturity and uterus receptivity.

During IVF procedures, sperm and oocytes are subject to close scrutiny and are handled delicately. This creates additional opportunities to improve the chances of conception. More sophisticated techniques involve injecting a sperm into the oocyte. Such techniques are technically difficult and are not available in all clinics. *Subzonal insemination* (SUZI) involves the injection of a single sperm cell between the outer coat of the egg and the cell's plasma membrane. *Intracytoplasmic sperm injection* (ICSI) delivers the sperm cell—mature or immature—directly into the cytoplasm of the oocyte. Both SUZI and ICSI require very delicate micromanipulation and have a relatively low success rate. If the technique can be improved, ICSI seems set to replace SUZI.

Gamete intra-Fallopian transfer (GIFT) is not, strictly, an IVF technique, but it uses many of the same procedures as IVF. In GIFT, sperm and oocytes are collected in essentially the same manner as in IVF, but instead of fertilization taking place outside the body, the eggs and sperm are placed in the woman's Fallopian tube, the natural site of fertilization. GIFT can be used only if at least one of the woman's Fallopian tubes is intact. GIFT's success rate is lower than that of IVF, but the cost per treatment cycle is considerably less.

Historical Development

In practice, IVF requires the management of biochemical and hormonal environments both within and outside the body. Developments in IVF drew upon efforts in many different fields of physiology and biochemistry. Mammalian reproductive hormones were first discovered in the 1920s. Controlled fertilization outside the body was depicted in Aldous Huxley's science fiction novel *Brave New World,* published in 1932. The fine structure of mammalian eggs and sperm was first revealed in the 1940s. During the late 1940s and 1950s, the wish to control reproduction in domestic animals, and so increase their breeding rate and produce strains with desired traits, resulted in major advances in reproductive technology. Among the achievements were the frozen storage of cattle sperm in 1949 and the transplantation of naturally fertilized sheep's eggs from one ewe to another in 1957.

In humans, development of the female contraceptive pill during the 1950s, and its approval for sale in the United States in 1960, was based on considerable research into the influence of hormones on fertility. In 1959, one of the principal investigators, M.C. Chang, reported the first mammalian birth resulting from IVF. He fertilized a rabbit egg in the laboratory using sperm recovered from the uterus of a second female. The resulting embryo was implanted in a third female and brought to term. By the early 1960s, progress in reproductive biochemistry and physiology was reaching the stage where IVF followed by implantation was fast becoming an achievable goal for humans as well.

In January 1961, Daniele Petrucci, a physiologist at the University of Bologna, Italy, announced that he had fertilized a human egg in vitro and had grown the resulting embryo for 29 days in the laboratory. He then destroyed the embryo, claiming that it was malformed. His work evoked a storm of protest from the Roman Catholic Church, and Petrucci abandoned further work in this area. The hostile reaction from church authorities deterred other researchers from publishing their work. Undoubtedly, a few had been engaging in similar studies.

In the early 1960s, Robert Edwards, a leading British reproductive physiologist, began work at Cambridge University. His move there was prompted by the refusal of the British Medical Research Council, a government-funded body, to support Edwards's transition from researching fertilization in pigs to studying the process in humans. At Cambridge, he was supported by the U.S.-based Ford Foundation. By the mid-1960s, Edwards was collaborating with British gynecologist Patrick Steptoe. In February 1969, they announced that they had achieved the fertilization of 13 human eggs (out of a total of 56) outside the body.

Between 1970 and 1978, Steptoe and Edwards were in the singular position of being funded to carry out research on human IVF and implantation at a time when others were barred from similar research. Such work was highly controversial in the United States, and was proscribed by the Medical Research Council in the United Kingdom. In 1974, a moratorium on publicly funded human embryo research was imposed in the United States. Steptoe and Edwards circumvented the problem by working in the United Kingdom with funding from private U.S. sources. The pinnacle of their achievement came in July 1978, when Louise Brown, the world's first test-tube

The IVF Process

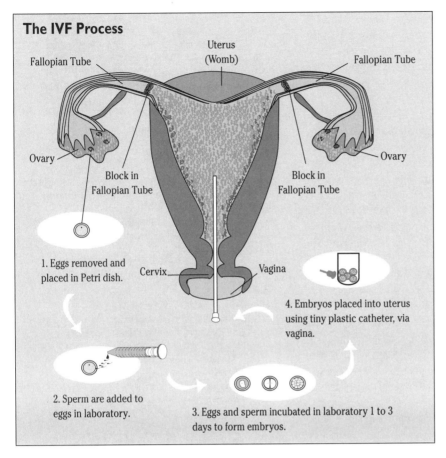

Fallopian Tube

Uterus (Womb)

Fallopian Tube

Ovary

Block in Fallopian Tube

Block in Fallopian Tube

Ovary

1. Eggs removed and placed in Petri dish.

Cervix

Vagina

4. Embryos placed into uterus using tiny plastic catheter, via vagina.

2. Sperm are added to eggs in laboratory.

3. Eggs and sperm incubated in laboratory 1 to 3 days to form embryos.

Adapted with permission from IGO Medical Group of San Diego.

baby, was delivered by cesarean section in the United Kingdom amid intense international media coverage.

Uses, Effects, and Limitations

IVF is a complex procedure whose success depends on the orchestration of events both within and outside a woman's body. British reproductive physiologist L.M. Baggot, writing in 1997, estimated that IVF failed for the following reasons: inability of the ovaries to produce follicles (15 percent of total failures), follicles inaccessible for oocyte collection (5 percent), oocytes not being fertilized (20 to 25 percent), embryos failing to develop normally (20%), and embryos failing to implant (over 30 percent). In a 1996 report in *Lancet,* success rates between 1991 and 1994 for IVF (live births per treatment cycle) in the United States were listed as 13.6 percent. Robert Winston, one of the world's leading experts on human fertility, has summed up IVF's limitations, stating: "IVF is still the most demanding, the most emotionally fraught, the most expensive and the least successful of all infertility treatments."

In the United States alone, well over 100 clinics offer IVF services. However, with an average success rate of about 14 percent for each IVF attempt, and with the cost of each IVF attempt averaging between $8000 and $12,000, the physical,

emotional, and financial costs of IVF can be prohibitively high, particularly where several IVF attempts fail. IVF is a method of last resort. It may result in a couple traveling many thousands of miles and spending many thousands of dollars to fulfill their desire to have a child. Many couples are more than willing to make this commitment.

Issues and Debate

Compared to less sophisticated methods of infertility treatment, the heightened control of IVF over the process of fertilization, together with the incubation of embryos outside the body, raises a great many legal and ethical issues as well as medical ones. Some religions forbid external fertilization for theological reasons, since the egg and sperm do not meet by natural sexual union and since the man's sperm is collected by masturbation.

In much current IVF practice, not all of the viable embryos produced in a single treatment cycle may be returned to the womb of the woman who provided the eggs. Deciding how many embryos to implant in the uterus is another contentious issue, because of the risk that more than one embryo will develop into a fetus. The more embryos a fertility doctor implants in a woman's uterus, the more likely it is that the woman will become pregnant, the goal of the procedure. But implanting three, four, five, or more embryos also increases a woman's chances of conceiving a multiple pregnancy (a pregnancy with three or more fetuses). Multiple pregnancies are medically inadvisable for many reasons, primarily because they greatly increase a woman's chances of miscarrying. To prevent multiple pregnancies, some countries, including Australia and the United Kingdom, limit to two or three the number of embryos that may be implanted in IVF procedures. This is not the case in the United States, where three or four, and sometimes more, are commonly implanted.

If more embryos are created than are implanted, surplus embryos create another set of problems. They may be frozen and stored, possibly for later use by the couple. But what if one or both members of the couple die before a decision is made as to the use of the frozen embryos? What will happen to the embryos then? The couple may wish to donate unused embryos to another infertile couple (embryos can be brought to term in the womb of an unrelated mother). This raises the possibility of frozen embryos—potential human beings—being bought and sold as a commodity.

Unused embryos may also be used for research into infertility, congenital disease, genetic testing and miscarriage prevention. But many pro-life groups campaign against such research on embryos and oppose IVF. Ultimately, the debate about embryo research, and whether it is morally acceptable, hinges on the issue of when life, or sentient life at least, begins. In the United States and United Kingdom, laws governing embryo research regard sentient life in humans as arising no earlier than in a 14-day-old embryo, and experiments are not permitted on IVF embryos older than 14 days. A 14-day-old embryo is less than a millimeter across and has yet to establish its central nervous system.

Manipulation of the embryo raises the possibility of selection for desired characteristics (or rejection of undesired ones) by GENETIC TESTING. The means does exist to determine the sex of an embryo prior to implantation, so a physician could control whether males or females are placed in the uterus. Both scientists and the wider public are concerned that greater manipulation of embryos could lead to "designer babies" and generate, in time, a culture in which embryos with "imperfect genes," and the babies that grow from such embryos, are seen as less acceptable.

IVF is expensive and both physically and emotionally demanding for the couple concerned. In the United States and United Kingdom, the issue of access to IVF is highly controversial, not least because national guidelines are lacking. In the United Kingdom, provision of IVF varies greatly from place to place. In some regions, the National Health Service, the nation's government-sponsored health care system, pays for IVF treatment for infertile couples if the couple meet certain criteria. These criteria vary, and may include age and marital status. In other regions, IVF is available only privately and at cost. In the United States, the provision of IVF varies greatly from state to state. As of 1998, only 12 states had laws requiring health insurers to offer coverage for some form of infertility diagnosis and treatment, although not necessarily IVF. Wealthier couples undoubtedly have much greater access to IVF wherever they reside.

—*Trevor Day*

RELATED TOPICS

Cloning, Gene Therapy, Genetic Testing

BIBLIOGRAPHY AND FURTHER RESEARCH

BOOKS

Baggott, L.M. *Human Reproduction.* Cambridge: Cambridge University Press, 1997.

Cooper, Susan L., and Ellen S. Glazer. *Beyond Infertility.* New York: Lexington Books, 1994.

Dunbar, R.I.M., ed. *Human Reproductive Decisions.* New York: St. Martin's Press, 1995.

Jansen, Robert. *Overcoming Infertility.* New York: W.H. Freeman, 1997.

Kaplan, Laurence J., and Rosemarie Tong. *Controlling Our Reproductive Destiny.* Cambridge, Mass.: MIT Press, 1994.

Pollard, Irina. *A Guide to Reproduction.* Cambridge: Cambridge University Press, 1994.

Turney, Jon. *Frankenstein's Footsteps.* New Haven, Conn.: Yale University Press, 1998.

Winston, Robert. *Infertility,* rev. ed. London: Vermilion, 1996.

Winston, Robert. *Making Babies.* London: BBC Books, 1996.

PERIODICALS

Cohen, Philip. "Dolly Helps the Infertile." *New Scientist,* May 9, 1998, 6.

Maranto, Gina. "Embryo Overpopulation." *Scientific American,* April 1996, 12.

"The Rights of the Many" (editorial). *New Scientist,* January 15, 1994, 3.

Wagner, Marsden. "IVF: Out-of-Date Evidence, or Not." *Lancet,* November 23, 1996, 1394.

Winston, R.M.L., and A.H. Handyside. "New Challenges in Human In Vitro Fertilization." *Science,* May 14, 1993, 932.

INTERNET RESOURCES

Frequently Asked Questions on Infertility
   http://www.asrm.org/patient/faqs.html

Gynecologic Web Library, Infertility Pages
   http://www.uc.edu./~pranikjd/infertil.html

International Council on Infertility Information Dissemination
   http://www.inciid.org/

Learning Center for Infertility
   http://www.womens-health.com/Infertility

LANGUAGE RECOGNITION SOFTWARE

Language recognition software enables computers to understand and respond to ordinary human language. Although computers process information using strings of binary numbers (zeros and ones), this form of communication makes little sense to humans. English-like computer languages were developed in the 1950s to make computers easier to use, but they have usually borne as much resemblance to mathematics and logic as to ordinary human language.

Today, computers can recognize speech, handwriting, and words printed in books and newspapers and convert them into natural language (the term computer scientists use to describe human languages, rather than computer languages and codes). Language recognition software has many applications, including palm-top computers that understand handwritten notes, advanced telephone exchanges that respond to the human voice, and computer systems that empower disabled people who are unable to use a keyboard and a mouse.

Scientific and Technological Description

Before human language can be processed by a computer, it must be converted into information that a computer can understand. Spoken words must be turned into a pattern of fluctuating electronic signals using a microphone and then *digitized* (converted to binary form) so a computer can

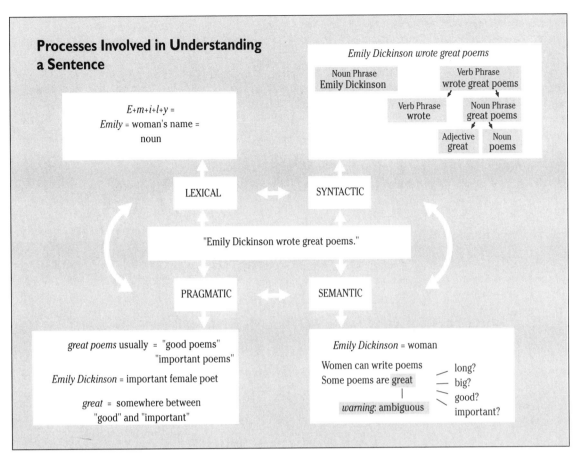

Processes Involved in Understanding a Sentence

$E+m+i+l+y$ =
Emily = woman's name = noun

Emily Dickinson wrote great poems

Noun Phrase
Emily Dickinson

Verb Phrase
wrote great poems

Verb Phrase
wrote

Noun Phrase
great poems

Adjective
great

Noun
poems

LEXICAL ⟷ SYNTACTIC

"Emily Dickinson wrote great poems."

PRAGMATIC ⟷ SEMANTIC

great poems usually = "good poems"
"important poems"

Emily Dickinson = important female poet

great = somewhere between
"good" and "important"

Emily Dickinson = woman
Women can write poems
Some poems are great
warning: ambiguous

— long?
— big?
— good?
important?

Four processes are at work when we understand a sentence: Lexical analysis interprets the letters and spaces and determines the words involved; syntactic analysis examines the syntax (grammatical structure) of the sentence; semantic analysis looks at the meaning of the sentence in isolation; and pragmatic analysis examines the sentence in the light of social conventions, context, and so on. The processes tend to overlap and feed into one another, so ambiguities in one process may be resolved with the help of others.

work with them. Similarly, written words must be digitized from a pattern of light and dark marks on paper. Once the computer is sure what words have actually been written or spoken, it can set about understanding what they mean.

Natural language can be input (entered) into a computer in a variety of ways. The simplest method is by typing a string of words on a keyboard. Other forms of input present harder problems. Printed words input using an optical scanner may be accurately recognized from a clean typewritten sheet if the scanning software recognizes the font (typeface), but ink spots or other marks, an unusual font, or too little space between letters can prevent the software from detecting reliably where one letter ends and another begins (see SCANNERS). When the input is handwritten instead of printed, there are extra problems. Not only does handwriting vary from one person to another, but even a single person seldom writes the same letters or words in precisely the same way. Most people connect their letters within words, which makes it hard for a computer to understand where one letter ends and another begins. Understanding speech adds a further degree of complica-

tion. There is much more variation in the way people talk than in the way they write (e.g., due to emotion). The handwriting of a person from Boston may be very similar handwriting to that of a person from London, but they probably speak the same words in quite different ways.

There are, then, essentially two stages in understanding language. The first, *recognition*, involves understanding what letters and words are being entered; the second, *processing*, involves understanding how these should be assembled into sentences, what those sentences mean, and how to respond to them. Printed or handwritten input is understood using pattern recognition software, which uses a combination of approaches. *Template matching* compares each scanned letter with a generalized model of each of the letters the machine recognizes, which are known as *prototypes*. *Feature matching* analyzes the components of the scanned letter; for example, it knows that two diagonal strokes that cross in the middle make an "x" whereas a horizontal stroke crossing the middle of a vertical stroke is a "+" sign.

Neural networks have also been used for pattern recognition (see ARTIFICIAL INTELLIGENCE). A neural network is a large, highly interconnected network of nodes or units that can recognize things if it is first "trained" by being presented with hundreds or thousands of examples. Another common approach to pattern recognition is based on statistical models that know the likelihood of one letter following another letter and one word following another word. For example, if the letters "p-l-e-a-s" have been recognized, the model can be fairly certain the next letter will be "e" or "u;" the word "thank" is often followed by the word "you."

Speech recognition software works in a similar way by identifying the 44 phonemes (distinct letter and letter-group sounds) of English instead of the 26 letters of the alphabet. The most advanced software uses statistical information to guess the most likely phonemes and words in the context of nearby phonemes and words. Recognizing written letters or spoken phonemes is only one part of language processing. The next steps (see figure) are lexical analysis (combining the letters into identifiable words), syntactic analysis (understanding the syntax of the sentence, which is the grammatical structure consisting of nouns, verbs, etc.), semantic analysis (what the sentence means), and pragmatic analysis (what the sentence means in the context of the wider text). Only when these stages have been carried out can the software begin to understand what the piece of language actually means, and only then can it carry out a request or formulate an appropriate reply (perhaps using artificial intelligence techniques).

Historical Development

Language recognition software got its start in the 1950s, when interest was growing in the parallels between computers and the human brain, which led to the closely related fields of artificial intelligence (the study of thinking machines) and cognitive psychology (a branch of psychology that investigates the human brain by speculating that it processes information like a computer). Both of these disciplines contributed greatly to the development of language recognition software. In particular, the branch of cognitive psychology known as *psycholinguistics* (the psychology of language) led to a better grasp of how people understand the structure (syntax) and meaning (semantics) of written and spoken language.

During the 1960s, artificial intelligence researchers tried to devise computer systems that could understand and respond to sentences in natural language. In 1966, Joseph Weizenbaum of the Massachusetts Institute of Technology (MIT) developed the ELIZA program, which could respond to typed sentences. A typical conversation might go:

ELIZA: Tell me about your problem.

PATIENT: I feel miserable.

ELIZA: I'm sorry to hear you feel miserable.

All ELIZA was really doing was picking out keywords and offering plausible rearrangements of the input sentences, but it contained the basis of syntactic analysis.

In 1972, Terry Winograd, also of MIT, developed a much more advanced program called SHRDLU, which could answer questions about a simple world made up of differently shaped building blocks. For example, it could understand a sentence as complex as "Does the shortest thing the tallest pyramid's support supports support anything green?" SHRDLU was capable of both syntactic and semantic analysis, and its impressive performance stemmed largely from the very limited nature of the building-block world it understood.

Neural networks, which had been pioneered in the 1940s, were investigated as a means of research into pattern recognition in the 1980s. In a neural network, different pieces of information are processed in parallel by a large number of interconnected units that work simultaneously. But since the 1990s, developers of language recognition software have concentrated mainly on using and improving statistical models that enable computers to recognize letters and language patterns. Another popular research strand in the 1990s investigates using artificial intelligence techniques to enable computers to build up semantic networks (detailed models of the information they read or hear and what it means) and progress from simply recognizing language to understanding and responding to it.

Uses, Effects, and Limitations

Language recognition software is already in widespread use. Inexpensive and accurate voice dictation software is widely available for personal computers. Those programs create documents in response to dictation. Palm-top computers and PERSONAL DIGITAL ASSISTANTS (PDAs) use handwriting recognition software. Telephone inquiry systems that feature voice recognition are already common. AT&T uses voice recognition to handle collect calls on its telephone network, and Bell Laboratories has developed a system called ShareTalk that can provide share price information on stocks in response to a spoken company name. Speech recognition is also finding important applications with people who are unable to use conventional computer keyboards and mice, including the disabled.

The simplest speech recognition software is known as *speaker dependent* and *discrete word;* this means that it recognizes only one speaker who must first train the system to the sound of his or her voice, and who must leave a short pause between each word. More advanced software is *speaker independent* and *continuous;* this means it can recognize potentially any speaker (because the manufacturers have already trained it to recognize a wide range of accents and emotions), and the speaker need not leave pauses between words.

There are, nevertheless, some important restrictions on what language recognition software can do. If it is speaker dependent, it is unsuitable for applications that might be used by any number of voices or accents, such as public information booths or automated telephone switchboards. Language recognition also requires considerable computer processor power and memory, which means that the miniature and portable computers that could benefit most from the technology are least likely to have the system resources to do so.

Although language recognition software will continue to make computers easier and more natural to use, it will not take over entirely from traditional methods of input. For example, high-security applications such as ATMs (cash-dispensing machines) will continue to require the privacy of the keyboard or touchscreen; voice recognition systems are unlikely to gain favor in open-plan offices even if they can filter out all the background noise. Any voice or handwriting system is eventually likely to encounter an accent or type of writing that falls outside the bounds of what it can recognize, so will need some method of calling on last-resort human assistance or asking for another kind of input.

Nevertheless, researchers have ambitious plans for systems based on language recognition software. Texas Instruments has developed a system that allows users to surf the Internet using voice commands alone. Elsewhere, artificial intelligence is being used to develop INTELLIGENT AGENTS that can answer and process telephone calls like a virtual secretary or search the Internet on the basis of a vague query and return a sensibly compiled report of information on a particular topic. And at Stanford University, language recognition software is forming the basis of Project SAGE, whose aim is to use speech recognition, multimedia computing, and speech synthesis to allow users to communicate with famous people who died long ago, such as Louis Pasteur and Ludwig van Beethoven.

Issues and Debate

Most people would agree that there have to be more efficient ways of getting information into a computer than typing it letter by letter. For some, the need is more pressing: Now that many jobs are computer-based, language recognition software could at last empower disabled people to compete in the workplace on equal terms with everyone else. Speech synthesizers, which can read computer documents out loud, allow blind people to use computers without the need for a display screen. British physicist Stephen Hawking, who suffers from a motor-neuron disease that has impaired his speech, uses a speech synthesizer to deliver lectures to his students at Cambridge University.

Voice recognition systems mean that paralyzed people who cannot use keyboards or mice can dictate words, enter computer commands, or guide a mouse pointer around the screen using only their voice. A 1998 remake of Alfred Hitchcock's film *Rear Window* featured *Superman* actor Christopher Reeve, who was paralyzed from the neck down following a riding accident, operating his computer using voice recognition alone. Instead of people having to make life easy for computers, in systems like this, computers are making life easier for people.

Much of the debate about language recognition software centers on the restricted way in which it has been applied so far. For example, handwriting recognition requires users to write on a slippery plastic tablet with a stylus. Speech recognition software often still requires that users sit next to their computers wearing a microphone headset and speaking, slowly and deliberately as though they were talking to a baby.

Some people embrace the idea of computers that can listen, understand, and talk back. Others, such as Ben Schneiderman of the University of Maryland, a pioneer of hypertext (a system of linking documents so that they can be accessed in a nonsequential manner rather than in a fixed order), argue that computers should be used not to mimic the things that people already do, but to help people do those things more creatively. They argue that a computerized bookstore should be designed, for example, so that people can find and order what they want as quickly as possible, rather than engage people in potentially time-consuming conversations such as one might have with a bookstore clerk. There is no real reason why computers of the future should not offer both options. Some users could be given a quick and easy-to-use menu system accessed by a keyboard or touchscreen; others may prefer to choose books by chatting at length to electronically reincarnated personalities such as Ernest Hemingway and Walt Whitman.

Many people find it difficult to use machines. Where people can be patient, flexible, and tolerant, machines can be harsh, logical, and unforgiving. But language recognition software bridges the gap between person and machine, opening up the possibility of a whole raft of devices that can talk or respond to spoken commands. In the future, doors may open simply by recognizing a person's voice; the small buttons on VCRs and cellular phones may be replaced by simple spoken commands; lights will probably be controlled by saying "lights on" and "lights off"; perhaps cars will even be driven by the commands "turn left" and "turn right." All this will revolutionize not just the ease with which people can use machines, but also the fundamental relationship between person and machine. As a result of language recognition software, people and machines will finally speak the same language.

—Chris Woodford

RELATED TOPICS
Artificial Intelligence, Intelligent Agents, Personal Digital Assistants, Scanners

BIBLIOGRAPHY AND FURTHER RESEARCH
BOOKS
Allen, J. *Natural Language Understanding*. Redwood, Calif.: Benjamin-Cummings, 1995.
Chomsky, Noam. *Language and Problems of Knowledge*. Cambridge, Mass.: MIT Press, 1988.
Crystal, David, ed. *The Cambridge Encyclopedia of Language*. New York: Cambridge University Press, 1997.
Edwards, Alistair, ed. *Extra–ordinary Human–Computer Interaction: Interfaces for Users with Disabilities*. New York: Cambridge University Press, 1995.

Gershenfeld, Neil. *When Things Start to Think*. New York: Henry Holt & Co., 1999.

Hofstadter, Douglas. *Gödel, Escher, Bach: An Eternal Golden Braid*. New York: Basic Books, 1979.

Negroponte, Nicholas. *Being Digital*. New York: Knopf, 1995.

Obermeier, Klaus. *Natural Language Processing Technology in Artificial Intelligence: The Scientific and Industrial Perspective*. New York: Ellis Horwood/Wiley, 1989.

Seureth, Russell. *Developing Natural Language Interfaces: Processing Human Conversations*. New York: McGraw-Hill, 1997.

PERIODICALS

"Computers that Listen" (editorial). *New Scientist,* December 4, 1993, 30.

Gibbs, W. Wayt. "Taking Computers to Task." *Scientific American,* July 1997, 64.

Holloway, Marguerite. "An Ethnologist in Cyberspace." *Scientific American,* April 1998, 20.

Lazzaro, Joseph. "Speech Enables the Common Desktop PC." *BYTE,* December 1997, 55.

"Profile: Humans Unite! Ben Schneiderman Wants to Make Computers into More Effective Tools—by Banishing Talk About Machine Intelligence" (editorial). *Scientific American,* March 1999, 21.

INTERNET RESOURCES

Home of Inquizit: a "Natural Language Understanding System" http://www.itpinc.com/

Microsoft Research Natural Language Processing http://www.research.microsoft.com/research/nlp/

New York Daily News Online Edition: "Your Voice Is My Command," by John Kosmer, December 6, 1998 http://www.nydailynews.com/1998-12-06/New_York_Now/Technology/a-12806.asp

Oregon Graduate Institute for Science and Technology http://cslu.cse.ogi.edu/

Stanford University's Applied Speech Technology Laboratory http://www-csli.stanford.edu/astl/ASTL.html

Textwise: a research and development company based at Syracuse University http://www.textwise.com

LASERS

Lasers are intense beams of light with special properties that allow them to be used for a wide range of applications, from microsurgery to computing and from communications to warfare. The word laser itself is an acronym for "light amplification by stimulated emission of radiation," a phenomenon discovered during the development of atomic physics in the early part of this century. Lasers have had a widespread effect on the modern world, and more applications are being discovered every day.

Scientific and Technological Description

A laser's unique properties arise from the fact that laser light is both *monochromatic* and *coherent*. In other words, all the light waves in a laser beam have exactly the same wavelength (and therefore color), and these waves all travel precisely in step with each other, so that the wave peaks and troughs reinforce each other, resulting in a very intense light source.

Normal light, such as sunlight, is a mixture of waves of different wavelengths, with each wave out of step with the others. Passing this light through filters can block out most of these wavelengths—resulting in a monochromatic beam, with a single color and a very limited range of wavelengths—but the waves remain out of step. This light cannot be coherent unless the individual photons of light were emitted simultaneously by the source, and even then, if the light rays have been allowed to diverge, they will fall out of step with each other.

Laser light therefore has to be produced using the only source of true monochromatic light: energy changes within individual atoms. Every atom of matter is composed of a heavy, positively charged nucleus, orbited in a series of fixed shells, or *orbitals,* by extremely light, negatively charged electrons. Each orbital corresponds to a different degree of electromagnetic attraction between the electron and the nucleus, and therefore a different *energy level.*

Normally, the electron orbits do not change: They remain as close as possible to the nucleus, keeping the atom's overall energy as low as possible. However, sometimes the atom is injected with additional energy (e.g., as a result of a collision with another atom). The additional energy given to the atom allows one of its electrons to jump to a higher energy level, becoming *excited.*

The atom can remain in this state for only a very short time before the electron drops back down to its previous level. However, as it does so, the excess energy is released as a photon of light. Because the energies of these *transitions* within each atom are always the same, the energy, and therefore the wavelength, of the photon of light released by a transition are predictable—a specific transition in an atom will release light of a specific wavelength.

A laser works by producing a huge number of identical atomic transitions simultaneously. This is possible because of a phenomenon called *stimulated emission*. Normally, an excited atom returns to its original state in less than one hundred millionth of a second, but specific atoms have *metastable states,* meaning that they can stay excited for up to a millisecond. If a photon of precisely the same energy as that produced by a transition strikes an atom in this metastable state, it can cause the atom to fall back to its lower-energy state early. The incoming photon is absorbed for an instant and then two identical photons are released by the atom. These photons can go on to cause similar stimulated emissions in other metastable atoms, eventually creating a cascade of identical, coherent photons—a laser beam.

However, for stimulated emission to work, a large number of metastable atoms must be produced for a long period. The creation of this *population inversion,* in which more atoms are in the metastable state than in their normal state, is the most difficult task in building a laser. Some types of laser can produce a population inversion for only a fraction of a second, resulting in a brief, intense flash of laser light (*a*

Solid-State Laser

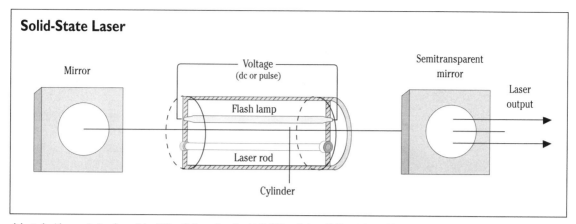

Adapted with permission from John Wiley & Sons, Inc. David Halliday, Robert E. Resnick, and John Merrill, Fundamentals of Physics, *John Wiley & Sons, Inc, 1992.*

pulsed laser), whereas others can sustain the population inversion indefinitely (a continuous-wave [C], laser).

Historical Development

The phenomenon of stimulated emission was predicted by Albert Einstein as early as 1917, but it took several decades for a working stimulated emission device to be produced. This was not a laser but a *maser* (microwave amplification by stimulated emission of radiation). The first maser was built in 1954 by Charles H. Townes and a team working at Columbia University. It used what we now know as the *laser principle* to generate a beam of coherent microwaves (a type of radio wave). Because microwaves have longer wavelengths and lower energies than light, they could be produced by comparatively small energy-level transitions in atoms, which made the production of a population inversion considerably easier. The pulses emitted by masers are so regular that they can be used as highly accurate timing devices, but they are most often used as amplifiers for weak radio signals (e.g., in radio telescopes).

Throughout the 1950s, physicists worked on the theory of producing a light maser and suggested various atoms and molecules with metastable states that could be used as *lasing media*. However, they had great difficulty in producing a working laser. The breakthrough was made by Theodore Maiman of the U.S.-based Hughes Research Laboratories in 1960, when Maiman produced a solid-state laser using a tube of crystalline ruby. Maiman's laser created a population inversion in chromium atoms in the ruby using a method called *optical pumping*. That method exposed the ruby to brilliant flashes of light from a white light source (similar to a camera flash) and thereby excited the chromium atoms to very high energy levels. As the atoms fell back to lower levels, they lingered in the metastable state, where stimulated emission could produce a cascade of photons with precisely the same energy. The laser beam was amplified by reflecting it back and forth between mirrors at either end of the ruby cylinder, and this also ensured that all the photons traveling in directions not perfectly parallel to the cylinder were absorbed by its walls.

Photons were allowed to escape by making one of the end mirrors semitransparent, and the result was a pulse of coherent, monochromatic and highly parallel photons with a wavelength of 694 nanometers (at the red end of the spectrum). This was the prototype *pulsed laser*.

Although pulsed lasers have found many uses, the more familiar and versatile today is the continuous wave laser. A CW laser is more difficult to produce than a pulsed laser, because some method must be found to produce and maintain a population inversion for long periods of time. Townes's group at Columbia, working with Arthur Schawlow's team at Bell Telephone Laboratories, announced the first CW laser just months after Maiman's ruby laser (the actual term *laser* was coined by Gordon Gould, a student of Townes'). Their helium–neon (HeNe) gas laser, built by Ali Javan at Bell Laboratories, found an ingenious way around the population inversion problem, using the metastable state of one gas as a reservoir of energy, and generating the actual lasing effect in a second gas.

The prototype gas laser used a mixture of 80 percent helium, 20 percent neon. Some of the helium atoms were split into negative electrons and positive ions by applying a high voltage across the gas mixture. When other helium atoms collided with these charged particles, they became excited and lingered in a metastable state, producing a population inversion in the helium. When these excited helium atoms collided with neon atoms, they transferred their energy, momentarily exciting the neon into an energy state close to that of the helium. By keeping a large proportion of the helium in the metastable state, it was possible to maintain a population inversion in the neon atoms as well, so that as they fell back to lower-energy states, they could trigger a cascade of stimulated emission events, producing laser light of wavelength 633 nanometers.

Uses, Effects, and Limitations

Lasers have uses across a wide array of fields. Industry and medicine take advantage of the intensity of lasers to use them as cutting tools. A focused laser beam can generate high tem-

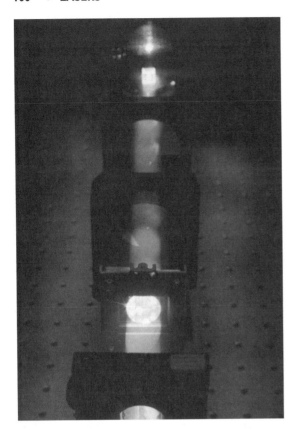

The latest development in laser technology is the first practical blue laser. Blue laser light has a shorter wavelength than red laser light and carries more energy. Blue lasers are expected to improve the storage capacity of compact disks and bring new precision to laser cutting, among other benefits (Corbis/Charles O'Rear).

peratures in a tiny area, thus can cut far more accurately than can a relatively blunt metal blade. The laser simply melts the material in its way while leaving its surroundings unaffected. Different intensities and wavelengths of beam are used for different tasks. For instance, in surgery, the long-wavelength infrared carbon dioxide laser is used as a cutting tool—the infrared light heats water in surface cells, causing it to boil, and destroys them. Shorter-wavelength lasers do not affect these surface cells and so can penetrate the body. Using two or more such laser beams meeting at a target inside the body, surgeons can kill dangerous cells such as cancer cells.

More complex uses for lasers arise from information technology. Perhaps the best-known application of lasers here is in HOLOGRAPHY—most familiarly used in the production of three-dimensional images on photographic plates but also useful for manufacturing optical instruments, analyzing materials, and storing data in a very compact form. A *hologram* is basically a photographic record of the interference pattern created by two light beams from the same source, one of which has been reflected off an object while the other has traveled directly to the photographic plate. The difference in the phase of the waves arriving at the plate creates interference fringes, which can be recon-

structed into the illusion of a three-dimensional image. However, the system works only with two absolutely identical coherent light beams, such as those produced by a laser.

Just as important, the fiber-optic cables that link many of the world's communications networks use lasers to carry information. Low-powered lasers, controlled by computers, fire invisible infrared laser beams along strands of optical fiber in extremely rapid pulses that encode digital information. Optical fibers are designed to reflect light off their walls with no absorption or loss of intensity, but become practical only when used in conjunction with lasers. A normal beam of light would not travel with its waves in parallel and so would still become weakened over long distances.

In fact, optical fibers and many other information technology applications rely on the most popular and widespread of all lasers, the semiconductor laser. These devices, invented in 1962, developed out of silicon chip and semiconductor technology (a semiconductor is a material, such as silicon, that conducts electricity). They use a layer of the semiconductor gallium arsenide to generate the laser, and amplify it by reflecting the light back and forth between two carefully cut crystal facets barely 0.2 millimeter apart. The resulting laser light spills out in a cone of rays from one edge of the semiconductor block; its intensity can be modified by altering the current applied to the semiconductor, allowing the laser beam itself to carry information. Another well-known application of semiconductor lasers is in compact disk players, where the laser is used to read the pits and lands on the disk surface (see CD-ROMS; DIGITAL VIDEO TECHNOLOGY).

Because of their unchanging properties, lasers are also ideal as a measuring device, in applications ranging from surveying to warfare. Because a laser beam will not spread out noticeably over even quite long distances, it can be aimed very accurately at a distant target. In many applications, this target is a reflector that bounces the laser beam back toward its source, where a detector is finely tuned to pick up that precise wavelength of light. By measuring the precise interval between the beam's departure and arrival, the distance it has traveled on its journey can be measured to a very high degree of accuracy. This principle has been used to measure the shifting of the San Andreas fault in California, and even the distance from the Earth to the Moon (using reflectors set up by NASA's *Apollo* astronauts).

The same basic principle of the reflected laser beam is used in military applications for targeting missiles. A laser beam fired at a target on the ground can be reflected back up into the sky and detected by a missile tuned to look for a specific wavelength of laser light. The missile can then fly down the reflected beam to zero in on the target. The beam itself can be fired either by a sniper on the ground or by an operator on board the bomber aircraft.

Recently, the Japanese company Nichia has pioneered the development of the first practical blue lasers. Blue laser light has significant advantages over traditional red laser light, arising from the fact that it has a shorter wavelength (enabling it to be focused more tightly) and carries more energy. Over the next few years, blue lasers are likely to create a second laser revolution, improving the storage capacity of compact disks and other laser-based storage devices, bringing new precision to laser cutting in industry and finding other uses that have not yet even been imagined.

Issues and Debate

Lasers are a fine example of "pure science" giving rise to a technology for which people then find uses. However, since their invention, lasers have had revolutionary effects on many aspects of modern life. Some applications simply use laser beams to replace earlier, more primitive techniques (such as in cutting, welding, and surgery), while others (particularly in the field of information technology) could not even have been thought of before the laser was a reality.

Most people would agree that the overwhelming effect of laser technology has been positive, but there is room for doubt in the field of warfare. The present applications of lasers on the battlefield are principally in improved targeting and are presented as a positive development, as they can help avoid the unwanted casualties and damage to civilian areas inevitably caused by less controllable bombing techniques. However, laser technology has also given rise to entirely new types of weapons. Since the 1950s, many military institutions have investigated the possibility of true laser weapons. At present, the U.S. Air Force has a fleet of attack laser aircraft on order for delivery in 2008. These are Boeing 747s fitted with high-energy gas lasers, capable of detecting and destroying enemy ballistic missiles at high altitude over ranges of several hundred miles.

A more ambitious laser program was the strategic defense initiative, developed in the 1980s, toward the end of the Cold War. This U.S. plan to protect the West with a network of satellite weapons platforms capable of destroying ICBMs (intercontinental ballistic missiles) launched from the Soviet Union would stretch even today's technology to its limits, and was eventually scrapped. If it had gone ahead, many analysts predicted it could have dangerously shifted the delicate nuclear stalemate of the time. This lesson, that perceived invulnerability is not necessarily a good thing, may prove just as applicable as these laser-dependent superweapons are introduced to the conventional battlefield.

—Giles Sparrow

RELATED TOPICS
CD-Roms, Digital Video Technology, Fiber Optics, Holography, Smart Weapons, Surgical Tools

BIBLIOGRAPHY AND FURTHER RESEARCH

BOOKS
Harbison, James P., and Robert E. Nahory. *Lasers: Harnessing the Atom's Light*. New York: W.H. Freeman, 1997.
Hecht, Jeff, and Dick Teresi. Laser: *Light of a Million Uses*. New York: Dover Publications, 1998.
Silfvast, William T. *Laser Fundamentals*. New York: Cambridge University Press, 1996.
Svelto, Orazio. *Principles of Lasers,* 4th ed. New York: Plenum Press, 1998.
Townes, Charles H. *How the Laser Happened: Adventures of a Scientist*. New York: Oxford University Press, 1999.
Whinnery, John R., and Jesse H. Ausubel, eds. *Lasers: Invention to Application*. Washington, D.C.: National Academy Press, 1989.

PERIODICALS
Laser Focus World

INTERNET RESOURCES
Rockwell Laser Industries RLI LaserNet
    http://www.rli.com/index.html

LIGHT-ACTIVATED DRUGS

Light-activated, or photoactivated, drugs are drugs that absorb or react with light to produce an intended effect. Plants, animals, and many bacteria have long made good use of light. Plants grow by trapping energy from light, for example. Now scientists are using light-absorbing molecules in their fight against cancer and other diseases, by taking advantage of the fact that energy from light causes some normally harmless molecules to become toxic.

Light-activated drugs have been used successfully to treat many conditions caused by tissues growing too fast. These include cancers and tumors, skin diseases such as psoriasis, and diseases of the eye that result in blindness. Doctors give patients light-activated drugs, then shine light on a part of the patient's body where unchecked cell growth needs to be stopped, such as a tumor. Although this type of therapy can cause unpleasant side effects, new light-activated drugs are being developed that cause fewer or less severe unwanted effects, and scientists are continually working to improve their performance. Some light-activated drugs are also being developed for more general use, for example to sterilize water.

Scientific and Technological Description

Light has energy, and this energy can be converted into other forms of energy, such as heat. Plants capture the energy from sunlight using a molecule called chlorophyll, and this captured energy is used to make sugar in a process called photosynthesis. There are many molecules that use the energy from light in a more harmful way. On a sunny day, if cows eat the yellow flowers of St John's wort (*Hypericum perforatum,* a plant that is commonly extracted and used to treat depression in humans), they become ill. Moving the cows to a dark barn allows them to recover. The reason they become ill only in sunlight is because the flowers contain a molecule called *hypericin* that absorbs the sunlight's energy and becomes toxic.

Exactly why light-activated hypericin is toxic is not well understood at the moment, but there are other light-activated molecules that scientists know more about. Some molecules hold onto the energy from the light and become

Light-Activated Drugs

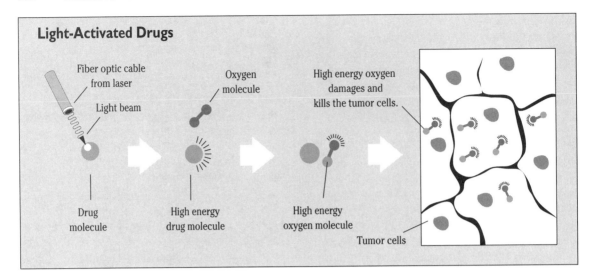

Fiber optic cable from laser

Light beam

Oxygen molecule

High energy oxygen damages and kills the tumor cells.

Drug molecule

High energy drug molecule

High energy oxygen molecule

Tumor cells

what are called *high-energy species*. Within a fraction of a second, these high-energy molecules bump into some other molecule, passing on their energy in the process. The receiving molecule is chemically changed (damaged) when it receives this energy: It might break into two parts or it might become joined onto some other molecule. Light-activated, high-energy species can cause enough damage to kill living tissues. Some light-activated molecules do not cause tissue damage directly themselves but cause the production of high-energy oxygen. It is the high-energy oxygen that actually causes tissue damage in these cases.

Light-activated molecules can be used as drugs. When someone suffering from a tumor is given a light-activated drug, the drug first travels harmlessly throughout the person's body. Sometime afterward, a laser is shone onto the tumor tissue. The drug in the tumor becomes photoactivated and the tumor tissue is killed as a result. The healthy tissue around the tumor is kept dark and so is not killed by the treatment. This type of therapy, involving light and light-activated drugs, is called *photochemotherapy* (see *Uses, Effects, and Limitations*). If the drug used causes the production of high-energy oxygen, the treatment is also called *photodynamic therapy* (PDT). The light used to photoactivate drugs is often provided by a laser, but only low levels of light are required and the light itself does not usually have any effect on the tissue.

Historical Development

Light-activated drugs and sunlight have been used to treat vitiligo for thousands of years. People with vitiligo have areas of unpigmented skin, which is particularly disfiguring for those with darker skin. In 2000 B.C. in Egypt, plant extracts containing light-activated drugs called *psoralens* were used to repigment the skin, and the treatment involved exposing the patients to sunlight. Psoralens were also used in India in 1200 B.C. to treat the same disease, although the Indians and Egyptians used different plant extracts in their remedies—

the Indians used the fruits of the plant *Psoralea corylifoli,* and the Egyptians used boiled *Ammi majus* seeds and plants.

The modern study of light-activated drugs can be traced back to 1900, when the German scientist Oskar Raab was assessing the toxicity of a dye called acridine. On a sunny day the acridine killed a single-celled organism called *Paramecium,* but not during a thunderstorm, when the sky was overcast. Raab realized the importance of light in the toxicity of the dye. Another major step came in 1948, when an Egyptian physician, A. M. El Mofty, isolated the pure psoralen compounds from *Ammi majus* and used them to treat vitiligo. This is the first real example of modern photochemotherapy using pure preparations of drugs. However, El Mofty didn't know why light was important in the treatment.

In 1974, J. A. Parrish and other dermatologists at Harvard University first suggested that light is needed to photoactivate the psoralens. They proposed the term *photochemotherapy,* and they were the first to use psoralens to treat psoriasis. Psoriasis is a skin disease caused by too much skin growth—skin hyperproliferation—which results in red, raised, scaly plaques over the skin. Photodynamic therapy for the treatment of cancer was developed in the 1970s and 1980s by Thomas Dougherty at the Roswell Park Cancer Institute in Buffalo, New York. This work resulted in the first licensed use of a light-activated drug, Photofrin, launched in the United States in 1996, for the treatment of certain tumors that could not be treated by surgery or conventional chemotherapy.

Uses, Effects, and Limitations

Photochemotherapy kills all living tissue, but it is most effective at killing tissues that are proliferating or growing, called diseases of hyperproliferation. To date, the hyperproliferative diseases that have been treated in this way include: tumors and cancers; blood vessels that are becoming blocked (angiopathies); blood vessels growing on the retina of eyes (a major cause of adult blindness); and the hyperproliferative skin disease, psoriasis. Most light-activated drugs in use today do not "home" in on diseased tissue; rather, tissues that are hyperproliferating tend to hold onto the drug longer than do other tissues. In theory there comes a point some time after giving the drug that normal tissues do not contain much drug but hyperproliferating tissues still contain enough to be severely damaged by light, although in practice all tissue that is exposed to light will be damaged to a degree. Even so, patients often have to wait for one to three days after receiving the drug before they are given the light treatment, which allows the drug to accumulate in their hyperproliferating tissues.

Two classes of drugs are now commonly used in photochemotherapy, the psoralens and the porphyrins.

Psoralens

The psoralens have been used to treat vitiligo, psoriasis, and a cancer of the blood called *cutaneous T-cell lymphoma* (CTCL). Psoralens are photoactivated by ultraviolet A light (UV-A), so this type of treatment is called *PUVA therapy.* Photoactivated psoralens cause damage to the genetic material (DNA) in cells, and this may partially explain how psoralens kill cells. Vitiligo (patches of skin lacking pigment) is not a hyperproliferative disease; psoralens somehow treat this disease by stimulating the production of the skin pigment melanin. Scientists do not fully understand how psoralens work to treat vitiligo. In PUVA treatment of psoriasis and vitiligo, the drug is applied directly to the skin, and then about two hours later the skin is exposed to UV light. For the treatment of CTCL, a patient's blood supply is actually treated outside the person's body with UV light, a process called *photophoresis.*

There are some side effects associated with PUVA treatments. Patients become photosensitized—that is, their skin becomes very sensitive to light and becomes badly

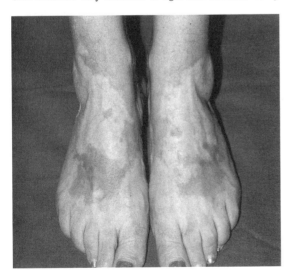

Light-activated drugs have been used to treat vitiligo, a disease in which people have areas of unpigmented skin, for thousands of years (Photo Researchers/Dr. P. Marazzi).

burned easily. Patients must stay out of sunlight or bright lights for several days after treatment. Many patients also feel dizzy or nauseated after treatment. The long-term effect of psoralen treatment is an increased risk of cancer, because both the photoactivated psoralens and the UV light cause DNA damage, which in turn can cause cancer. There is also an increased risk of cataracts (in which the lens of the eye becomes clouded over), caused by psoralens in the eye becoming photoactivated. New-generation drugs are being developed that do not have these problems. For example, 3-CPs (3-carbethoxypsoralens) do not cause DNA damage and are therefore not carcinogenic, and they also do not cause so much skin photosensitization.

Porphyrins

The porphyrins drugs are quite similar to the dye that makes our blood red. Unlike psoralens, which absorb UV light, porphyrins absorb visible light, Porphyrins cause the production of high-energy oxygen molecules, so treatments involving porphyrins are called *photodynamic therapies.* One class of porphyrins in particular is important in photodynamic therapy: the *hematoporphyrins* and their derivatives (HPDs).

Photofrin (sold by the Canadian company QLT Phototherapeutics) was the first light-activated drug licensed for the treatment of cancers and precancerous conditions. It is actually a mix of HPDs, and it is now used to treat many forms of cancer in which light from a laser can be directed onto the cancer. This is relatively simple for skin and mouth cancers, but light from a laser can also sometimes be directed to a cancer deep in the body by means of a fiber-optic cable (see FIBER OPTICS). Photofrin is given to patients intravenously, and then one to three days later the patient returns to the hospital to receive the laser treatment. Within a few days of treatment, the tissue exposed to the light becomes severely damaged and hopefully killed, while the surrounding healthy tissue remains unharmed.

The side effects of Photofrin are similar to those for psoralens. Patients are often nauseated, and because the drug causes skin photosensitivity, they must take care to stay out of strong light for as long as six weeks following treatment. Newer drugs in development, such as Levulan (aminolaevulinic acid [ALA]), reduce this time to one or two days. Levulan also can be given orally instead of intravenously, which is an advantage. Unlike psoralens, porphyrins do not cause DNA damage (because porphyrins do not get to where DNA is found in cells), so they do not have a tendency to cause cancers themselves.

In addition, the red light that is used to photoactivate porphyrins is much safer than the UV light used to photoactivate psoralens.

Other Drugs

Several new light-activated drugs are being developed to treat specific diseases. For example, QLT Phototherapeutics is developing one called Visudyne to treat choroidal neo-vascularization (the growth of new blood vessels in the retina of the eye). This disease is a common cause of blindness because the growing blood vessels can burst or leak, damaging the retina. Another new drug, Antrin, is being developed to treat blocked arteries (a major cause of heart attacks). This drug is taken up selectively by the cells that block arteries and so can be used to kill these cells while leaving the rest of the blood vessel unharmed. Because microorganisms can be killed by light-activated drugs, compounds that are not used medically have been attached to materials such as cloth and chopping boards, to make these materials self-sterilizing. These materials might be useful in places where hygiene is especially important, such hospitals and restaurants. There are other possible future uses for light-activated drugs, too: Some researchers are experimenting with putting light-activated drugs in toothpaste or using the drugs to sterilize water.

Issues and Debate

Photochemotherapy is a useful addition to the tools that doctors have to treat several medical conditions, although it is not yet used as an alternative to conventional treatments such as surgery. Both the drugs and the lasers required are expensive, but patients can be treated on an outpatient basis without the need to fill hospital beds for lengthy periods, which saves money, helping to contain rising health-care costs. An additional advantage is that the drugs can be used in cases where standard treatments such as surgery are not possible. As far as patients are concerned, light-activated drugs can greatly increase the chances that their lives will be prolonged. The drugs do have side effects (see above), but these are generally a small price to pay for the benefits of the treatment. Compared with today's light-activated drugs, the next generation of drugs will certainly have side effects that are less severe or shorter in duration.

The use of light-activated drugs for nonmedical uses—to sterilize water or special clothing, for example—might be cause for concern in cases where there is no real justification for sterilizing these materials in the first place. Some people have the impression that being exposed to microorganisms such as bacteria is always undesirable. Many products that feature antibacterial agents, such as liquid soap, are now being marketed and have been quite popular with consumers. But in fact, people benefit greatly from many such microorganisms, and our immune systems depend on being exposed to small numbers of microorganisms in order to fight serious infections effectively. Used sensibly, however, light-activated drugs could be an effective way of killing microorganisms when necessary.

—Matthew Day

RELATED TOPICS
Drug Delivery Systems, Fiber Optics, Surgical Tools

BIBLIOGRAPHY AND FURTHER RESEARCH

BOOKS
Foye, William, Thomas Lemke, and David William. *Principles of Medicinal Chemistry,* 4th ed. Baltimore: Williams & Wilkins, 1995.
Tannock, Ian, and Richard Hill. *The Basic Science of Oncology,* 3rd ed. New York: McGraw-Hill, 1998.

PERIODICALS
Barr, Hugh. "Gastrointestinal Tumours: Let there Be Light." *Lancet* 352(9136); October 17, 1998, 1242–1244.
Bown, Stephen G. "New Techniques in Laser Therapy." *British Medical Journal* 316; March 7, 1998, 754–757.
Moore, Pete. "Lethal Weapon." *New Scientist* 158(2130); April 18, 1998, 40–43.

INTERNET RESOURCES
"Blasting Cancer with Lasers," by Kelli Miller
http://www.aip.org/inside_science/scripts/109.htm
Cancer Treatment, Including Some General Information About Phytodynamic Therapy
http://www.phrma.org/guides/cancer/index.html
Information About Medical Lasers
http://www.medlight.com/
http://www.diomed-lasers.com/
Roswell Park Cancer Institute Web Site on Phytodynamic Therapy
http://rpci.med.buffalo.edu/departments/pdt/pdt.html
Sites Concerned with Biology and Light (Photobiology)
http://www.kumc.edu/POL/oth_site.html

MAGNETIC RESONANCE IMAGING

Magnetic resonance imaging (MRI) is a diagnostic medical technique that generates high-resolution images of the internal parts of the human body by utilizing the magnetic and spin properties of atoms, particularly those of hydrogen atoms. MRI scans yield cross-sectional or three-dimensional images that are particularly effective at revealing soft-tissue structure.

Since its introduction in the early 1980s, MRI has undergone very rapid development. MRI is still an expensive procedure compared with scanning techniques such as X-ray Imaging and Ultrasound. Nevertheless, MRI has certain advantages, particularly its apparent safety combined with its ability to discriminate fine detail of tissue structure and function. MRI is poised to become a much more important scanning methodology in the 21st century.

Scientific and Technological Description

Magnetic resonance imaging utilizes the magnetic and spin properties of atomic nuclei, particularly those of hydrogen atoms in water molecules. Soft tissues and body fluids contain abundant water. Since the amount of water, and its physical and chemical environment, vary from one tissue to another, MRI has the potential to discriminate between many types of body tissue and to distinguish healthy from diseased tissue. For example, cancerous tissues grow rapidly and divert blood flow to their vicinity; MRI can visualize this increased vascularization. Similarly, the gray mat-

A Magnetic Resonance Imaging Machine

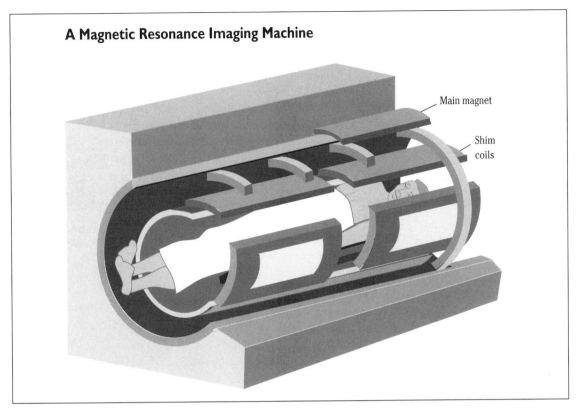

Main magnet

Shim coils

A full-body MRI scanner is depicted above. The main magnet produces a field that aligns the hydrogen nuclei in body tissues. The shim coils are smaller magnets. They create magnetic gradients that pinpoint the spatial arrangement of hydrogen nuclei. The radio-frequency (RF) transmitters and receivers are not shown. (Adapted from Durant et al., Encyclopedia of Science in Action, *Macmillan Publishers Ltd., 1995.)*

ter (tissue rich in cell bodies) of the central nervous system has a richer blood supply than does white matter (tissue rich in nerve fibers), so MRI can distinguish between the two types of tissue. More than this, MRI can pick up subtle differences in the physical and chemical environment of atomic nuclei, so is able to distinguish healthy from diseased tissue by detecting biochemical differences.

During an MRI procedure, the patient wears metal-free clothing, and any loose metal items are excluded from the examination room. The magnetic fields generated by an MRI machine are sufficiently high to cause metal objects to fly through the air like projectiles. The room is also screened from external radio waves.

When a patient is given an MRI scan, the body is subjected repeatedly to very powerful magnetic fields. Within the body tissues and body fluids, each hydrogen nucleus acts as a miniature bar magnet and aligns itself momentarily with the magnetic field. Radio-frequency (RF) energy is then applied to generate a second magnetic field, which causes the hydrogen nuclei to flip or rotate away from the static magnetic field. The term *resonance* refers to the fact that the atomic nuclei pick up energy from the radio waves and vibrate or resonate in harmony with them. The degree of rotation of the atomic nucleus depends on the amount of RF energy it absorbs. Once the RF field is switched off, the

hydrogen nuclei flip or rotate back and realign themselves with the magnetic field. This process is called *relaxation*. As the hydrogen nuclei relax, they emit the RF energy they had acquired. In fact, relaxation occurs in two stages, which means that two measures of relaxation time can be distinguished. The relaxation properties of diseased tissue typically differ from those of healthy tissue surrounding it.

An antenna in the MRI machine detects RF energies and relaxation times. The strength of the signal depends on the density of hydrogen nuclei present. Density is greater where there are abundant water or lipid (fat or oil) molecules. The MRI machine can be "tuned" to discriminate between the different types of relaxation effect, to detect particular kinds of tissue abnormality. Once detected, the signals are digitized, amplified, and finally, spatially encoded by computer. The resulting gray-scale images are displayed on a monitor and can be recorded as hard copy and stored on magnetic or optical digital media.

Magnetic resonance angiography is a means of visualizing blood vessels without the need to use contrast agents (substances administered to the patient that block or enhance the scanning medium and so better highlight internal structures). In time, it may replace conventional angiography, which uses potentially harmful X-rays as well as contrast media. Currently, almost all MRI is carried out without the use of contrast agents. Sometimes, however, a contrast medium containing gadolinium, a high-density metal, may be injected into blood vessels to enhance the visualization of tumors or to highlight regions in which there is inflammation, hemorrhaging, or some other vascular problem.

Most MRI is geared to visualizing hydrogen nuclei, or occasionally, to revealing metals such as sodium ions (Na^+) or the gadolinium in contrast agents. There is potential, however, to use *magnetic resonance spectroscopy* (the interpretation of nuclear magnetic resonance [NMR] signals that are "signatures" for different atoms) to investigate the presence of many elements and the molecules in which they occur. For example, the fate of phosphorus metabolites (among them, energized molecules that act as "metabolic fuel") could be explored in the brain or other organs.

Magnetic resonance elastography (MRE) is a technique, still in development, that combines MRI with ultrasound to reveal the elasticity of body tissue. It is hoped that MRE might disclose early tumors (conventionally, detected by a clinician's touch) that currently escape detection even with sophisticated imaging techniques such as computed tomography (CT; see X-RAY IMAGING) and standard MRI.

Functional MRI (fMRI) tracks changes in blood flow or tissue biochemistry over time. Researchers and clinicians then seek to match observed changes with organ function. Changes in the biochemical activity of the brain, for example, can be correlated with changes in brain activity as experienced by the patient. Like positron emission tomography (PET), a sophisticated technique in nuclear medicine (see SCINTILLATION TECHNIQUES), many brain researchers believe that fMRI will make great contributions in mapping the functions of various parts of the brain: in particular, those regions associated with higher mental processes and the control of movement.

Historical Development

In 1937, Austrian-born I.I. Rabi and German-American Polykarp Kusch revealed the nuclear magnetic resonance properties of atomic nuclei. NMR investigations were later to prove an invaluable means for determining the precise locations of hydrogen nuclei (protons) in molecules. Subtle differences in complex molecules could be distinguished by comparing their NMR spectra (electromagnetic signals of different frequencies). In 1944, Rabi was awarded a Nobel prize for his work on NMR.

MRI for use in medicine has its roots in NMR spectroscopy, an approach that has a tradition stretching back to 1946, when NMR was used for studying chemicals rather than living tissues. In that year, Americans Felix Bloch and Edward Purcell independently demonstrated magnetic resonance in solids and liquids. Between 1950 and 1970, NMR spectroscopy was applied to studying the atomic and molecular structure of a wide range of chemicals.

In 1972, X-ray-based computed tomography was introduced as a useful medical diagnostic technique. This system

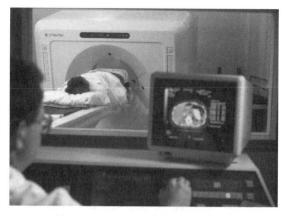

A technician performs an MRI scan, in which the patient's body is subjected repeatedly to powerful magnetic fields (Corbis/Ed Eckstein).

requires considerable computing power to generate a cross-sectional image of the body from X-ray data. CT demonstrated that complex computer-based diagnostic imagery was practical and that hospitals and health centers in the United States and Europe were willing to expend considerable resources on state-of-the-art diagnostic equipment.

In 1973, Paul Lauterbur, a chemistry professor at the State University of New York at Stony Brook, demonstrated how NMR could be used to visualize body tissues. By 1975, Swiss-born Richard Ernst, working for a private company in Palo Alto, California, had devised the systems for encoding NMR data that are used in MRI today. (In 1991, he was awarded a Nobel prize for that achievement.) By 1980 such systems were being piloted, and by 1982 they were available commercially. The term *nuclear magnetic resonance imaging* (NMRI) was dropped by the late 1970s and was replaced by *magnetic resonance imaging* (MRI); such were the negative connotations associated with the word *nuclear* at the time. Since 1980, MRI technology has developed rapidly. Early MRI body scans required the patient to stay still as long as five minutes. Today, scan times are down to less than one minute.

Uses, Effects, and Limitations

In diagnostic medicine, MRI is currently making its greatest contributions in visualizing the central nervous system—the brain and spinal cord—to detect suspected strokes, brain tumors, multiple sclerosis, and developmental disorders such as hydrocephalus, an abnormal accumulation of fluid in the brain. MRI reveals sufficient detail to plot the advancement of progressive neurological or muscular conditions such as multiple sclerosis and muscular dystrophy. MRI does not visualize bone's hard tissue well, but it does show up the soft tissue in and around bone, so is useful in evaluating damage to joints and abnormalities in muscle tissue. Fat, with its high hydrogen atom content, shows up well on MRI images. MRI scans of the heart following a heart attack (cardiac infarction) can evaluate the thickness of heart muscle and distinguish between healthy and dead or diseased muscle tissue. The growth of tumors in the abdomen and pelvic region can also be monitored with MRI.

MRI has several advantages over computed tomography, the X-ray technique that produces cross-sectional images of the body. First, contrast between soft tissues is greater using MRI. For example, MRI clearly distinguishes between white and gray matter, whereas CT does not. MRI does not use the potentially hazardous ionizing radiation associated with CT, and contrast media are used less often with MRI than with CT. When contrast media are used, the gadolinium-based media used in MRI are less likely to trigger adverse reactions than the barium- or iodine-based media used in CT. Most important, MRI scans in any plane, longitudinal (along), transverse (across), and oblique (at an angle), whereas most CT scans are restricted to sections across the long axis of the body (axial).

MRI is not suitable for people who have any electrical components implanted in their body. Patients with pacemakers, for example, cannot be examined with MRI. Patients with small prostheses that contain metal, and those who retain shrapnel from a previous injury, are advised not to have MRI scans. Those with larger prostheses, such as artificial hips, may in some cases have MRI, although any metal will most likely cause local disruption to MRI readings. There are some applications to which MRI is not well suited and for which CT is the technique of choice. CT detects fresh internal hemorrhaging better than MRI. CT is also much more effective at visualizing lung tissue and in discriminating between different bone disorders, where the degree of calcification (calcium content of the bone) needs to be assessed.

One of the few drawbacks associated with MRI is the tendency for some patients to feel claustrophobic when inside the MRI assembly. Scanning times until recently were quite long—several minutes—during which time the patient had to lie as still as possible. With each passing year, scanning times are dropping and new-generation machines are becoming less enclosed and tunnel-like than were earlier models. The cost of setting up and running MRI machines remains high. But while MRI machines are expensive, typically in excess of $1 million apiece, they have a long life and can be upgraded by improving their software. Nevertheless, the cost of an MRI scan is several times that of a CT scan.

MRI's latest applications are being seen on the operating table. In 1998, a multidisciplinary team working in Toronto, Canada, tested a small-scale MRI scanner that enables neurosurgeons to perform brain surgery while obtaining real-time MRI images. The surgeon's hands work within the MRI assembly itself.

Issues and Debate

MRI appears to be safer than other body imaging techniques, such as X-ray imaging and positron emission tomography (PET), which use ionizing radiation. However, the long-term effects of subjecting a person to the very powerful magnetic fields used in MRI have yet to be ascertained. In the interim, even if safety proves not to be a major issue, the high cost of MRI certainly remains so.

Although the utility and apparent safety of MRI suggest a flourishing future for this technology, its high cost raises ethical issues. Thousands of hospitals and medical centers across the world have MRI machines. Once installed, MRI machines and their operators have to justify their existence. In some cases, this situation has led to the overuse of MRI scans. In the United States, investigative journalists, public health organizations, and physicians' groups continue to expose the overuse of MRI imaging. It is easier for physicians to justify clinical use of MRI than use of some other imaging techniques: After all, MRI is apparently safe and has great utility. However, if the referring physician has a share in the imaging center to which he or she is directing patients, there is a potential conflict of interest. The high cost of MRI also means that most people in developing countries, and many in poorer economic circumstances in developed countries, will not have access to this apparently safe and effective technology. The cost of MRI scans will have to come down considerably before people from a range of income groups will be able to benefit from this advanced scanning technique.

—*Trevor Day*

RELATED TOPICS
Scintillation Techniques, Ultrasound, X-ray Imaging

BIBLIOGRAPHY AND FURTHER RESEARCH
BOOKS
Armstrong, Peter, and Martin L. Wastie. *Diagnostic Imaging,* 4th ed. Oxford: Blackwell, 1998.
Bushong, Stuart C. *Magnetic Resonance Imaging: Physical and Biological Principles,* 2nd ed. St. Louis, Mo.: Mosby, 1996.
Erkonen, William E., ed. *Radiology 101: The Basics and Fundamentals of Imaging.* Philadelphia: Lippincott-Raven, 1998.
Kevles, Bettyann H. *Naked to the Bone: Medical Imaging in the Twentieth Century.* New Brunswick, N.J.: Rutgers University Press, 1997.
Lisle, David A. *Imaging for Students.* New York: Oxford University Press, 1997.
McCormick, A.K., and A.T. Elliot. *Health Physics.* Cambridge: Cambridge University Press, 1996.
Smith, Robert C., and Robert C. Lange. *Understanding Magnetic Resonance Imaging.* Boca Raton, Fla.: CRC Press, 1998.
Weir, Jamie, and Peter H. Abrahams. *Imaging Atlas of Human Anatomy,* 2nd ed. London, U.K.: Mosby-Wolfe, 1997.

PERIODICALS
Barinaga, Marcia. "New Imaging Methods Provide a Better View into the Brain." *Science,* June 27, 1997, 1974.
Dixon, Adrian K. "Evidence-Based Diagnostic Radiology." *Lancet,* August 16, 1997, 509.
Gilman, Sid. "Medical Progress: Imaging the Brain." *New England Journal of Medicine,* March 19, 1998, 812.
Lentle, Brian, and John Aldrich. "Radiological Sciences, Past and Present." *Lancet,* July 26, 1997, 280.
McCone, John. "Maps of the Mind." *New Scientist,* January 7, 1995, 30.

McKinstry, Robert C., and David A. Feinberg. "Ultrafast Magnetic Resonance: A New Window on Brain Research." *Science,* March 20, 1998, 1965.

Schneider, David. "MRI Goes Back to the Future." *Scientific American,* March 1995, 42.

INTERNET RESOURCES

Basics of MRI
    http://www.cis.rit.edu/htbooks/mri
Bioethics Resources
    http://adminweb.georgetown.edu/nrcbl/
A Guided Tour of the Visible Human
    http://www.madsci.org/~lynn/VH/
Teaching Aids on Medical Imaging
    http://agora.leeds.ac.uk/comir/resources/links_c.html#teaching
The Visible Human Project
    http://www.nlm.nih.gov/research/visible/visible_human.html

MICROSCOPY

Microscopy is the art of making the invisible visible by observing it through a scientific instrument. Since the invention of the first microscopes in the 17th century, microscopy has changed our understanding of the world. Every field of science—from physics to biology to geology and ecology—has benefited from being able to see the world in miniature.

During this century the development of the electron microscope has allowed the viewing of large molecules and cellular structures. Microscopes developed during the last 20 years have revealed individual atoms. These microscopes are now being used to control the movement and placement of atoms, to construct tiny machines, and to store data.

Scientific and Technological Description

A microscope works by focusing waves to a point. In a light microscope the focusing is accomplished with glass lenses, while in an electron microscope magnetic fields bend the path of electrons. Light or electrons are supplied from a source: For a light microscope, this is a light bulb; and for electron microscopes, it is a beam of electrons, usually supplied by passing an electric current through a tungsten wire. In both cases dispersion, transmission, and absorption of the waves is measured by some sort of receiver, be it an eye, photographic film, or an electron detector.

The glass lenses of a light microscope can be either concave or convex. A concave lens (which bends inward to a thin center) spreads out light waves, whereas a convex lens (which bulges outward to a fat center) focuses light to a point. A single concave lens will make an object appear farther away and smaller; a convex lens will make the object appear closer and larger. In its simplest form, a modern light microscope has a concave lens that increases the size of the object under the lens, followed by a convex lens that pulls the light rays back to a point the eye can see.

The modern *compound microscope* is composed of many convex and concave glass lenses working in concert to focus light. This arrangement is designed both to magnify the object

and to overcome the problems that arise with magnification. Images can be blurred or distorted as they are continually refocused by the many lenses of a modern microscope. Also, imperfections in a lens can cause the image to blur.

One important property of all transparent materials is their *refractive index*. This number is a measure of the material's ability to slow the speed of light and alter its direction, focusing light. A refractive index of 1 means that there is no effect on light. Higher numbers indicate higher focusing powers. Water's refractive index is 1.33, glass's can vary from 1.46 to 1.96, and that of diamond is 2.42. So although a diamond lens will focus light more strongly than will a glass lens of equal thickness, the costs of diamond lenses are clearly prohibitive.

Electron Microscopes

An electron microscope uses a series of magnetic fields to focus electrons. (A solid object such as a glass lens would stop an electron in its path.) The magnification power of an electron microscope is determined by the voltage of the electrons. Higher voltages increase the speed of the electrons and shorten their wavelength, leading to greater resolution.

There are two different types of electron microscope: scanning and transmission. A *scanning electron microscope* (SEM) can image surfaces after they have been coated with a heavy metal such as gold or platinum. The most useful characteristic of SEMs is that they produce a three-dimensional image. A SEM beams electrons toward an object; these are scattered and reflected. After the paths of the reflected electrons have been detected, a computer can assemble a three-dimensional image of the object.

A *transmission electron microscope* (TEM) beams electrons through ultrathin slices of material. The electrons are deflected as they interact with the material's own electrons; the electrons that are able to pass through the sample are focused by magnetic fields into an image many times more detailed than that of a SEM.

Scanning-tunneling microscopes are a fusion of the two types of electron microscopes. These microscopes use a metal needle, usually made out of tungsten and only an atom wide at its tip, to pass electrons through a surface. As the tip passes close to an atom, an electrical current will "jump" or "tunnel" across to the material. The intensity of the electric current is directly proportional to the distance to the surface; mapping it gives a detailed three-dimensional image of the surface at the atomic level.

Atomic Force Microscopes

Atomic force microscopes (AFMs) offer even better imaging. Instead of measuring the electric current passed through the sample, these microscopes measure the electromagnetic force between the microscope and the atoms on the surface of the sample. They consist of a finely sharpened diamond, only one atom wide at its tip, attached to a flexible can-

The Atomic Force Microscope

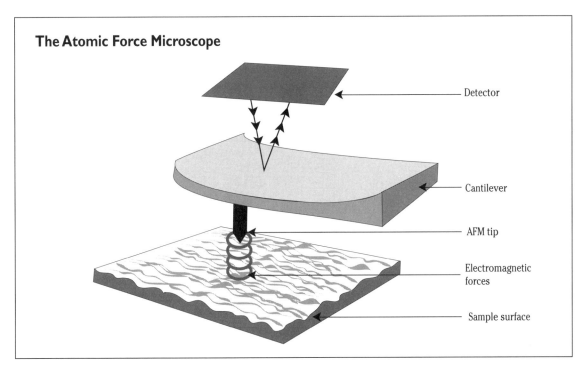

Detector

Cantilever

AFM tip

Electromagnetic forces

Sample surface

An Atomic Force Microscope (AFM) measures the electromagnetic force between the microscope and the surface of the sample being analyzed. The AFM tip, mounted at the end of a cantilever arm, is only one atom wide at its narrowest point. As the tip passes over the surface, forces of interaction cause the cantilever arm to move. Its motions trigger corresponding changes in a beam of laser light. The laser's movements represent the topography of the surface. (Adapted from an original illustration wtih permission from Almaden. Copyright by International Business Machines Corporation.)

tilevered arm. As the diamond tip passes over a surface, it is pulled toward the surface of the object by magnetic attraction as it approaches the peak of an atom. At the other end of the cantilever arm, the up-and-down motions are recorded. A higher amount of motion corresponds to a higher peak on the surface. This information can be used to create a three-dimensional map of the surface.

If the attractive force between the tip of an AFM and the surface is strong enough, individual atoms can sometimes be lifted off the surface. Manipulating atoms one at a time has long been a goal of scientists, and this type of microscopy is being investigated intensely. The most promising application of this technology is the development of tools that will assemble devices such as micromachines and computer chips one atom at a time. Knowing the location of each atom should lead to high-quality machine components. The ability to describe surfaces at the atomic level may also lead to better methods of data storage. Information that is currently stored in larger units on hard drives and compact disks may someday be stored at the atomic level.

Acoustic Microscopes

Acoustic microscopes are used in materials science, which deals with the structure of metals, plastics, ceramics, and other materials at the macro- and microscopic levels. These microscopes send ultrahigh frequency sound waves

through an object. (Sound waves that humans hear are measured in thousands of cycles per second, and ultrasound machines in hospitals use waves measured in millions of cycles per second; acoustic microscopes use frequencies of billions of cycles per second.) The reflections from the object are recorded and analyzed. A computer can reconstruct these reflections into a three-dimensional image that shows cracks, slippage between layers called *delaminations,* and holes in the material. Many of these faults cannot be detected by other microscopic means, or with other imaging devices, such as X-ray machines (see X-RAY IMAGING; ULTRASOUND).

A variation of acoustic microscopes uses lasers to detect vibrations set off by sound waves. After sound waves pass through the object, they strike a piece of thin gold foil, causing minute vibrations and wrinkles to form. The intensity of these wrinkles is measured by a laser, and the thickness, shape, and density of the object can be determined.

Historical Development

Before the first microscopes, it was thought that if something could not be seen with the naked eye, it did not exist. The first observations of magnification came from water: Many philosophers had noted that a drop of water could magnify the surface underneath. The same was also noted of precious gems, such as emeralds and diamonds.

In the late 13th century, English scientist Roger Bacon described the use of lenses to magnify printed words; around the same time two Italian scientists made the first pair of spectacles. All of these were simple microscopes composed of one convex lens. Microscopy with two glass lenses began at the end of the 16th century. Just who invented the first microscope is still a matter of contention among historians. Most believe that the Dutch lens grinders Hans and Zachariah Janssen built the first microscope, between 1590 and 1610. Soon afterward, the Italian astronomer Galileo Galilei invented a similar microscope with slightly different lens placement. Both of these designs were large—a typical microscope was over 2 feet long.

In 1611 the German astronomer Johannes Kepler first described a microscope containing two convex lenses, which sharpened and focused the image and also compacted the microscope design. No one knows who made the first microscope of this type, but the design was quickly adopted and produced in the Netherlands, Italy, and France. Since then many scientists have used the microscope to change and often revolutionize our understanding of the natural world. In the field of medicine, for example, soon after scientists were able to view the many kinds of microscopic organisms that live on Earth, they were able to come up with preventive measures that attacked disease on a microscopic level, such as sterilization and regular sanitation.

Light microscopes have not changed much since the end of the 19th century. At this time glass grinding was developed enough that the limiting resolution of light waves was reached. However, other scientists invented new techniques to improve the microscope, such as *fluorescence microscopy*, which uses the shorter wavelengths of ultraviolet (UV) light to increase resolution, and *phase-contrast microscopy*, introduced in 1935, which enhances small differences in the refractive indices of cellular structures. In 1953, Dutch physicist Frederick Zernike won the Nobel Prize in physics for this last invention.

After Louis de Broglie discovered in 1927 that electrons can behave as both particles and waves, it became obvious that the extremely short wavelength of electrons could be used in a microscope. The first electron microscope, finished in 1931 by German physicists Max Knoll and Ernst Ruska, had a magnification power of 17 times. This microscope, eventually known as the transmission electron microscope, was quickly improved and can now magnify up to half a million times.

The scanning electron microscope was initially designed in 1937, but a functional model wasn't completed until the 1960s by a team of scientists at Cambridge University. The scanning-tunneling microscope was developed in the early 1980s by two Swiss scientists working for IBM. They, together with Ernst Ruska, received the Nobel Prize for physics in 1986 for their contributions to microscopy. These scientists also helped develop the atomic force microscope.

Uses, Effects, and Limitations

Microscopes are used in all facets of science. The biggest users are biological and materials scientists. Biologists use microscopes to examine organs, tissues, cells, and their subcomponents. Materials scientists use microscopes to examine any substance for cracks, shears, breaks, and wear, as well as the structure of individual molecules.

Any microscope is limited by the wavelength of the type of radiation used to create images. In general, shorter wavelengths lead to better resolution. Light waves, with relatively long wavelengths, can be focused only so much before their magnifying power is exhausted. The same is true of the electrons used in electron microscopes, although the wavelengths of electrons are thousands of times shorter. Scanning-tunneling microscopes must deal with Heisenberg's uncertainty principle, which limits the precision in identifying a molecule's position. Atomic force microscopes must contend with heat vibrations in the cantilever arm, which may cause movements that indicate atoms that aren't there. Light microscopes have overcome the limitation imposed by long wavelengths by introducing techniques such as phase-contrast microscopy, which enhances subtle variations in refractive indices, and fluorescence microscopy, which takes advantage of radiation in the ultraviolet wavelengths, which are shorter than those of light and therefore offer greater resolution. However, light waves cannot be magnified much more than 2000 times.

Microscopes that utilize glass lenses suffer from three major types of distortion. The first is *chromatic aberration,* which occurs because different wavelengths are refracted by differing amounts through the same lens. For visible wavelengths, blue light (shorter waves) are focused to a point earlier than red light (longer waves). The result is that an image looks red at one focusing point and blue at another. To correct this, a lens with opposite effects is placed directly after the lens to correct the light waves' paths by averaging them out.

Spherical aberration is a problem caused by the diameter of glass lenses. Rays that enter at the lens's edges are refracted more than those that enter at the center. A resulting image may have two different focusing points, one for the outer objects and another for the inner. Another lens can be placed directly after the first lens to combat this problem, but it is almost impossible to correct perfectly.

Finally, *comatic aberration* is seen in objects that are not directly in the line of sight between the lenses, which results in objects that apparently have "tails." This is usually corrected by measures similar to those that combat spherical aberration.

A transmission electron microscope is hampered by the techniques required to prepare a sample for examination. The microscope works in a vacuum to ensure that electrons are not deflected by air and other contaminating molecules. But this vacuum also guarantees that biologists can never examine living tissue. Also, the samples must be dried, frozen, and covered with a resin. These extensive preparations can sometimes introduce *artifacts,* images that were never in the original samples. Finally, the sample must be sliced extremely thinly, which limits the amount of three-dimensional information that can be obtained. Still, the TEM can magnify a sample up to 500,000 times.

Light microscopes have recently found use by physicians for microsurgery. Microsurgery is used most often by vascular, eye, and neurosurgeons. Dealing with tiny body parts such as those in the eye and blood vessels requires extreme precision. Light microscopes have been developed specifically for physicians, to give them a three-dimensional view of the operating field. These microscopes are slightly out of line, and the surgeon's brain builds the two slightly differing images into a three-dimensional picture.

Issues and Debate

Even though light microscopes are microscopy's oldest technology, they still remain the most used. Tens of thousands of slides for light microscopy are made daily in hospitals worldwide. This is because light microscopes are relatively cheap and simple, yet continue to give rewarding knowledge. But for greater magnifications, the expenses associated with more powerful microscopes have made them somewhat rarer. Not every researcher or hospital has its own TEM or SEM. These machines, which are much more expensive and require skilled technicians to operate, are usually shared by an entire hospital or university campus. In addition, the chemicals used to prepare samples for investigation with a SEM or TEM, such as osmium, gold, and platinum, can be quite expensive if used frequently. These instruments are typically available on a fee per use or on a time-shared basis. Scanning-tunneling microscopes and atomic force microscopes are even more expensive, due to their increased complexity, and thus are generally available only within the largest research organizations.

In addition to their superb imaging capabilities, atomic force microscopes, may help increase the capacity of data storage in coming years. With their ability to "write" with molecules in paths only a few molecules wide, they can compress data far beyond the capabilities of today's compact disks and hard computer disk drives. Even so, someday these microscopes will become limited by the size of atoms themselves. Future data storage systems may make use of color, which will give scientists another variable that can be controlled and manipulated to store data.

—Philip Downey

RELATED TOPICS
Nanotechnology, Telescopy

BIBLIOGRAPHY AND FURTHER RESEARCH

BOOKS
Bradbury, Paul. *Introduction to Microscopy,* 2nd ed. Richmond, British Columbia: Steveston Scientific Publications, 1992.
Preiss, Byron, ed. *The Microverse.* New York: Bantam Books, 1989.
Stewart, Gail B. *Microscopes.* San Diego, Calif.:, Lucent Books, 1992.

PERIODICALS
Gross, Neil, and Otis Port. "The Next Wave." *Business Week,* August 31, 1998, 80.
Rotman, David. "Will the Real Nanotech Please Stand Up?" *Technology Review,* March 1999, 46.
Service, Robert F. "AFMs Wield Parts For Nanoconstruction." *Science,* November 27, 1998, 1620.

INTERNET RESOURCES
Almaden Research Center
    http://www.almaden.ibm.com
Introduction to Microscopy
    http://micro.magnet.fsu.edu/primer/index.html
Light Microscopy
    http://www.ruf.rice.edu/~bioslabs/methods/microscopy/microscopy.html

MICROWAVE COMMUNICATION

Microwave communication is a broad term that covers most aspects of modern wireless communication. Television and radio stations, cellular phones, citizen's band (CB) radios, and satellites all broadcast in the microwave region of the electromagnetic spectrum. Microwaves are superior to radio waves (the first form of wireless communication) for many physical reasons, including signal clarity and power.

In 1996 the United States began a program to auction off many of the publicly owned frequencies used to transmit wireless signals. Many private companies are currently jockeying for ownership of the frequencies needed to offer services such as digital television and radio, digital cellular phones, and satellite communications.

Scientific and Technological Description

Microwaves are a form of electromagnetic radiation, just like light or X-rays. All types of radiation have signature frequencies and wavelengths. Microwave wavelengths are fairly long and are measured in centimeters and meters. Microwave frequencies range from 3 kilohertz to 300 gigahertz (1 hertz [H_z] equals one cycle, or complete wave, per second; 1 kilohertz [KH_z] equals 1000 cycles per second; and 1 gigahertz [GH_z] equals 1 billion cycles per second). The most sought-after and commercially valuable microwave frequencies fall between 30 megahertz (MH_z; 1 megahertz equals 1 million hertz) and 30 GH_z. With the exception of AM radio, all modern microwave communications devices, such as television, cellular phones, and FM radio, operate in this range.

Different kinds of signals, such as a voice, computer data, or music, must all be transformed into a format that

How AM and FM Signals Are Sent

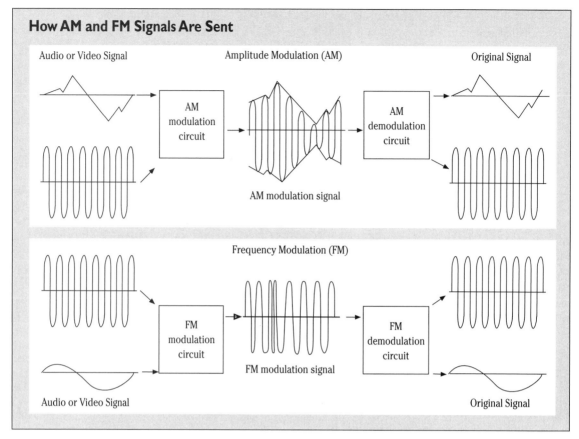

Before any signal—whether voice, computer data, or music—can be transmitted, it must first be changed to a format that allows it to be sent over microwaves. The original audio or video signal must be added to a "carrier wave," either by amplitude modulation (AM) or frequency modulation (FM). After the message has been modulated, it can be sent to a radio or television, for example. Those devices demodulate the message by separating the original signal from its carrier wave. (Adapted from Frank Baylin and Brent Gale, Satellites Today, Howard W. Sams, 1986.)

can be sent via microwaves. The two ways of doing this are amplitude modulation (as in AM radio) and frequency modulation (as in FM radio) (see figure). *Amplitude modulation* is done by altering the amplitude or power of the wave. *Frequency modulation* induces small variations in the transmission frequency to represent data. This means that a radio tuned to 102.1 FM is actually measuring small variations in the frequency of the signal, such as 102.1001 and 102.0999. The amount of variation is called the *bandwidth*. Except for AM radio, frequency modulation is used for all modern communications.

Radio waves were generated before microwaves, so the term *radio* has remained in use even though most modern transmissions are conducted with microwaves. Microwaves are preferable to radio waves for four reasons. First, microwave frequencies are much higher than those of radio waves, so more information can be packed into a signal. Second, microwaves have a shorter wavelength that allows them to be focused into small, narrow beams; consequently, smaller antennas and receiving dishes can be used. Third, microwaves are less susceptible to noise and static than are radio waves. Fourth, microwaves easily pass

through the upper atmosphere, where they are received by Earth-orbiting satellites.

Historical Development

Until the rise of the railroad in the 1830s, and the invention of the telegraph wire in 1844, the fastest means of communication were by horse or ship, both of which had not changed much in the previous 5000 years. From the beginning, the telegraph and the wireless radio, which was invented in 1896, were used for business, military, political, and personal communications.

After three failed attempts, the first transatlantic cable was laid underwater in 1858, from Ireland to Canada. Unfortunately, the cable worked for only 28 days before it broke. The British War Office figured that the first and only transatlantic communication it made saved it £50,000, by stopping the 62nd Regiment from sailing from Halifax, Nova Scotia, for India, where it was no longer needed. That amount was one-seventh of the privately raised money that funded the project. It was eight years before another could be laid. But this success encouraged others to attempt to lay more cables. Today, cables both underwater and overland circle the globe.

The first experiments with wireless communication were performed in the late 19th century. Developing a wireless telegraph was a goal of many scientists, engineers, and entrepreneurs throughout Europe who were devising ways of sending and deciphering signals through the air. The first person to succeed was Guglielmo Marconi, an Italian who sent a radio signal across 9 miles of English countryside in 1896. Five years later, in 1901, still experimenting with radio communications, he surpassed himself and sent the letter "S" in Morse code 2000 miles across the Atlantic Ocean, from Cornwall, England, to Newfoundland, Canada. Radio grew quickly and spread throughout North America and Europe.

By 1922, when there were just three radio stations in the United States, radio broadcasters and the public became engaged in a debate over who would pay for radio. Proposals ranged from attaching coin boxes to every radio, to renting telephone lines for transmissions, to providing a free system. AT&T solved the problem by introducing advertising to radio, and it has remained a fixture of telecommunications since then, allowing broadcasters to afford to transmit their signals without billing the recipient. One year later, in 1923, there were 595 radio stations. In 1926 the Federal Radio Commission, the predecessor to today's Federal Communications Commission, was formed to regulate this exploding industry. Its mandate was to "make available, so far as possible, to all people of the United States, a rapid, efficient, nationwide, and worldwide wire and radio communications service with adequate facilities at reasonable charges."

In the same year, the discovery of the ionosphere significantly changed our understanding of radio and microwave transmission. Until then it had been a mystery how a radio signal could be received by someone beyond the horizon. A signal should logically travel straight into space instead of following Earth's curvature. It was discovered that the ionosphere, an upper layer of the atmosphere that contains charged molecules called ions, reflected radio signals back to Earth. The reflective properties of the ionosphere allow us to receive transmissions from radio and television stations located below the horizon. The ionosphere is located 50 to 150 miles above the Earth. The amount of *ionization,* or number of charged particles, is affected by the energy output of the Sun. Sometimes, the ionosphere can be overloaded by bursts in the Sun's activity, disrupting Earth's communication, radar, and satellite networks.

A key turning point in wireless communication came in 1945, when science fiction writer Arthur C. Clarke, then a radar scientist with the British Royal Air Force, described a system of *geostationary satellites* that could receive signals from one point on Earth and retransmit them to other parts of the globe. But it wasn't until 1963 that rocket science could launch a satellite into the required orbit. Today, as Clarke predicted, geostationary satellites sit in a *geosynchro-* *nous orbit,* 22,247 miles above Earth. At this height they have the same rotational speed as the Earth's surface, therefore are always looking down at the same swath of Earth. Their great height also allows them to cover over 40 percent of the planet continuously with their signals. In theory only three supersatellites are needed to cover the entire planet; in practice there are several hundred, each dedicated to specific transmission frequencies and observation areas.

The first successful microwave communication in space didn't use satellites. In 1948 the U.S. Army Signal Corps bounced microwave signals off the Moon. In 1954 its researchers repeated the feat with voice messages. Two years later the branch created a successful radio link between Washington, D.C., and Hawaii. Of course, this system worked only when the Moon was simultaneously in view of both the transmitter and the receiver.

The first space system with an active transmitter instead of a passive reflector like the Moon was the NASA-sponsored SCORE (signal communication by orbiting relay equipment) satellite, launched by the U.S. Air Force on December 18, 1958. It received messages at 150 MH$_Z$, taped them, and rebroadcast them to Earth at 122 MH$_Z$. The satellite worked for 12 days, but on January 21, 1959, it fell back to Earth. On July 10, 1962, the United States launched its first true communications satellite, Telsat I. It worked for 226 days before a bombardment of cosmic rays shut it down permanently. It was the first television link between Europe and North America. The Syncom II satellite was the first to be placed in a geosynchronous orbit, on July 26, 1963.

Since then, hundreds of satellites have been placed into a variety of orbits, not all of which are geosynchronous. The majority satellites are in low-Earth orbits (LEOs), ranging from 100 to 1000 miles high, and whip around the Earth every two hours. It is much cheaper to place satellites into these orbits than into geosynchronous orbits. A string of satellites passing overhead continuously can achieve the same complete coverage as that afforded by geosynchronous satellites.

A handful of companies, including Teledesic, AT&T, Hughes, Motorola, and Loral, are setting up networks of hundreds of satellites that will relay voice, computer data, and other information to any point in the world via satellite instead of through earthbound wires and antennas. Motorola's Iridium telephone system has already begun operations, offering satellite telephone service anywhere in the world. Other companies are expected to join in the years 2000 to 2002. All of these companies have had to obtain broadcast rights at the same frequency from every country over which they wish to broadcast.

Uses, Effects, and Limitations

Millions of machines across the planet receive or transmit microwaves: baby monitors, analog and digital cellular phones, television sets, AM and FM radios, high-definition television receivers, cordless telephones, military communication systems, public safety services such as police and fire department radios, airport runway lights, and satellite transmissions. Long-distance communication, both wired or wireless, is one of humanity's most powerful tools. The news media bring messages and information from around the world to people's homes every day. Private citizens communicate by phone across the world by pressing buttons. Military forces and governments remain in constant contact with bases and operatives around the world. All of this is mediated by communication through the airwaves and through cables.

Microwave communication is affected by several factors. The first is *line of sight.* With low-powered transmitters, they must all be within sight of each other. Usually, this means elevating antennas to building tops and high terrain. This can create a simple network of transmitters separated by a few kilometers. The second factor affecting communications is the *ionosphere,* which reflects or passes radio signals depending on solar activity. In times of strong solar storms or flares, communication can be hampered and even shut down because of effects on the ionosphere and on satellite transmitters. Microwaves are less susceptible than radio waves to disturbances in the ionosphere. Microwave satellites in geosynchronous orbits are also limited by *latency,* the time it takes for the signal to travel to the satellite and back to Earth, which is about one-fourth of a second. This delay is too slow for certain military and financial communications. Therefore, ground-based wires and antennas will continue to be preferred in some arenas.

Issues and Debate

In 1996, the United States passed a sweeping Telecommunications Act, modernizing the outdated act of 1934. In 1934 all communication technologies were analog; a major goal of the 1996 law was to deal with current and future digital technologies. That law also initiated a program to auction the rights to broadcast to private enterprises on thousands of frequencies: those used by cellular phones, satellite communications, analog television and radio, and their future digital forms, cordless telephones, digital telephones, and Internet-like networks between satellites as well as those used in astronomical research. The frequencies to be sold will generate billions of dollars for the government and even more in eventual profits to industry.

Since the auctions began, the Federal Communications Commission (FCC) has raised more than $7 billion in the sale of the right to broadcast on digital cellular phone airwaves alone. Although vast sums of money are being spent in gaining access to the airwaves, many of these new ventures will probably have little difficulty recouping their investments. But the competition will be fierce and some will definitely lose out, since no single company has been permitted to dominate any set of frequencies, and because certain technologies will eventually be preferred over others by consumers. A current

example of this trend is the gradual loss of viewers by network television to cable broadcasters and the Internet.

Auctioning off a significant portion of the broadcast spectrum is going to take years. The *Guinness Book of World Records* has already called the auction for digital cellular phone airwaves the biggest ever. Eventually, the FCC will have auctioned off the broadcasting rights for the majority of the publicly available microwave bands. Some are being held in reserve for military, scientific, and public use, and in the future the entire system may have to be rearranged as new technologies are developed.

The program is also attracting criticism, because in the United States, citizens are considered the owners of the airwaves. To date, the government has acted as the people's agent in licensing the airwaves. Many of the technologies that will use them, such as digital television and digital cellular phones, are not yet even fully on the market. This makes it very difficult to estimate the value of the spectrum that these technologies will occupy. Some observers have argued that the frequencies up for auction are being sold too cheaply and that the companies getting in early are getting too good a deal considering the future billions that could be made. Some industries, such as digital television, may even be getting the rights to these airways at no cost (see HIGH-DEFINITION TELEVISION).

Allocation of microwave frequencies has been hotly contested in recent years. There is a limit to the number of frequencies there are to go around, since it is impossible for two stations to broadcast on the same frequency. In one sense this means that space on the broadcast spectrum is always limited; in another sense, however, the spectrum will always be around and cannot be depleted like such resources as forests. This means that the amount of time for which one owns the rights to a particular frequency will affect the use and value of that ownership.

Many frequencies are already set aside for exclusive use by the military, police, and fire departments, ATM (bank) machines, air traffic control, and other applications; the rest is shared between the government and the public. By law, 6 percent of the radio spectrum is set aside for the private sector, and 1.7 percent is for the government. In the remaining 92 percent, the private sector must always defer to the government sector. But this shared space is not evenly divided. According to a report by the Reason Foundation, just 10.6 percent of the shared spectrum is available to the private sector, while 81.6 percent remains in government hands. Given the rising importance of the flow of information and communications in today's economy, debate over proper allocation of the broadcast spectrum is likely to increase in the coming years.

—Philip Downey

RELATED TOPICS
Global Positioning System, High-Definition Television, Internet
   Protocol Telephony, Satellite Technology

BIBLIOGRAPHY AND FURTHER RESEARCH

BOOKS

Baylin, Frank, and Brent Gale. *Satellites Today*. Columbus, Ohio: Howard W. Sams, 1986.

Clarke, Arthur C. *Voice Across the Sea*. London: William Luscombe, 1974.

PERIODICALS

Lo, Catharine. "Get Wireless." *Wired*, April 1997.

Lo, Catharine. "Space Jam." *Wired*, October 1998.

Stephenson, Neal. "Mother Earth, Motherboard." *Wired*, December 1996, 97.

INTERNET RESOURCES

Federal Communications Commission
    http://www.fcc.gov

Internet Telephony
    http://www.internettelephony.com/

Lloyd's Satellite Constellations
    http://www.ee.surrey.ac.uk/Personal/L.Wood/constellations/

The Orbiting Internet: Fiber in the Sky
    http://www.byte.com/art/9711/sec5/art1.htm

Wireless Telecommunications Bureau Home Page
    http://www.fcc.gov/wtb/

MULTISPECTRAL IMAGING

Multispectral imaging is a remote sensing technique that involves the simultaneous collection of reflected and emitted radiation, over several bands of the electromagnetic spectrum, from a given scene.

Today many U.S. and foreign civilian satellite systems perform multispectral imaging. Its current applications involve imaging Earth—applications have rapidly evolved from government-sponsored uses in meteorology, cartography, and reconnaissance to more privately oriented uses in such diverse areas as resource evaluation, environmental monitoring, urban planning, disaster response, insurance claims adjustments, marketing, real estate, tourism, and precision farming. As satellite-derived multispectral imagery improves in resolution, not only will other interesting applications rapidly emerge in the 21st century, but so also will significant issues concerning data management, privacy, and international security.

Scientific and Technological Description

Multispectral imaging is a form of *remote sensing*, which is the technology of collecting information about an object at a distance (see REMOTE SENSING and SPACE-BASED REMOTE SENSING TOOLS). Specifically, multispectral imaging involves the simultaneous detection and measurement of photons of different energies emanating from a distant object or scene. Spatial and spectral features are then combined to reveal a variety of characteristics about the scene.

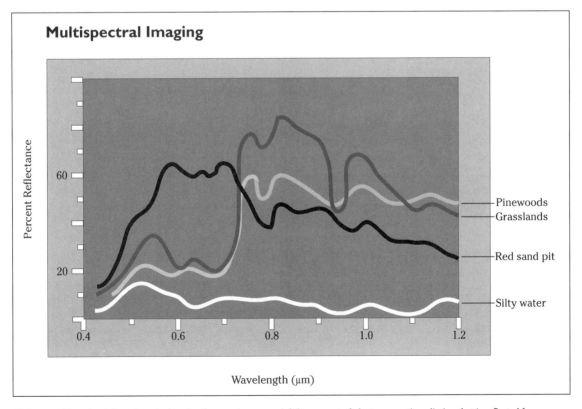

Multispectral Imaging

Multispectral imaging is based on the fact that for any given material the amount of electromagnetic radiation that is reflected from an object varies with wavelength. The figure shows how various natural objects (including pinewoods, grassland, red sand, and silty water) reflect sunlight at wavelengths ranging from the visible portion of the electromagnetic spectrum (e.g., 0.48μm wavelength is green visible light) up into the near-infrared region (1.2μm wavelength). This important property of matter allows for the possibility that different substances can be identified and distinguished by their "spectral signatures." (Adapted from an original illustration courtesy of NASA.)

Photons are elementary packets (*quanta*) of electromagnetic energy. These dual wave–particle objects have no mass and travel at the speed of light. The *electromagnetic* (EM) *spectrum* is a continuum of all electromagnetic waves (photons) arranged according to their frequency (μ) and wavelength (λ). Traditionally, the EM spectrum has been divided into several regions. These regions are (going from highest frequency to lowest frequency): gamma ray, X-ray, ultraviolet, visible light, infrared, and radio (which includes microwave). Visible light is composed of wavelengths ranging from 0.4 micrometers (μm; one millionth [10^6] of a meter) for blue light to 0.7μm for red light. Lying just beyond red light is the infrared region, which is especially important in multispectral imaging and is often divided into several subregions: near-infrared (about 0.7 to 1.0 μm), mid-infrared (about 1.5 to 3.0 μm), and thermal infrared (about 10.5 to 12.5 μm).

In satellite-based multispectral imaging sunlight illuminates objects on the surface. This incoming solar radiation (insolation) is confined primarily to the spectral interval between 0.2 and 3.4 μm, with the maximum solar radiation input peaking close to 0.48 μm wavelength (visible green light). Earth's atmosphere transmits, absorbs, and scatters solar radiation. The molecular constituents of the atmosphere can absorb EM radiation at certain wavelengths. Electromagnetic radiation at wavelengths shorter than 0.3 μm (X-rays and gamma rays) is completely absorbed in the upper atmosphere. Ultraviolet rays are also largely absorbed. Only microwave and longer radio wavelengths are able to penetrate the atmosphere with little absorption. Visible light and infrared wavelengths are generally transmitted through the atmosphere in special spectral regions of high transmittance called *atmospheric windows,* regions where multispectral imagery can be acquired by sensors on satellites.

On striking the land or ocean surface the arriving solar radiation experiences three types of interactive response: transmission, absorption, and reflection. First, some of the incoming solar radiation will penetrate a transparent surface, as sunlight penetrates water. Some of the incident radiation will also be absorbed by the medium, heating it. Finally, a portion of the incoming radiation will be scattered or reflected by the target. Most multispectral imaging sensor systems are designed to monitor this reflected radiation.

For any given material, the amount of incident solar radiation that is reflected will vary with wavelength. This important property of matter establishes the possibility that different surface materials can then be identified and separated by their *spectral signatures*. For example, at some wavelengths, sand may reflect more sunlight than green vegetation; but at other wavelengths, it may absorb more sunlight. Various kinds of surface materials (both natural and manufactured) are recognized and distinguished from each other by measuring these differences in relative reflectances.

The function of any multispectral imaging system is to detect reflected and/or emitted EM radiation signals, determine their spectral characteristics, derive appropriate signatures, and interrelate the spatial positions of the classes that these signatures represent. The multispectral image is a useful display that mirrors the reality of a particular surface in terms of the nature and distribution of the characteristic spectral features.

Historical Development

Remote sensing began with the appearance of the first aerial photographs. French balloonists made these initial photographs in the 1840s to help develop topographical maps. By World War I, cameras were mounted on aircraft and provided aerial views of reasonably large surface areas. From then until the early 1960s, the aerial photograph remained the single most important tool for portraying Earth's surface from a vertical or oblique perspective. Satellite-based remote sensing traces it origins to the early days of the space age in the 1950s, when both the United States and the Soviet Union pursued programs involving military photo-reconnaissance from space, including using an astronaut or cosmonaut to do photography.

The first multispectral photography performed from space was accomplished by American astronauts during the *Apollo 9* mission (March 1969). During this Earth-orbiting mission, four Hasselblad cameras were mounted in a special holder designed so that the astronaut could aim all four cameras at the same target point on Earth's surface. The system took four simultaneous photographs, including green-, red-, and blue-filtered images. The filters selected favored reflected light in the green, red, and blue portions of the EM spectrum (respectively), thereby providing an initial (although limited) ability to examine features on Earth's surface in a multispectral manner.

A six-camera multispectral system was flown on NASA's Skylab space station between May 1973 and February 1974. The value of such spacecraft-derived multispectral photography was quickly recognized by Earth-discipline scientists, especially geologists, hydrologists, agronomists, foresters, and others concerned with environmental monitoring and land-use assessment. Also during the mid-1960s, the idea of a civilian Earth resources satellite based on multispectral imagery was conceived by the U.S. Interior Department. Later, NASA embarked on an initiative to develop and launch the first Earth-monitoring satellite to meet the needs of natural resource managers and Earth scientists. The U.S. Geological Survey (USGS) entered into partnership with NASA in the early 1970s and assumed responsibility for archiving and distributing the multispectral data products that would result.

On July 23, 1972, NASA launched the first in a series of important satellites designed to provide repetitive multispectral imagery of Earth's landmasses on a global basis. Designated initially as the Earth Resources Technology

Satellite-A (ERTS-A), when operational orbit was achieved, the satellite was called ERTS-1. ERTS-1 continued to function beyond its designed life expectancy of one year and finally ceased to operate on January 6, 1978. But ERTS-1 created a revolution in multispectral imaging of Earth, a revolution that continues today. NASA launched a second spacecraft in this series on January 22, 1975, and renamed it Landsat-2. Three additional Landsats were launched in 1978, 1982, and 1984 (Landsats 3, 4, and 5). Then, a vastly improved system, called Landsat 6, failed to achieve orbit on October 5, 1993. However, on April 15, 1999, an even more sophisticated Landsat 7 spacecraft was launched.

Landsats 1 through 3 carried two sensors: *return beam vidicon* (RBV) and *multispectral sensor* (MSS). The RBV was essentially a television camera and did not achieve the popularity of the MSS sensor. The MSS sensor scanned Earth's surface with an array of detectors that ranged from the visible to the near-infrared (IR) portion of the EM spectrum. Landsats 4 and 5 carried both the MSS sensor and a new, more sophisticated multispectral imaging sensor called the *thematic mapper* (TM). The MSS and TM sensors detected primarily reflected radiation from Earth's surface in the visible and near-IR wavelengths, but the TM sensor, with its seven spectral bands, provided more detailed spectral information than the MSS sensor. The wavelength range for the TM sensor was from the visible through the mid-IR into the thermal-IR portion of the EM spectrum.

Multispectral imaging sensors such as the Landsat MSS and TM served as the prime Earth-observing sensors during the 1970s and into the 1980s. But these instruments contained moving parts, such as oscillating mirrors, that were subject to wear and failure. Therefore, another approach was developed to sensing numerous bands of reflected and emitted radiation simultaneously: the *pushbroom scanner,* which uses radiation-sensitive *charge-coupled devices* (CCDs) as the detector. The CCD is a versatile solid-state electronic device that contains an array of tiny sensors that emit electrons when exposed to light (or other types of EM radiation). CCDs are used to produce digital images in modern cameras, telescopes, and the like. The *enhanced thematic mapper+* (ETM+) on the newly launched Landsat 7 is a fixed pushbroom, eight-band multispectral scanning radiometer capable of providing high-resolution imaging information of Earth's surface in the visible and infrared portions of the EM spectrum. In the late 1990s, over 30 government (U.S. and foreign) and commercially sponsored remote sensing spacecraft are providing multispectral images of Earth at varying levels of spatial resolution, spectral resolution, and frequency of coverage to a wide variety of governmental, industrial, scientific, and private consumers.

Uses, Effects, and Limitations

Multispectral imaging, especially satellite derived, is a form of modern information technology that has now assumed a dominant place on the information spectrum, along with voice, data, and location information in the global marketplace. Great improvements in computer-based image processing have made satellite-derived multispectral imagery available to universities, resource-responsible agencies (such as the U.S. Interior and Agriculture Departments), small companies, and individuals.

The concurrent development and rise of *geographic information systems* (GISs) into widespread use has also provided an efficient means of integrating multispectral imagery data with other types of data. A geographic information system is a computer-assisted system that acquires, stores, manipulates, compares, and displays geographic data (e.g., land-cover information or vegetation status), often including multispectral imagery data sets from Earth observation spacecraft (see figure). The GIS approach is especially well suited to the storage, integration, and analysis of multispectral imagery data, leading to information that has a greatly enhanced practical value in many fields.

Almost three decades of continuous multispectral imagery from Landsat has supported creative uses in the following areas: agriculture, forestry and range resources, land use and mapping, geology, water resources, oceanography and marine resources, environmental monitoring, and global change research. As higher spectral and spatial resolution images become available at even lower cost, the list of innovative applications of multispectral imaging will continue to grow.

Of course, as multispectral imaging sensors improve in spatial and spectral resolution, the amount of data to be processed in a particular scene will also grow dramatically. The need to digitize more information-dense multispectral images efficiently creates new demands for faster processing procedures and for the creation of customized algorithms, which preferentially extract user-unique information from complex multispectral images. As the cost of multispectral imaging from space declines with international competition, more and more customers will begin to take advantage of the unique information content available in this powerful form of information technology. For example, precision farming is expected to become a standard form of agriculture in the 21st century. Scientists studying Earth system processes will also rely heavily on multispectral imagery to detect subtle natural and human-caused changes in Earth's surface, especially when such changes occur in remote or difficult-to-access terrain and when simultaneous observation of large portions of Earth's surface must be made without being hindered by political boundaries.

Issues and Debate

The first major issue involving multispectral imaging is the sheer quantity of the data becoming available and the efficient processing and archiving of this information. For example, Landsat's MSS archive now has over 630,000 scenes with a data volume of 20 terabytes (TB). One terabytes is roughly equal to the storage capacity of a half-million unformatted 2 megabyte floppy microdisks, the typical disks used on a personal computer. Storing that amount of information on 3.5-inch microdisks would require more than 10 million such disks. Meanwhile, the TM archive has over 300,000 scenes with a data volume of over 50 TB. Some advocates suggest that such data be made "freely" available (or perhaps at a nominal, low cost) to all users, as a government service, much as weather data are now provided. Others believe that such government-sponsored data should now be turned over to commercial organizations that would manage and operate the archive on a for-profit basis. *Multispectral imagery* (MSI) products from commercially sponsored remote sensing satellites are also entering the global marketplace. Their sponsors want these MSI products to compete with other imagery products at free market prices rather than being forced to compete with MSI products whose prices have been artificially lowered by heavy government subsidies.

The issue of privacy is also of concern. As the resolution (spatial and spectral) of commercial satellite-derived imagery continues to improve, vast quantities of information about surface objects become readily available for the first time to all willing to purchase the data. Individuals, organizations, and entire nations will begin to live and work in "glass houses." Previously protected information will be routinely collected by multispectral sleuths in the sky. Is such "sky spying" legal? Current international law allows states to take part in satellite remote sensing activities without advance notice and permits public dissemination of data without prior consent of sensed states. A United Nations treaty and resolution, both of which have been ratified by the United States, establish the current rule under which any nation or entity may collect imagery of Earth's entire surface, regardless of political boundaries, if it is technically capable of doing so.

Another interesting problem will arise when parties in a regional conflict seek to gain military advantage by acquiring commercially provided high-resolution imagery of their opponent's territory. Does the commercial company sell exclusively to one side and not the other, to both sides, or to neither? The obvious answer, perhaps, is neither. But that is not how the global arms business currently operates. Will multispectral imaging, a superb tool for intelligent management of Earth as a fragile ecosystem, also become an essential part of regional conflicts of the 21st century?

—*Joseph A. Angelo, Jr.*

RELATED TOPICS
Computer Modeling, Remote Sensing, Satellite Technology, Space-Based Remote Sensing Tools

BIBLIOGRAPHY AND FURTHER RESEARCH

BOOKS
Ginsberg, Irving W., and Joseph A. Angelo, Jr., eds. *Earth Observations and Global Change Decision Making, 1989: A National Partnership.* Melbourne, Fla.: Krieger Publishing, 1990.

Lillesand, T.M., and R. W. Kieffer. *Remote Sensing and Image Interpretation,* 3rd ed. New York: Wiley, 1993.
Sabins, Floyd F., Jr. *Remote Sensing: Principles and Interpretation,* 3rd ed. New York: W.H. Freeman, 1996.
Short, Nicholas M., Paul Lowman, Stanley Freden, and William Finch. *Mission to Earth: Landsat Views the World.* NASA SP-360. Washington, D.C.: U.S. Government Printing Office, 1976.

PERIODICALS AND REPORTS
Lang, Laura. "IRS Data Benefit Natural Resource Studies." *Earth Observation Magazine* (EOM), January 1996, 20.
NASA/Goddard Spaceflight Center. "NASA's Earth Observing System: EOS AM-1." NP-1998-03-018-GSFC, 1998.
Prins, Eric. "Remote Sensing Data for Biodiversity Management in Sahelian Africa." *Earth Observation Magazine* (EOM), August 1997, 18.

INTERNET RESOURCES
Commercial Satellite Imagery Company
    http://www.spaceimaging.com
Earth Observation Magazine Home Page
    http://www.eomonline.com/
EROS Center Home Page, USGS
    http://edcwww.cr.usgs.gov/eros-home.html
Landsat History
    http://geo.arc.nasa.gov/
Remote Sensing Tutorial
    http://rst.gsfc.nasa.gov/
USGS Earthshots (satellite images of environmental change)
    http://www.usgs.gov/Earthshots

MUSIC SYNTHESIZERS

Synthesizers are electronic musical instruments that can mimic the sound of any other instrument or generate a wide variety of entirely new sounds. Although primitive electronic instruments were developed around the turn of the 20th century, synthesizers were taken up by professional composers only in the 1950s, and became widespread in popular music only in the 1970s.

Although many musicians have embraced music synthesizers and have used them for both composing and performing, not everyone agrees that synthetic (computer-generated) music is a good thing. Some musicians believe that it lacks the warmth of sounds generated by traditional instruments; others are concerned that synthesizers may be replacing the skills of human virtuosos. Nevertheless, synthesizers have opened up a world of new options to musicians and appear to be here to stay.

Scientific and Technological Description

A conventional instrument produces musical notes by generating sound waves that take their shape from that of the instrument; for example, a long clarinet produces deeper notes than does a short piccolo. In a violin, sounds are made by plucking the strings so that they oscillate (vibrate), but in a synthesizer sounds are made using electronics alone. The heart of a synthesizer is a collection of electronic oscillators, each of which generates a regular sound wave of a specified frequency (pitch), amplitude (volume), and timbre (shape, or quality of sound). Because it can generate any kind of sound

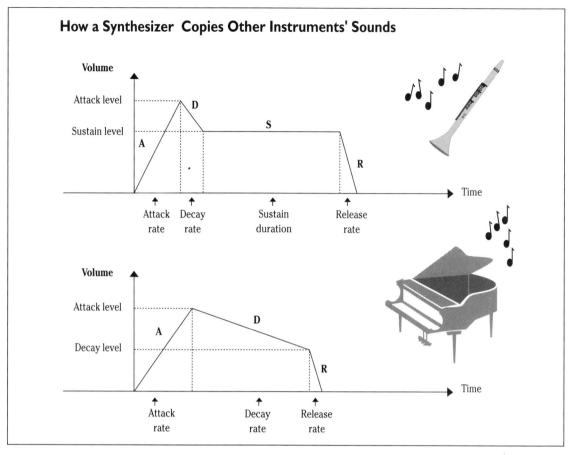

How a Synthesizer Copies Other Instruments' Sounds

A synthesizer copies other instruments partly by copying the way they change the volume of a note during its duration. They do this using a process known as Attack Decay Sustain Release (ADSR) envelope shaping. In a traditional wind instrument such as a clarinet (upper figure), the sound builds (attacks) quickly to its maximum volume, which decays to a lower volume. This is sustained for as long as the clarinet player keeps blowing, then decays rapidly to silence. In a piano (lower figure), the sound builds more gradually to its maximum volume and decays more slowly, before releasing to silence. There is no sustain unless the piano player presses the sustaining foot pedal. A synthesizer player can copy these effects by keying in numeric values or sliding switches that correspond to the different rates and levels of attack, decay, sustain, and release.

wave, a synthesizer can mimic the sound of any traditional instrument, but it can also generate sound waves that no ordinary instrument could produce. In other words, a synthesizer is much more than an orchestra in a box; it is a box containing every instrument that could ever be imagined.

A basic sound tone, which has an undulating pattern of crests and troughs known as a *sine wave,* can be turned into waves of other shapes, which have different timbres. These include waves with a sawtooth shape, which sound like brass instruments; square waves, which sound like harmonicas; and triangular waves, which sound like wind instruments. A single oscillator produces only a single tone, rather like a person humming at constant frequency. Additional oscillators add higher frequencies, known as *harmonics,* to produce a richer sound, like that of a barbershop quartet humming in harmony.

It is not just the presence of harmonics that makes one instrument sound different from another. A flute and a piano playing the same note do different things to the sound's vol-

ume over time. A note on the piano rises quickly to a maximum and falls off slowly when the key is released. A flute rises more sharply and remains at a sustained volume until the flautist stops blowing. A synthesizer copies the way that instruments change the volume of sounds by setting values for *attack* (the rate at which the sound builds to its maximum volume), *decay* (the volume to which the sound falls after the note reaches its maximum), *sustain* (the volume at which the sound continues after the note is released), and *release* (the rate at which the note dies to silence). This process is known as *ADSR envelope shaping.*

Synthesizers contain numerous other components for changing sounds. These include filters (e.g., for cutting off all frequencies below a certain level to produce a high-pitched sound) and modulators (for combining different frequency sounds to produce special effects such as tremolo—a vibrating effect——and the equivalent of electric guitar distortion). Synthesizers are abstract instruments whose design is inspired by complex theories of the physics of sound. To make them easier for musicians to relate to, they typically have piano-style keyboards, joysticks and dials for simple pitch-changing effects, "presets" (keys that instantly produce preprogrammed noises, such as piano or trumpet sounds), cartridges or disk drives for loading other preprogrammed effects, and a set of output connectors known as MIDI (musical instrument digital interface) ports for linking them to other synthesizers and computers.

Historical Development

Synthesizers have been inspiring musicians since around 1870, but only two early devices are still in regular use today. The Thérémin (invented in 1917) was a compact wooden box with a metal antenna, which produced a whistling noise that rose and fell in both pitch and volume as the player's hands moved near it; Thérémin effects are still used in today's synthesizers. Patented by French musician Maurice Martenot in 1922, the Ondes Martenot featured a keyboard and a slide control that could be used to produce eerie swooping noises. Several composers wrote works for the Ondes; notably, Frenchman Olivier Messiaen's epic *Turangalîla* symphony, composed in 1949, features extended Ondes solos. Exciting works such as this inspired avant-garde composers to experiment with electronic music in the 1950s. The pioneers of the new music worked in radio stations blessed with the latest equipment and technicians who understood how to use it. This period brought a number of electronic composers to worldwide attention, including Karlheinz Stockhausen in Germany and Milton Babbitt in the United States. However, composition was a time-consuming process that involved building up a piece by prerecording it. Stockhausen's 13-minute piece *Gesang* took 18 months to record, for example.

From the 1950s onward, the development of synthesizers was a partnership between electronic engineers and composers, each driving the other forward, each keen to use cutting-edge technology. In 1954 that technology was the RCA synthesizer developed by Harry F. Olsen. It produced sounds with a vacuum-tube oscillator, could be programmed using paper tape punched with holes, and could store sounds by cutting them into an internal, rotating lacquer disk.

During the 1960s, synthesizers made the transition from classical to popular music, thanks largely to Robert Moog. Moog had produced Thérémin kits in the early 1960s before turning his hand to full-scale keyboard synthesizers in 1964. They were an immediate success, not least because they could be played live like any other instrument and no longer required compositions to be recorded elaborately in advance.

By the 1970s, synthesizers were featuring prominently in hit records by musicians such as Frenchman Jean-Michel Jarre and the German group Kraftwerk. Groundbreaking instruments that appeared in the latter part of the 1970s included the Synclavier and the Fairlight CMI (Computer Musical Instrument), which were the first samplers (syn-

Music synthesizers began to be used in popular music in the 1960s and 1970s. Debate continues as to whether synthesizer effects such as drum machines are too artificial sounding. Above, composer Loren Rush tunes a drum using a synthesizer (Corbis/Henry Diltz).

thesizers that can copy sounds from the real world and play them back distorted in various ways).

The 1980s saw the beginning of a convergence between personal computers and synthesizers, thanks to MIDI, the standard method of connecting synthesizers that 100 synthesizer makers and electronics companies agreed on in 1984. One of the first machines to use it was the Yamaha DX7, which pioneered a technique known as frequency-modulated digital synthesis. This allows the timbre and volume of notes to be changed more precisely throughout their duration. Although the DX7 was a complex instrument that could be played in many new ways, it was simplified by plug-in cards carrying preset sounds. During the late 1980s and 1990s, computers gained sophisticated built-in sound cards, which meant they could work like synthesizers; and synthesizers incorporated sophisticated onboard computers. The difference between the types of machines had all but disappeared.

Uses, Effects, and Limitations

Synthesizers can effectively do only two things: copy existing sounds and generate new ones. Although a keyboard synthesizer can simulate any instrument in the orchestra, these are not always convenient for musicians who do not play piano. For this reason, synthesizers have been developed that resemble other types of instruments. These include synthesizer guitars, which have a keyboard in place of the strings; electronic drums, which are flat, hexagonal boxes connected to a synthesizer that produce a drumlike sound when they are struck; and wind synthesizers, which look like a clarinet or tenor saxophone but produce a synthetic noise when they are blown.

Because synthesizers are so versatile, electronic guitars, drums, and wind instruments do not necessarily have to produce the same sounds as their real counterparts. Thus, a classically trained flautist could use a wind synthesizer to produce the sound of a piano, a trumpet, or a completely new instrument. Just as synthesizers extend the capabilities of performing musicians, they can turn ordinary people into musicians of a different kind. Tod Machover of the Massachusetts Institute of Technology has developed a range of hyperinstruments, which can be played with no musical training at all. These include the digital baton (a kind of musical magic wand), the gesture wall (which makes sounds when people move toward it), and the sensor chair (which responds to movements of the seated occupant's legs and arms).

Machover's hyperinstruments are effectively virtual instruments, in which synthesizers are played using a VIRTUAL REALITY interface. Instead of pressing keys or plucking strings, the virtual musician simply moves around inside a virtual landscape and the synthesizer converts different movements into different sounds. Jaron Lanier, a computer scientist who coined the term *virtual reality,* is also one of the world's first virtual musicians. His instruments include the pianobeam, with which he generates piano harmonies simply by moving his hands through the air, and the flute lathe, with which he carves out computer representations of wooden forms as he plays a flute. Unlike real instruments, virtual instruments can be played by someone with relatively little skill or training, and could help many more people to discover the joy of making music for themselves.

Despite their versatility, there is one instrument that synthesizers have not yet been able to copy: the human voice. But it may not be long before they can do so. At Oxford University in England, scientist Ken Lomax is teaching synthesizers how to sing. This is much more difficult than simply copying a musical instrument. The human voice is more variable than a violin, say, and there is much greater variation between different human voices than between different violins.

The benefits of synthesizers include the ability to produce any sound of any pitch in the style of any instrument, real or imagined. They enable musicians to play a wider range of instruments, and nonmusicians to acquire musical skills more quickly. Sequencers, which can memorize tunes at slow speed and then play them back more quickly, enable relatively unskilled musicians to sound like dazzling virtuosos.

Issues and Debate

Synthesizers take their name from the Greek word *synthesis,* which means "to put together." But *synthesis* also gives us the word *synthetic,* which means "artificial and unnatural." Music synthesizers are often criticized for sounding precisely this way. Born in electronics and radio laboratories, they may seem closer to the world of computers than the world of composers. Synthesizer music is synthetic music, critics argue, lacking human subtlety and warmth. They charge that the virtuosity of musicians is replaced by the dull hand of machines, and that flowing emotional landscapes in the minds of composers become lifeless strings of numbers in the circuits of computers.

But purist musicians have always opposed new developments in music. Igor Stravinsky's symphony The *Rite of Spring* caused a riot when it was first performed in 1913 but is now recognized as one of the most inventive pieces in the classical repertoire. Creative musicians have always needed to push the boundaries of convention, and synthesizers may be just another example of this. Electronic-music composers argue that synthesizers take nothing away from music; on the contrary, they say, because they provide new ways to play instruments (e.g., allowing a keyboard player to make the sounds of a flute), they are opening up more musical possibilities to more people.

Some critics argue that synthesizers have replaced real musicians with computers. Instead of hiring an orchestra to add string effects to a pop record, many record producers now opt for a string synthesizer instead; it is cheaper than a dozen violin players, easier to control, and much more predictable. Where synthetic sounds were once highly artificial and thus easily recognized as such, many people now find them difficult to tell apart from the real thing.

Irrespective of their effect on existing instruments and musicians, synthesizers have introduced an entirely new musical vocabulary. Before synthesizers, no one had played an instrument just by sitting in a chair. No one had heard sampled sounds, such as familiar pieces of previously recorded music or speeches, incorporated into new songs. Composers such as Stockhausen and Babbitt, or more recently, John Adams, might never have been so creative without synthesizers and other electronic techniques.

Synthesizers are a controversial way of making music, perhaps because human beings resent computers intruding in an area that has traditionally been regarded as one of the peaks of human endeavor. As synthesizers become increasingly sophisticated and techniques such as virtual reality expose even more people to them, it may be that the skills of the human virtuoso are appreciated more, not less. It will be a long time before a computer can sing like Louis Armstrong or play piano like the great Canadian virtuoso Glenn Gould. Until then, comparisons between humans and synthesizers will make humans seem all the more remarkable.

—*Chris Woodford*

RELATED TOPICS
CD-ROMS, Digital Video Technology, High-Fidelity Audio, Virtual Reality

BIBLIOGRAPHY AND FURTHER RESEARCH

BOOKS
Bussy, Pascal. *Kraftwerk: Man, Machine, and Music.* London: SAF Publishing, 1993.
Crombie, David. *The New Complete Synthesizer: A Complete Guide to the World of Electronic Music.* London: Omnibus Press, 1986.
Darter, Tom. *The Art of Electronic Music.* New York: Quill/William Morrow, 1984.
Griffiths, Paul. *Modern Music: A Concise History.* London: Thames & Hudson,1994.
Hill, Brad. *Going Digital: A Musician's Guide to Technology.* New York: Schirmer Books, 1998.

Jourdain, Robert. *Music, the Brain, and Ecstasy*. New York: William Morrow, 1997.

Kurtz, Michael. *Stockhausen: A Biography*. Boston: Faber & Faber, 1992.

Noad, Frederick. *The Virtual Guitarist*. New York: Schirmer Books, 1998.

Pressing, Jeff. *Synthesizer Performance and Real-Time Processing*. Oxford: Oxford University Press, 1992.

Rich, Alan. *American Pioneers: Ives to Cage and Beyond*. London: Phaidon, 1995.

Russ, Martin. *Sound Synthesis and Sampling*. Oxford: Focal Press/Butterworth-Heinemann, 1996.

Young, Rob. *The MIDI Files*. Hemel Hempstead, Herts, England: Prentice Hall, 1996.

PERIODICALS

Geake, Elisabeth. "…and Hello to Playing Music Without Keys." *New Scientist,* August 14, 1993, 17.

Matthews, Mike, and Robert Myer. "Working Knowledge: Guitar Effects Pedals." *Scientific American,* July 1997, 84.

Neesham, Claire. "Digital Diva." *New Scientist,* September 7, 1996, 36.

Zatorre, Robert. "Sound Work." *Scientific American,* September 1997, 78.

INTERNET RESOURCES

120 Years of Electronic Music
    http://www.obsolete.com/120_years/

Beginner's Synthesizer Frequently Asked Questions
    http://www.tilt.largo.fl.us/faq/faqcont.html

Hyperinstruments at the MIT Media Lab
    http://brainop.media.mit.edu/Archive/Hyperinstruments/

Take a Look at Theremins
    http://www.ccsi.com/~bobs/theremin.html

The Virtual Synthesizer Museum
    http://www.synthmuseum.com/index.html

NANOTECHNOLOGY

Nanotechnology is the technology of the very small—technology on the scale of nanometers. A nanometer (nm) is one-billionth of a meter (0.000000001 m). This new technology can involve simple scaling down of normal machines to microscopic size (often also called microtechnology), or building up of new tools and machines atom by atom.

Nanotechnology's potential applications are nearly boundless—they range from computing to medicine to environmental engineering. But the obstacles to making those applications a practical reality are still huge. Nevertheless, nanotechnology is developing rapidly.

Scientific and Technological Description

The precise definition of nanotechnology varies. At its strictest, nanotechnology's meaning can be limited to refer purely to mechanical machines built up from individual atoms, in which case it is a technology still in its infancy. However, *microtechnology,* often included in definitions of nanotechnology, is far more advanced. In addition, the world is also full of natural nanomachines—molecules ranging from simple compounds to complex proteins and even deoxyribonucleic acid (DNA), which operate by chemical reactions. Although these are not actually examples of nanotechnology, organic molecules in particular

could be modified by nanotechnology to perform new roles as needed.

The broad scope of nanotechnology is often divided into two fields: top-down (microtechnology) and bottom-up (true nanotechnology). Top-down nanotechnology is simple miniaturization taken to the extreme, creating machines just a few microns (millionths of a meter) across.

Top-down nanotechnology often applies the techniques used in manufacturing silicon chips to making separate machine components. For example, components can be cut out of a wafer of raw material by first coating it with a *photoresist* material that hardens under ultraviolet (UV) light. UV is then shone onto the wafer through a computer-designed photographic mask, so that only the areas where the light shines through harden. Photoresist and raw material can be washed away from the other areas with various chemicals, and finally, the hardened photoresist can be removed, leaving a tiny, precisely shaped component, such as the wheels used to detect changes in speed in automobile accelerometers. More complex parts can be built up by repeating this process several times.

Bottom-up nanotechnology is a far more complex process, involving the construction of machines from individual atoms. The tools needed to do this already exist—the most important is the *atomic force microscope* (AFM). An AFM is a remarkably simple instrument that can create images of individual atoms and even rearrange them on a surface (see MICROSCOPY). It consists of a metal probe sharpened to a point that is at most a few atoms across. This probe is attached to a mirror that pivots as the probe moves up and down, and reflects a beam of laser light to a detector. As the probe moves across a surface, it moves up and down through the bumps and depressions caused by atoms in the surface material. These up-and-down movements are transformed into deflections of the laser beam, which are detected by the microscope and converted into a three-dimensional picture of the individual atoms on a surface. An AFM can also be used as a probe to push atoms around on a material's surface, joining them up in different ways, but this has so far resulted only in the creation of patterns from individual atoms, and nothing approaching even a simple machine.

The design complexities of bottom-up nanotechnology are daunting and mean that the simplest nanomachines would require many thousands of hours to manufacture, even with an accurate blueprint to work from. Because of this problem, most of the scientists who support nanotechnology advocate the development of machines called *assemblers,* nanomachines capable of manufacturing other nanomachines from raw materials with minimum human intervention.

Historical Development

Nanotechnology is a remarkable science because its origins can be traced back to a single event. In December

1959, noted U.S. physicist Richard Feynman (1918–1988) delivered a speech to the American Physical Society entitled "There's Plenty of Room at the Bottom," in which he laid out many of the basic principles of nanotechnology and outlined the huge range of fields to which miniaturization could be applied. By way of encouragement, Feynman announced that he was offering two prizes: $1000 for the first working motor occupying less than ¼₄ cubic inch (¼ inch on each side) and another $1000 for the first example of writing miniaturized to ¹⁄₂₅,₀₀₀ normal size. The first prize was claimed within just a few years of Feynman's speech, by an engineer who built a motor using traditional watchmaking skills, but the writing prize, which required a much greater degree of miniaturization, was not claimed until the 1980s. The original prizes have recently been replaced with two new Feynman prizes, an annual award for the team making the most significant contribution to nanotechnology, and a grand prize of $250,000 for the person or team that manufacturers both the first functional robot arm less than 100 nm long, and a simple functioning nanocomputer.

The techniques behind silicon chip manufacture, developed from the 1960s, were rapidly applied to the creation of micromachines. These developments enabled microengineering to become a reality by the 1970s, but it took considerably longer for nanoengineering to become a real possibility. The invention of the atomic force microscope at IBM's Zurich Research Laboratory was the spur to renewed interest in the 1970s. Then in 1989, IBM's Don Eigler was able to show an astonished world his company's name, IBM, written in just 35 atoms of xenon (although he missed the Feynman writing prize, which had already been claimed by more conventional means).

But writing made of atoms is still a huge way from functional nanomachinery, which most scientists now

think will arrive only with the development of an assembler. Assemblers are self-replicating robots, first proposed by Hungarian-born U.S. mathematician John von Neumann (1903–1957) long before the birth of nanotechnology. Von Neumann envisaged a robot that could process materials from the environment around it in order to create an identical copy of itself.

A functioning assembler has become the ultimate goal for nanotechnology research today. Once completed, assemblers already working at the atomic scale could do away with the need for the hundreds of laborious hours it would take to manufacture even the simplest of nanomachines. At present, most research is concentrating on the development of working designs for nanomachines. For instance, NASA researchers recently worked out how to

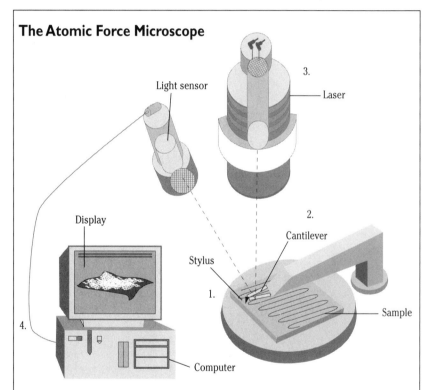

The Atomic Force Microscope

Light sensor

Laser

3.

Display

2.

Cantilever

Stylus

1.

Sample

4.

Computer

1. The AFM's stylus presses down on a sample surface.

2. The sample moves underneath the stylus, which is attached to a cantilever arm. Any bumps in the surface move the cantilever.

3. The movement of the cantilever reflects a laser beam, which is picked up by a sensor.

4. The laser's movements, which represent the topography of the surface, are fed into a computer. The computer displays a 3-D image of the surface.

The atomic force microscope, the most important tool in nanotechnology, is used to create images of individual atoms and rearrange them. (Adapted from Durant et al. Encyclopedia of Science in Action. *© 1995 Macmillan Publishers Ltd.)*

build gears that could be turned by a laser beam. The actual construction of such devices is still a long way off.

However, others are doing practical work that will have an impact on nanotechnology. The 1996 Nobel Prize for chemistry was awarded to Richard Smalley, Robert Curl, and Harold Kroto for their discovery of and experimentation with buckminsterfullerenes (carbon molecules in the shape of balls and tubes, which could be ideal building blocks for mechanical components in nanotechnology). In addition, U.S. software millionaire James Von Her recently set up the first company specifically founded to develop a nanotechnology assembler.

The development of nanotechnology has also been closely linked to advances in biotechnology in the past couple of decades, and specifically the ability to adapt and graft together parts of different molecules. Nature is already full of working nanomachines; proteins carry out mechanical functions in our bodies similar to those that a nanomachine might perform in a factory, and some nanoscientists suggest that the first generation of nanomachines could be developed most easily by adapting some of these natural machines (see SURGICAL ROBOTICS).

Uses, Effects, and Limitations

The possible uses of nanotechnology are almost limitless. Because of that, much of the literature on the subject deals with the implications of a nanotechnology revolution rather than with how to practically bring it about. One of the most exciting areas in which nanotechnology might be put to use is environmental engineering. Self-replicating nanomachines would be a fast and relatively easy way to repair the damage that human beings have done to the environment since the industrial revolution. Nanomachines could be produced, for instance, that bound onto and deactivated ozone-destroying chlorofluorocarbon (CFC) molecules. Others might even be able to transform oxygen in the upper atmosphere back into ozone. Nanomachines could also be used for clearing up chemical waste and other forms of pollution. In theory, they could return the Earth's environment to a preindustrial state. However, the dangers of releasing nanotechnology into the environment without proper controls are themselves considerable.

Even if nanotechnology might be too dangerous to release into the environment at large, it could certainly offer new and revolutionary methods of pollution-free manufacturing. Conceivably, everything from chemicals to large machine parts could be manufactured by nanotechnology, far more efficiently than by today's processes. All that would be required would be suitable supplies of raw materials and fuel for the tiny machines. Because the assemblers themselves work on the atomic scale, they would be able to construct individual molecules and larger materials, such as composite materials, according to preprogrammed designs. Trillions of nanomachines working together could literally "grow" complex machines such as engines. Each wave of

machines might perform one specific task before giving way to a new wave, building up the finished product step by step.

Most nanomachines would be quite simple devices, capable of just one task, or perhaps a limited range of tasks, depending on an external stimulus such as a chemical "signal." Some machines, however, will be expected to perform thousands of different tasks, and these will require some form of information-processing system—which today we call a computer. Computers as we know them do not function at the atomic scale, however; they rely on electronics (the movement of large numbers of subatomic particles called electrons), which work only on a bulk scale. Nanocomputers will have to work in a completely different way, which will allow them to convert the instructions they receive directly into mechanical movements in a nanomachine.

Some scientists have already found the solution to this problem—and it could also revolutionize computing as we understand it. Mechanical nanocomputers would use complex systems of atom-thick rods, sliding past each other, to mimic the electronic "1" and "0" signals of a normal computer. These rods would crisscross others, forming a *logic gate* (a component capable of performing a logical calculation—the basic building block of a computer). Strategically placed molecules grafted onto the rods would sometimes prevent, and sometimes allow, the signal rods to move, producing different "outputs" that could pass on to another component. The construction of logic gates that perform most common calculations is relatively simple, so the computer-aided design of "circuits" for a mechanical nanocomputer would be no more difficult than that for a normal integrated circuit.

Nanocomputers would have two important advantages over normal computers: size and speed. Component sizes would be at least 1000 times smaller than those of normal computers, and this would render the transmission time for the mechanical movements of the calculating rods to pass through the computer effectively instantaneous, compared to the small but measurable delay in the flow of electricity through a normal circuit. Nanocomputers would probably still rely on electronics for output and display devices, but the processors themselves would be almost immeasurably faster and more powerful than current personal computers.

One other extremely important area of nanotechnology applications is medicine. Nanomachines could be injected into the human body to carry out repairs in an advanced form of "keyhole surgery," akin to Isaac Asimov's novel *Fantastic Voyage,* in which a miniaturized submarine is injected inside a person to carry out surgery. In the longer term, nanomachines might even become permanent residents in our bodies, fighting infection and disease, repairing injuries, and perhaps even slowing the onset of aging.

Issues and Debate

Although nanotechnology is still almost entirely theoretical, it has already raised some fascinating and difficult issues.

On the positive side, if the predictions of futurists are correct, this new technology may become reality in the coming decades. If it does, it will, perhaps more than any other technology, actually transform our lives. The application of nanotechnology in medicine, industry, the environment, and many other spheres would radically change society itself. Some optimists have predicted an end to work, vastly increased (and even indefinite) lifespans, and the complete reshaping of the global economy. However, these predictions recall the earlier optimism of those who foresaw robots and computers bringing an end to workplace drudgery. The emphasis of work may shift, but it is unlikely that nanotechnology will entirely fulfill its potential to create a global utopia.

Indeed, some aspects of nanotechnology could turn it into a threat. Considerable caution will have to be exercised before nanomachines are released into the environment, for example. Although they might repair some of the damage done by pollution, they could equally create new and disastrous problems for our global ecosystem. Possible solutions would be to prevent these nanomachines from replicating, to give them a strictly limited lifespan, or to provide them with some sort of fail-safe "off" switch.

This sort of scenario reveals another worrying aspect of nanotechnology—its potential as a weapon. Nanomachines would doubtless become an important part of battlefield technology, but they could also present an insidious and invisible weapon of mass destruction, even more awesome than the nuclear bomb. As nanotechnology becomes more widespread, the possibility exists that terrorists could manufacture and release destructive self-replicating nanomachines into the environment.

One final ethical issue has already been explored in science fiction scenarios such as the *Star Trek* television series. In creating self-replicating assemblers, humans might in effect be producing ARTIFICIAL LIFE. Some nanotechnologists already say that the best way to make new nanomachines would be to allow them to breed and to evolve. In the future, humanity might be faced with a major moral dilemma if confronted with a nanotechnology that had become, to all intents and purposes, a new and independent form of life.

—*Giles Sparrow*

RELATED TOPICS
Artificial Life, Composite Materials, Microscopy, Smart Materials, Surgical Robotics

BIBLIOGRAPHY AND FURTHER RESEARCH
BOOKS
Drexler, K. Eric. *Engines of Creation.* New York: Anchor, 1987.
Drexler, K. Eric. *Nanosystems: Molecular Machinery, Manufacturing, and Computation.* New York: Wiley, 1992.
Regis, Edward. *Nano: The Emerging Science of Nanotechnology.* Boston: Little, Brown, 1996.

PERIODICALS
Nanotechnology

INTERNET RESOURCES
Foresight Institute: Preparing for Nanotechnology
    http://www.foresight.org
Nanotechnology Papers
    http://nanotech.rutgers.edu/nanotech/

NUCLEAR ENERGY

Nuclear energy is produced when the nuclei of atoms are split during fission or merged during fusion, two chemical reactions that create an enormous amount of heat, along with other forms of energy. The heat is used to make steam that powers turbines to generate electricity. Nuclear reactors can produce energy that is a million times more powerful than that produced by the classic burning of fossil fuels, but nuclear energy's potential to cause radiation poisoning, along with its other drawbacks, have engendered enduring controversy.

Scientific and Technological Description

Nuclear reactors most commonly use uranium for fuel—specifically, its isotope uranium 235 (although some use thorium or plutonium). Uranium 235 is important because it is *fissile,* it can split when hit by a relatively slow moving neutron. When the nucleus of an atom of uranium 235 fissions, it releases neutrons that can then split other nuclei, which release neutrons, and so on. A *nuclear chain reaction* forms, but its survival depends on the reactor becoming *critical;* that is, exactly one neutron from each fission event creates one more.

Several distinct steps lead from the uncovering of uranium to the production of electricity. First, the uranium ore must be mined and then converted chemically and enriched to be used in a power plant. Eventually, the fuel is molded into cylindrical pellets that are baked and hardened. The pellets are stuffed into long tubes called *fuel rods,* which are bundled together in fuel assemblies that make up part of the reactor core. In a reactor core, a release of neutrons triggers the chain reaction, but in common reactor designs, the resulting flyaway neutrons speed too quickly to sustain the chain reaction. They must be slowed by a moderator in the core. Several substances, including regular water, heavy water, graphite, and beryllium, serve as good moderators.

Moderation is not enough to control a nuclear reaction, so *secondary tubes,* or *control rods,* are inserted into the core. These rods are packed with chemical elements such as boron and cadmium that absorb renegade neutrons. Power plant technicians maintain the correct balance between fission and absorption by inserting or removing control rods. In the case of an emergency shutdown, or *scram,* the control rods are inserted fully. If the

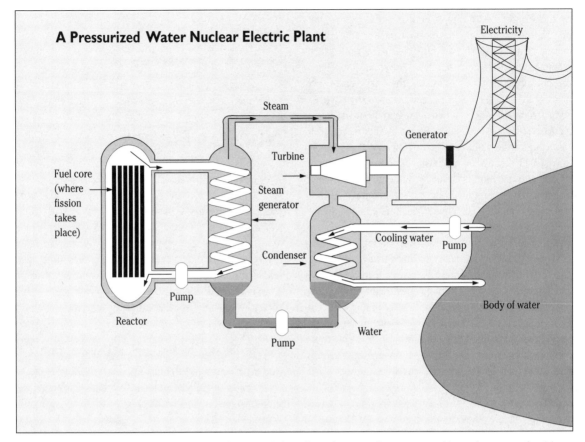

A Pressurized Water Nuclear Electric Plant

Just as in other types of power plants, in a nuclear plant, water is heated, creating steam that powers a turbine and generates electricity. (Adapted from Laurence Pringle, Nuclear Energy: Troubled Past, Uncertain Future, *Macmillan Publishing Company, 1989.)*

reactor's heat rises without hindrance, a meltdown in which the core liquefies could result.

When a nucleus fissions, or splits, it produces energy in varying forms. In common reactor designs, the heat that is generated is carried away from the reactor core by coolants such as water, heavy water, air, carbon dioxide, helium, and sodium. The heat from the coolants eventually powers turbines that create electricity. The exact method by which this is done depends on the reactor design. The most common design is a light-water reactor (LWR), of which there are two types: the boiling-water reactor (BWR) and the pressurized-water reactor (PWR). In a BWR, water circulates in the core, where it boils and turns to steam, which powers the turbines. In a PWR, the primary water system is kept under extreme pressure so that it does not boil. Instead, a second system of water absorbs the heat from the first and carries it to the turbine.

Another design, the heavy-water reactor (HWR), uses an isotope of water in which the hydrogen atom is replaced with deuterium. Gas-cooled reactors replace water as the coolant with gases such as helium and carbon dioxide. The gases carry heat away from the core to heat exchangers that produce steam for generators. Whereas most reactors use fuel to create electricity,

breeder reactors create more fuel than they use; they also sometimes produce electricity. In a breeder reactor, through several chemical processes, uranium 235 decays to plutonium 239. The plutonium can be used in making bombs as well as in nuclear reactors.

Historical Development

The world's first nuclear reactor was made by physicist Enrico Fermi in 1942 when he engineered a human-made sustained nuclear reaction. Tested successfully in December 1942, it ran for four and one-half minutes and produced one-half a watt of electricity. Earlier that year, Fermi had joined the Manhattan Project, the team of scientists brought together by the U.S. government to develop the first nuclear weapons. Fermi's work on the nuclear reactor paved the way for the first nuclear bombs, eventually dropped on the Japanese cities of Hiroshima and Nagasaki to end World War II in 1945.

The U.S. military became interested in nuclear energy shortly thereafter; officials had requested that the chemical company Du Pont build reactors in Hanford, Washington, during World War II. By 1948 the U.S. government wanted to explore postwar uses for atomic energy, and the Atomic Energy Commission (AEC) designated the Argonne National

Laboratory in Illinois as the nation's center for commercial reactor development.

In the early 1950s, Congress became eager for private industry to share in the expenses of nuclear energy development, and in 1953, President Dwight Eisenhower unveiled the "Atoms for Peace" program at the United Nations. This program helped to market nuclear energy, and to that end, Congress offered industrialists a partnership in the AEC. In 1955, the partnership produced the Power Reactor Demonstration Program, with the goal of creating prototype reactors.

Although interested in nuclear power, utility owners balked at the risk of investing millions of dollars in reactors that could go terribly awry. In response, Congress passed the controversial Price–Anderson Act in 1957. That law underwrote much of the risk of nuclear power production. It required that utilities purchase the maximum amount of insurance available, which in 1957 was $65 million, but it also guaranteed that the federal government would pay up to $560 million above that amount.

By the 1960s, competition for marketable reactor designs resulted in a breakthrough by General Electric, which had developed a boiling-water reactor. The design was less expensive and simpler than previous prototypes, which appealed to utilities. Westinghouse created a similar design, and between 1963 and 1966, the two companies sold 12 units to kick off the heady days for nuclear power in the United States. By 1970, over 100 reactors had been ordered.

The United States was not alone in its quest to develop nuclear power reactors. As early as 1945, France had established its atomic energy commission, called CEA. Its primary goal was to produce atomic bombs, but by 1973, the commission had overseen the development of gas-graphite reactors in that country. Meanwhile, the Soviet Union kept pace with the United States, and was one of the few countries that used nuclear energy to produce electricity by 1955. Also in 1955, the Japanese government awarded a multimillion-dollar budget to its Science and Technology Agency for nuclear research. In 1970, Japan opened its first commercial light-water reactor.

But much was to change in the next two decades. The Arab oil embargo of 1973 pinched the pockets of utility owners, who became cautious about heavy investments. By 1974, U.S. utility owners had canceled or deferred the construction of 253 nuclear and fossil-fueled (oil-, coal-, or natural gas–fueled) plants. Five years later, on March 28, 1979, an accident at the Three Mile Island nuclear station near Harrisburg, Pennsylvania, stirred international debate about the safety of nuclear energy. For the first time in the United States, a general emergency was declared at a commercial nuclear power plant. A series of malfunctions and human errors resulted in nearly one-third of the core in Unit 2 of the plant being exposed, releasing an estimated 17 curies (Ci) of radioactive and potentially damaging iodine 131 (on average, a reactor will release 10,000 Ci of radioactive gas throughout one year). More than 140,000 people fled the area, and the long-term physical effects on the residents who were exposed to the gas remain in dispute.

Seven years later, a far more damaging accident occurred in a reactor at the Chernobyl nuclear complex in the former Soviet Union. The core became supercritical, meaning that too many neutrons in the chain reaction were splitting. The tremendous heat created a steam explosion that sent a cloud of radioactive gas westward, contaminating major portions of Europe and parts of the Soviet Union. An estimated 100 million curies of radiation in total were released, but a little less than half were xenon gas, which is not considered a long-term threat. Seven million curies of iodine 131 gas were released, compared to the 17 Ci at Three Mile Island. Firefighters required 10 days to control the fire in the reactor core and another seven months to encase the unit entirely in concrete to block the radiation it emitted. Thirty-one people died as a direct result of the explosion and fires, 135,000 residents were evacuated, and 60,000 buildings required decontamination. Researchers continue to study the long-term effects of the radiation exposure on nearby residents. These accidents, coupled with high costs to maintain nuclear power plants, the problem of spent fuel storage, and other issues, have stalled construction of new plants in the United States, but other countries continue to plan and build nuclear power stations.

Uses, Effects, and Limitations

More than 100 licensed commercial nuclear reactors currently generate over 20 percent of the electricity in the United States, but the Nuclear Regulatory Commission (NRC) projects that the nation's nuclear capacity will almost halve over the next two decades, from 95,605 megawatts (MW) in the year 2000 to 49,217 MW in 2020. Not all countries can easily reduce their dependence on nuclear power, because many lack fossil fuels. Japan, which imports 80 percent of its fossil fuels, relies on nuclear power to produce 27 percent of the nation's electrical supply and plans to build 40 more reactors by 2010. France also lacks fossil fuels as a natural resource. The country began its nuclear power program in the 1970s amid tremendous protests, but a keen marketing campaign and respectable safety record have turned public opinion in favor of nuclear power. Today, France uses 54 reactors to produce 75 percent of the country's electrical output.

Nuclear power produces more than just electricity; it also creates waste that is more radioactive than is the original fuel. The storage, transportation, and disposal of spent fuel from nuclear reactors creates noteworthy risk, which limits the appeal of nuclear power. Part of that waste results from the disposal of spent fuel rods. On average, up to one-third of the fuel rods installed in a reactor need to be disposed of every year. These fuel rods emit high doses of radioactivity that must be blocked. In the United States,

France began its nuclear power program in the 1970s amid protests against the use of nuclear energy, but today, nuclear power is much more popular with its citizens. The country uses 54 nuclear reactors to produce 75 percent of the country's electrical output (Photo Researchers/Michel Baret).

most plants contain pools of water on-site in which the rods are immersed for years, but ultimately, the spent fuel must be stored or disposed of. Other countries recycle the uranium in spent fuel through a method called *reprocessing*.

Issues and Debate

Safety ranks among the greatest concerns about nuclear energy. In the United States, the NRC regularly inspects the plants it licenses. If safety inspectors find violations at a plant, they can and do issue hefty fines. In 1997, the NRC levied a fine of $650,000 for 70 violations found at a plant in Connecticut. The fine was the largest ever issued to a utility in New England and among the largest imposed in the country. The NRC also maintains a watch list of nuclear power plants with questionable past performances.

Regulation and inspection do not guarantee safety, as the accidents at Three Mile Island and Chernobyl attest, and the potential for disaster spurs fear in many people. Exposure to enough radiation, usually more than 100 rem (a measure of radiation dose as it affects human tissue), can produce radiation sickness. Exposure at levels above 400 rem can cause death (residents around Three Mile Island were on average exposed to 1 millirem [mrem] of radiation; 1 mrem is one-thousandth of a rem). For comparison purposes, people are exposed on average to 300 mrem of *background radiation* — the levels of radiation to which humans are exposed daily from sources such as space—every year.

Scientists continue to debate the long-term effects of the Chernobyl accident. An early study commissioned by the Energy Department projected that about 40,000 people would die of cancer-induced deaths related to the accident, but one year later that number was revised to 17,000. Scientists continue to study the health of residents near Chernobyl. No one is sure about the level at which low-level radiation becomes dangerous. The dangers of conta-

mination to humans extend beyond the health of the body to effects on air, food, and water sources. Following the Chernobyl accident, many European countries advised citizens not to drink milk or eat certain meats and vegetables until a certain amount of time had passed. Each country set its own limits.

Although most nuclear power plants have not had high-profile accidents, all of them do create radioactive wastes, and waste storage remains highly controversial. Waste from nuclear power plants will remain radioactive for tens of thousands of years, although levels will fall to background radiation limits within approximately 10,000 years. As a result, engineers must design disposal sites that will block direct radiation and prohibit radioactive particles from infiltrating food and water sources for millennia.

The United States plans to bury its waste in a deep geological disposal site but has yet to designate the location. Testing at Yucca Mountain in Nevada, a potential storage site, began in 1987, and scientists are researching the effects of water, bacteria, and seismic activity on waste storage, among other issues. The mountain is the only site being investigated currently, but this does not mean that it will be chosen. Whatever site is designated needs to hold approximately 84,000 tons of radioactive waste from commercial reactors as well as 10,000 tons of waste from Defense Department projects.

Issues of waste disposal also arise when a nuclear power plant is closed. The general life expectancy of a commercial nuclear reactor is 40 years, although U.S. plant operators can apply for a license renewal of up to 20 years. When a plant is closed, it is said to be *decommissioned*. Transportation also remains a significant issue in nuclear power. Trains, ships, and other means of transport carry uranium to enrichment plants and waste to storage areas. To avoid calamities in the United States, the NRC reviews approximately 100 new or amended containers designed for use in transporting radioactive materials each year.

With so many issues surrounding nuclear energy, one may wonder why any utility would want to build a nuclear power plant. The initial cost of building such a plant is staggering; one company in Michigan spent over $4.1 billion trying to erect a plant, which is not unusual. Once operational, a nuclear power plant offers some advantages over its competitors. Uranium 235 packs more energy per ounce than many classic fuels do. For example, 1 ounce of the isotope will provide 570,000 kilowatts (kw) of energy per hour, whereas 1 ton of coal gives only 7300 kw per hour. Also, as noted earlier, nations with few natural resources find harnessing atomic energy a practical means to gain independence in international energy markets.

—*Christina Roache*

RELATED TOPICS
Biomass Energy, Hydroelectric Power, Solar Energy, Wind Energy

BIBLIOGRAPHY AND FURTHER RESEARCH

BOOKS

Jasper, James. *Nuclear Politics*. Princeton, N.J.: Princeton University Press, 1990.

Medvedev, Grigori. *The Truth About Chernobyl*. New York: Basic Books, 1991.

Pringle, Laurence. *Nuclear Energy: Troubled Past, Uncertain Future*. New York: Macmillan, 1989.

Rhodes, Richard. *Nuclear Renewal: Common Sense About Energy*. New York: Penguin Group, 1993.

Wolfson, Richard. *Nuclear Choices: A Citizen's Guide to Nuclear Technology*. Cambridge, Mass.: MIT Press, 1991.

PERIODICALS

Booth, W. "Postmortem on Three Mile Island." *Science*, April 12, 1987.

Cobb, Charles E., and Karen Kasmauski. "Living with Radiation." *National Geographic,* April 1989, 403.

Edwards, Rob. "The World's Dustbin: Britain Wants to Cash In on America's Nuclear Waste." *New Scientist*, January 31,1998, 11.

Edwards, Rob. "No Smoke Without . . . (Nuclear Power Is Not the Green Solution)." *New Scientist,* September 12, 1998, 50.

Golay, Michael W., and Neil E. Todras. "Advanced Light-Water Reactors." *Scientific American,* April 1990, 82.

Greenwald, J., et al. "Meltdown." *Time,* December 5, 1986.

Kadak, Andrew C. "A New Look at Nuclear Issues." *Issues in Science and Technology,* Spring 1998, 7.

Kanin, Yuri. "Ukraine Will Close Reactors." *Nature,* November 7, 1991, 8.

Kerr, Richard A. "Yucca Mountain Panel Says DOE Lacks Data." *Science,* February 26, 1999, 1235.

Saegusa, Asako. "Japan Reacts Uneasily to Nuclear Plan for Meeting Carbon Cuts." *Nature,* July 2, 1998, 3.

Slovic, Paul. "Perceived Risk, Trust, and the Politics of Nuclear Waste." *Science,* December 13, 1991, 1603.

"Time to Take the Plunge at Yucca Mountain." *Nature,* December 10, 1998, 497.

"Twilight of Forgotten Dreams? The Plan to Close Britain's First Commercial Nuclear Plant Should Not Be Misread (and It Has Been) as a Further Nail in the Coffin of the Nuclear Industry Worldwide." *Nature,* July 21, 1998, 183–184.

INTERNET RESOURCES

International Atomic Energy Agency
http://www.iaea.or.at

Office of Nuclear Energy, Science and Technology, U.S. Department of Energy
http://www.ne.doe.gov

Reports on the Yucca Mountain site
http://www.ymp.gov

U.S. Nuclear Regulatory Commission
http://www.nrc.gov

ORGAN TRANSPLANTATION

By the late 1990s, technological and surgical expertise had reached the stage where up to six of the body's major organs could be replaced by donated organs. Organ rejection, rather than surgical technique, remains the major limitation to transplant success. Tissue matching and immunosuppressant drugs minimize the problem of rejection, but side effects of medication make the body more liable to infection.

Nevertheless, organ transplantation is a late 20th-century surgical success story, and the demand for donated organs far outstrips supply. Various strategies are being considered to alleviate this shortfall. In the short term, this includes better public education about organ donation and transplant practices; in the longer term, research is under way to make organs available by engineering them in tissue culture or possibly by xenotransplantation (using nonhuman organs). Many medical, social, legal, and ethical issues affect organ transplantation, and their complexity is likely to increase markedly in the future.

Scientific and Technological Description

In the context of this article, a transplant refers to an organ or part of an organ that is transferred from one part of the body to another or from one person to another. When donor and recipient are the same, the transplant is an *autograft;* when they are different individuals of the same species, the transfer is an *allograft* or *homograft;* when of different species, the transplant is a *xenograft* or *heterograft.*

In essence, organ transplantation shares the same difficulties as those of TISSUE TRANSPLANTATION, plus additional ones because of the three-dimensional nature of organs and their additional complexity. To transplant an organ successfully, the replacement needs to be connected to the appropriate blood vessels in the recipient. Microsurgical techniques are used to connect the severed ends of arteries and veins, so making leakproof connections between the donor organ and the recipient's circulatory system.

In humans, an organ is transplanted to replace or supplement a damaged organ. Organs that are currently transplanted are the kidneys, the pancreas (with or without a kidney), lungs, the heart (with or without lungs), the liver, and parts of the intestine (see figure). In most cases, the organs are obtained from a recently deceased person. However, in the case of a kidney, a lung, or part of a liver or intestine, a living person, typically a relative of the recipient, can donate the organ or part without experiencing a debilitating loss of function.

Two major barriers limit organ transplantation as currently practiced: availability and acceptance. As medical technology and surgical sophistication advance, so the demand for organs for transplantation far outstrips the supply. In the United States, the United Network for Organ Sharing (UNOS) is the national body responsible for allocating and matching donor organs with recipients. On March 31, 1999, slightly more than 62,000 patients were on the UNOS waiting list for donor organs. In 1997, about 20,000 organ transplants were performed in the United States using organs from about 5500 dead and 3800 living donors.

The second barrier to transplantation, acceptance, has been largely overcome by a combination of *tissue matching* and the use of *immunosuppressants* (drugs that reduce the activity of the immune system that would otherwise cause the rejection of transplanted organs as foreign). Most body organs and tissues have chemical markers (antigens) on their cell surfaces. These are recognized by the person's immune system as *self,* or if they enter the body of someone else, that person's immune system recognizes the tissue as *foreign.* Foreign tissues are immediately attacked by white blood cells, especially lymphocytes. The affected tissue becomes inflamed and is ultimately killed. The foreign tissue has been rejected.

Tissue matching is a means of minimizing the problem of rejection. The major cell surface antigens that trigger rejection are *histocompatibility antigens* and the most important of these are *human leukocyte antigens* (HLAs). Where possible, transplant surgeons seek to match three major groups of HLAs—HLA-A, HLA-B, and HLA-DR—between donor and recipient. In addition, a match between compatible blood groups (on the ABO blood group system) is necessary. In practice, tissue matching is usually best between close relatives, and best of all between identical twins. However, as organ donors are frequently those that are deceased, and viable organs can be stored for only a matter of hours, relying solely on organs donated from a close relative is not feasible. In most cases, organs are transplanted from a person who is unrelated to the recipient.

The use of immunosuppressant drugs is the second line of attack to minimize tissue rejection. Corticosteroid drugs limit the activity of the immune system overall, whereas cyclosporine has a more specific effect on the lymphocytes responsible for tissue rejection. Of the several side effects associated with immunosuppressants, lowered immunity to attack by infectious microbes, in particular fungi and protozoa, is one of the more serious. Under current transplant regimes, most organ recipients have to take immunosuppressant drugs for the rest of their lives if they are to avoid a potentially fatal rejection of the transplanted organ.

Historical Development

In the 1910s, French-born surgeon Alexis Carrel and his American colleague Charles Guthrie pioneered blood vessel *anastomosis* (stitching the ends of cut vessels together to form a leakproof join). This breakthrough made organ transplantation feasible. Carrel successfully autografted animal organs such as the kidney and liver. However, when he attempted animal *allografts* (transplants between different individuals of the same species), the transplanted organs were rejected.

The first attempts to transplant entire human organs are credited to Serge Voronoff, a Soviet surgeon. Anatomically, the kidney is the easiest major organ to be transplanted. In the 1930s, Voronoff tried unsuccessfully to transplant kidneys from cadavers into six seriously ill patients. Tissue rejection was the biggest impediment. It was not until 1954 that Joseph Murray and J. Hartwell Harrison at Peter Bent Brigham Hospital in Boston carried out the first successful human organ transplant. They bypassed the problem of rejection by transplanting a kidney from one man to his identical twin, who was suffering kidney failure.

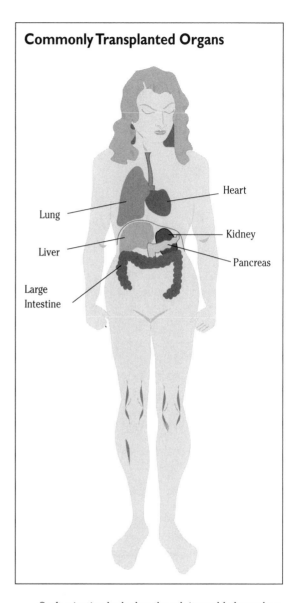

Commonly Transplanted Organs

Heart

Lung

Kidney

Liver

Pancreas

Large Intestine

Graft rejection looked as though it would always be a major obstacle in allografts. Between the 1940s and the 1960s, major advances were achieved in unraveling the causes of rejection and its control. British researcher Peter Medawar, working with burn victims during World War II, discovered that skin allografts were rejected because donated skin was genetically incompatible with that of the recipient. The donor skin was recognized by the recipient's immune system as foreign and attacked. During the 1950s, various studies by Medawar and others showed that corticosteroids could be administered to limit the recipient's lymphocyte attack on a donor organ. However, lymphocyte-produced *antibodies* (active chemicals that attack antigens) would still attack donor tissue, and it was not until the late 1950s and 1960s that a class of drugs—those that selectively inhibited DNA synthesis in lymphocytes—were applied to the problem of tissue rejection. *Azathioprine,* the first of these drugs, was developed by

British researchers Roy Calne and J.E. Murray in 1959 and was in clinical use by 1962. *Cyclosporine,* a more powerful and selective immunosuppressant, was tested in 1978 by Calne and colleagues and was marketed by 1983. Nowadays, a cocktail of immunosuppressant drugs—often cyclosporine in combination with azathioprine and corti-costeroids—is administered to transplant patients after surgery. *Monoclonal antibodies,* antibodies formed following the laboratory fusion of lymphocytes with cancerous cells, can be manufactured to specifically attack some of the recipient's lymphocytes and then administered to control the recipient's rejection response.

During the 1950s, the development of life-support technology had a major impact on organ transplantation. First, such technology forced surgeons to develop more highly refined methods for establishing death. At this time the concept of *brain-stem death* was introduced, in which those parts of the brain responsible for the control of basic life processes such as breathing and circulation are irreparably damaged. The brain is no longer able to maintain the life of the body. Brain-stem death is now the benchmark used to establish a person's death. (Prior to this, lack of physical responsiveness to certain stimuli and the absence of vital signs such as breathing and heartbeat were used to establish death.) Nowadays, subtler means for determining the presence or absence of brain activity may be employed in certain cases. Second, using life-support machines, the organs of brain-dead patients could be maintained alive within the body until the point at which the organs were required for transplantation.

With the advent of immunosuppressants, in combination with the development of tissue typing, the 1960s and 1970s saw a great increase in the number and type of organs being transplanted. By the mid-1960s, dozens of kidney transplants were being performed each year. In June 1963, U.S. surgeon James Hardy performed a lung transplant on a man dying of lung cancer. The new lung, transplanted from a heart attack victim, was accepted and functioned for 18 days before the patient succumbed to unrelated kidney damage. In the same year, Denver surgeon Thomas Starzl attempted a liver transplant: The 3-year-old patient died on the operating table. It was only a matter of time before the heart would be transplanted.

Heart transplants require a heart–lung machine to maintain the patient during the operation. On December 3, 1967, South African surgeon Christiaan Barnard transplanted the heart of a brain-dead young woman into 53-year-old terminally ill Louis Washkansky. He died of pneumonia 18 days later. In the following 10 years, heart transplants garnered considerable funding and media interest, but it was not until the early 1980s, with the introduction of cyclosporine as an immunosuppressant, that the success rate of heart transplants (previously less than 30 percent survival after one year) reached anything like acceptable levels.

With the ever-greater shortfall of donor organs, strategies other than allografting are now being actively investigated. One approach is to engineer organs (see Artificial Organs) that are grown from the recipient's own tissues or are cultured from donated *stem cells* (cells that are initially unspecialized but can be induced to differentiate into various tissues). The use of Genetic Engineering (also see Genetic Testing) raises the possibility that such tissues could be stripped of marker molecules so that they are not recognized as foreign, or that the marker molecules could, in some way, be masked. Another approach, *xenotransplantation,* involves using genetically engineered domestic animals, particularly pigs, as potential donors of organs. However, because pig cells are so genetically distinct from human cells, and because they contain proteins and other chemicals that are different from the equivalent molecules in humans, such organs may prove to be incompatible even if the problems of tissue rejection can be overcome. Nevertheless, xenotransplantation is a very active area of research and some researchers consider it the best long-term solution to the critical shortage of human donor organs.

Uses, Effects, and Limitations

Nowadays, organ rejection can be more or less controlled therapeutically, so that barring other complications, if the surgical procedure is successful, the functional survival of most organ transplants is about 80 percent at one year. With well-matched kidneys, functional kidney survival after five years is better than 90 percent. There are patients scattered across the world who have lived with an allografted kidney for more than 25 years. Heart transplant patients have survived more than 20 years, and many pancreas recipients, the beneficiaries of a more recent surgical procedure, have so far survived for more than 10 years. Recipients of organ transplants, who previously had life-threatening medical conditions, can now lead healthy, active lives for many years.

Matching a donor to a recipient needs to be timely, because once an organ is removed for donation, it must be implanted within a matter of hours. Many countries use computer databases to help facilitate the matching of organs with donors. The delivery of an organ for donation is often a rushed affair, with a helicopter or speeding ambulance ferrying the cooled organ from deceased donor to living recipient. To maintain an organ in as pristine a state as possible, it is immersed in cold sterile fluid and perfused through its blood vessels with cooled, preservative fluids. Treated in this manner, a kidney is usable for 48 to 72 hours, a liver or pancreas for 12 to 24 hours, and the heart and lungs for 4 to 6 hours.

Despite considerable progress in immunosuppression, rejection and infection are still the principal causes of organ transplant failure. Immunosuppressant drugs lower immunity to infection, and research is continuously under way to improve the selectivity of immunosuppression while minimizing unwanted side effects. Interest is being rekindled in the possibility of inducing immunological tolerance in humans. By a combination of bouts of drug immunosuppression, interspersed with the introduction of cells from a potential donor, there is the possibility of "educating" the recipient's immune system to tolerate the foreign cells of a potential donor: in effect, to fool the immune system into recognizing specific foreign cells as self. If this could be achieved, patients would no longer need to take immunosuppressant drugs in the long term, and so could avoid the risks associated with this medication.

In time, other major organs and limbs or parts of limbs will be allografted. By the late 1990s, surgeons in the United States, Canada, and the United Kingdom were carrying out intestinal transplants. In September 1998, an international team of surgeons operated on New Zealander Clint Hallam at a hospital in Lyons, France. They attached the hand from a recently deceased man to Hallam's handless arm. As of early 1999, the procedure was a success, and similar operations were being performed in the United States.

Issues and Debate

Many social, legal, medical, and ethical issues surround the use of donated organs. On the supply side, the ethics of living donors being paid to supply organs, or of the legal estates of dead donors being paid for organs supplied, is vigorously debated in legal and medical circles. Currently, payment for organs is prohibited by the International Transplant Society and most industrialized societies adhere to this. Stories of people being abducted and their organs removed are greatly exaggerated. However, in a few societies, organs from executed criminals are sometimes harvested for use in transplants; and there are perhaps justifiable fears of a black market trade in human organs.

Legal organ allocation is also a thorny issue. What criteria should be used to decide who should receive the highest priority on a waiting list to receive a donor organ? In the United States alone, thousands die of organ failure each year before they receive a replacement organ, but organ allocation is at least moving in the direction of a system that is more utilitarian and more equitable. However, balancing conflicting demands is no easy matter. Organ transplants could be targeted at those for whom the procedure is likely to be most cost-effective, has the highest chance of success, and is likely to offer the greatest number of additional years of patient life. However, such a priority system would necessarily discriminate against older persons and those in minority groups. On statistical grounds alone, patients in small minority ethnic groups may be harder to tissue match with available donors. Also, some minorities have a cultural or religious bias against organ donation, so suitable tissue-matched donors may be unavailable. A point system that balances fairness against utility criteria may be one way of achieving a more desirable system. Devolving the responsibility for organ allocation to panels of representatives from many parts of the community, such as a scheme that has been pioneered in Oregon, is one way forward.

The U.S. system of organ allocation seeks to be blind to wealth or social status. However, in a study published in 1998 by G.C. Alexander and A.R. Segal in the *Journal of the American Medical Association,* access to kidney transplantation was found to be lower for blacks than for whites, lower for women than for men, lower for poor than for rich patients, and varied by geographic region. However, the reasons for this were many and complex and affected all stages of transplant allocation, from a patient's initial inquiry to the final operation itself.

In the United States, organ transplantation is, in many respects, at the forefront of medical and bioethical practice. With people living longer and with new kinds of transplant operations coming into use, the ethical and philosophical dilemmas surrounding organ transplantation are likely to increase. As it is, many more people could donate organs than currently do so. Increased public awareness and reassurance about the principles and practice of organ donation are called for.

—Trevor Day

RELATED TOPICS
Artificial Organs, Artificial Tissue, Fetal Tissue Transplantation, Genetic Engineering, Genetic Testing, Tissue Transplantation

BIBLIOGRAPHY AND FURTHER RESEARCH

BOOKS

Caplan, Arthur L. *Am I My Brother's Keeper?* Bloomington, Ind.: Indiana University Press, 1997.

Cohen, Lloyd R. *Increasing the Supply of Transplant Organs: The Virtues of an Options Market.* Austin, Texas: R.G. Landes, 1995.

Crigger, Bette-Jane, ed. *Cases in Bioethics,* 3rd ed. New York: St. Martin's Press, 1998.

Durrett, Deanne. *Organ Transplants.* San Diego, Calif.: Lucent Books, 1993.

Fox, Renée C., and Judith P. Swazey. *Spare Parts: Organ Replacement in American Society.* New York: Oxford University Press, 1992.

Institue of Medicine. *Xenotransplantation: Science, Ethics and Public Policy.* Washington, D.C.: National Academy Press, 1996.

Kimball, Andrew. *The Human Body Shop,* 2nd ed. Washington, D.C.: Regnery Publishing, 1997.

McCartney, Scott. *Defying the Gods: Inside the New Frontiers of Organ Transplants.* New York: Macmillan, 1994.

Roitt, I.M. *Essential Immunology,* 8th ed. Oxford: Blackwell, 1994.

Weatherall, D.J., J.G.G. Ledingham, and D.A. Warrell, eds. *Oxford Textbook of Medicine,* 3rd ed. Oxford: Oxford University Press, 1996.

PERIODICALS

Alexander, G. C., and A. R. Segal. "Barriers to Cadaveric Renal Transplantation Among Blacks, Women and the Poor." *Journal of the American Medical Association,* October 7, 1998, 1148.

Concar, David. "Hands Today, Faces Tomorrow." *New Scientist,* October 3, 1998, 13.

Concar, David. "The Organ Factory of the Future?" *New Scientist,* June 18, 1994, 24.

Lin, H.-M., H.M. Kaufmann, and M.A. McBride. "Center-Specific Graft and Patient Survival Rates: 1997 United Network for Organ

Sharing (UNOS) Report." *Journal of the American Medical Association,* October 7, 1998, 1153.

Milford, Edgar L. "Organ Transplantation: Barriers, Outcomes and Evolving Policies." *Journal of the American Medical Association,* October 7, 1998, 1184.

INTERNET RESOURCES

M.O.R.E. (Multiple Organ Retrieval and Exchange) Program of Ontario
http://www.transplant-ontario.org/

National Foundation for Transplants
http://www.otf.org/

National Institute of Transplantation
http://www.transplantation.com/

Organ Transplantation and Donation
http://www.transweb.org/

United Network for Organ Sharing (UNOS)
http://www.unos.org/

PARALLEL COMPUTING

Traditional computers process one programming instruction or piece of information (data) at a time, a method known as serial computing. When complex tasks or large amounts of data are involved, this means that even the largest computers take an inordinate amount of time to complete their work. A more advanced technique known as parallel processing enables a computer to do different jobs or analyze different pieces of data at the same time.

Parallel processing is used in the world's fastest computers (known as SUPERCOMPUTERS) and is widely regarded as the best method of extending computer power beyond the limitations of current technology. However, parallel computing is both much more complex and much more expensive than serial computing.

Scientific and Technological Description

An ordinary personal computer (PC) uses a single processor chip to run a single program instruction on a single piece of data at a time. This approach is known as *serial processing* (more precisely, single instruction, single data [SISD]). Parallel computing aims to make computers quicker and more productive by doing more than one thing at a time. There are numerous ways of going about this, but they break down into two main approaches: data parallelism and functional parallelism.

Data parallelism involves carrying out the same operation at the same time on more than one piece of data (sometimes known as *single instruction, multiple data* [SIMD]). It includes a method known as *vector processing,* which enables a single instruction (a computing operation such as a calculation) to be carried out simultaneously on a long list of numbers (known in computing terminology as a *vector*). Vector processing is used when large amounts of highly regular data are being processed, such as in scientific and engineering calculations. It is commonly used in the fastest supercomputers, which are designed entirely around the task of processing vectors.

Functional parallelism splits up a computer's work and does different jobs in different places. This is similar to the way that jobs are split up on an assembly line in a factory. Different parts of the computer carry out different operations on different data at the same time (sometimes known as multiple instruction, multiple data [MIMD]).

Functional parallelism includes techniques known as multiprocessing and massive parallelism. In *multiprocessing* more than one processor is working inside a single computer; each processor has its own memory (known as *distributed memory parallel* [DMP]) or shares a common memory (known as *shared memory parallel* [SMP]). *Massive parallelism* is a method of combining hundreds or thousands of individual processors and memories to do different jobs simultaneously. A massively parallel supercomputer known as the Connection Machine, built by Thinking Machines Corporation, has no fewer than 64,000 individual processors. Sometimes processors are connected in a line, but variations include tree shapes, stars, and even complex multidimensional shapes known as hypercubes. The connections between processors and memories are known as *interconnects.* A type of interconnect known as a *crossbar switch* enables any processor to be connected to any memory.

Functional and data parallelism are most often used with vast supercomputers, but there is another method of using parallel techniques to achieve the same computing power. Multiple PCs can be connected over a local area network (LAN), spanning an office block, or a wide area network (WAN), spanning the globe. In the most ambitious examples of this approach, sometimes known as *hypercomputing* or *distributed processing,* hundreds of people connect their PCs together over the Internet (a type of WAN) to form temporary supercomputers that can tackle complex mathematical and scientific problems. There are standard methods of getting computers to work together in this way, including a common communications method called the *message passing interface* (MPI) and a technique of sharing memory over many linked computers called *distributed shared memory* (DSM).

Historical Development

Parallel computing has so far been used most notably in the world's most powerful computers and its development is a reflection of theirs (see SUPERCOMPUTERS). The first computer to make use of parallel techniques was the ENIAC, a 30-ton electronic calculator constructed in the early 1940s at the University of Pennsylvania. Later in the 1940s, computer scientists began to explore the similarities between computers and human brains. This led to the field of artificial intelligence, which aims to produce computers that share some of the characteristics of thinking human beings. It also led to the idea that computers might use parallel processing in a way similar to the human brain. In 1948, British mathematician Alan Turing, who later became famous for his research into artificial intelligence, published a scientific paper entitled "Intelligent Machinery." This outlined the concept of a computer called the O-machine that used a highly interconnected set of parallel processing units now known as a *neural network*.

The first large-scale parallel supercomputer, ILLIAC IV, was designed in the late 1960s and completed in the early 1970s. It used a type of functional parallelism in which 64 processors were linked in an array, each with its own 16-kilobyte memory. The entire machine had 1 megabyte of memory, which is considerably less than that of today's PCs. It worked at a speed of 25 megaflops (25 million flops), a *flop* being a measure of computing speed equal to one calculation performed in a second. Today's supercomputers are up to 100,000 times faster.

During the 1980s, parallel processing ideas first aired by Turing and others 40 years earlier became a fashionable avenue of research in artificial intelligence (where they spawned neural networks), psychology (where they led to neural network or *connectionist* models of the brain), and supercomputing. In 1983, Thinking Machines Corporation launched its massively parallel Connection Machine CM-1, which contained 64,000 individual processors using the SIMD technique. The same year, the world's best-known supercomputer maker, Cray Research, launched the Cray X-MP, which was the world's first *parallel vector processing supercomputer* (a machine in which each processor uses vector processing independently). It used four separate processors, worked at 941 megaflops, and cost up to $16 million.

The 1990s marked the end of the Cold War and a drastic reduction in defense spending. This had a serious effect on supercomputer makers, for whom defense had been a sizable market. At the same time a relatively new technique known as *reduced instruction set computing* (RISC), which speeds up computers by paring down the set of instructions they have to understand, was enabling high-performance workstations such as the IBM RS/6000 to achieve near-supercomputer performance. The logical conclusion was the disappearance of several supercomputer manufacturers in the mid-1990s and the merger of Cray Research, a leading supercomputer maker, and Silicon Graphics, its equivalent in the RISC workstation world.

Today, parallel processing is found not just in supercomputers and RISC workstations. The Microsoft Windows NT operating system also includes support for multiple processors, and many PCs can be linked to form a temporary supercomputer using software known as *parallel virtual machine* (PVM), which was developed in the early 1990s at Oak Ridge National Laboratory, Tennessee.

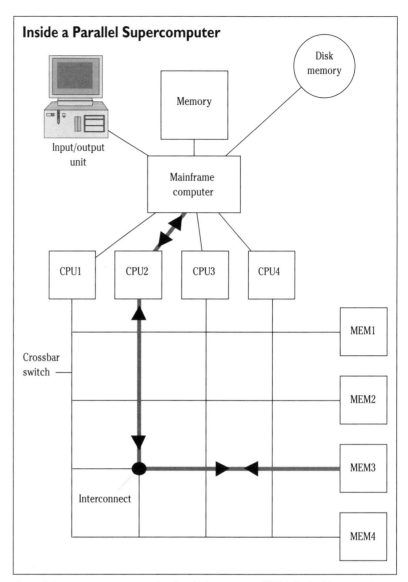

Inside a Parallel Supercomputer

A parallel supercomputer may consist of multiple processors (CPUs). Each of these may be connected to its own memory by a gridlike system of connections known as a crossbar switch. Work is coordinated between the different processors by a large mainframe computer, which also communicates with users via an input/output unit, such as a keyboard and monitor.

Uses, Effects, and Limitations

Parallel processing is typically used in traditional supercomputing applications such as weather forecasting, space and defense research, and high-energy physics. It is an effective method of increasing the performance of supercomputers so that they produce in days or weeks results that would otherwise take months or years. One early supercomputer, the Cray-1, could work at a speed of up to 160 megaflops. Today, the state-of-the-art, massively parallel Intel ASCI Red TFLOPS machine, which links together over 9000 ordinary 200-megahertz (200-megaflop) Pentium-Pro microprocessors, can achieve a speed of 1.8 teraflops (1.8 trillion flops). One of the most spectacular demonstrations of the power of parallel processing is a project known as the Search for Extra Terrestrial Intelligence (SETI) based at the University of California in Berkeley, which processes data from radio telescopes in an attempt to locate intelligent life in space. A million ordinary PC users have downloaded software called SETI@home, which effectively turns their humble desktop PCs into small components of a massively parallel supercomputer connected by the Internet. Thanks to parallel processing, the SETI@home project had completed over 50,000 years of computer processing in just four months after its launch in May 1999.

Initiatives such as this highlight not just the trend toward greater speed, but also the way in which parallel processing has moved from Cape Canaveral and the Defense Department to ordinary homes and offices. A corporation that has a LAN or WAN connecting tens, hundreds, or thousands of PCs potentially has its own supercomputer. The movie *Toy Story* was produced not with a supercomputer but using 117 powerful workstations, each of which contained 300 processors running in parallel. In a few years' time, similar feats may be possible using ordinary PCs.

But the benefits of parallel computing are not just in speed and performance. The computer company Tandem carved out a market niche by producing *fault-tolerant computers* for critical business applications, such as electronic banking. These systems feature two identical high-specification mainframes (large computers), which run in parallel. One machine exactly mirrors the other's performance. If a fault occurs, the backup machine immediately takes over. Fault tolerance is something the human brain is particularly good at; two similar lobes and "processing" distributed throughout the brain structure make humans capable of surviving a surprising degree of brain damage with relatively unimpaired performance. Distributed processing offers computers similar benefits; the Internet is a good example. It was originally developed by defense researchers as a computer system to enable communication between people in different locations, in which processing abilities were distributed (spread around) to enable the network to survive hostile attack even if parts of it were disabled.

For all the benefits it offers, parallel computing is not without its drawbacks. At the supercomputer end of the market, there is no such thing as off-the-shelf software; everything has to be written from scratch. The complexity of parallel computer systems and the wide variety of ways

in which parallelism is implemented makes developing software a costly and time-consuming business. Transferring software between machines is notoriously difficult, although the convergence between supercomputers and high-specification workstations, which use standard software, has eased this problem.

Another difficulty is more subtle. By their nature, parallel computers can solve only those problems that can be neatly broken up into subproblems or calculations involving large sets of data that can be divided between different processors. Not all problems fall into this category. For some applications, parallel computing actually takes more time than serial computing.

Issues and Debate

For corporations frustrated by time-consuming chores such as processing vast amounts of similar data, parallel techniques that bring the power of supercomputers to desktop PCs should be very good news. But although parallel processing can be much faster than serial processing, it can also be more complex and is much harder to implement. Until PCs and workstations can use parallel techniques in a way that is both simple and invisible to users, the benefits are likely to be limited. Also, not everyone will welcome the prospect of faster and more powerful computers.

The "computing economy" is based on developing and selling ever-faster computers, which can run more advanced software, which needs more memory, which leads people to buy faster computers, and so on. From the biggest computer companies such as IBM and Microsoft to the countless enterprises that gain their competitive advantage from high-technology products, millions of workers depend on this trend. But there is a problem. Current computer technology is based on making electrons flow quickly through silicon (a constituent of sand) in devices known as semiconductors. However, the speed of silicon semiconductors is limited; the laws of physics set a speed limit even on computers, which is rapidly being approached. Other semiconductor materials, notably gallium arsenide, thought to offer computing at higher speeds, have so far failed to fulfill their promise. Radically different forms of computing, such as using light instead of electricity in optical computers, may provide a very long term solution. But in the short term, one of two things will happen. Either the sprinting progress of the computer market will slow to a crawl or some new means of perpetuating progress will have to be found. Parallel computing potentially offers a way out.

One interesting consequence of parallel computing has been the discovery that large computational chores can be divided between computers in different cities or continents, a technique known as *distributed processing*. This will make possible greater collaboration among scientific researchers; global research efforts such as the Human

Genome Project, in which geneticists are mapping out the entire human genome, which contains the blueprint for human life, are already benefiting from the technique. World Wide Web inventor Tim Berners-Lee has championed such global collaborations as a particularly creative and profitable direction in which the Internet might develop.

There are drawbacks, of course. Some have argued that distributed systems such as the Internet, with their lack of centralized control, make terrorists, criminals, and political extremists harder to track down. Others counter that distributed processing is a libertarian model of computing in which centralized control gives way to autonomy and personal responsibility. If centralized computer systems are the instruments of what writer George Orwell described as the "Big Brother" state, distributed systems go some way toward redressing the balance.

—*Chris Woodford*

RELATED TOPICS
Artificial Intelligence, Supercomputers

BIBLIOGRAPHY AND FURTHER READING

BOOKS
Duncan, Ralph. "Parallel Computer Construction Outside the United States." in *Advances in Computers*, Vol. 44, Marvin V. Zelkowitz, ed. San Diego, Calif.: Academic Press, 1997.
Fountain, Terence. *Parallel Computing: Principles and Practice.* New York: Cambridge University Press, 1994.
Hsu, Jeffrey, and Joseph Kusnan. *The Fifth Generation: The Future of Computer Technology*. Blue Ridge Summit, Pa.: Windcrest, imprint of TAB Books, 1989.
Kuck, David. *High-Performance Computing: Challenges for Future Systems*. New York: Oxford University Press, 1996.
Sima, Deszö, Terence Fountain, and Peter Kacsuz. A*dvanced Computer Architectures: A Design Space Approach*. Harlow, Essex, England: Addison-Wesley, 1997.
Stork, David, ed. *Hal's Legacy: 2001's Computer as Dream and Reality*. Cambridge, Mass.: MIT Press, 1997.

PERIODICALS
Alves, Alexandre. "Parallel Computing Windows Style." *BYTE*, May 1996, 169.
Copeland, B. Jack, and Diane Proudfoot. "Alan Turing's Forgotten Ideas in Computer Science." *Scientific American*, April 1999, 77.
Gibbs, W. Wayt. "Taking Computers to Task." Scientific American, July 1997, 64.
Pountain, Dick. "Parallel Processing in Bulk." *BYTE*. November 1996, 71.
Sharp, Oliver. "The Grand Challenges: The Supercomputer Makers." BYTE, February 1995, 65.
Thompson, Tom. "The World's Fastest Computers." *BYTE*, January 1996, 44.

INTERNET RESOURCES
IEEE Computer Society ParaScope
    http://computer.org/parascope/
Intel Technology Journal (TFLOPS issue)
    http://developer.intel.com/technology/itj/q11998.htm
National Energy Research Scientific Computing Center
    http://www.nersc.gov/
San Diego Supercomputer Center
    http://www.sdsc.edu/
UCLA Parallel Computing Laboratory
    http://pcl.cs.ucla.edu/

PCS PHONES

PCS (personal communications services) is a mobile phone service that first became available in the United States in 1995. Unlike earlier phone services, PCS offers some extra benefits to phone users, including greater security and compatibility with mobile phone systems used in other countries.

With the arrival of PCS, there are now several incompatible mobile phone systems in the United States (unlike in Europe, where there is a single standard system). Although PCS has been highly successful, with users expected to reach 15 million by the year 2000, concern over the possible health risks of using mobile phones could frustrate the growth of all types of mobile telecommunications, including PCS.

Scientific and Technological Description

Mobile phones are an example of wireless telecommunications—they are much more closely related to radios than telephones. (For a description of how a traditional telephone works, see INTERNET PROTOCOL TELEPHONY.) A mobile phone contains a small low-power radio transmitter and receiver. When a call is made, the transmitter in the phone sends out a signal, which is picked up by the antenna on a nearby base station. This routes the call to a central handling center, a *mobile telephone switching office* (MTSO), which uses fiber optics to pass the call to the conventional telephone network, the *public switched telephone network* (PSTN). Calls from a traditional phone (a *land line*) to a mobile phone simply route in the opposite direction, from the PSTN to the mobile via the MTSO. As a mobile phone moves around with its user, its call is passed or "handed off" from one antenna to another, which usually ensures an unbroken signal (see figure).

PCS is a digital system, which means that speech is converted from an undulating sound wave (like the up-and-down vibration of a plucked guitar string) into a series of numbers before it is transmitted. This makes a digital phone much less susceptible to background noise than the other main type of phone technology, analog, which simply transmits sound waves as they are without converting them. It also means that the numerical signal can be scrambled for greater security by the sending phone and unscrambled by the receiver (using the mathematical coding technique known as CRYPTOGRAPHY).

PCS is also a cellular system. This means that each antenna, mounted on a high building or freestanding tower, handles all the calls in a small geographical area known as a *cell*. A cell could be anything from a block or two of Fifth Avenue in New York City to most of a small town in a state with large rural areas such as Oregon. If a cell regularly

Making a Call with a PCS Phone

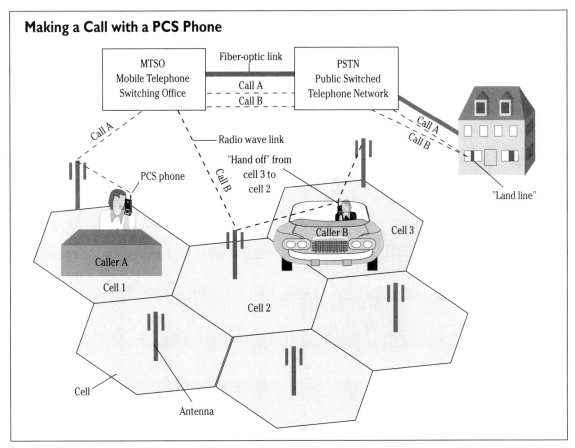

When caller A calls home on a PCS phone, it sends a radio signal to an antenna in his local cell (Cell 1). This transmits the call to the ordinary phone network (the Public Switched Telephone Network or PSTN) via an intermediate station known as a Mobile Telephone Switching Office or MTSO. When caller B crosses the boundary from Cell 3 to Cell 2 while making a call from her mobile phone, the call is "handed off" from the antenna in Cell 3 to the one in Cell 2 with no disruption in conversation.

becomes swamped with calls, the telecommunications company can simply divide it into smaller cells.

Mobile phones, pagers, television sets, radios, and similar devices use invisible beams of electromagnetic radiation, such as radio waves and microwaves, but there are only so many frequencies (bands of the broadcast spectrum) available for all these devices (see MICROWAVE COMMUNICATION). This problem is partly overcome using cells. Both phones and cell antennas use relatively low-powered transmitters, which means that a cell can use the same frequencies as nearby (although not neighboring) cells without causing interference, a technique known as *frequency reuse*. There are other solutions to the frequency shortage problem. Analog cellular systems transmit calls using the frequency-division multiple access (FDMA) technique, in which different phone calls are transmitted on different frequencies to prevent them from interfering with one another. PCS systems use one of two other (incompatible) techniques to squeeze more calls onto the limited frequencies. Time-division multiple access (TDMA) alternates rapidly between the calls from several phones, giving each call a small time share of a particular frequency. Code-division multiple access (CDMA) tags different calls with numeric codes, so that many calls can be transmitted on the same frequencies but still kept apart. TDMA offers at least three times more capacity than analog networks; CDMA claims 10 to 20 times more capacity.

Historical Development

Mobile phones have been around since the 1940s. The first commercial mobile phone service was established by AT&T and Southwestern Bell in 1946 in St. Louis, Missouri, mainly for vehicles such as taxis. In 1947, AT&T began a long-distance radio telephone service between New York and Boston, another key milestone for the technology. In 1948, the Richmond Radiotelephone Company started up a fully automatic mobile telephone service in Richmond, Indiana. By today's standards, these were cumbersome and unsophisticated systems, not unlike citizen's band (CB) radio. Signals could be sent only over a fixed number of frequencies (channels), which limited the number of calls that could be made at any one time and meant that other users tuned to the same frequency could eavesdrop on calls.

Since their arrival in the U.S. in the early 1990s, cellular phones have radically changed the way people do business and conduct their personal lives, making it possible to reach phone owners virtually anytime and anywhere (Photo Researchers/David Weintraub).

These problems would later be solved by cellular, digital cellular, and PCS phones.

Due partly to delays by the Federal Communications Commission (FCC), which oversees the regulation of U.S. telecommunications, in making enough frequencies available for mobile telephony services, and partly to the lack of microelectronic technology, which would ultimately make small-scale portable devices possible, it was another 30 years before cellular mobile phones became generally available in the United States. Indeed, it was not until 1978 that AT&T and Bell launched the popular analog system known as *advanced mobile phone service* (AMPS). It was tested the following year in Chicago, with 2000 customers spread over an area of 2100 square miles covered by 10 cells, and was launched commercially in 1983.

The next major development in mobile phone technology was already well under way in Europe, where 26 phone manufacturers had been jointly developing a world standard method of digital cellular telecommunications since 1982, which was known as GSM (originally meaning Groupe Speciale Mobile and later renamed Global System for Mobile Communications). The idea behind GSM was to devise a single standard system for mobile telephony that would work throughout Europe, using digital technology to ensure the greatest possible call-carrying capacity with the most efficient use of frequencies. Other benefits would include clearer calls, encryption to protect privacy,

the ability to transmit and receive digital computer data, and lower power consumption (which meant longer battery life). Although GSM was rapidly adopted throughout the world (about 100 countries now use it), U.S. phone companies were less enthusiastic about the standard, partly because of their considerable investment in existing analog systems.

Digital cellular phones arrived in the United States in 1991, initially using TDMA, the method of increasing the capacity of mobile phone frequency channels by time sharing. But U.S. phone companies remained out of step with the worldwide GSM standard until PCS became available in 1995. That year, the FCC announced an auction of mobile phone frequencies, and U.S. telecommunications companies used the opportunity to launch their own PCS systems. In the biggest auction in history, $7 billion was raised by selling off 2958 PCS operating licenses, which were divided between companies using rival—and only partly compatible—CDMA, TDMA, and GSM technologies. In other words, unlike GSM, PCS is not a single standard with which all manufacturers and telecommunications companies comply.

Uses, Effects, and Limitations

PCS offers many benefits over traditional cellular phone systems. Its digital technology provides call security through cryptography, longer battery life, and clearer call quality, and means that PCS can be used to send short text messages from one phone handset to another. Unlike other cellular systems, PCS phones use plug-in SMART CARDS, which contain details of the phone user's identity, a sizable personal phone book, and the types of services to which the user has access. PCS users can instantly transfer their phone number and account details to another handset (perhaps a rented phone in another country) simply by removing the smart card from the old phone and plugging it into the new one.

PCS phones can also be used for an array of purposes other than transmitting voice conversations. Digital technology allows computers connected to PCS phones to send and receive data, faxes, or e-mail; to browse the Internet; or even to provide connections between a mobile worker's laptop and his or her main office computer network while the worker is on the move.

Cellular phone systems such as PCS offer another very important benefit. Mobile phones are frequently used at the scene of accidents and emergencies, and because each call is made through a particular cell, it is possible for the emergency services to locate roughly where calls are coming from. In the future, mobile phones are expected to use a type of satellite navigation called the GLOBAL POSITIONING SYSTEM (GPS), so that the origin of emergency calls can be located with pinpoint accuracy. The same technology could also be used to provide directions to people while they are traveling or to help blind people negotiate unfamiliar areas.

For all their benefits, mobile phones are not without their problems. PCS offers poorer service coverage than that of traditional digital cellular phones and is often more expensive; calls can cost several times as much as those from an ordinary telephone and tend to be more expensive than digital cellular. Although PCS offers longer battery life than digital cellular, limited-life batteries are still a problem. Attempts to make batteries last longer include power amplifier chips made from gallium arsenide semiconductors, which consume less electricity, and different power-supply technologies, including fuel cells (see ALTERNATIVE AUTOMOTIVE FUEL TECHNOLOGIES; SOLAR ENERGY). Loss of service is caused not only by battery failures, but also by the limited capacity of phone cells.

Mobile phones are notorious for interfering with other sensitive electrical equipment; their use is often banned in hospitals, on aircraft, and at gas stations. But it is not just phone handsets that cause these sorts of problems. Cell transmitters can sometimes interfere with the locking systems used on automobiles, immobilizing cars if they are parked nearby. Satellites that transmit mobile phone signals also regularly disrupt space telescopes. Scientists at the Giant Meterwave Radio Telescope (GMRT) at Pune, India, and the Arecibo Radio Telescope in Puerto Rico have negotiated with mobile phone company Motorola to switch off its satellites to enable at least a few hours of "quiet time" for research each day (see TELESCOPY).

Issues and Debate

Mobile phones have brought benefits to millions of people who need to keep in touch while they are on the move and have become a fixture of modern life, but they are not without controversy. In the United States, the arrival of PCS has left the country with several different—and only partly compatible—mobile phone systems. PCS is compatible with the worldwide GSM standard, so U.S. travelers can use their phones abroad. However, unlike GSM, there is no strict definition of PCS, and not all telecommunications companies have adopted PCS in the same way. Mobile phone manufacturers have found themselves forced to supply devices that can operate using analog, digital cellular, and PCS, creating more expense and confusion for customers.

PCS and GSM are now worldwide standard methods of transmitting voice calls and data between mobile phones over long distances, but other standards have recently been proposed for transmitting data between appliances such as palm-top computers and mobile phones over shorter distances. These include a system known as Bluetooth, from a consortium of companies including IBM, and a rival system called Wireless Knowledge, from Microsoft. Attempts to marry these various systems, and pressure from various manufacturers to see their own proprietary systems adopted worldwide, could lead to a confusing variety of wireless communications "standards" in the early part of the 21st century.

PCS and other mobile phone technologies also face a potentially more serious hurdle. Possible health risks from mobile phones (which involve placing a microwave transmitter very close to a person's brain) and cell transmitters (which are sometimes sited next to homes and schools) continue to produce scare stories in both the popular media and the science press. No evidence has been found to confirm that there is any danger, but electromagnetic radiation from devices such as phones has been shown to affect the behavior of animals in laboratory experiments. Although the U.S. National Research Council concluded in May 1999 that there was no health risk posed by mobile phones and no need even for further study, other official scientific bodies are not so sure. The World Health Organization and the Japanese Science and Technology Agency are both conducting large-scale studies of the health effects of electromagnetic radiation.

With regard to health and safety concerns, cell-phone manufacturers may find the cigarette industry's experience instructive. Denying the health risks of cigarettes stored up tremendous legal problems for their manufacturers in the 1980s and 1990s. In the 1960s and 1970s, lawyers advised cigarette makers not to market "safe cigarettes," because by implication it suggested that ordinary cigarettes were dangerous. This controversial policy prevented the true health risks of cigarettes from being understood for many years.

Mobile phone manufacturers appear to have learned from that lesson. In response to health concerns, some companies are now selling low-radiation phones; others have developed cases that incorporate radiation shields or "hands-free" devices that enable phones to be used at a much safer distance from the head. Technology can usually solve the problems it creates, given time, and it seems unlikely that mobile telephony, one of the fastest-growing technologies of the 20th century, will be stalled by even this problem.

—*Chris Woodford*

RELATED TOPICS
Fiber Optics, Global Positioning System, Internet Protocol Telephony, Microwave Communication, Personal Digital Assistants

BIBLIOGRAPHY AND FURTHER RESEARCH

BOOKS
Harte, Lawrence, Steve Prokup, and Richard Levine. *Cellular and PCS: The Big Picture.* New York: McGraw-Hill, 1997.
Lee, William. *Mobile Cellular Telecommunications.* New York: McGraw-Hill, 1995.
Muller, Nathan. *Mobile Telecommunications Factbook.* New York: McGraw-Hill, 1998.
Ricci, Fred. *Personal Communications System Applications.* Upper Saddle River, N.J.: Prentice Hall, 1997.

Stetz, Penelope. *The Cell Phone Handbook: Everything You Wanted to Know About Wireless Telephony (but Didn't Know Whom or What to Ask).* Newport, R.I.: Aegis Publishing, 1999.

Tisal, Joachim. *GSM Cellular Radio Telephony.* New York: Wiley, 1997.

Tsakalakis, John. *PCS Network Deployment.* New York: McGraw-Hill, 1997.

PERIODICALS

"Cell Phones Threaten Radio Telescope." *Science,* November 28, 1997, 1569.

Emmerson, Bob. "Wake-up Call." *BYTE,* May 1996, 228.

Emmerson, Bob, and David Greetham. "GSM's Extraordinary Growth." *BYTE,* March 1996, 21.

Hills, Alex. "Terrestrial Wireless Networks." *Scientific American,* April 1998, 74.

Jerome, Marty. "Air Wars." *BYTE,* August 1997, 93.

LaPedus, Mark. "Korea to Install New CDMA Phone System." *BYTE,* October 1996, 4.

Mackenzie, Dana. "Motorola to Limit Signals over Arecibo." *Science,* March 27, 1998, 2032.

Pelton, Joseph. "Telecommunications for the 21st Century." *Scientific American,* April 1998, 68.

"ScienceScope: Power Surge." *Science,* May 21, 1999, 1245.

"Special Investigation: Mobile Phones." *New Scientist,* April 10, 1999 (a special issue of *New Scientist* devoted to the safety of mobile phones).

Ward, Mark. "Havens of Peace." *New Scientist,* May 23, 1998.

Ward, Mark. "Mobile Emergencies." *New Scientist,* August 30, 1997.

INTERNET RESOURCES

AT&T Digital PCS Demonstration
http://www.attws.com/general/pcs_demo/index.html

"Can PCS Dethrone Cellular and Become the Wireless King?" by Charles Eickmeyer and Nicole S. Davison
http://fiddle.ee.vt.edu/courses/ee4984/Project1997/eickmeyer_davison.html

"Digital Wireless Basics," by Tom Farley
http://www.privateline.com/PCS/PCS.htm

"Hello, This Is Dick Tracy," by Alan Hall, *Scientific American Online,* May 3, 1999
http://www.sciam.com/exhibit/1999/050399dicktracey/index.html

"Technical Trends in Wireless Telecommunications," by Dale N. Hatfield
http://tap.gallaudet.edu/hatfield.htm

World Health Organization EMF Project
http://www.who.int/peh-emf/

PERSONAL DIGITAL ASSISTANTS

Personal digital assistants (PDAs) are small, hand-held computers that offer greater mobility than desktop or laptop computers, yet are more limited in their information processing and storage capabilities. For data input, PDAs offer either a small keyboard or a touch-sensitive screen (touchpad) that responds to penstrokes and can interpret the user's handwriting.

When first launched, PDAs were criticized for their poor handwriting-recognition capabilities, among other weaknesses. Over the past several years, PDAs have greatly improved in handwriting recognition as well as in computing power and storage capacity. A majority of users employ PDAs to keep track of their schedule and store address book information. Newer PDAs offer e-mail and paging services. Although the latest PDAs are far more useful than the original models, *there is still debate about their future—whether they will remain computing accessories or become powerful and versatile enough to replace laptop or other computers.*

Scientific and Technological Description

All computers are conceptually the same: They have some way of receiving information (input), a way of processing that information, and a way of feeding out the results (output). PDAs are much smaller than desktop and laptop computers, and this means that the input, processing, and output of a traditional computer have to be miniaturized or simplified in a PDA. Where a typical desktop PC has a 2-foot-wide 105-key keyboard, a typical PDA has a miniature keypad or a touch-pad that responds to handwritten input (see LANGUAGE RECOGNITION SOFTWARE). Where a typical desktop machine has a 17-inch monitor, a PDA has a liquid-crystal display (LCD) screen that is perhaps only 3 or 4 inches across. Where a desktop has a powerful microprocessor (central processing chip) and memory, a PDA typically has a slower processor, a much smaller memory, and no hard-disk drive.

Like desktop computers, PDAs have two kinds of memory: RAM (random access memory, which stores temporary data created by users or by application programs, such as diaries and word processors, and is wiped clean when the power is switched off) and ROM (read-only memory, which is preprogrammed in the factory and cannot be overwritten by users). Because they lack a hard disk, PDAs have much less available storage, so application programs are usually preprogrammed into ROM memory. This can make PDAs faster to use (because accessing RAM and ROM memory is faster than reading data from a hard disk) and less error-prone (because applications are stored in virtually indestructible ROM memory, they cannot be damaged or overwritten by users).

Few computers are used in isolation: Most desktop machines are connected either to private corporate networks or to the Internet. Partly because they have such small memory and limited processing power, connectivity is a particularly important feature of PDAs. Most PDAs can be connected to a desktop PC using a standard connection cable, which enables data to be copied across from one machine to the other. In this process, known as *data synchronization,* whichever machine contains the most up-to-date information copies its data onto the other machine, overwriting the older version of the data. Some PDAs can transmit and receive information using an infrared port. This is a small plastic transmitter built into the side of the case that beams information in the form of infrared radiation (electromagnetic waves with a shorter wavelength than visible light) to a similar infrared port on a printer or a desktop PC. The most advanced PDAs have built-in modems, which enable them to be connected to the Internet for sending and receiving e-mail or browsing the World Wide Web. PDAs are sometimes combined with cellular phones in a single device that can function as a mobile phone, a stand-alone PDA, or a PDA connected to the Internet using the phone to send and receive its data; these combined units tend to have a tell-tale antenna protruding from their case (see PCS PHONES).

PDAs come in various shapes and sizes, ranging from small shirt-pocket or palm-sized devices, such as the Palm Computing organizers and Psion Series V, to hand-held PCs (HPCs) slightly smaller than a traditional laptop, such as the Phenom Express by LG Electronics and the Hewlett-Packard Jornada. Different models have different combinations of features, such as voice-recognition capabilities and built-in modems, but generally speaking, the smaller machines use touch-sensitive screens and infrared data synchronization, and the larger models operate more like laptops, with keyboards, high-resolution color screens, and cable connections.

Historical Development

In 1993, Apple Computer introduced the first PDA, the Newton MessagePad, which had been the pet project of John Sculley, then Apple's chairman. Developed at a cost of $200 million, it was the result of research into simpler, more intuitive ways of human–computer interaction based on handwriting analysis, voice recognition, and artificial intelligence, conducted by the Apple Advanced Technology Group from the mid-1980s to the mid-1990s. The first Newtons received almost universally poor reviews for their design flaws, the most significant of which were unreliable handwriting recognition and a lack of application software. Although there were two later versions, which offered an improved operating system (the main control software, such as Microsoft Windows, that organizes how different programs use a computer's resources), more processing power, and a bigger display with greater resolution, Apple announced in February 1998 that it had discontinued development of the Newton.

Apple pioneered the PDA concept, but other companies have been much more successful in exploiting the marketplace, notably Palm Computing (which has subsequently been acquired by the modem maker U.S. Robotics, itself taken over by network equipment manufacturer 3Com). In 1996, Palm Computing introduced what was regarded as the first truly useful palm-sized PC, the PalmPilot, which offered users up to 512K of memory, enough to store large amounts of data and run powerful applications. But it was the PalmPilot's usability that made it such a success. Its designer, Jeff Hawkins, realized that desktop machines running Windows were becoming increasingly complex and overfeatured at a time when most computer users seemed to want exactly the opposite: greater simplicity and ease of use. In the PalmPilot, Hawkins adopted a "less is more" approach, reducing the product's functions to exactly those that its users would want, promising only features that he knew he could deliver, and producing a machine that was reliable and dependable.

Later models of the PalmPilot built on this success. In 1997, the PalmPilot Professional Edition came with backlight and contrast controls, buttons that quick-started specific applications, an improved operating system, 1 megabyte (MB) of memory, and a HotSync cradle to synchronize data on the PDA with data on a desktop or laptop PC. In 1998, Palm Computing introduced the Palm III, organizer, with 2 MB of memory, a further improved operating system, and a new infrared device that allowed Palm organizers to "beam" information to each other.

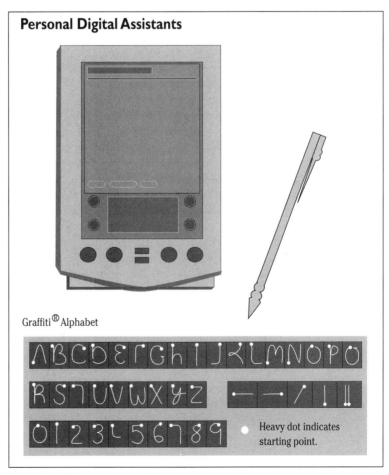

Personal Digital Assistants

Graffiti® Alphabet

Heavy dot indicates starting point.

The smallest PDAs have a touchpad that responds to penstrokes to receive input. Users write on the screen with a special stylus. They must also use a modified form of handwriting so that the computer will recognize letters and numbers correctly. The Graffiti Alphabet is a popular handwriting-recognition software program used with many PDAs. (The Graffitti Alphabet provided courtesy of Palm Computing.)

The latest Palm devices are the Palm V organizer and the Palm VII organizer. The Palm V device is half the thickness of previous units, provides a sharper screen, and comes with rechargeable batteries and 2 MB of memory. The Palm VII device has a built-in wireless communication service (effectively a modem and mobile phone transmitter/receiver seamlessly integrated into the PDA), which can be used to send and receive e-mail or to browse the Web. Because the Palm VII device's small screen is not really suited to displaying full Web pages, Palm offers a Web clipping service that delivers news updates, stock information, and sports scores (minus graphics, links, and other things that would slow down data transmission) from a variety of popular Web sites.

In fall 1996, Microsoft responded to the release of the first Palm Computing PDAs by introducing a new operating system for hand-held machines called Windows CE. This provides a similar experience to that of the Windows software that runs on desktop machines, yet uses the limited power of PDAs. Among the first companies to incorporate Windows CE into their PDAs were Casio, Goldstar, Hewlett-Packard, and NEC. Early reviews of Windows CE–based PDAs recommended them over Palm Computing organizers and Apple's Newton, but criticized their inefficient power use, a significant problem for mobile users. Over the past several years, competition between Palm PDAs and Windows CE–based PDAs has escalated. Currently, analysts note that since both types of devices have varied uses in a large market, both will continue to be feasible choices.

Uses, Effects, and Limitations

People use PDAs primarily to keep track of diaries and schedules and to maintain personal address books. Different PDAs vary in the amount of information they can store, but all can synchronize stored information with a laptop or desktop PC, for example, so that an appointments diary kept on the desktop machine will be the same as a diary kept on the PDA. Additionally, many PDAs offer synchronization software to allow them to link up to industry-standard applications running on desktop PCs, such as Lotus Organizer, Microsoft Outlook, and Symantec Act personal information managers.

The restricted processing power and memory of PDAs makes them unable to run the same powerful operating systems as desktop machines, and this in turn means that they cannot run the same powerful applications (such as the Microsoft Word word processor or the Lotus 1-2-3 spreadsheet). PDAs have simplified operating systems, such as Windows CE or Psion EPOC, which have built-in simplified versions of common desktop applications. Windows CE includes "pocket" versions of Microsoft's popular word processor (Word), spreadsheet (Excel), and presentation graphics tool (PowerPoint). Other software includes time, expense, and mileage tracking programs, and mapping and navigation applications.

The Palm VII handheld computer is the latest personal digital assistant from Palm Computing, featuring built-in two-way wireless communication. The Palm VII organizer can send and receive e-mail and retrieve information from the Internet (Courtesy of Palm Computing).

Although PDAs are highly portable, that portability commands a price: namely, limited memory and processing power. (Memory in PDAs ranges from 1 to 32 MB of RAM; by comparison, memory in desktop PCs ranges from about 32 to 128 MB.) For this reason, the scaled-down Pocket PowerPoint supported by Windows CE hand-held devices will allow users only to display presentations, not create or edit them. Additionally, the Windows CE versions of Word and Excel lack most of the advanced features of the corresponding desktop software. For this reason, PDAs are not really an alternative to desktops or laptops; they are better suited for limited data handling, such as jotting down notes, creating tasks, and sending e-mail while traveling.

Some PDAs are specifically designed as communication devices; indeed, the distinction between PDAs and cellular phones is becoming increasingly blurred (see PCS PHONES). Connected to the phone network through a modem, or using a built-in wireless communications system, PDAs can serve as pagers, devices for sending and receiving e-mail, PC-based fax machines, and Web browsers (see INTERNET AND WORLD WIDE WEB). While hand-held PCs (with their laptop-like screens) accomplish these tasks much like desktop PCs, the small, LCD screens of palm-top PDAs are not really suited to displaying Web pages. A great deal of scrolling may be needed to view a single Web page and, because of the limitations of the screen, the Web browser may be unable to display images.

Issues and Debate

The greatest issues and debate regarding PDAs address the role they will play in the future of personal computing, considering the current limitations of their capabilities. Industry analysts debate over growth projections for the PDA market. Optimistic projections suggest that Windows CE–based PDAs will proliferate so much that within the next decade, they will hold equal market share with other Windows-based computers. If these projections are borne out, people will be doing much more of their computing while in transit or away from their desks. Most analysts, however, predict more moderate growth.

Another issue that users debate is whether they will want to bring a laptop computer or a PDA when traveling outside the office in coming years. The short answer is that people will probably need both. Each serves separate purposes, although both offer some level of portability. PDAs offer greater portability; they are smaller and lighter than laptop computers. In addition, because their software is stored entirely in ROM memory, PDAs start up more rapidly than do notebook computers and offer greater battery life. Generally, battery life on hand-held PDAs runs from 8 to 12 hours, although some machines may work for several weeks on a single set of batteries. Notebook computers, by contrast, run for 2 to 4 hours. However, laptop computers give users advantages in other areas. For instance, with laptop PCs documents can be written and edited using full-scale versions of word processing, spreadsheet, and other application software. In addition, laptop computers offer greater processing power to make the applications run more quickly.

—Christine Doane

RELATED TOPICS
Language Recognition Software, PCS Phones

BIBLIOGRAPHY AND FURTHER RESEARCH

INTERNET RESOURCES
About.com: Introduction to Palmtop/PDA/Handheld Computing
http://palmtops.about.com/library/weekly/aa051699.htm?pid=2764&cob=home
"Getting a Handle on Handhelds," *PC Week,* July 26, 1999
http://www.zdnet.com/pcweek/stories/news/0,4153,410138,00.html
Handheld PC Magazine
http://www.hpcmag.com
"How to Buy Handhelds," *Computer Shopper Online*
http://www.zdnet.com/computershopper/edit/cshopper/frames/online.html
Microsoft Windows CE Page
http://www.microsoft.com/windowsce/default.asp
"Palmtop PCs Grow Up," *PC World Online* by Michael S. Lasky, Harry McCracken, and Vince Bielski
http://www.idg.net/crd__79985.html
"The PalmPilot's Creator Reflects on Good Design," *Business Week,* September 1998
http://www.businessweek.com/bwdaily/dnflash/sep1998/nf80929b.htm
PDA Central
http://www.palmpilotzone.com/
ZDNet PDA User
http://www.zdnet.com/products/pdauser/index.html

POLYMERS

A polymer is any material or fabric made of repeating molecular units linked together into a long chain. Polymers, which include synthetic materials such as plastic, vinyl, nylon, polyester, and Teflon, and natural materials such as DNA, silk, cotton, starch, and cellulose, are used in our lives every day.

Scientists began to control and manipulate polymers in the 19th century when they were looking for ways to improve natural rubber. In this century, chemists have created hundreds of durable synthetic polymeric materials from just a few simple building blocks. Scientists are experimenting today with polymers that enhance drug delivery to specific locations, exert mechanical forces, and emit light.

Scientific and Technological Description

Polymers are giant molecules composed of simple small units. The word *polymer* comes from the Greek words *poly,* meaning many, and *mer,* meaning unit. A *monomer* is a single unit; a *dimer* is composed of two; a *polymer* can have ten to millions of individual units. These units do not have to be identical: Polymers with two or more different monomers are called *copolymers.* The most common forms of copolymers are *random copolymers,* which contain different types of monomers in random order, and *block copolymers,* which contain stretches of different monomers: for example, five A monomers followed by five B monomers. Nylon is a typical copolymer that is composed of two repeating units, a diacid molecule and a diamine molecule. Polymers can be named for the bond that joins them (ester bonds are polyesters, amine bonds are polyamines) or for the monomer that makes them, such as polyethylene or polyvinyl chloride (PVC).

Polymers are created in nature and by mechanical processes. Common natural polymers are silk, cellulose, DNA, and starch. A polymer's characteristics are defined by

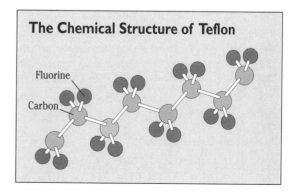

The Chemical Structure of Teflon

Fluorine

Carbon

Teflon, or polytetrafluoroethylene, is a linear polymer formed by the repeating molecular unit CF_2—a carbon molecule and two fluorine molecules. Teflon's special properties—it is uniquely nonadhesive and creates very little friction—make it ideal for nonstick coatings on pots and pans. It is also used to insulate wires, cables, and motors.

a small number of features. The first is the length of the polymer, which can run from 10 units to millions of units. At the molecular level the chains of molecules can be branched, unbranched, or cross-linked to other strands; they can be aligned or unaligned; and they can be flexible or inflexible. Changing any of these characteristics affects the macroscopic properties of the material, such as its melting point, flexibility, rigidity, and elasticity.

There are surprisingly many polymers that have been made from a few simple starting units. The five most common starting molecules are ethylene, methane, propylene, benzene, and butylene. All of these are obtained from crude oil and natural gas. These molecules are the lightest fractions of crude oil, which is a rich mixture of various hydrocarbons (molecules made of only carbon and hydrogen). Other chemicals derived from crude oil include gasoline, naphtha, kerosene, diesel fuel, and waxes.

The most common way of creating polymers is through *addition polymerization,* a process divided into three phases: initiation, addition, and termination. In *initiation,* a chemical must form a *free radical,* which is a molecule with an unpaired electron. These molecules are unstable and therefore very reactive. Once a monomer with an unpaired electron is formed, *addition* begins as the free radical attacks and reacts with another monomer. This reaction creates another free radical, which starts a chain reaction that feeds on the next-closest monomer. Many reactions add tens of thousands of monomers in a second. *Termination* of the reaction occurs when two free radical *tips* encounter each other and bond together to form a larger molecule, or when a free radical reacts with the midsection of a different chain. A contaminating molecule such as water or alcohol can be added to stop the reaction.

Once polymers have been formed in a reaction, there are a few common ways of manipulating them to form a final product. The first way is to heat the chemicals until they melt or soften. Then they are poured, pressed, or blown into a cold mold, which cools and hardens the polymer to a desired shape. Second, strips of plastic can be formed by passing the molten polymer through a small die or hole, which forms a long, thin fiber of the polymer. This product can then be pulled and stretched to form fabrics such as nylon. Other methods involve using hot presses to stamp polymers into desired shapes.

Historical Development

The science of polymers began in the 19th century. Natural *rubber* from the South American tree *Hevea brasiliensis* had been imported to Europe since the mid-15th century. The natives called it *caoutchouc.* They used it to mold shoes around their feet, to waterproof fabrics, and to make toy balls. Joseph Priestley, an English chemist, called it India rubber because it came from the West Indies and was good for rubbing out pencil marks. Rubber was not a perfect

material. In cold weather it would crack and come apart, and in hot weather it would begin to sweat and stick. The first commercial application of rubber was created by Charles Macintosh, a Scotsman who dissolved rubber in naphtha (an inflammable oil) and sandwiched it between two layers of cloth. Waterproof jackets made from this fabric are still called macintoshes in many parts of the world.

Rubber wasn't very useful until 1839, when the American Charles Goodyear discovered the process of *vulcanization.* After an accident where he spilled rubber and sulfur on a hot stove, he found that the new compound wasn't sticky, retained its shape, stayed flexible in cold temperatures, and resisted damage from chemicals that degraded raw rubber. Rubber was eventually discovered to be a polymer of the molecule *isoprene,* which was repeated 40 or 50 times to make a single rubber molecule. Adding sulfur to rubber created cross-links between the individual strands of rubber, locking them in place. When stretched, vulcanized rubber returns to its original shape. Natural rubber molecules slip around each other and don't spring back to their original positions.

Other researchers were also experimenting with *cellulose,* a natural polymer of glucose molecules that is created by all plants. Cellulose has been used to build homes of wood, make clothing from cotton, and create paper from wood pulp. Christian Schönbein added sulfuric and nitric acids to cotton in 1846 and created guncotton (cellulose nitrate), an explosive that was much more powerful and gave off less smoke than gunpowder. Later that year, Louis Menard treated paper with the same acids. After treating the product with ether and alcohol, he created a hard, transparent, colorless plastic called *collodion.* At the time, nobody appreciated its uses. It wasn't until 1863, when the American company Phelan & Collender offered a $10,000 prize to anyone who could make a synthetic billiard ball, that a use was found. John Wesley Hyatt took up the challenge, tried his best, and found that adding camphor to collodion made a plastic that softened in heat and could be rolled, spread, or molded into almost any shape. He called his plastic *celluloid.* It was useless as a billiard ball but great for applications such as combs, nail files, men's shirt collars, photographic film, and table-tennis balls.

Other scientists were also busy experimenting with collodion and cellulose nitrate. Alfred Nobel (who later founded the Nobel prizes) combined collodion and nitroglycerin in 1875 to produce blasting gelatin, the first explosive that could be transported safely. In 1884, Paul Vielle invented smokeless gunpowder from colloided cellulose nitrate. Later, collodion also found use as a synthetic silk fabric. Unfortunately, it was rather flammable; a single spark could cause a dress to disappear very quickly in a puff of smoke.

In 1907 the first completely synthetic polymer was discovered by Leo Baekeland. He created Bakelite by mixing

Scientists are trying to either find or develop bacteria that can degrade and metabolize plastics. While plastics and other polymers remain nonbiodegradable, recycling continues to be an essential part of plastics disposal (Corbis/Joseph Sohm; Chomo Sohm, Inc.).

and heating phenol and formaldehyde. This resulted in a hard, clear, amber-colored plastic that when hot could be poured and easily molded into a permanent shape. Bakelite is still used today. As an electrical insulator, it can be dyed or colored, and is strong but light. It is used in everything from buttons to car parts to radios and knife handles.

Polymer research exploded in the 1930s when the U.S. company Du Pont Chemicals chose Wallace Hume Carothers to head up its research division. In 1934 his laboratory discovered nylon, which was marketed as being stronger and more elastic than natural silk or rayon. In May 1940, women lined up to purchase the first stockings "made of coal, air, and water": Four million pairs were sold in four days.

Shortly afterward, World War II forced the Allies to look for a source of synthetic rubber, since the natural supply had been cut off by the Japanese in the South Pacific, and Germany was already producing synthetic rubber. During the war, U.S. researchers made a synthetic rubber, Buna S, from butadiene, sodium (atomic symbol Na, which initiated the polymerization reaction), and styrene. Since then, polymer chemists have created thousands of polymers, ranging from Styrofoam and Teflon to Kevlar (used in bulletproof vests) and Lexan (used in bulletproof windows).

Uses, Effects, and Limitations

Polymers have thousands of uses in modern society. Some common polymeric materials and their uses include: phenolics, used in wiring devices and handles; unsaturated polyesters, used in surface coatings, fake marble, pipes, and toys; epoxies, used in adhesives, moldings, and flooring; polyethylene, used in food wraps, bags, disposable diapers, drink bottles, and storage drums; polyvinyl chloride, used in pipe, tubing, electrical insulation, and floor tiles; polystyrene, used in videocassettes and disposable kitchen utensils and plates; polyesters, used in blankets, carpeting, clothing, parachutes,

and lingerie; fluoropolymers such as Teflon, used in nonstick coatings—the list could go on indefinitely. It is impossible to imagine the course this century would have taken without photographic film, nylon, glue, polyester, Teflon, Styrofoam, plastic wraps, acrylic paints, and hundreds of other compounds used daily by millions of people around the world.

There is such a variety of polymers in the world that it is hard to find an area of industry or home life that they have not penetrated. The auto industry uses plastic to reduce the weight of cars, which makes them more fuel efficient and able to meet environmental emissions standards. Plastics such as Tupperware are used every day in kitchens to store and preserve food, and nonstick Teflon pans make cooking and cleaning up much easier. Kevlar is the major component of the bulletproof jackets worn daily by police officers and soldiers.

Many other uses are also under investigation. Researchers are looking at the use of polymers in medicine to make better sutures for closing wounds and to deliver drugs more efficiently by coating them with fast-dissolving polymers (see DRUG DELIVERY SYSTEMS). Polymer pastes may someday deliver drugs to normally inaccessible spots such as the inner eyelid and the nasal passages. Polymers that expand or contract when stimulated by light or heat may someday be used to generate precisely controlled mechanical forces.

Most polymers have high electrical resistivity. This means that they conduct almost no electricity and therefore make good safety coverings for electrical wire and other electric appliances. This is also the reason that nylon carpeting can build up a static electricity charge and shock those walking on it when they later touch a material that does conduct electricity. Polymers that themselves conduct electricity are also being studied. Electrically conducting polymers have been around for the last 20 years. These were created by doping polymers with metal. Now, scientists have created metal-free conducting polymers. These may be used in displays (called *all-organic displays*) for cellular phones and hand-held computers, to generate energy in a fashion similar to photosynthesis in plants, and to shield or dissipate electrical charges in machines.

Issues and Debate

Environmental and economic issues have arisen from the burgeoning use of polymers in the latter half of this century. This is particularly true with respect to the disposal of long-lasting durable plastics, but fabrics such as nylon and polymeric materials such as celluloid are also difficult to degrade safely. Natural polymers such as wood and sugars degrade much more readily. Americans currently use around 60 billion tons of plastics each year. The simplest way to dispose of them is by burning them; unfortunately, this can release many poisonous chemicals into the environment. Plastics disposal has therefore gotten a bad reputation, mostly because plastics are nonbiodegradable. Many plastics and polymers take hundreds of years to degrade by natural processes. Most people believe that landfills are filled with plastics, but plastics currently make up only about 8 percent of the total weight of U.S. landfills and 20 percent of their volume. The most common material in landfills is actually paper.

Since plastics have been around only for the last 100 years, bacteria have not yet evolved mechanisms to degrade and metabolize them. Some scientists are looking for naturally occurring bacteria that can do this; they are also searching for ways of genetically engineering bacteria to degrade plastics. Recycling and disposing of plastics and other polymers has therefore become a hot topic. Recycling programs for papers, plastics, and metals have sprung up worldwide in the last decade, and many fast-food chains and some manufacturers have switched from using containers made from Styrofoam to paper wraps and other biodegradable forms of packaging.

Recycling is currently the most important and best way of disposing of plastics. Many plastics, such as those used in housing, cars, and appliances, are designed to last for years and don't need to be recycled often. The more disposable types, such as plastic wraps, bags, and bottles, need to be recycled. This is not without its problems. Although many household plastic products are recyclable, they must be carefully sorted to be reclaimed properly. The recycling symbol with a number that is found on the bottom of many bottles and containers is a good start but is not a complete solution. The number allows recyclers to sort the plastic containers by type, but impurities in the bottles, such as dyes or other colorants, and things like labels and glues, make a recycler's job difficult, and any impurity can drastically reduce the efficiency of recycling. Mixtures of different plastics are difficult to separate, and after being cleaned, melted, and reshaped, are not usually as strong or durable as they once were.

Although polymer products present society with certain disposal problems, their manufacturers are seeking ways to create new plastics that can either be more easily recycled or more easily broken down. Many bottles and bags that are recycled are reclaimed and used in different forms in car parts, for example. But more can still be done: Car manufacturers are clamoring for the plastics industry to improve methods of recycling the plastic in cars. Polymer researchers hope that their advances will help solve some of today's waste disposal problems while creating new efficiencies and savings for manufacturers. Industy packaging practices may continue to change as well, as members of the environmental movement continue to pressure manufacturers to use biodegradable, reusable, or recyclable materials whenever possible.

Since plastics and polymers are an inextricable part of modern life, their efficient use through recycling and reuse programs is essential to keeping the environment healthy. The most promising programs will find efficient chains and systems of reuse and recycling that will pass plastics from one industry to the next. This will place less strain on the environment and will also make financial sense compared to the cost of continually creating new plastics from raw resources.

—*Philip Downey*

RELATED TOPICS
Composite Materials, Smart Materials

BIBLIOGRAPHY AND FURTHER RESEARCH
BOOKS
Alper, Joseph, and Gordon L. Nelson. *Polymeric Materials.* Washington, D.C.: American Chemical Society, 1989.
Mark, Herman F. *Giant Molecules.* New York: Time, 1966.
PERIODICALS
"The Briefest History of Fibers." *Whole Earth,* June 22, 1997, 14.
Cox, Jack. "Plastics Unpredictable Partner in the Life of an Online Recycler." *Denver Post,* December 29, 1998, E-01.
Mapleston, Peter. "Auto Sector's Recycling Goals Keep Plastics in the Hot Seat." *Modern Plastics,* May 1995, 48.
Rathje, William, and Cullen Murphy. "Five Major Myths About Garbage and Why They're Wrong." *Smithsonian,* June 1992, 113.
Stone, Robert F., and Ambuj D. Sagar. "Recycling the Plastic Package." *Technology Review,* July 1992, 48.
INTERNET RESOURCES
Du Pont
　http://www.dupont.com
Polymers and Liquid Crystals
　http://abalone.phys.cwru.edu/tutorial/enhanced/main.htm
Polymers Dot Com
　http://www.polymers.com
Recycler's World
　http://www.recycle.net/recycle/index.html

RADIOISOTOPE THERMOELECTRIC GENERATORS

A radioisotope thermoelectric generator (RTG) is a compact, long-lived, highly reliable electric power system that uses the natural decay of a radioisotope to provide thermal energy (heat), which is then directly converted into electricity by the process of thermoelectric conversion.

Since the early 1960s, RTGs have been used successfully by the U.S. government in a variety of electric power applications involving remote locations on Earth as well as in outer space. The U.S. National Aeronautics and Space Administration (NASA) has conducted successful missions to the surface of Mars (Viking), the polar regions of the Sun (Ulysses), and the giant outer planets (Pioneer 10 and 11, Voyager 1 and 2, Galileo, Cassini), using RTGs as a source of spacecraft electric power. Yet despite its proven performance, the use of RTGs in future space missions is now the subject of considerable public debate, due to concerns about safety.

Scientific and Technological Description

The radioisotope thermoelectric generator is an inherently simple device for generating electric power. It consists of two basic components: a radiosiotopic heat source and a thermoelectric converter that transforms this heat directly into electricity by means of the thermocouple or Seebeck effect (see figure). As a direct energy conversion device, the RTG contains no moving parts.

A *radioisotope* is an unstable isotope of an element whose nucleus decays spontaneously, emitting nuclear particles (*alpha particles* or *beta particles*) or very high energy electromagnetic radiation, called *gamma rays*. *Radioactive decay* involves the decrease in the amount of radioactive material with time due to the spontaneous (but statistically predictable) transformation of one isotope into a different isotope or sometimes into a different energy state of the same isotope. The process results in the emission of nuclear radiation (alpha, beta, or gamma radiation), and subsequently, the generation of heat as the emitted nuclear radiation is absorbed in the radioactive material or its surroundings. Alpha particles do not travel far and can easily be stopped by the thickness of a sheet of paper. Therefore, radioisotopes that emit only alpha particles are especially desirable for use in RTGs, since all the decay heat is released in the fuel itself, and nuclear radiation shielding requirements are minimal.

The *half-life* of a radioactive substance is the time required for the disintegration of half the number of atoms originally present. The half-life varies for specific radioisotopes from millionths of a second to billions of years. The shorter the half-life, the more *radioactive* the substance: that is, the greater the number of nuclear disintegrations per unit of time. For a given quantity of a specific radioisotope, nuclear engineers use the *law of radioactive decay* to calculate its heat output precisely as a function of time. This heat is harnessed in an RTG by the *Seebeck effect,* which holds that if two dissimilar metals are joined at two places and each of these junctions is kept at a different temperature, an electric current flows through the circuit. The magnitude of this phenomenon, also called the *thermoelectric effect,* depends on the nature of the two materials and the temperature difference. A modern RTG (using semiconductor thermoelectric materials) generally has an overall thermal-to-electric energy conversion efficiency of between 5 and 7 percent. This means that for each 100 watts (W) of thermal energy generated by the decay of the radioisotope source, between 5 and 7 W of electric power is ultimately produced. The remaining thermal energy must be rejected to the surroundings as waste heat.

All RTGs flown in space by the United States have used the radioisotope *plutonium-238* as their nuclear fuel. Plutonium-238 is a non-weapons-grade isotope of the transuranic element plutonium. It is an alpha emitter, characterized by a long half-life (87.7 years) and a favorable thermal power density. However, in metallic form, any isotope of plutonium poses a significant radiological and toxicological hazard, especially if the substance enters the body as a tiny particle. To help overcome this drawback, NASA and U.S. Department of Energy–sponsored research has resulted in a plutonium dioxide RTG fuel, a ceramic compound that sig-

nificantly reduces the overall risks and environmental consequences of an inadvertent fuel release.

Historical Development

RTGs have depended on the weaving together of two threads of scientific discovery. Thomas Seebeck observed the thermoelectric effect in 1821, and radioactivity was co-discovered between 1896 and 1898 by Henri Becquerel and Pierre and Marie Curie. In 1913, the English physicist Henry Moseley (1887–1915) constructed the world's first nuclear battery, a device that can be considered as a distant ancestor of the modern RTG. As late as 1945, the Moseley model of the nuclear battery influenced other researchers in their attempts to generate electricity directly from nuclear radiation emissions.

Following World War II and the development of the nuclear reactor, large quantities of radioisotopes became available, including those suitable as thermal power sources for modern RTGs. In 1954, at Mound Laboratory in Ohio, two scientists, Kenneth Jordan and John Birden, hit upon the idea of applying the thermocouple principle to create a device (a *thermopile*) that uses the decay heat of a radioisotope to generate electricity directly. By cleverly combining the thermoelectric effect with a radioactive heat source, they avoided all the difficulties previously encountered by other researchers who tried to create useful nuclear batteries. They patented the concept and it remains the basis for all radioisotope thermoelectric generators.

Cold War pressures further stimulated development of the modern RTG. In the late-1950s, the U.S. Air Force and the Atomic Energy Commission (AEC) performed studies concerning nuclear power supplies for military spacecraft. The AEC gave the space nuclear program the title Systems for Nuclear Auxiliary Power (SNAP). One unusual event helped significantly to change the direction of the early SNAP program. On January 16, 1959, a SNAP-3 RTG unit (fueled by the radioisotope polonium-210) was demonstrated on President Eisenhower's desk in the White House. Impressed, the president became eager to share this interesting technology with the world and sought ways to express openly the peaceful uses of nuclear energy. This event provided NASA with a unique opportunity to propose several ambitious civilian space missions that could utilize the potential of an RTG device.

As the electric power demands of ever more sophisticated space exploration missions increased, RTG technology improved to satisfy those needs. For example, early SNAP units established the technical antecedents necessary for the more sophisticated units deployed on the lunar surface by *Apollo* astronauts (1969–1972). Mission-enabling SNAP-19 units powered the *Viking 1* and *2* lander spacecraft as they searched for life on Mars (1975) and the *Pioneer 10* and *11* spacecraft as they performed the first flyby missions of the giant planets Jupiter and Saturn (1972–1979).

A new generation of more powerful RTGs, called simply the *multihundred-watt* (MHW) *system,* allowed the far-travel-

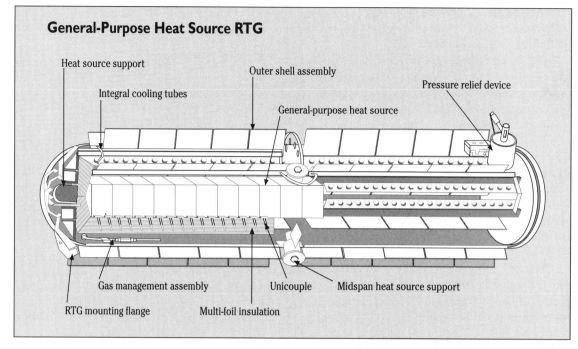

General-Purpose Heat Source RTG

Heat source support

Integral cooling tubes

Outer shell assembly

General-purpose heat source

Pressure relief device

Gas management assembly

Unicouple

Midspan heat source support

RTG mounting flange

Multi-foil insulation

Adapted from an original illustration courtesy of NASA.

ing *Voyager 1* and *2* spacecraft to make a grand tour of all the gaseous giant planets. Jupiter and Saturn were visited by both spacecraft (between 1979 and 1981), while *Voyager 2* encountered Uranus (1986) and Neptune (1989). Today, the *Voyager* spacecraft are still functioning and transmitting scientific data as they leave the outer regions of our Solar System and venture out into the void of interstellar space.

At the dawn of the 21st century, an even more advanced radioisotope-fueled system, the GPHS-RTG (general-purpose heat source RTG), now supplies electric power to NASA's *Galileo* spacecraft as it explores the Jovian system, to the joint NASA/European Space Agency *Ulysses* spacecraft as it ventures through interplanetary space observing the Sun's never before studied polar regions, and to NASA's *Cassini* spacecraft as it heads for a scientific rendezvous with Saturn and its intriguing system of moons in the year 2004.

Uses, Effects, and Limitations

The RTG is most suitable for missions operating beyond the main asteroid belt, missions involving hostile environments (e.g., intense radiation belts or frigid regions of a planetary surface), or where long-lived, reliable operation independent of orientation to or distance from the Sun is important. Engineering limitations suggest that RTGs be used on future missions that require up to a few hundred watts of electric power (or less) for periods of five years or more. NASA is currently studying several future space missions between 2002 and 2015 that will probably use RTGs. These include the *Pluto-Kuiper Express* (a flyby of Pluto and its moon, Charon), the *Europa Orbiter,* the *Europa Lander* (a mission that will search for possible life in the suspected liquid

water ocean beneath the frozen surface of Jupiter's intriguing moon, Europa), a *Neptune Orbiter,* a *Titan Organic Explorer,* and an *Interstellar Probe.*

Consistent with federal regulations and law, no RTG has been or will be launched into space by the U.S. government without presidential approval, granted only after extensive safety review and analysis by panels of experts. This process imposes an important administrative limitation on the use of RTGs that inherently ensures an appropriate, safe application of this technology.

Safety is of prime concern in the deployment of any RTG. As a result, U.S. government agencies have invested over 30 years in the engineering, safety analysis, and testing of these systems. Not only are important safety features incorporated in the RTG design, but extensive testing is also performed to demonstrate that the system (as designed) can withstand the physical conditions encountered during normal operation as well as all credible accidents.

NASA is using the GPHS-RTG on board several current spacecraft. Each GPHS-RTG assembly contains approximately 10.9 kilograms of plutonium dioxide fuel and uses advanced silicon–germanium (SiGe) thermoelectric materials to accomplish the direct conversion of thermal energy into electricity. This modern RTG contains a stack of 18 GPHS modules. The outer layer of each module is a high-strength graphite aeroshell, or heat shield, designed to withstand reentry heating in the event of an accident. Each aeroshell contains two graphite impact shells that provide additional reentry protection as well as impact protection. Together, the two impact shells cradle four fuel pellets. Each pellet is enclosed in its own metal capsule of iridium and is about the size of a spool of thread. This system of multilayered protective materials is designed to prevent or minimize fuel releases under any operational or anticipated accident condition.

Issues and Debate

The great contemporary debate associated with the use of plutonium-fueled RTGs in future space missions involves public safety and risk. Stimulated by recent "techno-catastrophes" such as the *Challenger* accident (January 1986), in which seven astronauts were killed, and the Chernobyl nuclear plant disaster (April 1986), which contaminated major portions of the former Soviet Union and Europe, people have generally become more concerned about our ability to control and use advanced technologies. The RTG risk and safety debate is further influenced by a public apprehension about the use of nuclear power, even for peaceful applications.

Within the international space community, especially through the work of the United Nations' Committee on the Peaceful Uses of Outer Space (COPUOS), it is recognized and accepted that RTGs "may be used for interplanetary missions and other missions leaving the gravity field of the Earth. They may also be used in Earth orbit if, after conclusion of the operational part of their mission, they are stored in a high orbit." The international space community further recommends containment of the radioisotope fuel under all operational and accident conditions.

Despite international acceptance and decades of cautious application of RTGs, it has been difficult to develop a popular consensus as to whether the benefits offered by RTGs outweigh their risks. For the purposes of risk assessment, NASA and the Energy Department have suggested that the risk of using plutonium-fueled RTGs be defined as the probability (per unit radiation dose) of producing, in an individual or population, a radiation-induced detrimental health effect, such as cancer. Government risk analyses performed prior to the *Cassini* mission concluded that an early launch accident with plutonium dioxide release had a probability of occurrence of 1 in 1400 and would cause 0.1 fatality, adding up to an overall risk factor of 0.00007. A later launch accident or spacecraft reentry with plutonium dioxide release had a probability of 1 in 476, would cause an estimated 0.04 fatality, and represented an overall risk factor of 0.00008.

Not everyone agrees with the risk-assessment approach, however. They maintain that government risk-assessment studies are, by virtue of their sponsorship, untrustworthy because they are biased toward a pro-technology outcome. It is also important to recognize that no matter how unbiased and objectively performed, any risk-assessment study provides only a quantification of a particular risk. The acceptability of that particular risk (no matter how numerically insignificant on various comparative scales) is still a very personal, subjective judgment that can never be forced by mathematical arguments alone.

To some observers, the unique and enabling benefits of RTG-powered, deep-space exploration spacecraft far outweigh the small calculated risks they pose. These people accepted and supported the RTG-powered *Cassini* mission. To others, however, such projected risks are unacceptable. Although the *Cassini* safety debate still raged, the launch was approved. On October 15, 1997, a Titan IV rocket lifted off from Cape Canaveral Air Force Station and placed the RTG-powered *Cassini* on a gravity-assisted flight trajectory that will bring the spacecraft to Saturn in 2004.

—Joseph A Angelo, Jr.

RELATED TOPICS
Reusable Launch Vehicles, Robotic Space Exploration, Space Shuttle

BIBLIOGRAPHY AND FURTHER RESEARCH

BOOKS
Angelo, Joseph A., Jr., and David Buden. *Space Nuclear Power.* Malabar, Fla.: Orbit Book Co., 1985.
Engler, Richard E., ed. *Atomic Power in Space (A History).* Washington, D.C.: U.S. Department of Energy, 1987. (Available from U.S. Department of Commerce, NTIS as NTIS-PR-360.)
Morrison, David. *Voyages to Saturn.* NASA SP-451. Washington, D.C.: U.S. Government Printing Office, 1982.
NASA. *Viking: The Exploration of Mars.* NASA EP-208. Washington, D.C.: U.S. Government Printing Office, 1984.

PERIODICALS AND REPORTS
"The Future of Space Exploration." *Scientific American Quarterly* 10(1), spring 1999.
Salisbury, David. "Radiation Risk and Planetary Exploration." *The Planetary Report,* May–June 1987.
U.S. Department of Energy (DOE). "GPHS-RTGs in Support of the Cassini Mission: Final Safety Analysis Report," May 1997.
U.S. Government Accounting Office. "Space Exploration: Power Sources for Deep Space Probes (Report to the Honorable Barbara Boxer, U.S. Senate)." GAO/NSIAD-98-102. May 1998.

INTERNET RESOURCES
NASA Space Link
    http://spacelink.nasa.gov/
U.S. Department of Energy Homepage
    http://www.doe.gov/
Jet Propulsion Laboratory Homepage
    http://www.jpl.nasa.gov/
Cassini: Mission to Saturn
    http://www.jpl.nasa.gov/cassini/rtg/
NASA Homepage
    http://www.nasa.gov/

RATIONAL DRUG DESIGN

For centuries humans have used naturally occurring substances as sources of drugs without knowing how drugs work in the body. Even today many drugs are discovered and used without any understanding of how they work.

The last 20 years have seen the rise of rational drug design, a field in which researchers use computers and their own knowledge of biology to design new drugs without testing large numbers of molecules hoping to find one that has a

How a Drug Fits into a Protein

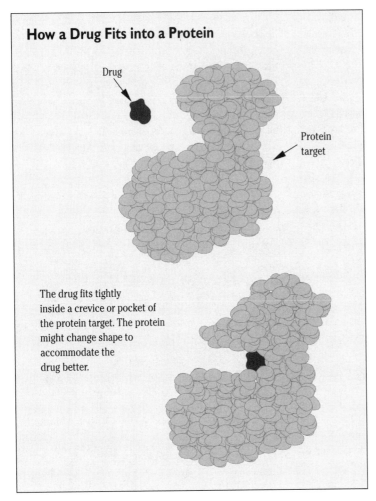

Drug

Protein target

The drug fits tightly inside a crevice or pocket of the protein target. The protein might change shape to accommodate the drug better.

body. Biologists look for an Achilles' heel, a target molecule from either the virus or the infected host that plays a vital role in the virus's life cycle. Biologists must also decide how a drug should modulate the function of their target molecule: Should the drug prevent the target from functioning, or should the drug make it function unusually efficiently?

The next step is to work out the *atomic structure* of the target molecule, how the atoms of the target molecule are arranged in three-dimensional space. Drug designers need to know this if they are to design a drug that can interact with the target. One of the most common ways to do this is using X-ray crystallography, in which a crystal of the target molecule is bombarded with X-rays. The target scatters the X-rays as they pass through the crystal, and researchers can determine the three-dimensional atomic structure of the target from the way the X-rays are scattered.

Once the structure of the target molecule has been established, drug designers decide which part of the target they want their drug to interact with. To look at a molecule's three-dimensional structure, *molecular graphics* computer programs are used to visualize molecules on a computer screen, which is far easier and faster than building physical models out of wood or plastic. Drug designers consider molecules that might interact with the target in the way they want; then to see how well these molecules should interact with the target, molecular dynamics simulations are performed. In *molecular dynamics simulations*, computer programs estimate the forces that different molecules or parts of molecules exert on each other, which can show, for example, that two different molecules would actually push each other away rather than interact tightly with each other. Researchers often use the term *molecular modeling* to describe the process of looking at molecules on computer screens and carrying out molecular dynamics simulations.

An important requirement in designing a drug is that the drug's structure complement the structure of the target, like a glove complements the shape of a hand or the way two right hands complement each other's shapes in a handshake. The better the fit between the drug and target, the more time the drug will stay bound to its target (as opposed to being in the body but not bound to its target) and therefore the more effective the drug will be. Often, the target molecule has a cavity and a drug can be designed to fit snugly into this cavity. Some parts of the target might behave like tiny magnets, and the drug will need a complementary magnetic region if it is to interact tight-

beneficial effect. Although there are only a few examples of commercially available drugs that have been designed in this way, the techniques of rational drug design are likely to become increasingly useful for producing new drugs in the future.

Scientific and Technological Description

A drug is any molecule that perturbs a biological system, to the benefit or detriment of an organism. To do this, a drug must interact physically with that system, and therefore drugs must bind to (i.e., interact with) other molecules, called *targets*, inside the organism (see figure). Drugs cause their effect either by interfering with how well a target molecule carries out its normal function or by making it do something it doesn't normally do. Rational drug design is the design of new drugs using knowledge about a particular biological system. A drug is designed to have a particular biological effect and the drug's designers know how the drug will work even before they have invented it.

Rational drug design involves several key steps. The first is *target identification*. Drug designers need to know which target molecule their drug should bind to and how the function of the target molecule needs to be affected. For example, biologists might study how a virus invades the human body or how the virus reproduces (increases in number) inside the ly with the target. Although the designed drug and the target will never become irreversibly joined chemically, the two can be made to stick together so tightly that the drug will only rarely "drop off" its target, if that is what the designer wishes. Finally, chemists must synthesize the designed drug for further testing.

Historical Development

For many centuries drugs were crude preparations that contained multiple ingredients and were usually produced by a single person, but over the last 100 years this picture has changed dramatically. The development of chemistry as a science in the 19th century led to the production of pure compounds, both by distilling compounds from complex mixtures such as biological samples (i.e., without actually synthesizing the compound at all) and by chemical synthesis (i.e., by synthesizing a new compound from compounds already available to chemists). Anesthetics (such as ether and chloroform) and analgesics (pain-relieving compounds such as aspirin, morphine, and codeine) were discovered in the 19th century and produced commercially in large quantities.

As chemistry developed, biologists started asking how chemicals affect living organisms. At the turn of the 19th century, Paul Ehrlich (1854–1915) and co-workers developed the *receptor theory* for drug action, in which a drug binds to a particular receptor molecule on or in the cells that make up an organism, and the drug can then alter how this receptor functions. This theory could explain, for example, how the beating of a heart cell could increase by 50 percent when only 0.001 percent of its surface was covered by the drug acetylcholine.

Rational drug design has its origins in the 1960s, when the structures of large biological molecules such as proteins began to be determined. The first molecular structure determined by X-ray crystallography was that of penicillin in 1949. The first protein structure (of myoglobin, a protein similar to the hemoglobin that carries oxygen in our blood) was not determined until the 1960s.

In the 1970s, the development of powerful, low-cost computers began to have a profound effect on the biological science upon which rational drug design is built. Over the past 20 years, computers have allowed an explosion in the determination of the structure of biological macromolecules. Computers also permitted molecular dynamics simulations, first applied to proteins in 1977. In the last 10 years, computers have become sufficiently powerful to allow the design of completely new molecules that interact very specifically with their targets. There has also been a great increase in the scientific understanding of biological systems, such as how an entire organism is built up from a limited collection of molecules and how different types of cells and organs function. This has given drug designers more potential targets for drugs.

Uses, Effects, and Limitations

To date there are only a few cases in which rational drug design has been the sole technique used to develop a new drug. The techniques of target identification, target structure determination, molecular graphics, and molecular dynamics simulations have all improved greatly over recent years, but it takes many years for a drug, once designed, to pass through government regulatory procedures and reach consumers. It will be some time before drugs designed by a rational approach will become more commonplace.

The AIDS drug Norvir recently brought to the market owes much to the methods of rational drug design. This drug prevents the action of the human immunodeficiency virus (HIV) protease, a protein made by HIV that is vital for its function after it has infected host cells. The atomic structure of the HIV protease was worked out using X-ray crystallography, and Norvir was designed using molecular dynamics simulations to inhibit the activity of the protease.

Another example of the value of a rational approach to drug design is the development of cimetidine (better known under the commercial names Tagamet and Peptol), which was designed to reduce the secretion of acid into the stomach by specifically inhibiting the action of molecules called H_2 *receptors*. Knowledge of the structure of H_2 and that of similar receptors allowed the design of a drug that did not interact with those similar receptors.

The biological sciences are now identifying large numbers of potential targets for new drugs. Every time a new biochemical pathway is uncovered or a molecular peculiarity of a disease state is found, potential drug targets are also discovered. For example, researchers studying cancer have found that some cancer cells produce proteins that are not normally found in noncancer cells. Designed drugs that interact with these proteins might result in new anticancer treatments.

Many drug targets are proteins, and an organism's *genome* (its complete genetic blueprint) contains sequences of coded information about every protein an

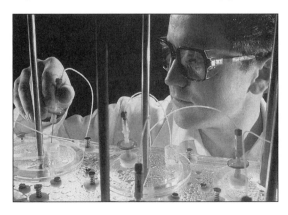

In traditional approaches to drug development, researchers test large numbers of molecules, hoping to find one that has a beneficial effect. In rational drug design, by contrast, the drug's designers know how the drug will work even before they have invented it (Photo Researchers/Geoff Tomkinson/Science Photo Library).

organism can produce. The complete genomes of several organisms, including that of the bacterium that causes tuberculosis, have now been sequenced (i.e., the sequences of coded information have been "read"), and it will take only a few more years before the complete human genome is sequenced as well. From these genomes, researchers can deduce the amino acid sequence of every protein in that organism, and in principle it should be possible to design drugs that interact with each of these proteins. It will take many years to analyze fully (*decode*) all of this genomic information, but many new drugs should result from this analysis.

It is important to put the role of rational drug design in context with other methods of drug discovery. Rational drug design requires knowledge about a potential drug target and its molecular structure so that how a rationally designed drug works is understood. But many modern drugs have been discovered and used without any real understanding of how they work. One way this can happen is by luck and through observation. A recent example is Viagra, which was tested to see if it would help angina (heart pain), until some men given the drug reported increased erectile function. The drug is now used to treat impotence.

A new drug to treat a particular disease can also be found by *random screening,* in which thousands of compounds are tested in an attempt to find one that has some desired effect. Chemists are developing techniques to synthesize large numbers of molecules simultaneously to use in drug screening trials (see COMBINATORIAL CHEMISTRY). Once a potential drug has been found, similar molecules can be synthesized to see if they improve the original compound's performance. Computers are used to analyze the results of these random screens, using *QSAR* (quantitative structure–activity relationships). QSAR compares the effectiveness of a large number of compounds, identifying features that appear to be required for a drug to have a particular biological effect.

Rational approaches have been used in drug design to improve drugs that have already been identified by random screening. Once a potential drug has been identified, perhaps using QSAR, it is sometimes possible to identify quickly the target molecule on which the drug acts to cause its therapeutic effect. If the atomic structure of the target is then worked out, molecular graphics and molecular dynamics simulations can be used to improve the drug so that it interacts better with its target. Generally, the better the fit between the drug and the target, the less likely the drug is to cause side effects by binding to other, undesirable target molecules.

The techniques of random screening, combinatorial chemistry, and QSAR can be applied to develop drugs, whether or not the drug target molecules are known, but rational drug design can be used only if a target and its structure are known. The process of rational drug design is not necessarily more likely than random screening and QSAR to result in a new drug, although drug designers hope that if they fail to identify potential new drugs by random screening and QSAR, they might have more luck with a rational design process. In practice, drug development companies use all the techniques at their disposal.

Issues and Debate

Rational drug design itself does not raise ethical concerns, as drug designers are simply using the computer tools of modern biology to develop new and improved drugs. Rational drug design might even reduce the need for animals in research by reducing the need for the random screening of thousands of compounds using tests that require fresh animal tissue. There is some hope that rational drug development can reduce the cost of drug discovery, by avoiding the time and expense of carrying out random screening. The development of a new drug costs around $200 million on average and can take over 10 years, and advancements in rational drug design will probably help to reduce those costs overall. Yet a substantial part of drug-development costs result from the work required by complex government regulations to prove that a drug is effective and safe for use. Even rapid improvements in rational drug design will not be able to reduce the costs associated with that part of the process.

Some industry observers believe that rational drug design is the only way forward; others maintain that random screening and QSAR remain much more effective. Most experts believe that the two approaches should go hand in hand, however.

—Matthew Day

RELATED TOPICS
Combinatorial Chemistry, Gene Therapy, Light-Activated Drugs

BIBLIOGRAPHY AND FURTHER RESEARCH
BOOKS
Lambright Eckler, J. A., and J. M. Stimmel Fair. *Pharmacology Essentials.* Philadelphia: W. B. Saunders Co., 1996.
Smith, H. J., ed. *Introduction to the Principles of Drug Design,* 3rd ed. Amsterdam: Harwood Academic Publishers, 1998.
PERIODICALS
Hunter, W. N. "Rational Drug Design: a Multidisciplinary Approach." *Molecular Medicine Today* 1, April 1995, 31.
Lane, David. "The Promise of Molecular Oncology: The Promise of Cancer Research and Treatment." *Lancet* 351(suppl. II), May 1998, 17.
INTERNET RESOURCES
General Information on Rational Drug Design
    http://www.techfak.uni-bielefeld.de/bcd/ForAll/Introd/drugdesign.html
Introduction to Molecular Modeling
    http://cwis.nyu.edu/pages/mathmol/quick_tour.html
Pictures and Information about Drug Molecules
    http://www.chem.ox.ac.uk/mom/

REMOTE SENSING

Remote sensing uses electromagnetic radiation to obtain information about an object without making physical contact with it. From surface, aircraft, or space-based platforms, special instruments acquire reflected or emitted radiation from selected portions of the electromagnetic spectrum.

Up until the 1960s, remote sensing was limited primarily to aircraft-derived photographic images within the visible and near-infrared portions of the spectrum. Spacecraft now provide an unobstructed view of Earth and the Universe. Advances in detector technology and information processing have produced versatile instruments that operate well beyond the narrow region of the electromagnetic spectrum associated with visible light and the human eye. However, remote sensing systems also generate large amounts of data, and this presents a major challenge to analysts, who must process, interpret, and manage vast quantities of information.

Scientific and Technological Description

Remote sensing uses *electromagnetic radiation* (EMR) to acquire information about an object, phenomenon, or event without having the sensor in physical contact with the object. EMR consists of oscillating electric and magnetic fields that propagate through space with the speed of light. Under some conditions EMR behaves like a wave, under other conditions it behaves like a stream of particles, called *photons.*

When sunlight passes through a prism, it throws a rainbowlike array of colors onto a surface. This display from red to violet, called the *visible spectrum,* represents the narrow spectral band to which the human eye is sensitive. The EM spectrum actually includes much more than meets the eye, however. It is a continuum of wavelengths, going from very long wavelength radio waves to very short wavelength gamma rays. The EM spectrum includes radio waves, radar waves, microwaves, infrared radiation, visible light, ultraviolet radiation, X-rays, and gamma rays.

All objects in the universe emit, reflect, and absorb EMR in their own distinctive ways. The way an object does this provides scientists with a *signature,* or set of special characteristics, that can be detected by remote sensing instruments. Many factors influence the characteristic EMR signature of an object. These factors include the physical properties of the object (such as its temperature and surface conditions) and the wavelength region over which the sensor operates. Any intervening (transparent) medium, such as a planetary atmosphere, can also influence the radiation signal collected by a sensor.

Unlike photographic cameras, which operate using the visible spectrum, modern remote sensing instruments collect radiation signals from many different portions of the EM spectrum. Careful processing and interpretation of such nonvisible light images often provide interesting new information about an object. For example, a thermal image describes an object's heat signature and energy content.

Remote sensing systems are divided into two categories: passive sensors and active sensors. *Passive sensors* collect the sunlight reflected by an object or characteristic radiation emitted by an object. *Active sensors* (e.g., an imaging radar system) provide their own source of illumination (EMR) on the target. Both active and passive remote sensing systems obtain images of an object. Passive sensors operating in the visible and near-infrared portions of the spectrum need sunlight to acquire an image. A thermal imaging system collects the thermal (or heat) signature of an object and operates well in darkness. A radar imaging system (see figure) uses its antenna to emit a pulse of microwave radiation that illuminates an object. Depending on the surface properties of the object, most, some, or none of this microwave energy is reflected back toward the radar system. Any backscattered radiation is received and recorded by the antenna. A *radar image* is created as the antenna alternately transmits and receives pulses at particular microwave wavelengths. Radar imaging systems can operate day or night and make images of the target even when clouds are present.

A *multispectral image* is produced by simultaneous detection and measurement of EMR at different wavelengths as radiation is reflected or emitted by the object (see MULTISPECTRAL IMAGING). The multispectral image is then processed and interpreted to discover any special characteristics of the object that are revealed by both its spatial features and by various combinations of its spectral features. Typical imaging systems create a multispectral image by simultaneously sweeping an array of detectors (each sensitive to a different portion of the EM spectrum) across the scene. The latest scanning systems use charge-coupled devices (CCDs). The CCD is a versatile solid-state electronic device that consists of an array of EMR-sensitive detector elements that emit electrons when exposed to certain wavelengths of radiation. This emitted charge is collected by attached electronic circuits, and the resulting signals are readily available for digital processing. Modern multispectral scanners generally collect up to about 10 spectral bands simultaneously, with each spectral band being some 0.10 micrometer (μm; a wavelength) wide.

Hyperspectral imaging systems collect 100 or more very narrow spectral bands (on the order of 0.01 μm wide in each band) simultaneously and generate very detailed visible and near-infrared images of a target. Analysts are just beginning to unlock the power and potential of this new class of scanning instrument.

There are other, nonimaging types of remote sensing instruments. Some of these instruments measure the total amount of energy (over a certain portion of the EM spectrum) that falls within the field of view of the sensor, but do not create an image of the object being viewed. For example, a *spectrometer* measures the intensity of EMR emitted or reflected by a distant object as a function of wavelength. A *radiometer* quantifies the amount of thermal energy coming from an object. A *scatterometer* emits microwave radiation and then senses the total amount of microwave radiation scattered back over a wide field of view. This instrument is often used by meteorologists to measure wind speeds and direction. A *radar altimeter* sends out a narrow pulse of microwave energy toward a surface and then accurately times the arrival of the return pulse reflected by the surface. This instrument provides a precise measurement of the altitude of an aircraft or spacecraft above a planet's surface. Similarly, a *lidar* (light detection and ranging) *altimeter* emits a narrow pulse of visible or infrared radiation toward a surface and then accurately times the arrival of the light pulses reflected back from the surface.

Historical Development

Modern remote sensing started in the 1840s when French balloonists took the first aerial photographs of Earth's surface to assist in mapmaking. The invention of the aircraft (1903) and the improved airplanes of World War I (1914–1918) provided additional support for aircraft-based remote sensing. During World War II (1939–1945), extensive aircraft-based photoreconnaissance operations helped mature the science of photograph interpretation. Remote sensing was pushed beyond the visible spectrum when near-infrared sensitive film was used to detect camouflaged installations such as gun emplacements, tanks, aircraft, bunkers, and factories. The near-infrared photos showed fake vegetation very clearly; the same camouflage could not easily be detected in visible photos.

The initiation of the Cold War between the United States and the Soviet Union in 1945 created an urgent need for expanded information gathering and triggered a revolution in remote sensing technology. New instruments, based on rapid advances in solid-state physics and microelectronics, received heavy support from the mili-

tary. For example, the need for improved missile surveillance stimulated major advances in the detection of infrared radiation. The superpower missile race also helped bring about the development of a new remote sensing platform—the satellite. Whereas rockets had been used to carry photographic cameras before 1957, the arrival of the satellite provided a new, vibration-free platform with an unobstructed view of Earth. The development and use of spy satellites provided the technical heritage that made possible expanded image processing and interpretation, advanced remote sensing instruments, and sophisticated scientific and civilian spacecraft.

The first civilian application of satellite-based remote sensing of Earth took place in the 1960s with the development of the early weather satellites. The technical progeny of these early satellites now carry a variety of instruments designed to help meteorologists understand and predict the weather. Similarly, the first civilian satellite designed to monitor Earth's resources, called *Landsat-1*, was successfully launched by NASA on July 23, 1973. The *Landsat* tradition of providing high-quality multispectral imagery data to a broad user community continues with *Landsat-7*, launched on April 15, 1999.

Planetary scientists and astronomers have also enjoyed the use of new remote sensing instruments and platforms to observe celestial objects in the infrared, ultraviolet, X-ray, and even gamma-ray portions of the spectrum. For example, NASA's latest orbiting observatory, called the Chandra X-Ray Observatory, launched in July 1999, is cur-

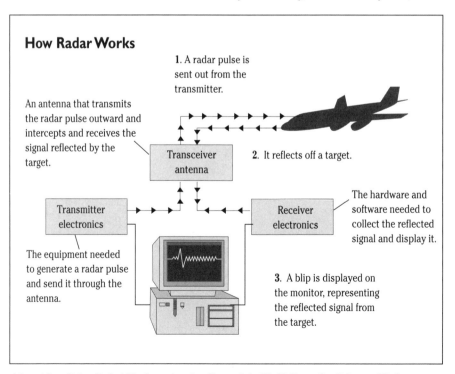

How Radar Works

An antenna that transmits the radar pulse outward and intercepts and receives the signal reflected by the target.

Transceiver antenna

The equipment needed to generate a radar pulse and send it through the antenna.

Transmitter electronics

Receiver electronics

1. A radar pulse is sent out from the transmitter.

2. It reflects off a target.

The hardware and software needed to collect the reflected signal and display it.

3. A blip is displayed on the monitor, representing the reflected signal from the target.

Adapted from *Robert Buderi*, The Invention that Changed the World: How a Small Group of Radar Pioneers Launched a Revolution, *Simon & Schuster, 1996.*

rently providing detailed X-ray images of a wide variety of astrophysical objects.

Uses, Effects, and Limitations

Remote sensing provides military, governmental, scientific, industrial, and individual users with large quantities of high-quality data that make it possible to perform a variety of important tasks. For example, weather satellites, carrying both imaging and nonimaging instruments, have become an indispensable part of the science of meteorology. Such spacecraft provide meteorologists with continuous coverage of environmental conditions around the planet and warn against severe weather conditions such as devastating hurricanes.

Military applications include expanded technical intelligence collection, better indication and warning of pending threats, the location of hostile forces and facilities that seek to operate undercover, and prompt and efficient battle-damage assessment. Industrial and environmental applications include mineral prospecting, crop monitoring, and surveillance of national forests. The telecommunications industry uses all-weather radar imagery to assist in the design of wireless networks. There are several other areas in which scientists and government planners use remote sensing: planetary science and exploration, astronomy and astrophysics, geology, geography, oceanography, hydrology, land use and mapping, water resource management, urban planning, and natural disaster recovery.

The medical profession currently uses high-resolution thermal imagery to perform noninvasive diagnosis of chronic pain and soft tissue injuries. Archaeologists use radar images to detect the remains of lost cities, ancient roads, and early trading routes. Environmental scientists depend on multispectral imagery from space, aircraft, and ground platforms to conduct intensive global change studies. Information specialists use computer-based geographic information systems to merge remote sensing imagery data with other key data to produce powerful specialized information collections.

The use of remote sensing technology is physically limited by the spatial and spectral resolution of the instrument, as well as by the effects of any intervening media, such as a planetary atmosphere. The intervening atmosphere can block or absorb certain portions of the EM spectrum, so remote sensing instruments are often limited to viewing a distant object in selected wavelength bands called *atmospheric windows*. Another major technical limit to the successful application of remote sensing technology is the ability to process and interpret vast quantities of information. Advanced computational methods, intelligent sensors (i.e., sensors that "know" when they have something interesting to report), and creative data fusion activities (in which multiple data streams are automatically blended and evaluated using ARTIFICIAL INTELLIGENCE) will be required to overcome

this limitation and to unleash the full power of remote sensing technology in the 21st century. The use of remote sensing could also be limited by concerns for privacy.

Issues and Debate

In the past, dramatic advances in remote sensing technology were promoted by Cold War military needs. Often, new remote sensing instruments gave the U.S. government and its military forces unique informational advantages over all potential adversaries. Much of the Cold War–era remote sensing technology has now been diffused into the civilian sector. The spatial and spectral quality of commercially available aircraft and satellite-derived imagery data is approaching levels that could give rogue nations and terrorist groups notable advantages in regional conflicts against information-poor neighboring states. This circumstance has created a national and international debate concerning any government's right to limit the quality (i.e., resolution) of commercially available remote sensing instruments and products. Some argue for tight controls in the interests of international security and global stability. Others demand that free-market conditions prevail, with the needs of commercial end-users driving the remote sensing marketplace.

A parallel issue is access to and the cost of remotely sensed data. For example, meteorological data have always been regarded as a free service provided to all citizens by the government—but what about detailed multispectral imagery of a person's hometown? Such imagery might be collected to look for unhealthy vegetation, which would indicate illegal toxic waste dumps, or to aid urban planners in studying population growth and environmental stresses. There is considerable debate about how much a government should charge for such imagery. As commercial companies move into the business of collecting and selling their own satellite-derived imagery, competition between government-sponsored and commercially provided imagery data will expand. Should governments compete with private companies in this area? Some say yes; others argue vehemently for fair and open commercial competition.

Another very important issue has also emerged concerning privacy and the constitutional protection of an individual against unfair search and seizure. Can law enforcement authorities use nonvisible remotely sensed data (such as thermal imagery) to conduct a search operation without probable cause? Or should a court order be required before a government can apply the power of modern remote sensing in the surveillance of its citizens? This very important personal freedom issue is currently being debated by legal experts.

—Joseph A. Angelo, Jr.

RELATED TOPICS
Multispectral Imaging, Space-Based Remote Sensing Tools, Telescopy

BIBLIOGRAPHY AND FURTHER RESEARCH

BOOKS

Ginsberg, Irving W., and Joseph A. Angelo, Jr, eds. *Earth Observations and Global Change Decision Making, 1989: A National Partnership*. Malabar, Fl.: Krieger, 1990.

Pebbles, Curtis L. *The Corona Project: America's First Spy Satellites*. Annapolis, Md.: Naval Institute Press, 1997.

Sabins, Floyd F. *Remote Sensing: Principles and Interpretation*, 3rd ed. New York: W.H. Freeman, 1999.

PERIODICALS AND REPORTS

Brewster, Alice, and Heather Monday. "Airborne Imagery Used in Wetlands Restoration." *Earth Observation Magazine* (EOM), August 1994, 57.

Corbley, Kevin P. "Wireless Communication Tunes in with Satellite Imagery." *Earth Observation Magazine* (EOM), August 1997, 21.

NASA/Goddard Spaceflight Center. "NASA's Earth Observing System: EOS AM-1." NP-1998-03-018-GSFC, 1998.

INTERNET RESOURCES

Earth Observation Magazine Online
http://www.eomonline.com/

NASA/Earth Observing System (Terra) Home Page
http://terra.nasa.gov/

National Oceanic and Atmospheric Administration (NOAA), Meteorological Satellites Home Page
http://www.noaa.gov/

National Reconnaissance Office: U.S. Spy Satellites
http://www.nro.odci.gov/

Radarsat Interactive: Applications of Radar Imagery
http://www.rsi.ca/

Remote Sensing Tutorial
http://rst.gsfc.nasa.gov

U.S. Geological Survey: Images of Environmental Change
http://www.usgs.gov/Earthshots

RETINA AND IRIS SCANNING

The retina and the iris are unique to each person. The retina, at the back of the eye, has a unique pattern of blood vessels, whereas the iris, the colored part of the eye that surrounds the pupil, is a complicated weaving of many different connective tissues. Given these two tissues' properties, they can be used as markers to identify almost anybody.

The human iris and retina can be rapidly scanned and checked to confirm a person's identity, much as fingerprinting has been used in the past. A digital picture of a person's retina or iris can be checked against a stored record every time a person wants to withdraw money from a bank or automated teller machine, access secure information over a computer network, or enter a secure building. In light of the development of these and other identification techniques that rely on parts of the body or bodily fluids, many people have expressed concern that people's privacy will be invaded by these kinds of scanners. They fear that their body parts may be used to identify them without their knowledge or consent.

Scientific and Technological Description

Biometrics is the generic term for using the body as identification. Body parts available for use as identification can include fingerprints, hand shape, the retina and iris, face shape, and DNA sequences; behaviors, such as signatures, keystroke patterns, and voice recognition, can also be used. All biometric systems do their work by comparing a stored record of a biometric with a current record.

Retina and iris scanning depend on the uniqueness of individual body parts. The retina is the lining at the back of the eye that contains the cells that sense light. Most important for biometrics is the fact that the blood vessels that supply the retina with nutrients and oxygen form a unique pattern in every person. The iris is the ring at the front of the eye that gives our eyes their color and acts to limit how much light enters the eye before it arrives at the retina. The iris is a complex interweaving of connective tissue—a tough mesh of fibers that, like the retinal tissue, is unique in pattern.

The first stage of any biometric system is *enrollment*. At this time the biometric is recorded and preserved in a computer database. For iris scanning, a simple black-and-white picture of the iris is taken from about a foot away. The position of the light and dark areas in the picture are measured. These contrasting areas and their locations are digitized and stored as a long string of numbers and letters. Usually, a few pictures are taken and the best result is chosen for storage. In a retinal scan, the subject must place his or her head in a headrest so that the eye is just a few inches from a camera that takes a picture of the blood vessels lining the retina. As with the iris scan, the best picture is digitized and stored.

After the enrollment stage, a biometric scanner is now set up to recognize that person. When that person next enters a biometric sensor's receptive field, it can go to work at the identification process. The first step is to take a picture of the iris or retina. Most systems use infrared light, which is emitted continuously by any living body; using infrared light also allows the system to work at nighttime without needing any light, such as overhead lights or a camera flash to illuminate the object. For an iris scanner to operate properly, the person needs to be within about 3 feet of the camera to be scanned. The person doesn't need to hold still for the iris to be imaged. In a retina scan, the person must approach the sensor, hold still for a few seconds, and be less than a foot away so that the sensor will have a view of the inner eye.

Once the system has taken an image of the iris or retina, this pattern is converted into a digitized form, just as in the enrollment stage. Once this digitization is complete, a computer program can compare the current data with the stored data. The stored data must now be retrieved for the comparison process. The database that holds this information can be a centrally located computer or even a hand-held card that could be inserted into an automated teller machine (ATM). To compare the two images, the computer analyzes the strength of the match between the two numbers obtained from the encoded pictures.

After this analysis is complete, the computer must decide whether or not to recognize the person being scanned. If the match is strong enough—say 99.9%—the system will grant the identification. If not, the person will be refused. The degree of exactitude required in matching can easily be varied.

The most important aspect of any biometric system is its *accuracy*. The two most important measures of accuracy are the false-acceptance rate and the false-rejection rate. *False acceptance* occurs when an impostor is accepted mistakenly. *False rejection* occurs when an authorized user is denied access to a system. Finally, other concerns, such as vulnerability to fraud, also called *barrier to attack*, and *long-term stability*, which concerns infrequent users being rejected, must also be considered.

Historical Development

Commercial biometric systems are just now beginning to enter their prime, in the late 1990s. Many of the systems, although they were hypothesized decades ago—such as iris and retina scanning, keystroke and voice recognition, and odor detectors—needed the power of modern computers and data networks before they could become technologically or commercially feasible. Fingerprinting systems have been in place since the end of the 19th century and have also benefited by merging with computers. The first modern

biometric scanner measured hand geometry and finger length in the 1970s at the Wall Street investment bank Shearson Hamill. The public's imagination was first caught by retina and iris scanners in the 1980s, when James Bond movies showed images of lasers passing over eyes.

The unique pattern of blood vessels in the retina of the eye was first described in 1935, but it was four decades before this knowledge was put to any use in biometrics. Robert (Buzz) Hill, an American electronics engineer, was helping his father, Robert Hill, an ophthalmologist, take photographs of the retina of the eye to detect disease. Hill Jr. noticed that the retinas were different and after some research, began developing a plan to create an identification scanning system. He patented his idea in 1975 and founded a company, Eyedentify, Inc., to develop the product in 1976. In 1984, the first retina scanner came on the market. In the 1990s the company's scanner was featured in the movies *Mission: Impossible, True Lies,* and *Demolition Man.*

The first person to note the uniqueness of the iris was Alphonse Bertillon, a French physician. Bertillon devised a system of biometrics that measured the features of the head and hands for identifying criminals in the 1880s. He also developed the first formal fingerprinting system in the 1890s. The idea of using iris patterns for personal identification was originally proposed in 1936 by Frank Burch, an

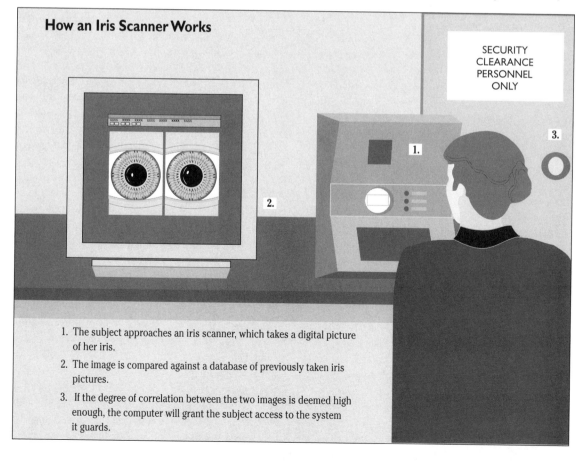

How an Iris Scanner Works

SECURITY CLEARANCE PERSONNEL ONLY

1. The subject approaches an iris scanner, which takes a digital picture of her iris.
2. The image is compared against a database of previously taken iris pictures.
3. If the degree of correlation between the two images is deemed high enough, the computer will grant the subject access to the system it guards.

ophthalmologist. Iris scanning was again postulated in the 1980s, when two ophthalmologists, Leonard Flom and Aran Safir, noted that there were many unique features in the iris. They developed the idea along with a Cambridge University computer scientist, John Daugman, then at Harvard University. The New Jersey company IriScan, founded by Daugman, Flom, and Safir, now holds worldwide patents on the technology used in iris scanning.

The technology for iris and retina scanning has been developed and licensed by many other companies that are trying to get a share of this new market. Its most widely used application will probably be in ATMs. Banks have long sought a better system than the current one involving bank cards and oft-forgotten personal identification numbers (PINs). Studies have shown that about one person in five has been unable to access money due to a forgotten PIN. Companies developing iris recognition systems are planning to introduce them in the United States for testing at ATMs early in the 21st century. Tests have already begun in England. One study at an English bank involving 1000 customers found that 90% preferred iris scanning to PIN cards. Other studies have found a wide international variation in willingness to accept biometric scanners, with Great Britain at the top, France at the bottom, and the United States and Germany in between.

Uses, Effects, and Limitations

There are many potential uses for biometric identification systems. Banks and businesses are always looking for more effective ways to control security. Banks want to be as sure as possible that they are giving money to the correct person and not an impostor. Businesses and government agencies that routinely deal with sensitive information—nuclear secrets, business plans, or other confidential information—also must control and record who is accessing what information.

Banks could use biometric systems for any in-person transaction, whether it is with a teller or with an ATM. Businesses and governments could use them for access to stored computer information or for physical access to a filing cabinet drawer, a specific room, or an entire building. A biometric sensor could be placed at every worker's terminal; each day all workers would have to be scanned before being allowed access to their workstations. Or a person might have to be scanned every time that he or she entered a certain room or requested specific data. Some government laboratories already have biometric systems in place for building and room access. Biometric systems can also eliminate "buddy-punching," a practice at some factories where workers punch their late (or absent) friends' timecards, so they appear to be showing up for work on time and get paid for hours for which they were not present. One Australian supermarket chain used fingerprint scanners for its timecard system and estimated

that the scanners generated an overall savings of 2 percent on payroll expenses.

Iris Scanning

Iris scanners are powerful identification tools, because there are many features available for use as markers in the iris. One company in the iris-scanning field claims that there are 266 unique features of the iris, including coronas, pits, filaments, crypts, striations, and radial furrows, that can be used. By contrast, standard fingerprint recognition uses about 35 individualized features. Iris scanning can uniquely identify around 1 million people before overlaps in individual features will begin to confuse the system. There are also other problems. The iris must still be picked out of a picture by computer programs, which are not 100 percent reliable. As well, systems have occasionally been fooled by pictures of irises. One advantage is that unlike retina scanners, iris scanners feel less invasive to a user since they do not require a subject to place his or her head close to a camera. A South African bank and its customers found iris scanning systems most suitable after previous attempts with fingerprint and retina scanners failed because the fingerprint pad became excessively dirty and no one liked the invasiveness of retina scanners.

The iris doesn't change much during life, which makes it a reliable, permanent biometric. Hands, faces, and fingerprints can all change with time, due to natural aging, weight gain or loss, and accidents. The pattern of the iris is set around the seventh month of gestation. Six months after birth, cells called *chromophores* give the eyes their color. The iris can change in two rare circumstances; moles can grow on irises just as they do on the skin, and certain cancer treatments with prostaglandin drugs can also affect the iris. The retina is also a very stable tissue. But it, too, can be damaged in accidents or by disease. It is not uncommon for the retina to undergo a process called *macular degeneration,* which leads to fuzzy vision or blindness in about 25 percent of the elderly and many diabetics.

Retinal Scanning

To date, Eyedentify retina scanners have never been fooled into admitting impostors. However, they have occasionally rejected the proper person. The retina scanner images around 30 to 40 of the eye's thickest, centrally located blood vessels; these have thicknesses of 100 to 250 microns. About 320 points of data are obtained after creating a digitized picture.

The U.S. government has already investigated using a retinal scanning system to prohibit truckers who have their licenses suspended in one state from obtaining a new license somewhere else. At the end of the 1980s, retinal scans began to replace fingerprint comparisons for identification purposes in prison systems in Dade County, Florida, which processes over 100,000 prisoners each year. In correctional facilities, prisoners may trade identities as favors or through intimidation. When a prisoner's name is called for release, the prisoner has traditionally needed to know only his or her social security number and mother's maiden name, information that can easily be traded. Checking fingerprints took hours: First the paper documents had to be found and then the comparison had to be performed. Often this was done with the naked eye by untrained guards. Retinal scanning takes only a couple of minutes and has elevated the security of the system.

A major problem with these technologies has long been their expense. Until quite recently, computers, memory storage space, and data networks were very expensive to develop, operate, and maintain. Now that computers and networks are becoming ubiquitous, and their manufacturing costs and retail prices have decreased considerably, the time may be right for biometrics. Banks seem to believe in this technology, even though it adds considerable cost to a single ATM. One unit already costs between $20,000 and $50,000; the addition of a biometric scanner increases the cost by another $8000 to $10,000.

Issues and Debate

The biggest issue concerning all biometric devices is their ability to track a person throughout his or her life. No identification is more certain than the features of a person's own body. Unfortunately, as many critics and privacy advocates have pointed out, these types of identification could also become a lifetime brand. In some of the scenarios they envision, no one would ever be free of the past as recorded by biometric scanners. Insurers could refuse to insure someone by scanning that person's retina and using it to check for future diseases. Tests available today can already identify AIDS, glaucoma, and diabetes by scanning the retina.

Representatives of the companies making these devices swear that the information they obtain will remain private and will not be given or sold to anyone else. But computers have been hacked into (accessed by unauthorized users) before and many privacy advocates remain leery of a centralized database containing biometric information. Also, what if the government requested these data for its own purposes? In a now-famous episode of evangelist Pat Robertson's *The 700 Club* TV program, biometrics was savaged as an invasive tool of the government and a harbinger of the Apocalypse. Other groups have also expressed their concerns, somewhat less stridently.

Instead of storing information in centralized computer databases, one option is to store a person's biometric data on a small card, similar to a modern bank card. But this essentially defeats the purpose of biometrics in the first place, which is to rid us of the need for identifying items such as keys, bank cards, magnetic cards, and PINs. A stolen or lost biometric card—which would be useless to anyone else—would bring the user back to square one, unable to access money or data.

Losing an ID card or forgetting a PIN number is a problem in today's society, even though there are mechanisms in place to handle them, but what will happen when biometric systems

mistakenly reject the correct person? An error rate of just 1 percent could exclude thousands from a system handling 1 million people. Alternatively, admitting impostors at a rate of 1 percent could lead to significant problems with fraud, theft, or other criminal activities. This is a problem that will someday have to be addressed by the biometrics industry. Commercial pressures will probably be the major force of change: Signature recognition systems have been around for years, but the credit card industry has yet to find one with a small enough error rate. If biometrics companies can not satisfy the operators, let alone the users, they may not survive long.

One final problem that is being dealt with slowly is compatibility: Of the wide variety of biometric system models, few can talk to each other. An organization called the Biometric Consortium has been formed to deal with this and to develop other standards for this growing industry.

—Philip Downey

RELATED TOPICS
DNA Fingerprinting, Genetic Testing

BIBLIOGRAPHY AND FURTHER RESEARCH
BOOKS
Jain, Anil, Ruud Bolle, and Sharath Pankanti, eds. *Biometrics: Personal Identification in Networked Society.* Dordrecht, The Netherlands: Kluwer Academic Publications, 1999.
PERIODICALS
Davis, Ann. "The Body as Password." *Wired,* July 1997, 132.
Malkin, Elizabeth. "Optical Conclusions: 'Eyedentify' System Is Faster Than Fingerprints." *Los Angeles Times,* June 14, 1987, 8.
Randall, Neil. "Biometric Basics." *PC Magazine,* April 6, 1999, 193.
Tan, Clarissa. "Hi-Tech ATMs: Cash with a Wink of an Eye." *Business Times,* June 18, 1999, 10.
INTERNET RESOURCES
A National ID Card: Is Big Brother on the Way?
    http://www.the700club.org/newsstand/stories/980715b.asp
Biometrics Consortium
    http://www.biometrics.org
Eyedentify, Inc.
    http://www.eyedentify.com
IriScan, Inc.
    http://www.iriscan.com
John Daugman
    http://www.cl.cam.ac.uk/~jgd1000/
Retinal Scans Always Get Their Man
    http://www.autoidnews.com/technologies/concepts/retinal.htm
Sensar, Inc.
    http://www.sensar.com
Your Body IS Your PIN!
    http://webreference.com/ecommerce/mm/column3/index.html

REUSABLE LAUNCH VEHICLES

A reusable launch vehicle (RLV) is an aerospace vehicle that incorporates functional designs and fully reusable components to achieve airline-type space flight operations. Although no completed RLV yet exists, the RLV as designed will use advances in space technology to provide low-cost access to

RLV Designs

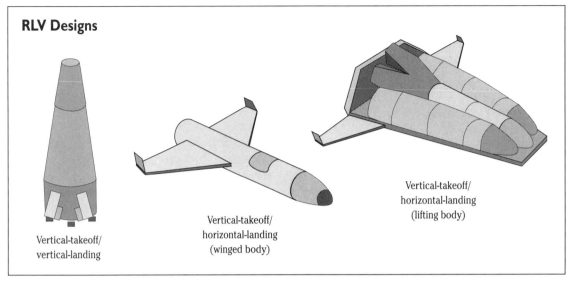

Vertical-takeoff/
vertical-landing

Vertical-takeoff/
horizontal-landing
(winged body)

Vertical-takeoff/
horizontal-landing
(lifting body)

Adapted from an original illustration courtesy of U.S. Congress, Office of Technology Assessment.

space, thereby significantly expanding commercial and scientific opportunities.

Aerospace experts now consider it possible to build an RLV. However, achieving the performance levels necessary to reach orbit with a useful payload, for example a satellite, will require a host of engineering advances that improve fuel efficiency and lower vehicle mass without compromising structural integrity. Completely reusable launch vehicles are technologically much more difficult to achieve than expendable ("throwaway") rockets because all of the RLV's components must be capable of resisting deterioration and surviving numerous launches and reentries without requiring refurbishment between flights.

Scientific and Technological Description

With the exception of NASA's Space Shuttle, all modern rockets used to lift payloads into space fall into the category of *expendable launch vehicle's* (ELVs). This means that all of a vehicle's flight components, such as its engines, fuel tanks, and support structures, are discarded after just one use. In the case of *staged rockets* (which have two or more rocket units stacked one on top of another), expended rocket units are discarded (jettisoned) in steps or increments during the flight, making the later stages function more efficiently because they have less "useless" mass to carry into orbit. This practice makes access to space with ELVs quite expensive.

Getting into space requires the use of a powerful rocket vehicle capable of lifting itself, large amounts of propellant, and any useful payload to velocities of about 8 kilometers (km) per second (18,000 miles per hour). One measure of success used by rocket scientists is the *mass fraction,* which is defined as the mass of propellant the vehicle needs to acquire a specific orbit divided by the gross liftoff mass of the vehicle (including its structure, propellant supply, and payload). The smaller the mass fraction, the better the per-

formance of the vehicle. As a point of reference, the U.S. Air Force's mighty Titan IV expendable rocket has a mass fraction of 0.87, which means that at launch 87 percent of this multistaged vehicle's total mass is propellant. In fact, most rockets are really flying "gas cans." The RLV is no different.

Aerospace experts now believe that the United States has the appropriate technology to design and produce a single-stage-to-orbit (SSTO), fully reusable launch vehicle with sufficient payload capacity to meet most government and commercial space transportation requirements. *Fully reusable* means just that. Between space missions, the vehicle would simply undergo inspection, refueling, and payload processing. Much like a commercial airliner, there would be no major components to be reassembled or refurbished completely after every flight. Since the SSTO RLV has no expendable components, it needs to carry more fuel than would be necessary if it were discarding mass by dropping expended stages during ascent.

There are three candidate SSTO configurations. Each poses unique technical obstacles but also offers distinct advantages for reducing the cost of accessing space. These configurations are the vertical takeoff/vertical landing conical vehicle; the vertical-takeoff/horizontal-landing winged-body vehicle; and the vertical-takeoff/horizontal-landing lifting-body vehicle. The *conical* configuration has the advantage of a low-mass airframe that does not require enormous wings and provides a simple aerodynamic shape. However, this vehicle's rocket engines must be restarted after reentry into the Earth's atmosphere to accommodate a vertical landing and its payload volume is limited. The *winged-body* SSTO offers the advantage of simple fuel tanks and easy maneuverability during reentry. However, it also has a limited payload volume and involves a high landing speed. The *lifting-body* SSTO offers low reentry temperatures, a low landing speed, and a low mass design. It does, however, represent a more complicated airframe. In 1995, NASA, in partnership with the Air Force and private industry, started an RLV technology program in which all three of these SSTO designs were considered. The lifting-body configuration was eventually selected for further development and is now called the X-33 program. The X-33, which will measure about 20 meters in length and will have a dry mass of about 28,350 kilograms (kg), will be flight tested at Edwards Air Force Base in California starting in mid-2000.

VentureStar is Lockheed Martin's potential commercial follow-on to the X-33. It will be similar in design to the X-33 but twice its size and about eight times its launch mass. Lockheed Martin hopes to operate this RLV at a flight rate of 40 launches per year, leading to launch costs of approximately $2000 per kilogram. NASA and Lockheed Martin are also studying the accommodation of crew missions on VentureStar. One option is to launch this vehicle as a cargo-only craft initially, with crew-capable modules phased in later.

NASA's X-34 is designed to be a suborbital technology (the vehicle will not actually go into orbit) testing ground for the RLV program. The goals of the X-34 are to achieve a maximum speed of Mach 8 (eight times the speed of sound) and to reach altitudes of up to 80 km. In June 1996, NASA awarded a contract to Orbital Sciences Corporation to design, develop, and test the X-34. The X-34 design features a cylindrical body with delta wings. The vehicle measures 17.7 meters in length and has a dry mass of 19,500 kg. The X-34 will be launched from an L-1011 carrier aircraft and is powered by a new liquid oxygen (LOX)/kerosene rocket engine called Fastrac. Flight testing at NASA's Dryden Flight Research Center in California (starting in 2000) will focus on RLV operations such as 24-hour turnaround, landing in adverse weather conditions, and safe abort procedures.

Historical Development

From its establishment in 1958, NASA has studied the concept of the reusable launch vehicle. In 1963, NASA joined the Air Force in research toward the development of a reusable, crewed vehicle that could go into orbit and return, taking off and landing horizontally. In 1969, an Air Force–NASA Space Task Force Group examined *post-Apollo* era space objectives for the United States. With a view toward the 21st century, this group recommended (among other things) the development of an economical, reusable space transportation system of interplanetary scope. However, the extensive program recommended by the Task Force Group was chopped down by extreme budget pressures to just a single new program—a recoverable, reusable new launch vehicle, eventually called the SPACE SHUTTLE.

Between 1969 and 1971, the aerospace industry studied a reusable space transportation system. However, these studies were greatly influenced by continued budget pressures, so much so that an early design for a two-stage, fully reusable liquid-propelled vehicle eventually became a triangular-winged, aircraftlike vehicle mounted on a large expendable external tank and supplemented in thrust by two large, reusable solid rocket boosters. This austerity-driven final design compromised the original visions of a completely reusable launch vehicle and led to a contemporary Space Shuttle vehicle that is only partially reusable, requires extensive refurbishment between launches, and is extremely costly to operate.

A 1995 partnership with the Air Force and private industry enabled NASA to create the Reusable Launch Vehicle Technology Program. The main objective of this program is to develop and demonstrate new technologies for the next generation of fully reusable space transportation systems that can radically reduce the cost of accessing space. An all-rocket-powered SSTO RLV is the goal of the program, which is now focusing on two demonstrator vehicles: the X-33 and the X-34. NASA's Future X program will continue with research on advanced RLV concepts. NASA initiated the X-33 program to develop a test for integrated RLV technologies, paving the way for full-scale development of a SSTO reusable launch vehicle that would be contracted for government and commercial use. The X-33 is targeted to reach high hypersonic speeds and demonstrate SSTO and autonomous operations capabilities. In July 1996, NASA selected Lockheed Martin's design, a lifting-body SSTO.

Uses, Effects, and Limitations

The development of the RLV is driven by the desire to reduce launch costs. A significant reduction in costs is anticipated because the RLV can be reflown with minimum refurbishment after each mission rather than having the entire vehicle replaced. This reduces long-term production costs. Since the RLV is designed for quick-turnaround operations that support a higher flight rate than can be achieved with today's expendable or partially reusable vehicles, the cost per flight will also be reduced. Some studies, such as NASA's 1994 *Commercial Space Transportation Study,* suggest that if an operational RLV is capable of reducing launch costs in the early decades of the 21st century, many new space missions and commercial business ventures will also emerge. A suc-

A reusable launch vehicle delivering a scientific experiment payload to the International Space Station, circa 2008 (Digital image courtesy of NASA).

cessful first-generation-design RLV should reduce the cost of accessing low Earth orbit to about $1000 per kilogram (or less). A more advanced RLV, perhaps emerging out of NASA's Future X program, could reduce the cost to as little as $100 per kilogram of payload delivered into LEO.

Issues and Debate

Space enthusiasts proclaim the RLV as the 21st century space transportation system that will provide routine, affordable access to space for many different customers. Although no one truly opposes the idea of getting into space more cheaply, less enthusiastic engineers point out that many obstacles must be overcome before a working, reliable, and affordable RLV can actually take a payload into space at low cost. Only focused research programs and flying demonstrators will resolve this debate.

Reliance on the Space Shuttle, only a partially reusable aerospace vehicle, can be reduced if RLVs are developed. Payload hauling with the Shuttle is even more expensive than with expendable launch vehicles, because of the high cost of maintaining the fleet and the extensive refurbishing of each vehicle between flights. Current launch costs using both ELVs and the Space Shuttle range between $5000 and $20,000 per kilogram or more, with Shuttle-delivered payloads dominating the high end of this range. The cost per unit mass is used so that different vehicles, systems, payloads, and launch sites can be compared against a common aerospace industry figure. The exact delivery bill depends on such factors as desired orbital inclination, characteristics of the launch vehicle, and the possible need for an upper propulsive stage (should the payload have to travel to a higher orbit or through interplanetary space).

The RLV will be capable of placing payloads into orbit more cheaply through the use of fully reusable components that incorporate advanced technologies with uncomplicated, functional designs. Innovative operational techniques, including *automated vehicle checkout*—the use of sophisticated computer systems to perform vehicle inspection and launch readiness checks during the launch process—as well as efficient payload processing, will provide an airline-like operational environment that does not require the presence of a large and expensive launch complex. The first generation of RLVs promise to carry payloads into low Earth orbit (LEO), about 200 km (125 miles) above the Earth, at an estimated cost that is about 90 percent less than the current cost of accessing space. Aerospace engineers further predict that the second generation of RLVs (available perhaps by the year 2020) will reduce this cost by another 90 percent, so it may eventually cost only $100 to $200 to deliver a kilogram of cargo to LEO.

In addition to providing the desired low-cost access to space, a successful RLV will also raise some interesting issues. Who will regulate this greatly expanded access to space? Because of its operational simplicity, an SSTO RLV can be launched from just about any place on Earth and requires only a minimal launch complex infrastructure. Nations that lie along or near the equator could suddenly become highly competitive "spacefaring" countries, due to the favorable rotation of the planet, which gives a "free" kick to equatorial launches. Rogue nations might acquire instant access to space, possibly with specific hostile intentions toward the military space systems of more powerful countries. Finally, without proper transition planning, free-market initiatives could make today's existing government space agencies and their large, billion-dollar space complexes essentially obsolete, at least from the perspective of space transportation.

—*Joseph A. Angelo, Jr.*

RELATED TOPICS
International Space Station, Satellite Technology, Solar Power Satellites, Space Alerting and Defense System, Space-Based Materials Processing, Space Shuttle

BIBLIOGRAPHY AND FURTHER RESEARCH

BOOKS
Angelo, Joseph A., Jr. *The Dictionary of Space Technology*. New York: Facts On File, 1999.
Associate Administrator for Commercial Space (AST). *1999 Reusable Launch Vehicle Programs and Concepts*. Washington, D.C.: Federal Aviation Administration, January 1999.
Heppenheimer, Thomas A. *Countdown: A History of Space Flight*. New York: Wiley, 1997.
Neal, Valerie, Cathleen S. Lewis, and Frank H. Winter. *Spaceflight: A Smithsonian Guide*. Toronto, Ontario: Macmillan, 1995.

PERIODICALS
Austin, Robert E., and Stephen A. Cook. "SSTO Rockets: Streamlining Access to Space." *Aerospace America,* November 1994, 34.
Berkowitz, Bruce D. "NASA's X-33." *Air and Space,* October–November 1996, 35.
Kistler, Walter P. "Humanity's Future in Space." *The Futurist,* January 1999, 43.
Robertson, Donald F. "Shuttling into the 21st Century." *Astronomy,* August 1995, 32.

INTERNET RESOURCES
Aeronautics and Space Transportation Technology
    http://www.hq.nasa.gov/office/aero/
Dryden Flight Research Center
    http://www.dfrc.nasa.gov/
Marshall Spaceflight Center
    http://www.msfc.nasa.gov/
Venturestar
    http://www.venturestar.com

RNA: ANTISENSE TECHNOLOGY

Antisense RNA technology is a way of treating diseases by stopping proteins from being produced in living cells. This technology can also be used to create genetically modified organisms. Proteins are molecules that carry out most of the chemical reactions in cells, and proteins can be considered to be directly responsible for almost all diseases and illnesses. Over the past 20 years, scientists have developed and used antisense technology in biological research to study what particular proteins do.

Drug developers have now designed a new generation of drugs called antisense therapeutics, and the first such drug, Formivirsen, was licensed for use in 1998. Scientists have also produced genetically engineered plants, such as the Flavr Savr tomato, that have altered properties because they permanently produce antisense molecules. Although some uses of antisense technology have caused concern about possible ethical or environmental problems, antisense technology will probably become increasingly useful in the development of drugs and genetically engineered organisms.

Scientific and Technological Description

All living organisms contain molecules called *proteins,* which carry out almost all the chemical reactions necessary for life. Proteins are made according to the genetic instructions contained in an organism's deoxyribonucleic acid (DNA). To make a particular protein, DNA containing the instructions to make the protein is first copied into RNA (ribonucleic acid which is a molecule similar to DNA), which is then used to make the protein. DNA consists of two molecules, called *strands,* that associate tightly with each other. One of the strands, called the *sense strand,* is converted into RNA; the other strand is called the *antisense strand.*

If the antisense DNA strand is converted into RNA as well as the sense strand, the sense and antisense RNA strands

stick together to form a double-stranded molecule (similar to the original DNA molecule). Proteins cannot be made from double-stranded RNA, so if a scientist can somehow get some antisense RNA into an organism, or somehow get the organism to produce its own antisense RNA (from antisense DNA), he or she can stop the organism from producing a particular protein (see figure). Because proteins are responsible for almost all chemical processes, good and bad, that occur in living organisms, scientists would like to manipulate which proteins an organism produces so as to, for example, cure diseases and change the properties of our foods. For diseases caused by the lack of some protein, GENE THERAPY is a possible cure. Antisense therapy can be thought of as gene therapy, using antisense DNA or RNA. Conventional gene therapy uses the sense strand of DNA to cure a disease by *causing* the production of a particular protein, not by preventing it.

Although antisense RNA can stop protein production just by binding to its sense RNA partner, there is a way to make the antisense RNA far more effective at blocking proteins that are being produced inside organisms. Antisense RNA molecules can be attached to special RNA called *RNAse H,* which can physically cut up and destroy other RNA molecules. When an antisense RNA–RNAse H hybrid molecule forms a double-stranded molecule with some sense RNA, the RNAse H cuts the sense RNA. So not only is no protein produced from the sense RNA because it associates with the antisense RNA, but also the RNAse H destroys the sense RNA so that it can not be used to

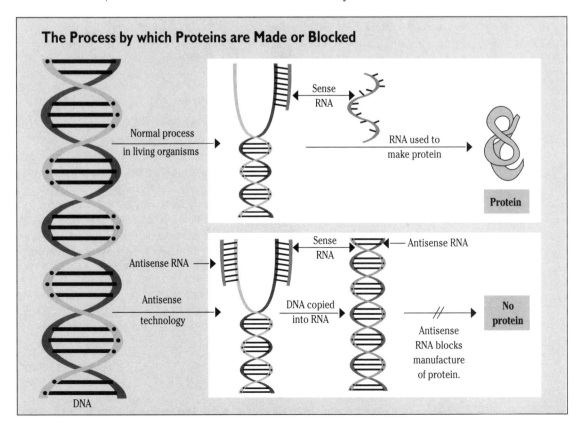

The Process by which Proteins are Made or Blocked

produce protein even if the antisense RNA happens to drop off. Many drug developers are using RNAse in their antisense therapeutics.

RNA is a very unstable molecule, and it is destroyed very quickly inside an organism so that an RNA drug would not survive as a drug long enough to be effective when given to a patient. To solve this problem, drug developers are trying to find the best way to modify RNA chemically so that it can survive much longer inside patients and still work as it is supposed to.

Historical Development

In 1953 the chemist Francis Crick and the biologist James Watson correctly guessed the structure (the three-dimensional shape) of DNA. In the years that followed, Crick and other researchers then worked out how DNA contains the information to make proteins and how DNA is copied into RNA before protein can be made. From this knowledge, researchers were able to theorize about how antisense DNA or RNA might be used to affect which proteins living organisms produce.

The first experiment with antisense technology was by Paul Zamecnik in 1978, in which he found that antisense RNA could reduce the growth of a chicken virus (the rous sarcoma virus) in chick embryos. Then in the 1980s, researchers became increasingly able to make short sections of DNA and RNA in the laboratory, and they also found ways to modify RNA chemically so that it is not destroyed quickly inside cells. These two advances meant that researchers could start to think seriously about using antisense RNA for therapeutic uses. Also in the 1980s, researchers developed many ways to cut up molecules of DNA and rejoin them. By cutting out a section of DNA from a larger DNA molecule and then putting it back "back to front," researchers could create DNA that would produce antisense RNA. Today, researchers have two ways to make antisense RNA: They can chemically synthesize it or they can make organisms produce it themselves.

The first commercial use of antisense RNA came in 1994, when the Flavr Savr tomato went on sale in the United States. This tomato was genetically modified by Calgene Inc. of California (now part of the Monsanto Corp.) to produce antisense RNA to prevent overripening that can occur during storage and transportation. In 1998 the first antisense drug, Formivirsen, was licensed for sale to AIDS patients in the United States to treat eye infections of a virus called cytomegalovirus. Formivirsen was developed by Isis Pharmaceuticals in California.

Uses, Effects, and Limitations

In recent years great expectations have arisen about what antisense therapeutics can deliver. Most diseases and illnesses are caused by proteins, and some diseases are caused by having too much of a particular protein, or by having a malfunctioning protein. In these cases it is hoped that antisense therapy will reduce the production of these unwanted proteins and so cure the disease. The key point about antisense technology is that it has the potential to be very specific: Only the disease-causing protein is affected and none others. Conventional drugs almost always work by binding to a protein and inhibiting (or stimulating) that protein's function. Drugs cause side effects when they affect proteins other than the one that is causing an illness, and researchers hope that this will not occur nearly as much with antisense RNA drugs.

Formivirsen (see above) is not a miracle cure for cytomegalovirus infections—it is not 100 percent effective—but it does show that antisense RNA can be used as a drug. Other antisense drugs are being tested in people to see if they too are effective at treating diseases. In particular, there is a great deal of experimentation in using antisense for antiviral and anticancer therapies. There is, however, some controversy about whether antisense RNA actually works as researchers hope. Some researchers argue that the beneficial effects of antisense RNA molecules don't always result because the molecules interact specifically with sense RNA molecules. These researchers say that the RNA can act by other mechanisms to interfere with chemical reactions inside cells. Ultimately, most researchers agree that it is not so important how a drug acts, but rather, that it does have some beneficial effect (without causing unwanted side effects). One of the biggest problems with using antisense RNA as a drug is how to get RNA into the cells from which organisms are made (see DRUG DELIVERY SYSTEMS; GENE THERAPY). Researchers have come up with a number of ways to do this, but none of them yet work as well as might be hoped.

Antisense technology has been used very successfully in biomedical research and plant GENETIC ENGINEERING. Biomedical researchers can use antisense RNA to stop some particular protein from being produced in cells as a way of trying to find out what that protein does. The genetic engineering of plants to produce antisense RNA has also been very successful, and the Flavr Savr tomato is a good example of this. Researchers have also taken the plant from which 60 percent of the world's coffee is made, *Coffea arabica,* and engineered it genetically using antisense technology so that it doesn't produce caffeine. In so doing, there will be no need to decaffeinate coffee beans after they are harvested, a process that costs money and affects the flavor of the coffee.

Issues and Debate

Because antisense technology is so closely related to genetic engineering and gene therapy, the ethical questions concerning the use of genetic engineering for the treatment of human disease or the creation of new animals or plants also apply to antisense technology. Some people argue that it is immoral or dangerous to use genetic engineering to change a person's or other organism's DNA permanently, as well of that of their offspring. Others think that fears about the safety of genetic engineering are unfounded as long as all work is carried out properly and the results are carefully monitored.

The first antisense drug, Formivirsen, does not raise any ethical concerns on the basis of its being an antisense drug. The antisense RNA in Formivirsen is directed against a virus, not a human cell, and the chance that the antisense RNA could somehow get incorporated in the patient (and worse, the patient's offspring) permanently is vanishingly small.

The Flavr Savr tomato has not been accepted as readily. Many researchers argue that because antisense technology "switches off" genes rather than adds new genes to an organism (as conventional genetic engineering does), antisense technology is inherently safer than conventional genetic engineering. This argument has been used to allay fears about the Flavr Savr tomato, but it does not convince everyone, on the grounds that one can never tell what all the effects of switching off a protein will be. Some undesirable change to the organism might occur that could not be predicted. Not many people (scientists or otherwise) are prepared to say that it is impossible that a plant genetically engineered with antisense technology could damage either the environment or human health. For this reason, it is important that commercial applications of antisense technology be monitored properly, so that even quite small effects in people or the environment do not go unnoticed.

—Matthew Day

RELATED TOPICS
Cloning, Drug Delivery Systems, Gene Therapy, Genetic Engineering, Genetic Testing, Rational Drug Design

BIBLIOGRAPHY AND FURTHER RESEARCH

BOOKS
Aldridge, Susan. *The Thread of Life: The Story of Genes and Genetic Engineering.* London: Cambridge University Press, 1996.
Grave, Eric S. *Biotechnology Unzipped: Promises and Reality.* Washington, D.C.: Joseph Henry Press,1997.
Klug, William S., and Michael R. Cummings. *Essentials of Genetics,* 3rd ed. Upper Saddle River, N.J.: Prentice Hall, 1995.

PERIODICALS
Ault, Alicia. "First Antisense Compound Clears U.S. Regulatory Hurdle." *Lancet* 352(9125), August 1, 1998, 377.
Day, Stephen. "Switching Off Genes with Antisense." *New Scientist* 124(1688), October 28, 1989, 50–55.
Gavaghan, Helen. "To Kill a Superbug." *New Scientist* 161(2173), February 13, 1999, 34–37.

INTERNET RESOURCES
General Background and Overview
    http://www.enzo.com/therapeutics/antisense_primer.html
    http://www.hybridon.com/graphic_version/antisense/ antisense.html
Isis Pharmaceuticals
    http://www.isip.com/antisens.htm
Science of Antisense RNA Technologies
    http://www.godriva.com/~jkimball/BiologyPages/A/ AntisenseRNA.html

ROBOTICS

A robot is a machine designed to perform a broad variety of tasks and functions. Robots are often ideally suited for going where human beings cannot go. Extremes of heat, cold, or nuclear radiation, and hostile environments such as war zones, the ocean deeps, or the surface of Mars, allow a well-designed robot to operate where people would be unable to function, or would be killed.

Robots have been used in cleaning up environmental disaster sites such as the Chernobyl nuclear reactor. They are also used in mining exploration, land-mine detection, bomb disposal, and to attack enemy troops and vehicles. Teleoperated robots can be controlled by people from a remote location; autonomous robots can act on their own or cooperate with others. Given the wide variety of beneficial applications that people can develop for robots, the field of robotics has undergone tremendous growth in recent years; this growth will undoubtedly continue. Nevertheless, as the possible applications of robotics are expanded, some observers are concerned that these nonhuman "actors" may take on too much power and that people may not be able to control them.

Scientific and Technological Description

What separates a robot from a simple machine is its versatility, flexibility, and adaptability. While a simple machine repeats a single task over and over, robots can usually be reprogrammed with new instructions to perform different tasks. Some robots also have the ability to "learn" and to reprogram themselves to accommodate changing situations.

All modern robots have a few features in common. The first is a central processing unit that controls the robot and serves as its brain. This unit sends the commands that control the robot's movements, interpret its sensor data, and communicate with humans. Computer programs control all of these actions. All robots also have some kind of body that can be almost any size or shape: Insectoid, vehicular, humanoid, and canine models are presently being made. In addition, there must be some kind of interface—through the physical contact of wires or fiber-optic cables or the remote control of radio and microwave transmissions—through which people can communicate with the robot to extract its recorded data or send it new instructions. Finally, robots generally have a tool or sensor to manipulate or record data about an environment.

Robots typically have two methods of locomotion: wheels or legs. Robots with wheels or tanklike tracks are typically very stable on flat ground, but they quickly run into difficulties on uneven terrain, such as rocky landscapes or even a simple set of stairs. Legged robots can negotiate uneven terrain, but they are often extremely slow, and even more often have tremendous balance problems, even on relatively flat terrain. Working robots have been developed featuring all

A Robotic Arm and Hand

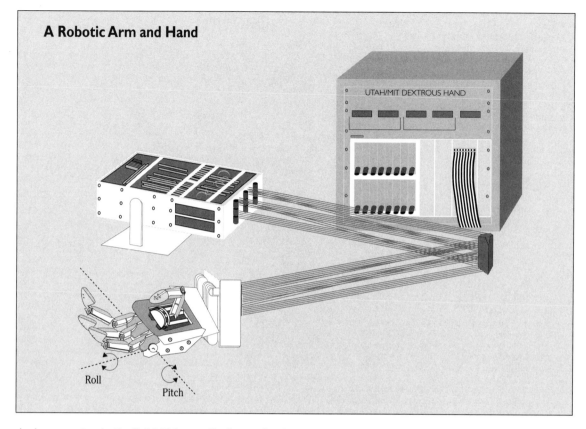

UTAH/MIT DEXTROUS HAND

Roll

Pitch

A robot arm equipped with a Utah/MIT Dextrous Hand, a specific robot-hand design that was invented in 1984 by S. C. Jacobson and others and has been used in research projects since 1991.

manner of locomotion, from a one-legged robot that must hop continuously to eight-legged spiderlike models.

Robots can be made in almost any size. Today, feasible robots—from land rovers that measure a few centimeters to tanks and unmanned aircraft measuring tens of meters—are being produced or actively researched. They can be shaped into any mechanically feasible form. Robots' prime purposes are to accomplish a specific task, and they can easily be designed to carry and use any kind of tool. They can be equipped with guns, lasers, shovels, drills, video cameras, water cannons, radar, or scientific equipment. No robot carries all of these tools, of course: It would quickly become overly complicated and heavy.

Robots have three different degrees of independence. At one extreme, *teleoperated robots* are controlled from a remote location by human beings who remain in constant communication with them. At the other, *autonomous robots* are almost entirely independent of human control. They have almost complete independence in deciding how a task will be accomplished or a problem solved. In between, many robots combine properties of both independence and reliance on people. Teleoperated robots remain in contact with human controllers by radio contact or guide wires. These robots are typically used in strange environments where novel situations would befuddle an independent

robot. The mobile robotic rover Sojourner, of the 1997 *Pathfinder* mission to Mars, is an example of a teleoperated robot that was guided by Earthbound humans. The robot was equipped to analyze Martian soil and rock samples. The samples themselves were chosen by ground controllers on Earth. Autonomous robots are largely independent and must be carefully programmed to accomplish their goals. Simple tasks for a person, such as finding the way across a crowded room of moving people, are often complex and demanding tasks for an independent robot.

Historical Development

Robots and other animated creatures have held places in human literature and folklore for hundreds of years. For the most part they have been viewed as evil destroyers. The first modern robot was suggested in Mary Shelley's *Frankenstein,* an 1818 novel about a monster constructed of human cadavers who runs amok among humans. Since then, animated machines and robots have often been viewed with trepidation by the public. The word robot was invented in 1921 by the Czech playwright Karel Capek in his play *R.U.R.,* which described a society in which robots designed to do humans' work rebelled and took over the world. Previously, all robots had been called *automatons.* In 1926, the Fritz Lang movie *Metropolis* depicted a robot that was designed to pacify and be worshiped by the common workers; instead, they revolt and riot.

The first positive portrayals of robots began to appear in the 1930s, when Isaac Asimov and others began publishing science fiction stories depicting robots that were for the most part under the control of human masters. Asimov also coined the term *robotics*. This more positive portrayal led to the helpful and servile 1950s' movie star Robby the Robot, and in the 1970s, the friendly and efficient C3PO and R2D2 of the *Star Wars* movies. Since then robots have been viewed with a mixture of admiration of their sometimes awesome abilities and fear of their future capabilities. Still, machines like HAL 9000, the artificial intelligence of the 1968 movie *2001: A Space Odyssey,* were able to bring back the image of the malevolent machine.

The first working robot was a mechanical duck invented by Jacquard de Vaucanson in 1738. This duck could swim, drink water, quack, eat grain, and digest and excrete it. De Vaucanson toured Europe with his duck, where many spectators came to gaze at the marvel. After selling his duck, de Vaucanson went on to invent the Jacquard loom, the first automated weaving system. Most important, it could be reprogrammed to weave different patterns. Other inventors followed with automated scribes, tigers, and animated clocks.

The first patent for an industrial robot was issued in 1954 to George C. Devol, Jr. He developed a control and memory system called *universal automation* to direct machines in factories. Joseph F. Engelberger took this concept and developed it into Unimation, Inc., which was established in 1958 and led the field of robotics for decades. Unimation produced a wide variety of robots for industrial uses.

In the 1950s, ARTIFICIAL INTELLIGENCE (AI), a branch of computer science that creates machines that can "think," learn, and solve problems, got its start. The first university department devoted to studying this was the Artificial Intelligence Lab at MIT, founded by Marvin Minsky and John McCarthy in 1959. Many universities quickly followed suit, creating their own AI labs. Since then robots have been developed for a variety of uses outside the factory. With the beginning of the exploration of our solar system in the 1960s, robots have had a role to play in exploring other planets (see ROBOTIC SPACE EXPLORATION). Robot landers have been used repeatedly on the surfaces of Mars and the Moon for exploration. Robots designed for handling extreme environments, such as the interior of the Chernobyl power station in the former Soviet Union after its 1986 explosion, began to be developed after the Three Mile Island nuclear accident in Pennsylvania in 1979. Bomb disposal robots were created in the 1980s. Bomb disposal has been a growing concern since World War II, when the Nazis first dropped time-delayed bombs on London.

Deep-sea exploration has also exploited robots for commercial and archeological purposes (see DEEP-SEA VEHICLES; TELEOPERATED OCEAN VEHICLES). Teleoperated robots built to withstand the crushing pressures of the deeps have made possible the following discoveries: a 2700-year-old Phoenician ship in the Mediterranean Sea; the Israeli submarine *Dakar,* which sank in 1968, was found in 1999; and most famous, the *Titanic,* which sank in 1912 and was rediscovered in 1985 with the *Alvin* submersible explorer.

In the 1990s, robots have been turned to home entertainment purposes. The Sony Corporation introduced its robot dog AIBO (Japanese for "companion" or "pal") in 1999. Each dog has a "personality" that will develop over time. A set of programming instructions gives the robot behaviors unique to its environment and tasks. Sony claims that AIBO has six "emotions": happiness, sadness, fear, anger, dislike, and surprise. It also has instincts for love, search, movement (or play), and hunger (it needs a recharge). The Matsushita Electric Industrial Company is also working on an electronic cat, and the car company Honda has developed a 5-foot 2-inch humanoid robot called P3 that can walk up and down stairs and over bumps, open and close doors, shake hands, and bow. At MIT, researchers have developed Cog and Kismet. These robots, with their humanoid heads and eyes, observe people moving across rooms and meet them eye to eye, an experience that people describe as being eerie.

Uses, Effects, and Limitations

Robots are used for two principal reasons: to replace human labor for economic or for safety purposes. The four major fields in which teleoperated and autonomous robots are used are industrial manufacturing, environmental cleanup, bomb disposal, and exploration of hostile environments. The U.S. government is the largest American user of robots for environmental cleanup. Fifty years of purifying the radioactive elements tritium and plutonium for use in nuclear weapons has left the United States with some extremely polluted environments and facilities. Cleaning up these spots is a job that no person could long tolerate. This was learned in painful detail during the Chernobyl meltdown, when 28 workers struggling to contain the radioactive fires at the plant lost their lives due to radiation injuries, and 134 were diagnosed with acute radiation syndrome. Overall, the 5000 workers at Cheronbyl each absorbed around 25 rem of radiation, five times the recommended yearly dosage. In 1990, the first robots sent in to clean up the Chernobyl reactor lasted about 7 minutes before being fried by radiation. In this instance, the usefulness of robots was clear: Officials estimate that robots were able to save 500 workers from entering Chernobyl.

Tougher robots, built by RedZone Inc. and U.S. government laboratories, have managed to last for short periods of duty—hours—inside a reactor core. They have also been the first to return. Equipped with jets that shoot blasts of pressurized water at over Mach 2—2000 kilometers an hour—the robots carve the radioactive slag, called *corium,* into manageable chunks that can be packed into protective

The Sony Corporation introduced the AIBO robot dog in 1999 (Courtesy of Sony Electronics, Inc.).

containers the size of a shoebox. These boxes can now be removed safely from the reactor. Of course, disposal of this radioactive waste presents an entirely new problem.

Bomb disposal is the second-largest use of robotics. This job requires mainly teleoperated systems that can examine and contain unexploded ordnance (UXO). The U.S. Defense Department defines a UXO as unexploded explosives such as dynamite, land mines, and time bombs purposely left behind, or accidentally unexploded ammunition such as bombs, mortar shells, and fuel storage containers. The robots used have the ability to examine suspicious packages and bombs with portable X-ray scanners to determine their contents and components. They are also equipped with manipulators to contain and sometimes inactivate the UXO. One common technique is use of a high-powered water cannon to simultaneously short-circuit the UXO and tear it apart.

Robots are also being used in hostile and battlefield environments as security and defensive systems. Teleoperated tanks have already been developed for combat. Using these still requires a human operator, but one who can remain in safety hundreds of kilometers away. If the robot is destroyed, the operator can simply start up another robot, without loss of human life. Robot systems may someday be used by police to flush out fugitives and criminals from fortified positions.

The biggest problem with all robots is getting them to learn from experience. This is the prime problem faced by all artificial intelligence researchers. Robots must somehow be programmed to learn from their mistakes, anticipate problems, and come up with novel solutions. Computer scientists try to make robots mimic human thought processes, but since these aren't fully understood either, robots are still falling short of human ideals. Two well-understood methods of problem solving are the top-down and bottom-up methods. *Top-down processing* involves breaking a problem down into smaller prearranged parts, solving the small problems, and reassembling the whole. *Bottom-up processing* involves making vari-

ations in the entire setting to see their effects on the final results. A robot using this method must be imbued with creativity, a difficult programming task. Each problem-solving method has its merits, but robots must learn to determine which are appropriate for use in a certain situation, something that human beings do unconsciously.

A problem related to learning is *coevolution*. This is when roboticists try to get many robots to work together and cooperate at solving problems. Each robot—they can be the same or different models—can learn through trial and error and hopefully, specialize at performing a single task that will further the solution to the overall problem. For this to occur, each robot must be a flexible learner, able to cooperate with others, and perhaps give and receive orders. One example of the progress in creating successful cooperation among robots is the Robot World Cup, or Robocup, which began in 1997. In that international event, held most recently in July 1999, artificial intelligence research groups from competing universities build teams of soccer-playing robots. The robots then meet in a championship to determine which university group has created robots with the best soccer skills. The robots pass the ball back and forth, guard one another, and shoot for and defend the goal. There is also the Robot Wars competition, established in 1994, in which teleoperated robots of wildly different designs have the sole goal of destroying one another completely.

Issues and Debate

Since the introduction of machines, people have always worried about being replaced by mechanized labor. This fear is still with us today. Robots and other machines have encroached upon manual labor in all areas of human endeavor. Car factories use robots for assembly, COMPUTER ANIMATION has replaced hand-painted animation cels, and MUSIC SYNTHESIZERS mimic human musicians. Not all of these things are necessarily bad. Robots can perform dangerous tasks in a factory, and computers have opened up entirely new avenues of expression in the creative arts.

Now that artificial intelligence researchers are giving robots rudimentary learning skills, some observers wonder what will happen if they suddenly "evolve" and increase quickly in intelligence. Even worse, what if the robots can replicate themselves? It is not difficult to imagine a scenario where these sorts of robots go out of control quickly. In the end it will be up to scientists to create machines with "failsafe" mechanisms that give us a final degree of control over all robots. A general method for doing this, one familiar to all robotics researchers and creators,which may someday save us from our own creations, is Isaac Asimov's Three Laws of Robotics, as laid out in his science fiction stories:

1. A robot may not injure a human being, or, through inaction, allow a human being to come to harm.

2. A robot must obey the orders given it by human beings except where such orders would conflict with the First Law.

3. A robot must protect its own existence as long as such protection does not conflict with the First or Second Law.

Yet there is a tension inherent in robot development—the very abilities that researchers most want to develop in robots are those that elicit fear in many observers. If robots become "smart" enough and have enough autonomy to make their own decisions, and if they adapt, evolve, and work in teams, critics ask, will it not be possible for them to act in ways that are beyond human control? In the end it seems we will need a secure fail-safe strategy that will allow us to keep control of our robots.

—Philip Downey

RELATED TOPICS

Artificial Intelligence, Deep-Sea Vehicles, Nanotechnology, Robotic Space Exploration, Surgical Robotics, Teleoperated Ocean Vehicles

BIBLIOGRAPHY AND FURTHER RESEARCH

BOOKS

Asimov, Isaac, and Karen A. Frenkel. *Robots: Machines in Man's Image*. New York: Harmony Books, 1985.

Malone, Robert. *The Robot Book*. New York: Jove Publications, 1978.

PERIODICALS

Anderson, Mary Rose. "Ecological Robots." *Technology Review*, January 1998, 22.

Blank, Jonah. "But Did Any of the Wine Survive?" *U.S. News & World Report*, July 5, 1999, 52.

Crawford, Mark. "DOE Plan Would Boost Robotic Applications." *Energy Daily*, October 8, 1998.

Crossman, Craig. "Robotic Dog Is the Cat's Meow." *Palm Beach Post*, September 4, 1999, 3E.

Graham-Rowe, Duncan. "Booting Up Baby." *New Scientist*, May 22, 1999, 42.

Hovertson, Paul. "Robot to Map No Man's Land in Chernobyl." *USA Today*, April 23, 1998, 10A.

Johnson, R. Colin. "Moving Beyond Analysis, Systems Visualize Their Internal World." *Electronic Engineering Times*, May 17, 1999.

Kunii, Irene M. "This Cute Little Pet Is a Robot." *Business Week*, May 24, 1999, 56.

INTERNET RESOURCES

Adaptive Systems Group at the Navy Center for Applied Research in Artificial Intelligence
http://www.aic.nrl.navy.mil/~schultz/research/

AIBO Home Page
http://www.world.sony.com/robot/top.html

Artificial Intelligence Laboratory at MIT
http://www.ai.mit.edu

Intelligent Robots: Do We Need Them and Can They Be Built?
http://www.ornl.gov/ORNLReview/rev26-1/net526.html

NASA Space Telerobotics Program Home Page
http://ranier.oact.hq.nasa.gov/telerobotics_page/telerobotics.shtm

RedZone Inc.
http://www.redzone.com

Robots Come to the Rescue of Chernobyl
http://www.llnl.gov/automation-robots/chern.html

Robot Wars
http://www.robotwars.com

Soviet Official Admits That Robots Couldn't Handle Chernobyl Cleanup
http://www.the-scientist.library.upenn.edu/yr1990/jan/anderson_p2_900120.html

ROBOTIC SPACE EXPLORATION

Robotic space exploration refers to the use of robotic space-craft to explore the solar system. In just four decades, increasingly sophisticated exploring machines have visited every planet, save Pluto, and have produced an explosion in scientific knowledge unmatched in human history.

But despite all the spectacular discoveries made by exploring machines, a subtle rivalry has appeared within the U.S. space program. This rivalry ultimately pits human astronaut against robotic exploring machine. Risk, cost, and net scientific return are the major points of contention in deciding whether humans or robotic spacecraft should take on exploratory missions.

Scientific and Technological Description

ROBOTICS involves the theory and practice of merging versatile machines and powerful computers to automate tasks normally performed by humans. A *robotic spacecraft* is essentially a computer-controlled machine that can travel and function in the hostile environment of space or on an unknown planet and perform its mission of scientific investigation successfully. Several major types of robotic spacecraft are used for exploration: the *flyby*, the *orbiter*, the *atmospheric probe*, the *lander*, and the *surface rover*. Aerospace engineers often combine robotic systems: for example, an orbiter and atmospheric probe or a lander and rover.

Robotic spacecraft come in all shapes and sizes; each is custom designed to satisfy the demanding objectives of a specific exploration mission. In addition to carrying a collection of specialized scientific sensors, a robotic spacecraft usually contains the following functional subsystems: structural, thermal control, data handling, spacecraft clock, data storage, telecommunications, navigation, attitude and articulation control, and power.

An orbiter spacecraft might also contain an onboard propulsion system that enables the robotic explorer to perform several energy-intensive orbital maneuvers during a planetary encounter. A lander spacecraft and rover spacecraft should possess some type of "soft" landing system. The lander and/or rover spacecraft might have a mechanical arm with which to probe, poke, or pick up soil and rock specimens. Each could also contain an onboard laboratory in which to conduct experiments on specimens collected during surface exploration.

A robotic spacecraft may function automatically (i.e., without any direct human control) or semiautomatically (i.e., under some type of limited human-in-the-loop supervision through a process called *teleoperation*). However, as the round-trip telecommunications distance between the spacecraft and Earth exceeds tens of light minutes (a *light minute* is the distance traveled by light in 1 minute; it takes

The Voyager Spacecraft

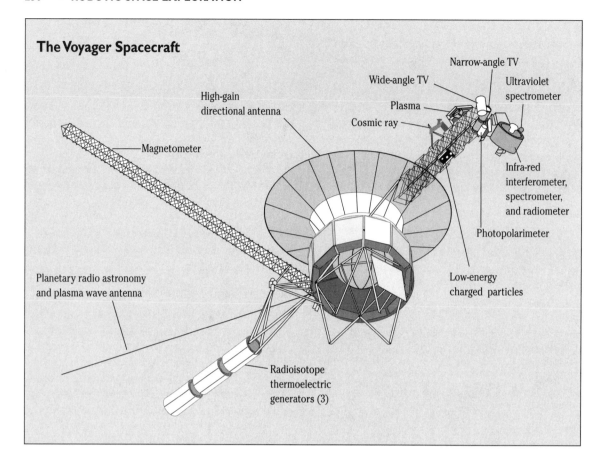

about 8 minutes for light from the Sun to reach Earth), any type of real-time human supervision becomes essentially impossible. Therefore, robotic explorers must be quite (machine) intelligent, reasonably independent in their actions, and even fault tolerant.

The flyby spacecraft follows a continuous trajectory past a target planet and is not captured into an orbit around the planet. This robotic explorer must have the ability to use its onboard sensors and scientific instruments to observe a rapidly passing target. It must also be capable of transmitting encounter data back to Earth at high rates of speed, as well as storing some data onboard. NASA uses this type of robotic explorer during an initial, or reconnaissance phase of exploration.

An orbiter spacecraft is designed to travel to a distant planet and go into orbit around that object. This type of robotic explorer must contain a significant propulsive capability to decelerate it at just the right moment so that it can achieve the desired orbit. Because an orbiter spacecraft experiences regular periods of shadowing (temporary obscuring of the Sun) by the planet, it must be designed for periodic interruptions in its communications with Earth, for wide variations in thermal conditions, and for maintaining its onboard electric power levels when direct sunlight is not available. NASA uses this type of spacecraft as part of its second, in-depth study phase of exploration.

Some exploration missions involve the use of atmospheric probes. These probes separate from the carrier spacecraft prior to closest approach to a target planet or moon in order study the target's gaseous atmosphere as the probes descend through it. After probe release, the carrier spacecraft usually performs a trajectory correction so that it does not follow the same path taken by the probe into the planet's atmosphere. Following this corrective maneuver, the carrier spacecraft can serve as either a flyby spacecraft or even become an orbiter spacecraft. In either event, it is available to record data from the atmospheric probe and then relay these data back to Earth. NASA uses atmospheric probes during in-depth exploration missions.

A lander robotic spacecraft is designed to reach the surface of a planetary body, make a soft landing, and then function at least long enough to acquire some useful scientific data and transmit these data back to Earth. A lander spacecraft is generally used to acquire panoramic imagery of the landing site, measure local environmental conditions, and make preliminary investigations of local soil composition.

A rover robotic spacecraft is carried to the surface of a planet or moon, soft-landed, and deployed. The rover then begins to operate and explore the surface, collecting technical data and acquiring imagery. The rover can function in an independent mode, semiautonomous mode, or fully controlled mode. How far the rover can wander from the original landing site is a function of many variables, including physical conditions at the site, rover design and power supply, available telecommunication links, and the mission's science objectives. Data collected by the rover can be transmitted back to Earth through one of several pathways (or "pipelines"): first, by means of the lander spacecraft; second, via an orbiting mothership; or third, directly from the rover vehicle itself. Sometimes robotic exploring teams are used in which a lander spacecraft delivers a rover or an orbiter an atmospheric probe.

Historical Development

In the early part of the 20th century, space visionaries boldly suggested the use of rockets to hurl humans and machines into interplanetary space. However, it was not until the development of powerful rocket vehicles at the end of World War II and during the Cold War missile race between the United States and what was then the Soviet Union that such space missions could be seriously considered. Parallel technology developments, also stimulated by the Cold War, helped accelerate the arrival of robotic spacecraft. The miniaturization of electronics, the availability of efficient portable power sources, the development of versatile digital computers, the rise of modern telecommunications, and the appearance of a great variety of useful remote sensing instruments all contributed to a technical infrastructure that enabled the rapid emergence of progressively more capable robotic spacecraft.

Another important stimulus also encouraged the development of versatile robotic spacecraft. It was a deeply political one, the "space race" between the United States and the Soviet Union. On October 4, 1957, the Soviet Union surprised the world when it successfully launched the first artificial satellite, *Sputnik 1*. A complacent American public went into nothing short of "technoshock."

Throughout the 1960s, any progress in space was linked directly to national prestige. This intensely competitive environment, highlighted by the race to explore the Moon with human beings, set the stage for the rapid development of some of NASA's most interesting initial robotic spacecraft. When *Apollo 11* astronaut Neil Armstrong became the first human to set foot on another world (July 20, 1969), the U.S. political objectives in the space race were essentially satisfied. Under growing fiscal pressures, government planners immediately began trimming the space program. Fortunately, the technical heritage of machine-based exploration quietly carried forward into the 1970s and 1980s and set the stage for some of the most spectacular episodes of discovery ever performed by any society.

Voyager 1 was launched from Cape Canaveral on September 5, 1977. This robotic explorer made its closest approach to Jupiter on March 5, 1979, and then used Jupiter's gravity to swing itself to Saturn. On November 12, 1980, it flew successfully through the Saturnian system and

was then hurled up out of the *ecliptic plane* (the plane of the Earth's orbit in its annual motion around the Sun) by Saturn's gravity on an interstellar trajectory.

The *Voyager 2* spacecraft lifted off from Cape Canaveral on August 20, 1977. Even though it left earlier than its companion, this spacecraft encountered the Jovian system on July 9, 1979 (closest approach) and then used the gravity assist technique to follow *Voyager 1* to Saturn. A gravity-assist maneuver results in a desired change in a spacecraft's direction and velocity achieved by a carefully calculated flyby through a planet's gravitational field. In effect, the planet's gravity "pulls" the spacecraft along and then "whips" it around in a new direction. On August 25, 1981, *Voyager 2* performed a flyby mission of Saturn and then went on (again using gravity assist) to encounter Uranus (January 24, 1986) and Neptune (August 25, 1989). Space scientists consider the end of *Voyager 2's* encounter of the Neptunian system as the end of a truly extraordinary epoch in planetary exploration.

While the nation mourned the loss of the Space Shuttle *Challenger* and its seven astronauts in January 1986, it also witnessed the incredible "extended" mission of the robotic spacecraft *Voyager 2* to Uranus (1986) and then Neptune (1989). Other robotic spacecraft, such as NASA's *Magellan* mission to Venus, performed detailed scientific investigations of planetary bodies. Our knowledge of the solar system increased exponentially despite setbacks in human space exploration.

At the close of the 20th century, an armada of robotic explorers (including *Mars Pathfinder, Mars Global Surveyor, Mars Surveyor 98,* and *Mars Surveyor 2001*) have been targeted on the mysterious Red Planet. Previous spacecraft explorations have indicated that Mars once possessed liquid water on its surface. Where did that water go? Did life ever exist on Mars? If so, is it still there, perhaps clinging in microscopic form in some fragile subsurface biological niche? A

The Sojourner robotic rover landed on Mars on July 4, 1997, as part of NASA's Mars Pathfinder *mission. The Sojourner roamed across the planet's surface, collecting an array of scientific rock and soil samples. In September 1997, the robot stopped communicating with Earth because of a power failure (Courtesy of NASA/Jet Propulsion Laboratories).*

prime mission for this new wave of Martian robotic explorers is to help resolve this complex question. Other sophisticated robotic spacecraft, such as NASA's *Galileo* (Jupiter) and *Cassini* (Saturn), are performing equally important, detailed scientific investigations of the giant planets.

Uses, Effects, and Limitations

Within the past four decades, through the creative use of robotic spacecraft, we have visited all the planets known to the ancients (plus two, Uranus and Neptune, that were unknown until fairly recent times). Only distant Pluto has yet to be visited by a robotic explorer, and NASA's *Pluto-Kuiper Express* mission is slated to remedy that situation in the next decade. These worlds have now become familiar celestial objects that we can see and touch—if not yet in person, at least through the "electronic sight" and other senses provided us by sophisticated robotic spacecraft. Such missions of scientific discovery represent some of the most remarkable human achievements of all time.

The successful use of robotic spacecraft to explore the solar system is perhaps best illustrated using NASA's *Voyager* spacecraft and their Grand Tour mission. These robotic exploring machines are actually a gangly pair of instrument-laden machines that represent the technical heritage of all the NASA space robots that flew before them (see figure). Over the course of two decades, these robotic explorers have drawn back the curtain of mystery on nearly one-half of the solar system. One by one, they brought the faces of the four giant outer planets into sharp focus in a series of highly successful flyby missions. They also revealed an interesting assortment of companion moons and important new facts about each system. Both spacecraft are now headed (in different directions) toward the outer boundary of our solar system. They are destined to wander through the Milky Way, robotic emissaries from Earth that each carry a recorded message from their human creators—a message aptly entitled "The Sounds of Earth."

Issues and Debate

One debate involving the use of robotic spacecraft centers around the basic question: Should humans or machines be used to explore other worlds? The ideal response is that both would be used in partnership, as was done during the Apollo Project when robotic spacecraft served as precursors to the human lunar landing missions. There is no real debate about using robotic spacecraft to explore deep space beyond the main asteroid belt. However, when it comes to the detailed exploration of Mars in the 21st century, the aerospace community is generally split. One faction wants to fully explore the Red Planet—including the search for microscopic Martian life forms (current or extinct)—with an increasingly sophisticated family of robotic explorers, eventually bringing rock and soil specimens back to Earth for detailed analysis. The other faction accepts the role of robotic explorers in pre-cursor missions but emphatically demands the use of astronauts. In fact, this group suggests that Mars cannot be truly explored until human beings set foot on its surface.

The major point of contention is overall cost, since a human expedition to Mars will cost between $40 billion and $80 billion. NASA must answer the following question: Is machine exploration sufficient to inspire the human spirit, or must a human explorer be personally at risk and directly involved in the mission to create the same level of societal support? Actually, human beings are always involved in robotic space exploration, since they construct the robotic systems, operate the systems, and receive the data from the systems. At the center of this important debate is economics and where the human explorer should be located.

—*Joseph A. Angelo, Jr.*

RELATED TOPICS
International Space Station, Radioisotope Thermoelectric Generators, Reusable Launch Vehicles, Robotics, Space Shuttle

BIBLIOGRAPHY AND FURTHER RESEARCH

BOOKS
Angelo, Joseph A., Jr. *The Dictionary of Space Technology,* 2nd ed. New York: Facts On File, 1999.

Beatty, J. Kelly, and Andrew Chaikin, eds. *The New Solar System,* 3rd ed. Cambridge, Mass.: Sky Publishing and Cambridge University Press, 1990.

Fimmel, Richard O., Lawrence Colin, and Eric Burgess. *Pioneer Venus.* NASA SP-461. Washington, D.C.: U.S. Government Printing Office, 1983.

French, Bevan M., and Stephen P. Maran. *A Meeting with the Universe.* NASA EP-177. Washington, D.C.: U.S. Government Printing Office, 1981.

Gulkis, S., D. S. Stetson, and E. R. Stofan. *Mission to the Solar System: Exploration and Discovery.* NASA/JPL Publication 97-12. Pasadena Calif.: Jet Propulsion Laboratory, 1997.

NASA. *Origins: Roadmap for the Office of Space Science Origins Theme.* NASA/JPL Publication 400-700. Pasadena Calif.: Jet Propulsion Laboratory, 1997.

NASA. *Viking: The Exploration of Mars.* NASA EP-208. Washington, D.C.: U.S. Government Printing Office, 1984.

NASA. *Voyager: The Grandest Tour.* NASA/JPL Publication JPL 400-445. Pasadena Calif.: Jet Propulsion Laboratory, 1991.

NASA. *Voyages to Saturn.* NASA SP-451. Washington, D.C.: U.S. Government Printing Office, 1982.

PERIODICALS
"The Future of Space Exploration." *Scientific American,* Spring 1999.

INTERNET RESOURCES
NASA Homepage
    http://www.nasa.gov/
NASA Space Link
    http://spacelink.nasa.gov/
Mission to the Solar System Roadmap
    http://sse.jpl.nasa.gov/roadmap/
Planetary Sciences at the National Space Science Data Center
    http://nssdc.gsfc.nasa.gov/planetary/planetary_home.html
The Jet Propulsion Laboratory: Planets
    http://www.jpl.nasa.gov/planets/
Cassini: Mission to Saturn
    http://www.jpl.nasa.gov/cassini/rtg/

SATELLITE TECHNOLOGY

Any object placed in orbit around another one is an artificial satellite, but the term is usually restricted to uncrewed spacecraft that orbit Earth. Satellites can be used to study the Earth, explore space, and even assist navigation.

In the four decades since the first satellites were launched, they have had a revolutionary effect on many aspects of modern life, giving us a fuller understanding of the way our planet works, revolutionizing international politics, and turning the Earth into a global village with intercontinental communications.

Scientific and Technological Description

A satellite's purpose is to support a cargo (the payload) in orbit. It can be seen as a number of subsystems that provide a suitable environment and allow the payload to operate correctly. Satellites are carried into orbit by a rocket-powered launch vehicle, usually composed of several rocket stages, each with its own engines and fuel tanks. The stages fire in a preset sequence during launch, then jettison and fall away to reduce the vehicle's weight as it gains altitude and speed. By the time the last engine cuts out, the satellite is typically at an altitude of about 125 miles (200 kilometers [km]) and traveling so fast that the pull of Earth's gravity is exactly balanced by the satellite's tendency to shoot off into space on a straight path—it has achieved a *low earth orbit* (LEO).

Because the forces acting on it are exactly balanced, the satellite behaves as if it were weightless. An *orbit* is in fact any path where the satellite's speed and gravity are balanced in this way, and is usually elliptical (a circular orbit is just a specific kind of ellipse). LEOs above the equator are a very common orbit for satellite operations, because launch vehicles can be given an additional boost by the speed of the Earth's rotation. The SPACE SHUTTLE, crewed space stations (see INTERNATIONAL SPACE STATION), and many scientific satellites all operate in this kind of orbit. Other satellites that study the Earth itself are often placed in a highly inclined LEO called a *polar orbit,* which is tilted so that it passes over or close to the poles. As the Earth rotates, the satellite can eventually pass over and study any area on the planet's surface.

Many other satellites operate at higher altitudes. Weather satellites may be positioned farther from Earth, where they can view an entire hemisphere, and communications satellites often occupy the much higher geostationary orbit (22,300 miles [35,900 km] above the equator) (see figure). To reach these higher orbits, a satellite can be equipped with its own rocket engine, called an *inertial upper stage.* Careful firing of this rocket can boost the satellite into a *transfer orbit,* an elongated path between two more circular orbits at different altitudes.

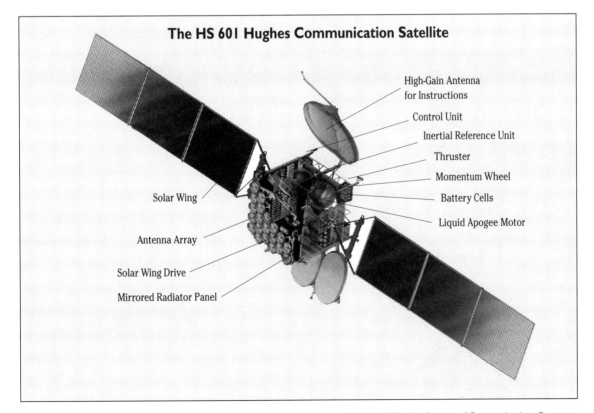

The HS 601 Hughes Communication Satellite

High-Gain Antenna for Instructions
Control Unit
Inertial Reference Unit
Thruster
Momentum Wheel
Battery Cells
Liquid Apogee Motor

Solar Wing
Antenna Array
Solar Wing Drive
Mirrored Radiator Panel

Courtesy of Hughes Space and Communications Company.

A satellite has five chief subsystems in addition to its payload: structure, power, communications, thermal control, and attitude sensing and control.

Structure and Power

To reduce their weight and launch cost, satellites are usually designed and constructed utilizing simple structures based on the shape of boxes or drums, simple solids that can be built up from a framework of just a few supports. These supports can be made in lightweight alloys of metals such as aluminum and titanium. Much of the satellite's strength is provided by a thin, lightweight casing around the framework, often a sandwich design with a rigid honeycomb of interlocking hexagonal cells between two layers of carbon fiber or another composite material.

Satellite power systems frequently use chemical batteries during launch and deployment before switching to solar power once the launch vehicle casing falls away. Depending on the payload's power requirements, solar panels may be mounted around the satellite or on large expanding arrays that unfold in orbit. A very few satellites rely on nuclear power systems such as RADIOISOTOPE THERMOELECTRIC GENERATORS, although these are generally confined to space probes venturing far from the Sun.

Communications and Controls

Communications systems relay data and instructions between the satellite and its ground station by radio, while thermal control systems regulate the satellite's temperature using reflective insulation or heat conductors to transfer heat from the searing sunlit side of the satellite to the side in the freezing shade. Many satellites use light sensors to find their orientation in space. Sun sensors are relatively inexpensive but crude; star sensors are more accurate but expensive. Some satellites have an *inertial guidance system,* a series of detectors that monitors the forces acting on the satellite in different directions from the moment of launch. Attitude corrections (corrections to the satellite's position or orientation) are often made using rockets, either small gas thrusters that release pressurized gas, or ion engines that use a magnetic field to push out an electrically charged gas. Because rocket engines will eventually use up propellant, satellites with longer life spans are fitted with *reaction wheels,* weighted disks spun by electric motors. According to *Newton's third law of motion,* also known as the *law of action and reaction,* as the disks spin one way the satellite reacts by spinning slowly in the opposite direction.

Historical Development

Like virtually all space technology, satellite technology arose from the early days of the Cold War between the United States and the Soviet Union, but its theoretical background dates back to the time of British scientist Isaac Newton. Newton, who discovered the laws of gravity (published in 1687), first recognized the possibility of an artificial satellite. The idea was adopted by several science fiction writers from the 19th century onward, but did not become a practical proposition until the first liquid-fueled rocket flights of the 1920s and 1930s. Wernher von Braun (1912–1977) and the team that built the German V2 (a rocket-powered missile that paved the way for most of today's rocket technology) in the 1940s also proposed a larger rocket capable of launching a satellite into orbit. The U.S. Defense Department studied several options for a satellite program as early as 1948 but failed to adopt any of them.

The development of nuclear weapons at the end of World War II led to a race for a new weapons-delivery system, one that was capable of striking at targets halfway around the world. Both the United States and the USSR used German rocket technology and scientific experts captured after World War II to develop the *intercontinental ballistic missiles* (ICBMs). As it happened, many of the scientists involved in the military programs were also advocates of space exploration. It soon became clear that rockets powerful enough to propel weapons long distances could very likely launch satellites into orbit.

In 1954, both the United States and the USSR announced intentions to place a satellite in orbit for the forthcoming International Geophysical Year 1958. The possible applications for satellites were becoming apparent, but above all a satellite launch would be a powerful demonstration of rocket (and therefore missile) technology and capability. The USSR stole an early lead by developing launch vehicles directly from its ICBMs; this enabled the Soviets to launch *Sputnik 1,* the first artificial satellite, on October 4, 1957. Meanwhile, the United States attempted to develop a civilian launch vehicle from small scientific rockets rather than relying on military technology. After the launch of *Sputnik,* they too fell back on ICBM technology, which allowed them to launch the *Explorer 1* satellite in January 1958.

Early satellites were very simple: *Sputnik* was an 83.6-kilogram (kg) (184-pound [lb]) aluminum ball with antennas sticking out of it and some basic instrumentation to measure inside temperature and pressure. *Explorer* carried far more instrumentation, including a radiation detector, but weighed only 14 kg (31 lb.), most of which was accounted for by the upper rocket stage that housed the satellite.

As the space race accelerated, satellite payloads became more important than the satellites themselves. Early satellite applications such as communications were developed for military or propaganda purposes. The first broadcast from space, in late 1958, was a Christmas greeting from then-U.S. president Eisenhower, intended as a clear symbol that America was keeping up with the USSR. Spy satellites were another obvious goal for a space program driven by the military.

From the early 1960s onward, the satellite program was revolutionized by commercial pressures. *Telstar 1* (launched in 1962) was the first satellite used for live television relays, and also the first privately funded satellite. Although scientific and military satellites remained the province of government agencies, communications satellites became an international business. In 1964, Intelsat, the International Telecommunications Satellite Organization, was founded to promote the development of internationally owned communications satellites, and it is now a cooperative owned by over 100 countries. Individual countries have launched their own communications satellite networks, as have large telecommunications companies and broadcasting organizations. Intelsat, meanwhile, has served as a model for other organizations, such as Inmarsat (the International Maritime Satellite Organization), which launches maritime navigational and communications satellites.

Uses, Effects, and Limitations

As can be seen from the history of their development, satellites have a vast range of different applications. Several of these are covered in detail in other articles (see GLOBAL POSITIONING SYSTEM, REMOTE SENSING, and SOLAR POWER SATELLITES). Other uses include communications (see PCS PHONES), space science, and intelligence. One of the first satellite applications, and one still in use, is intelligence gathering. Early spy satellites used photographic film that had to be returned to Earth in a heat-shielded capsule. This technique was soon replaced with satellite cameras that sent images back to Earth in electronic form. Since spy satellite technology is covert, it is difficult to say how rapidly it has advanced, but modern satellites are probably capable of distinguishing details down to about 8 inches (20 centimeters) across. However, images are only one form of intelligence. An equally important role for spy satellites is listening in to radio communications.

Explorer 1 was a space science satellite gathering information about the environment near the Earth, but astronomers soon realized that satellites beyond the atmosphere were a unique platform for looking farther into the universe. Specially designed orbiting telescopes can study radiation with wavelengths longer and shorter than those of visible light, such as infrared radiation and ultraviolet radiation, that are usually blocked by Earth's atmosphere. The information gathered by such telescopes has revolutionized astronomy, but the most spectacular success of all has been the Hubble Space Telescope, an orbiting optical observatory.

If one use of satellite technology has changed our everyday lives, however, it is its application in communications. The principle of the geostationary communications satellite was outlined in 1945 by science fiction writer Arthur C. Clarke, who pointed out that a satellite nearly 22,500 miles (36,000 kilometers [km]) above the Earth's equator would orbit in exactly 1 day, remaining above the same point on the Earth. This would enable it to act as a relay station for radio signals (although Clarke thought the station would have to

be manned). Current telecommunications satellites indeed utilize this principle but are in fact packed with microcircuitry that allows them to perform all the tasks of an Earthbound telephone exchange automatically.

Modern communications satellites are capable of handling tens of thousands of telephone calls simultaneously, receiving them from one ground station, amplifying them, and sending them back to another ground station. However, geostationary satellites have one major limitation—they appear low on the horizon from high latitudes, so their signals have to pass through thick layers of atmospheric interference. To overcome this problem, the Soviet Union developed the *Molniya* satellites from the 1960s onward. These have highly elliptical, inclined orbits that carry them high into the sky over northern regions and move slowly across the sky. A series of these satellites is needed to ensure full communications coverage at all times.

Issues and Debate

Few would deny that the effects of satellite technology have been overwhelmingly positive. Satellite communications systems such as pagers, global positioning systems, and satellite phones have brought the world closer together, making business communications more efficient, keeping families together across continents, and bringing news to our television screens from all parts of the world as it happens. Intelligence satellites, meanwhile, have made it more difficult for aggressive nations to launch sudden attacks on their neighbors

But these benefits have sometimes come at the cost of individual privacy. Satellite intelligence is only as good as our ability to search through the huge amounts of data a satellite collects, so satellites tend to monitor specific areas of military interest. However, the planned launch of commercial spy satellites that will provide images to anyone willing to pay for them has stirred significant controversy. Countries without their own spy satellite programs are expected to be major purchasers of commercial reconnaissance images, and as yet there are no regulations to prevent images falling into the hands of aggressive governments or individuals. On the other hand, commercial spy images will enable some people to monitor their own governments more effectively, and will end the duopoly on satellite imaging shared by Russia and the United States.

Satellite technology also poses some physical dangers, arising primarily from the sheer amount of material being launched into orbit. Tens of thousands of satellites have been launched since 1957, and in addition to the satellites themselves, their launch vehicles have carried huge amounts of debris into space, ranging in size from discarded rocket stages to flecks of paint. Most of these fast-moving objects are still in low Earth orbit (about 125 miles [200 km] above the Earth), so that space relatively close to Earth is beginning to get crowded. At least one satellite is thought to have been destroyed by a collision with debris, and an early Space Shuttle flight returned to Earth with a half-inch crater in one reinforced window due to a head-on collision with a paint fleck.

Although low Earth orbit is beyond Earth's atmosphere, tenuous traces of atmospheric gases still linger there, and the gradually braking effect from collisions with gas particles causes many objects to fall back to Earth each year. Most of these small objects burn up during their fall due to friction from the atmosphere, which assists in a gradual sweep-out of space near Earth. Objects that are large and robust enough to survive reentry to the atmosphere are a further cause for concern, however. The chances of a satellite actually hitting a populated area are small but not negligible, and it can be very difficult to predict a satellite's reentry path in advance. In addition, some satellites can carry dangerous payloads, such as the radioactive material scattered across Canada by the orbiting Soviet nuclear satellite Cosmos 954 in 1978.

Satellite technology is still prohibitively expensive for many organizations—launch costs are typically $10,000 per pound of cargo ($22,000 per kilogram). Different groups are working to lower these costs, and in the coming decades, satellites may become even more ubiquitous. REUSABLE LAUNCH VEHICLES are planned that could cut launch costs by 90 percent, and microsatellites, which pack all the technology of a large satellite into a very small, lightweight vehicle, will make space even more accessible. Satellites will find many new applications in the future, some of which we cannot even imagine today.

—*Giles Sparrow*

RELATED TOPICS

Global Positioning System, Multispectral Imaging, Radioisotope Thermoelectric Generators, Remote Sensing, Reusable Launch Vehicles, Robotic Space Exploration, Solar Power Satellites, Space-Based Remote Sensing Tools, Space Shuttle

BIBLIOGRAPHY AND FURTHER RESEARCH

BOOKS

Davis, J. K. *Space Exploration* (Chambers' Compact Reference). New York: W&R Chambers, 1992.

Gatland, Kenneth, ed. *The Illustrated Encyclopedia of Space Technology,* 2nd ed. New York: Orion Books, 1989.

Gavaghan, Helen. *Something New Under the Sun: Satellites and the Beginning of the Space Age.* New York: Copernicus Books, 1998.

Luther, Arch C., and Andrew Inglis. *Satellite Technology: An Introduction,* 2nd ed. New York: Focal Press, 1997.

PERIODICALS

"The Future of Space Exploration." *Scientific American,* Special Issue, March 1999.

INTERNET RESOURCES

NASA History Office
   http://www.hq.nasa.gov/office/pao/History/history.html

NASA Homepage
   http://www.nasa.gov

Small Satellite Home Page
   http://www.ee.surrey.ac.uk/EE/CSER/UOSAT/SSHP/

Space Transportation Association
   http://www.spacetransportation.org/

SCANNERS

A scanner is an array of photosensitive silicon cells that takes a picture of a page and measures the light reflected off documents. From this information, a scanner renders the image it captures into digital form for editing and display on a computer. Scanners are available in various forms and sizes, depending on their uses, which range from scanning simple photographs and text documents to high-end graphics work. Although scanners were originally tools for graphic artists, with price tags customarily exceeding $1000, many less-sophisticated computer users now purchase scanners for as little as $100 for use at home.

The proliferation of scanners has been accompanied by the periodic abuse of these devices; people have used them to violate copyright laws, especially for commercial reuse. Specific copyright and other laws governing scanning prohibit the infringement of traditional copyright law on items such as photographs and prohibit the use of scanners for creating counterfeit money.

Scientific and Technological Description

A scanner's purpose is to capture documents and images and render them in digital form for computer editing and display. The most common form of scanner is a flatbed scanner, which is a long, flat unit that resembles a small photocopier. Users place photos and documents flat on the glass bed of the scanner. An array of photosensitive silicon cells shines on the document and measures the light reflected off the document. This light source is normally embedded in a carriage that is pulled along the document by a drive belt.

The light collected by the array is reflected through a series of mirrors and then passes through a lens, which reduces the image and focuses it onto a charge-coupled device (CCD). The CCD measures the light intensity and converts the information into voltage. If it is a color scanner, the CCD measures the intensity of each of three colors—red, green and blue (RGB)—before converting this information into voltage. An analog-to-digital converter will then convert the voltage's analog signals into digital signals. From the digital information, the computer attached to the scanner can recreate the image on the screen.

Scanners range in technical capability, which is measured in two ways: brightness and resolution quality. First, the extent of brightness that a scanner can capture is measured in bit depth, which depends on the sensitivity of the scanner's sensor and the capability of its analog-to-digital converter. The bit depth refers to the number of levels of color a scanner can detect. Greater bit depth will offer better image reproduction. The minimum bit depth needed to produce a decent-quality scan is 24 bits. Even this produces a color range that is inferior to that seen in color photographs. For slides, negatives, or transparencies, the minimum is 30 bits.

Scanners also range in two types of resolution quality: optical resolution and interpolated resolution. *Optical reso-*

lution refers to the amount of information the scanner's CCD can capture. *Interpolated resolution* is the maximum resolution the scanner can achieve by using optical enhancements and algorithms to fill in between pixels (the dots that make up the image). The optical resolution determines how much real information the scanner can capture. A good minimum resolution measure is 300 dots per inch (dpi). This usually proves sufficient for documents, and more than sufficient for images used for the World Wide Web. For slides and negatives, it is best to have a scanner with an optical resolution of 1200 dpi or higher, since slides and negatives are smaller in original size.

Historical Development

At the core of scanner technology are photosensitive silicone cells. These measure light from images and emit electronic currents that are then translated into digital information that a personal computer (PC) can use to render an image. Photosensitive cells are derived from technology introduced in the phototube, which contained two electrodes enclosed in a glass tube: an anode and a light-sensitive cathode. A battery or other voltage source was connected to the electrodes to release current in an amount proportional to the intensity of the light. Scientists have developed many types of applications for photocells, including use as electric eyes that sense light and operate as light-actuated counters, automatic door openers, intrusion alarms, and digital scanners.

The first digital scanners became available in the mid-1980s. Originally, they were used by graphic designers who needed a way to store and edit images, including photographs and art. The first scanners resembled photocopying machines—with a long, flat glass bed in which designers would place the documents and images to be scanned. However, these units didn't produce photocopies—they produced digital images of the documents. The original scanners debuted for approximately $1500, about ten times the price of a flatbed scanner today.

Today's scanners, produced by companies such as Hewlett-Packard, Visioneer, Canon and Epson, are also about half the size and weight of original models. Scanners are no longer solely the tools of graphic artists. Many scanners are now purchased by home consumers, who use them to store family photos and documents. In the late 1990s, manufacturers have begun to incorporate scanning capabilities into multi-function devices that also print, fax, and copy documents. Most notable are the new flatbed systems, which can accommodate document-feeding systems that allow stacks of documents to be scanned automatically rather than having to be placed on the scan bed one by one.

Uses, Effects, and Limitations

Scanner Types

Flatbed scanners are the most commonly used. Although they are the largest and most cumbersome, they are also the most convenient for scanning a large number of documents or for handling fragile documents or photographs. Professional graphic scanners, including drum scanners, are more expensive than basic flatbed scanners because they provide high-quality optics and sensors, which creates a better image. In addition, professional graphic scanners are available in a wider

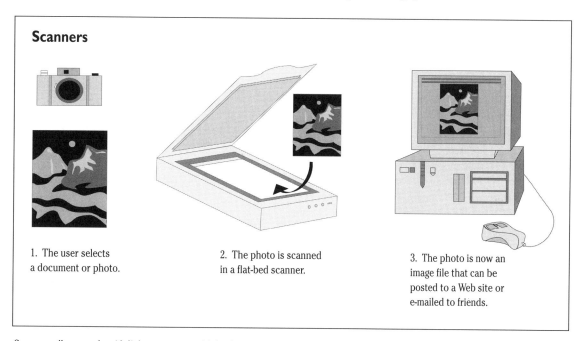

Scanners

1. The user selects a document or photo.

2. The photo is scanned in a flat-bed scanner.

3. The photo is now an image file that can be posted to a Web site or e-mailed to friends.

Scanners allow people with little computer sophistication to convert photographs into computer images that they can then manipulate and use in many ways.

range of shapes and sizes to accommodate very large or small originals. These types of scanners also tend to include higher-quality software, including color-calibration tools.

Sheetfed scanners are smaller and more compact than flatbed scanners. With sheetfed scanners, the scanning array remains still while the document travels through the scanner. Although they are more compact than flatbed scanners, sheetfed scanners are not a good option for small or fragile documents or photographs, since the feeder could crumple them. In addition, sheetfed scanners generally produce poorer image quality. Sheetfed scanners are the most common type of scanner found in multifunction devices, which operate as printers, fax machines, copiers, and scanners rolled into one machine. Some high-end multifunction devices currently incorporate flatbed scanners with fax/copier/printer capabilities, but the majority still use sheetfed scanning technology.

Hand-held scanners are even smaller still. They are best suited for scanning small documents, such as business cards, although they are a convenient option for scanning any document. Their size can be a detriment if documents are large or lengthy, since the scanner must be dragged across the documents.

Specialty Uses

For specialty uses, scanners have peripheral devices that enhance the scanner's functionality. One such device is a transparency adapter, which is useful for scanning negatives, slides, and transparencies. Normal flatbed scanners reflect light off a document to capture its image. With transparencies, however, the light passes through, so a transparency adapter must backlight the original.

Some scanners are also compatible with an automatic document feeder. These units allow documents to be fed into the scanner in large quantities rather than being removed and placed in the scanner individually. In addition, many scanners come equipped with optical-character-recognition software (OCR), which can translate digital images of text documents into digital text files that can be edited in a word processing program. Although these applications are not 100 percent accurate, many do a good job of translating most text accurately and retaining document formatting such as borders, headlines, and special text boxes.

Scanner Speeds

Many low-cost scanners use a parallel-port connection (a very basic connection that uses a cable to hook up to most, if not all, PCs), with a pass-through connection for a printer. This means that the scanner also has a port that allows it to be hooked up to a printer, so that with just one parallel port on the back of the PC, the two peripheral devices can be connected. The advantage to this setup is that it is inexpensive, and most PCs have parallel ports, so there are no problems with incompatible hardware. However, parallel-port scanners are generally slower than scanners with universal serial bus (USB) ports in sending scanned images to the PC. Parallel-port connections are steadily being replaced by USB ports, which offer better speed and reliability.

Mid- to high-end scanners—those with more than 600 dpi and a color depth of 36 bits or more—are likely to have a small computer system interface (SCSI) connection. SCSI connectors can handle high-bandwidth traffic between scanner and PC, resulting in faster transmissions.

Issues and Debate

Although scanners are fairly straightforward devices, a few people have used them to violate copyright and usage laws by illegally copying and selling or disseminating images and literature that have been published under copyright law. For instance, in 1997 a photographer named Francesco Sanfilippo was sued for violating *Playboy* magazine's copyright license, which protects the images in the magazine. Sanfilippo had purportedly scanned in several images from *Playboy* and was selling them on CD-ROM. He advertised his product on his company's Web site. In April 1998, a judge awarded *Playboy* $3.74 million in damages for copyright infringement, the largest Internet-related damage award to date.

Other scanner-related copyright laws relate to scanning money and other sensitive documents, such as birth and death certificates or social security cards. Those who scan such documents risk seizure of their computer equipment, up to $25,000 in fines, or up to 15 years in prison. However, analysts say that it would take millions of dollars of equipment and technology to be able to create extremely realistic counterfeit money through scanners, which would not be worth the investment.

Not all societal issues relating to scanners are negative, however. The ability to "take a picture" of a photograph or document, which can then be manipulated at will, has enabled many relatively inexperienced people to do a range of new things with their computers. With scanners, proud parents can easily trade baby pictures over e-mail, for example. Similarly, people wishing to sell antiques or other collectibles over the Internet may scan and place photographs of the items on a personal site or on an auction site such as Ebay. As scanner technology evolves and improves in the future, there will no doubt be both an upside and a downside for society.

—Christine Doane

RELATED TOPICS
Digital Libraries, Electronic Books, Language Recognition Software

BIBLIOGRAPHY AND FURTHER RESEARCH

PERIODICALS
"It's Done with Light and Mirrors." *New York Times,* March 11, 1999.
Jantz, Richard. "Fast, Easy Scanners from HP, Umax, Visioneer." *PC World Online,* 1999.
Napoli, Lisa. "Judge Awards Playboy $3.74 Million in Copyright Case." *New York Times,* April 3, 1998.

INTERNET RESOURCES
Copyright and Other Laws Governing Scanning
    www.infomedia.net/scan/
Definitions of Technology Terms: Scanner
    http://www.whatis.com/scanner.htm

SCINTILLATION TECHNIQUES

Scintillation techniques are specialties within nuclear medicine (the use of radioactive substances in medical diagnosis, monitoring, and treatment). Scintillation techniques such as single-photon emission computed tomography (SPECT) and positron emission tomography (PET) utilize radioactively labeled substances that are ingested, inhaled, or injected. Radiation detectors trace the fate of these substances within the body and visualize their location as an image on a monitor screen. Although SPECT and PET show less detail than other body imaging techniques, their great diagnostic strength lies in providing information about tissue function rather than structure.

Scientific and Technological Description

Scintillation techniques are body imaging methods that employ *radionuclides* (atoms that are unstable and undergo radioactive decay). A natural radionuclide is used, or one is made by placing a pure sample of the stable element inside a nuclear reactor core or in the path of positively charged atomic particles accelerated by a cyclotron (an atomic particle accelerator). The element's atoms are destabilized and form radionuclides that will eventually return to the element's stable state. As they do so they emit nuclear radiation (alpha particles, beta particles, or gamma rays). The radionuclide has the same chemical properties as those of the normal element. It can be incorporated in other chemical substances by conventional chemical reactions, and once incorporated, it acts as a radioactive *label*. The radioactively labeled substance is metabolized by the body in exactly the same manner as the unlabeled substance, but radiation-detecting devices can trace its fate within the body.

For body imaging, the radionuclides selected are those that yield gamma radiation, since this type of radia-

tion can be detected outside the body and its path backtracked to its source. Radionuclides are chosen that have a short *half-life* (the time taken for the radioactive decay of half of the atoms in a pure sample). Choosing to use short-half-life radionuclides reduces the radiation dosage a patient receives. Typically, the radioactively labeled substance is injected into the patient's blood or is inhaled or swallowed by the patient. Inside the body it accumulates at sites where it is stored or is involved in metabolic activity. Outside the body, either a gamma camera or an array of scintillation detectors discloses the radiation emitted by the radionuclide. A computer analysis "backtracks" the source of radiation to a precise location in the body, revealing hotspots of metabolic activity. The radionuclide decays fairly rapidly, and the patient remains faintly radioactive for only a few hours after treatment.

By selecting an appropriate radioactively labeled substance, the nuclear medicine specialist can test a hypothesis as to the cause of a patient's observed signs and symptoms. For example, radioactive iodine (^{123}I) is used to investigate thyroid gland function, since iodine accumulates in this gland and is incorporated in the thyroid gland hormone, thyroxin. Technetium (^{99}Tc), a radioactive element, can be attached to substances such as phosphates that are incorporated in bone. Glucose labeled with radioactive fluorine (^{18}F) is commonly used to visualize sites of high metabolic activity in the brain.

Scintillation detectors disclose photons of gamma radiation (discrete amounts of electromagnetic energy) as they emerge from the body. When radiation passes through

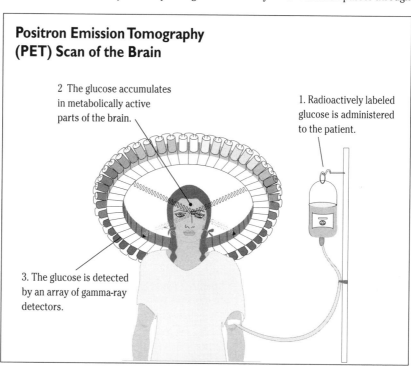

Positron Emission Tomography (PET) Scan of the Brain

2 The glucose accumulates in metabolically active parts of the brain.

1. Radioactively labeled glucose is administered to the patient.

3. The glucose is detected by an array of gamma-ray detectors.

Adapted from Durant et al., Encyclopedia of Science in Action, *Macmillan Publishers Ltd., 1995.*

a detector, it interacts with a light-generating substance, most commonly sodium iodide. The resulting flash of light from the striking gamma ray is directed into a photomultiplier, which converts the light flash into a stream of electrons. The electron stream is amplified so that it will register as a detectable current and can be analyzed by computer. A scintillation detector is typically a cylindrical crystal with a diameter up to about 7.5 centimeters (cm) in PET machines (where there are many) or up to 39 cm across in SPECT machines. SPECT machines normally have the appearance of a single large circular lens assembly mounted on an arm (see figure).

The point of origin of the gamma ray is displayed as a dot on a monitor screen. The dot is then photographed along with many other dots representing other points of origin. Alternatively, the data from photomultipliers are collated as digital information by a computer and are stored and displayed. Computer-generated color is often used to distinguish regions of differing metabolic activity.

SPECT Scans

Single-photon tomography uses radionuclides, such as technetium (^{99}Tc) or iodine (^{123}I), that emit gamma-ray photons one at a time. These are detected by a gamma camera. Commonly, the camera is moved in an arc around the body region under investigation. The camera incorporates a very large sodium iodide crystal that captures gamma rays over a large area. The position of entry of a gamma ray is computed from the output of many photomultipliers arrayed above the sodium iodide crystal. The series of images from various gamma-camera positions is processed and analyzed by computer. Usually, these data are used to computer-generate a series of cross-sectional slices through the patient's body, in which case the technique is called *single-photon emission computed tomography* (SPECT).

PET Scans

Positron emission tomography (PET) utilizes radionuclides that produce two positrons at a time. Such positron-emitting forms include radioactive isotopes (alternative forms of an atom) of carbon (^{11}C), nitrogen (^{13}N), fluorine (^{18}F), and oxygen (^{15}O). Commonly in PET scanning, one of these isotopes is incorporated into a metabolically active substance that is injected into the bloodstream. Inside the body, the nucleotide label releases positrons that immediately encounter electrons. Positrons and electrons are effectively the same forms of subatomic particles, but they carry opposite charges. A positron is the antimatter equivalent of an electron. When they meet, they annihilate each other, giving rise to two gamma-ray photons that move in exactly opposite directions. This "double emission" is detected by a ring of scintillators, and after computer analysis, these data can trace the rays back to their point of origin.

SPECT or PET scans are typically used to generate two- or three-dimensional false-color representations that show regions of differing metabolic activity for the process under investigation. For example, they can indicate the oxygen consumption of different tissues. Typically, oxygen consumption will be lower than normal in damaged or dying areas and higher than normal where there is uncontrolled cell growth characteristic of cancers. SPECT offers poorer image resolution than PET, but because it utilizes readily available radionuclides, it is much more affordable than PET, which relies on the use of costly radionuclides. Until recently, PET has been restricted largely to use in brain function research, but in some medical facilities PET has been adopted as an almost routine diagnostic procedure for certain kinds of brain abnormality and for imaging heart tissue.

Historical Development

Scintillation technology in medicine is based on the use of radioactively labeled substances that act as *tracers* within the body. The notion of using a radioactive substance as a label was first devised, and the technique first demonstrated, by Hungarian György Hevesy in 1913. Using the tracer principle, Hevesy was able to show that living organisms, including humans, have a characteristic turnover of chemical elements and molecules. Such studies were later to show that even seemingly permanent tissues, such as bone, exist in a dynamic, ever-changing state. In 1943, Hevesy was awarded a Nobel prize for his work on radionuclides.

Following the development of the nuclear reactor during World War II, radioactive isotopes of most elements became available by the late 1940s. The first generation of scintillation detectors for body imaging were rectilinear scanners, developed initially in 1950. These were insensitive and slow. They worked by tracking the scanning assembly back and forth across the body in a series of adjacent lines, gradually building up a picture of the distribution of the radionuclide within the body. Rectilinear scanners created a two-dimensional image usually based on a horizontal scan of a prone patient. Because scanning times were long, the distribution of nucleotide material in the body often changed markedly in the time it took for the scan to be performed.

The gamma camera, first developed by Hal Anger in the United States in 1957, represented a considerable advance on the rectilinear scanner in terms of speed and sensitivity. Using a single large crystal, the gamma camera captured simultaneous gamma ray emissions over a much wider area. The first SPECT machines were introduced in the late 1960s. Positron emission tomography was not developed until the 1980s. By that time, there was sufficient confidence in scintillation technology to attract the high level of funding needed to develop PET scanning machines, and PET was able to draw upon the proven technology of SPECT and computed tomography (CT). Cyclotrons, the main source of the short-life radionuclides needed in PET, were also more widely available.

Uses, Effects, and Limitations

PET and SPECT have advantages over most other body scanning methods because they reveal the function of body tissues, not just their structure. PET and SPECT can detect changes in function long before changes in structure become apparent. Early accurate diagnosis as a result of PET and SPECT scans leads to a higher likelihood of success in treatment of several conditions. PET scanning is particularly valuable in investigating brain function. Radioactively labeled glucose, for example, which accumulates in metabolically active parts of the brain, can reveal a region where a fast-growing tumor is located or where epileptic activity might originate. PET and SPECT can be used to detect abnormalities in blood flow or biochemical balance or change in brain disorders such as Alzheimer's disease, Parkinson's disease, or schizophrenia. PET is an important research tool used extensively for mapping the functions of various parts of the brain in healthy volunteers. With the right choice of radionuclide and carrier, PET is able to visualize biochemical changes in the brain and highlight specific kinds of nerve cells with a clarity that no other technique can match. PET is also employed to distinguish among dead, damaged, and healthy heart tissue following a heart attack (cardiac infarction). This information can be invaluable in directing the best forms of treatment.

Like PET, SPECT is used to study blood flow in the brain and is employed to assess damage to heart tissue following a heart attack. SPECT is also used to evaluate the function of the thyroid gland, the lungs, liver, spleen, and bone. For such purposes, it may be used to shed light on structural changes revealed by X-ray or ultrasound techniques. The radioactively labeled substances (radiopharmaceuticals) used in PET and SPECT studies are costly to make. They must be chemically pure, their radioactive properties must be precisely matched to a particular use, and they must be suspended in a sterile medium that matches the pH of body fluids. In the case of PET, the radiopharmaceuticals have short half-lives and so must be manufactured on-site or nearby using a cyclotron. SPECT does not require the highly trained personnel needed to operate a cyclotron, and the SPECT scanner assembly itself is simpler, making the technique much less costly and more readily available than PET. However, SPECT scans do not have the higher resolution level of PET scans and so are not as discriminating as PET when used in heart and brain studies.

As with X-rays and other forms of ionizing radiation, gamma radiation is potentially damaging to living tissue. Very high doses can kill cells immediately, and moderate doses can increase the likelihood of cancers. Even quite low doses received by germ line cells (cells that give rise to sperm or egg) may cause genetic mutations that can be passed on to offspring. At present, the damaging effects of ionizing radiation are generally regarded to be proportional to their dosage over time. However, this assumption has yet to be thoroughly tested, particularly for low radiation dosages. In general, SPECT studies subject a patient to lower doses of ionizing radiation than do X-RAY IMAGING techniques such as computed tomography. Levels of exposure to ionizing radiation in PET scans of the brain are comparable to, or higher than, those experienced in CT brain scans.

Issues and Debate

The costs of a PET scanner and its cyclotron source of radionuclides are on the order of several million dollars. Moreover, a minimum of three highly paid specialists is involved in the stages between radionuclide production and generation of a completed PET scan. PET is costly and thus has been used almost exclusively by research specialists until the 1990s. More recently, however, some hospitals and medical centers have adopted PET for routine or almost-routine diagnosis of a fairly narrow range of medical conditions. The high cost of the technology means that availability is, to a large degree, limited by geographic and economic considerations. In the United States, for example, a high-income family with good health insurance living in a large city is likely to have access to PET for the diagnosis of certain brain disorders. A low-income family living in a rural area is very unlikely to have such access. SPECT, being a lower-cost technology, is more widely available and is often an effective complement to other forms of medical imaging. For heart and brain investigations, SPECT has poorer resolution than PET, but may be the only option on the basis of cost and local availability.

In her 1997 book, *Naked to the Bone*, Bettyann Kevles notes a double standard in the use of PET for research. Healthy adult male volunteers are encouraged to be subjects in PET studies of brain function, on the assumption that such studies are not injurious to patients' health. Researchers, however, and the community at large, are reluctant to allow minors and women of childbearing age to participate in such studies. The risks of PET have yet to be properly evaluated in long-term studies. Although using PET to diagnose debilitating or life-threatening conditions can be justified, using PET on healthy volunteers for the purpose of more-or-less pure research raises moral questions. The use of standard PET protocols is unlikely to be injurious to health, but long-term follow-up studies of PET volunteers have yet to run their course.

The availability of PET scanning is limited by its high cost, but by providing information about living tissue that cannot be obtained in other ways, PET scans are likely to become more widespread in the coming decades. In the United States, the Food and Drug Administration is carefully evaluating each new radiopharmaceutical as PET scanning's applications shift from pure research purposes to clinical diagnostic use.

—*Trevor Day*

RELATED TOPICS

Magnetic Resonance Imaging, Ultrasound Imaging, X-ray Imaging

BIBLIOGRAPHY AND FURTHER RESEARCH

BOOKS

Armstrong, Peter, and Martin L. Wastie. *Diagnostic Imaging,* 4th ed. Oxford: Blackwell, 1998.

Erkonen, William E., ed. *Radiology 101: The Basics and Fundamentals of Imaging.* Philadelphia: Lippincott-Raven, 1998.

Kevles, Bettyann H. *Naked to the Bone: Medical Imaging in the Twentieth Century.* New Brunswick, N.J.: Rutgers University Press, 1997.

Lisle, David A. *Imaging for Students.* New York: Oxford University Press, 1997.

McCormick, A. K. and A. T. Elliot. *Health Physics.* Cambridge: Cambridge University Press, 1996.

Patel, P. R. *Lecture Notes on Radiology.* Oxford: Blackwell, 1998.

Weir, Jamie, and Peter H. Abrahams. *Imaging Atlas of Human Anatomy,* 2nd ed. London: Mosby-Wolfe, 1997.

PERIODICALS

Barinaga, Marcia. "What Makes Neurons Run?" *Science,* April 11, 1997, 196.

Dixon, Adrian K. "Evidence-Based Diagnostic Radiology." *Lancet,* August 16, 1997, 509.

Gilman, Sid. "Medical Progress: Imaging the Brain." *New England Journal of Medicine,* March 19, 1998, 812.

Lentle, Brian, and John Aldrich. "Radiological Sciences, Past and Present." *Lancet,* July 26, 1997, 280.

McCone, John. "Maps of the Mind." *New Scientist,* January 7, 1995, 30.

Powledge, Tabitha M. "Unlocking the Secrets of the Brain (Part 2)." *Bioscience,* July–August 1997, 403.

Watzman, Haim, and Peter Aldous. "Have the Mind Mappers Lost Their Way?" *New Scientist,* May 4, 1996, 12.

INTERNET RESOURCES

Bioethics Resources
http://adminweb.georgetown.edu/nrcbl/

A Guided Tour of the Visible Human
http://www.madsci.org/~lynn/VH/

How PET Works
http://csee.lbl.gov/cup/Su98/Tony/index.html

Teaching Aids on Medical Imaging
http://agora.leeds.ac.uk/comir/resources/links_c.html#teaching

The Visible Human Project
http://www.nlm.nih.gov/research/visible/visible_human.html

SMART CARDS

Smart cards resemble credit cards but possess a microprocessor instead of a magnetic strip. The chip allows smart cards to store a variety of information, including financial information and medical and health records. First popularized in Europe, smart cards have gained acceptance in the late 1990s in the United States to pay for gasoline and road tolls with great convenience, as well as serving other purposes.

Although the cards provide consumers with convenient options in some scenarios, they have yet to prove themselves as more convenient than existing debit and credit cards in several situations, such as in the purchase of groceries. Critics of smart cards are concerned that hackers can break the securi-ty of smart card encryption to tamper with financial or other personal information.

Scientific and Technological Description

A smart card possesses a small microprocessor chip that tracks and stores a variety of information, such as bank account information or the amount of "digital" cash left on the card. Smart cards can be used as digital cash in which customers prepay to "fill up" their cards, or consumers can use them as debit or automated teller machine (ATM) cards. Smart cards can also store secure information such as medical records.

Conventional debit and bank cards use a magnetic strip that identifies the card and account number, but smart cards are more complex. Their chip allows them to store information in the same manner that a computer can keep track of bank account information. This is why smart cards are unique in their ability to act as digital cash; they can deduct money from the card's balance each time it is used. Debit cards rely on another computer to make that calculation. For instance, in a grocery store, a debit card relies on the grocery store's debit machine to contact the bank and debit the account. A smart card calculates and retains that information on the card itself. This is useful for situations in which there is no other computer to call the bank account, such as when using a vending machine.

There are three types of smart cards: contact, noncontact, and combination. *Contact smart cards* must be inserted into a smart card reader. For instance, an ATM could read the financial information on a digital cash smart card. *Noncontact smart cards* have an antenna coil as well as a chip embedded on the card. The antenna can transmit and receive information when in proximity with a compatible smart card reader/transmitter. For instance, a noncontact card could be used to pay a toll fare—the driver wouldn't have to stop to pay the toll. The toll booth would simply read the card and subtract the fare as the car maintained highway speeds. A *combination card* is both a contact and a noncontact card. It possesses the reliability of a contact card and the convenience of a noncontact card.

Historical Development

Smart cards owe their existence to two key electronic inventions: the integrated circuit (IC) and the microprocessor. An IC comprises thousands or even millions of separate electronic components deposited by a chemical process onto a square silicon chip smaller than a fingernail. Invented in 1959 by Jack Kilby of Texas Instruments and Robert Noyce of Fairchild Semiconductor, it led ultimately to Noyce's invention, in 1971, of the first microprocessor, the Intel 4004, which was a complete miniature computer on a single silicon chip.

Inspired by this revolution in miniature electronics, French journalist and amateur inventor Raymond Moreno conceived an entirely electronic payment system in 1974.

Moreno envisioned people wearing rings containing computer chips that could store electronic money, which could be debited as purchases were made in stores or refilled with new credit at banks and other outlets. By September of that year, the ring had become an epoxy card, and by the following spring, a credit-card-style smart card. Moreno patented his invention in 1975 and went on to found a pioneering smart card company called Innovotran, which licensed the right to use its invention to other European companies, including Honeywell Bull, Schlumberger, and Philips.

France continued to pioneer smart card technology in the years that followed. By the early 1980s, a consortium of French banks had begun to look into using smart cards as a system of electronic payment. A number of computer companies were invited to participate in early consultations, and trials in the French cities of Lyons, Blois, and Caen began in the spring of 1982. The following year, the French government announced trials of smart cards for storing medical records, and the Schlumberger company began installing smart card pay phones throughout the country.

With their complex electronics, smart cards were expensive to produce and took a long time to catch on. But by the mid-1980s, they had spread to other countries. Customer trials were held in Norway and Italy in 1984 and by 1985, the giant credit card company MasterCard International had announced its intention to develop a smart card system for the United States following a detailed study of the French system. Rival company VISA announced its own smart card the same year. The 1980s also saw smart cards being used in other ways, for example, to secure personal information (such as account details and personal identification numbers) in the GSM (global system for mobile communications) cellular phone system pioneered in Europe (see *Uses, Effects, and Limitations*).

More recent developments have been driven by London-based Mondex International (now 51 percent owned by MasterCard), which announced a much more powerful type of smart card in 1994 and tested it in the English town of Swindon from 1995 through 1998. Users charge up their cards at a modified ATM and can then use them to make payments in stores, on buses, in public telephones, and in electronic commerce (Internet trading) transactions (see ELECTRONIC COMMERCE). They can also transfer cash to or from Mondex cards held by other people, using either a special electronic wallet that accepts both cards or by making a phone call, which means that Mondex transactions can be carried out cheaply without a direct computer link to a bank, an important advantage that makes them much more like ordinary cash. Tamper-resistant chips and built-in encryption (digital coding) technology are designed to prevent fraudulent transactions from taking place (see CRYPTOGRAPHY). Currently, more than 450 com-

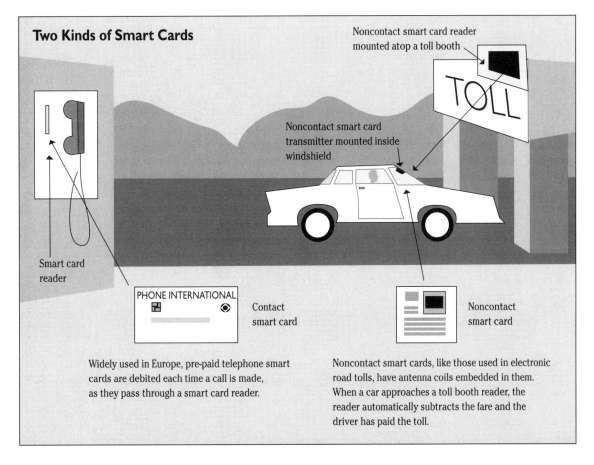

Two Kinds of Smart Cards

Noncontact smart card reader mounted atop a toll booth

Noncontact smart card transmitter mounted inside windshield

TOLL

Smart card reader

PHONE INTERNATIONAL

Contact smart card

Noncontact smart card

Widely used in Europe, pre-paid telephone smart cards are debited each time a call is made, as they pass through a smart card reader.

Noncontact smart cards, like those used in electronic road tolls, have antenna coils embedded in them. When a car approaches a toll booth reader, the reader automatically subtracts the fare and the driver has paid the toll.

panies in 40 countries license the Mondex technology to make smart card systems.

The latest smart cards are nothing less than credit-card computers that can be used for a number of different applications. No one buys a new desktop personal computer (PC) every time he or she wants to run a new application but simply loads the new application into an existing PC. Similarly, no one wants a wallet full of smart cards. For a single smart card to work in the same way as a desktop PC, it cannot run just a single preset program like smart cards used to. Instead, it must run an operating system, a piece of central control software such as Windows that enables multiple programs to run inside a single computer by sharing its memory, processor, and so on.

In 1999, Microsoft announced a trimmed-down operating system called Smart Card for Windows, and rival systems are offered by Sun Microsystems (Java Card) and MasterCard (Multos). Unlike traditional smart cards, which are limited to a single application (working a pay phone or storing health records, for example, but not both), these advanced cards can be used for numerous applications (e.g., storing financial and health records, acting as a customer loyalty card at the supermarket checkout, and storing electronic cash). Each application effectively runs as a separate program inside the smart card's computer, and new applications can be added as needed.

Uses, Effects, and Limitations

Smart cards were conceived originally as a kind of electronic wallet into which money could be deposited from a bank and extracted to pay stores or other cardholders, and now have many financial applications. In the United Kingdom and elsewhere in Europe, prepaid telephone smart cards are sold with a fixed amount of credit, which is reduced gradually as calls are made at public phones; when the card runs out of credit, it is thrown away.

Noncontact smart cards, with their built-in antennas, can be used to make payments without putting the cards into a machine. The best-known application is electronic road tolls, in which a smart card fitted to a car windshield is debited each time it drives along a toll bridge, tunnel, or road. The toll system, proposed in 1991 and introduced on the Verrazano-Narrows Bridge in New York City in 1995, is now used throughout the world to speed traffic through tolls without stopping.

Electronic commerce has introduced another important new application for smart cards. The key problems of Internet trading are security (preventing fraudulent transactions) and authentication (ensuring that Internet customers are who they say they are). Smart cards, which can carry digital signatures (lengthy numeric codes that identify a person as uniquely as their handwritten signature), may be used to identify people securely to electronic commerce systems before they can make Internet purchases or other sensitive transactions.

Smart card support is now built into Web browsers such as Netscape Navigator and Microsoft Internet Explorer.

Another major application is granting or restricting access to computer and telecommunications systems. The development of the GSM cellular phone system in Europe from 1982 onward saw smart cards known as Subscriber Identification Modules (SIM cards) fitted inside phone handsets. Each card is programmed with account billing information, personal preferences for how the phone should work, a personal phone book, and other information. The card (and all the information it contains) can be removed from one handset and plugged into another one, instantly transferring the information to the new device. Some corporations with highly secure computer networks (such as banks and insurance companies) now use smart cards to grant employees access to their systems, instead of the traditional (and relatively insecure) technique of entering a user name and password. In August 1999, public libraries in Michigan introduced smart cards on their public Internet terminals to enable parents to restrict their children's access to unsuitable Web sites.

Smart card pioneers realized early on that if the devices could store cash, they could also store other sensitive information. One popular application is using smart cards to store medical records and patient details. Unlike paper cards, which can be stolen and used fraudulently, smart cards can be used to provide secure access to prescription drugs. They are particularly useful in accidents and emergencies where

The E-ZPass automatic toll collection system, shown here at the George Washington Bridge in Fort Lee, New Jersey, is a form of non-contact smart card. As cars outfitted with E-ZPass cards inside their windshields pass by the E-ZPass readers, the toll is automatically deducted from drivers' pre-paid accounts (AP/Michael Sypniewski).

their holders may be unable to reveal their blood type, allergies, next of kin, or other vital information.

Despite their many benefits, smart cards are not without their limitations. They are not entirely independent of the traditional cash system: They still need to be filled up regularly with money. Although built-in security devices help to prevent fraudulent use, they cannot protect against accidental loss or damage. This problem has become more of a concern with the latest generation of multifunction cards, which may store a whole range of personal information, cash, and other sensitive details. Losing a smart card of this kind is equivalent to losing an entire wallet and may cause the holder much more personal inconvenience than simply losing a credit or debit card.

Issues and Debate

Although smart cards offer convenience of use, they are also vulnerable to cryptographic attacks—attempts to break their codes and to gain access to their information—since they use microcircuitry. The problem with microcircuitry is that it can betray the operations of a card and its secret code. Research teams have cracked the smart card's digital security code by measuring fluctuations in the amount of electrical power it consumes. When a card is in use, its microchip performs operations, each of which requires a different amount of power to execute. By studying the energy fluctuations, researchers can uncover patterns that give clues about the card's security code. After analyzing this, code crackers can eventually piece together the encryption key to the card. With an encryption key, anyone can tamper with a smart card. For instance, criminals could use the card to obtain access to a person's bank account to refill the card automatically with cash. A code break like this does not require sophisticated equipment. Realistically, it can be performed with a few personal computers and several thousand dollars of electronic equipment. Thus it is feasible that a number of criminals could exploit the weaknesses in smart card technology.

To make smart cards more secure, smart card makers, such as financial corporations Citicorp and the Chase Manhattan Corporation, have developed several techniques. One technique involves the creation of *digital noise,* meaningless random calculations that would consume random amounts of current. This would thwart the attempts of someone trying to analyze energy fluctuations produced by a microchip. According to MasterCard, this security method has already been implemented in their Mondex cards. Another defense against smart card fraud varies the order of the software's operations. This would make it difficult for attackers to deduce patterns in the card's consumption of power. Finally, since smart cards must be used with other appliances, such as ATMs; banks and financial institutions could monitor for unusual behavior and put a stop on any card with unusual activity. This

would be a last-defense option for preventing the criminal abuse of smart cards, which is becoming more plausible with advances in code cracking.

There are also concerns that smart cards will not be adopted by the public. An experiment conducted by Citibank and Chase Manhattan Bank revealed that when those companies gave smart cards to more than 100,000 people living on the Upper West Side of Manhattan, most people did not use the cards consistently. Participants in the experiment were instructed to use the cards to transfer cash from their bank accounts onto the cards at ATMs and spend the money at 500 area merchants. This would purportedly make small purchases faster and more convenient. Although people followed these instructions initially, most did not load the cards up (from their bank accounts) a second time.

As a result, Citibank and Chase Manhattan have canceled the smart card experiment. Other financial companies, such as VISA and MasterCard, have discovered similar consumer reactions. One of the problems with the smart cards in these experiments is that they didn't offer any additional convenience. Most people chose to use the smart cards in grocery stores and convenience stores, which already accept debit and credit cards. They were less likely to use the cards at newsstands and with other merchants who generally deal in cash. In addition, the smart cards didn't work with pay phones or vending machines, devices that require change.

Smart card proponents argue that there are several benefits to a smart card over a magnetic stripe card. For instance, smart cards offer far more information storage capacity. Smart cards are also more reliable, perform multiple functions, and are more secure. Ultimately, consumers will decide whether smart cards offer them enough advantages to balance out their drawbacks. Currently, there are more than 1 billion smart cards in use.

—Christine Doane

RELATED TOPICS
Cryptography, Electronic Commerce, PCS Phones

BIBLIOGRAPHY AND FURTHER RESEARCH

BOOKS
Henning, Kay. *The Digital Enterprise*. London: Random House, 1999.
Miller, Ben, ed. The 1997 *Advanced Card and Identification Technology Sourcebook*. Bethesda, Md.: Personal Identification News, 1996.
Rankl, Wolfgang and Wolfgang Effing. *Smart Card Handbook*. New York: Wiley, 1997.

PERIODICALS
Hansell, Saul. "Got a Dime? Citibank and Chase End Test of Electronic Cash." *New York Times,* November 4, 1998.
Kalish, David E. "High-Tech Group Releases Standards in Bid to popularize 'Smart Cards.'" *New York Times,* May 16, 1997.
Peterson, I. "Power Cracking of Cash Card Codes." *Science News,* June 20, 1998, 388.

INTERNET RESOURCES
Definitions of Technology Terms—Search for Smart Card
http://www.whatis.com

How Smart Card Technology Works
   http://www.mondex.com/mondex/cgi-bin/printpage.pl?style=
   noframescash&fname=../documents/tech1.txt&doctype=genp
"Mondex: A House of Smart-Cards? With e-Cash Privacy Is Illusory
   and Security Is Questionable," by David Jones, *Convergence,* July
   12, 1997
   http://insight.mcmaster.ca/org/efc/pages/
Mondex International: Reengineering Money
   http://www.msb.edu/faculty/mchenryw/mgmt681-f98/mondex/
   mondex.htmlCards
"MTA Bridges and Tunnels E-ZPass" by Michael C. Ascher, MTA
   President
   http://www.nyu.edu/wagner/transportation/journal/MTA.htm
"Port Authority Brings E-ZPass to the World's Busiest Bridge"
   http://www.panynj.gov/pr/97-97.html
Smart Card Cyber Show/Smart Card Museum
   http://www.cardshow.com/museum/
"Swindon Sniffs at Smart Card," *Electronic Telegraph,* July 4, 1995
   http://insight.mcmaster.ca/org/efc/pages/media/
   e-telegraph.04jul95.html

Smart Highways: Intelligent Vehicle Highway Systems

Intelligent vehicle highway systems (IVHSs), also known as smart highways, describe an array of technologies developed by engineers to improve traffic management. The systems grew from the field of telematics, which describes the integration of communications and information technologies. When applied to roadway development, telematics is called IVHS (and in Europe, road transport informatics or advanced transport telematics).

Predicted to be revolutionary in its safety advance and traffic relief, some critics remain dubious about IVHS, wondering if the disparate technologies involved will work together feasibly. Others question who will pay for the IVHS implementation and who will be able to afford the vehicles capable of taking advantage of these technologies.

Scientific and Technological Description

Because it draws from many broad fields, IVHS encompasses a startling number of tools, but a few can be identified as major components. Fundamental to the field are computers, which play key roles in both the vehicles that roll on the roadways and the roads themselves. Experts consider onboard computers to be the primary building block of IVHS because they are multifunctional. Specialized computers can control specific functions, such as speed, and displays can download and transmit data in real time. One computer can be used to access a database of maps, then check on the locations of other vehicles, then transmit its own location to tracking satellites.

The information held in onboard computers comes from sensors such as loops that are embedded under highway surfaces or video cameras that are mounted along roads. The information they gather is sent to centralized computers that disperse it to the vehicles; more commonly, the data are sent to electronic signs on highways that give drivers messages about upcoming traffic or road conditions. Sensors also help control the timing of traffic signals. Historically, traffic signals were set on timers that were unrelated to actual traffic flow. Under such a system, drivers could find themselves waiting at red lights without a single vehicle present to pass through the intersected green lights. Sensors under surfaces register the weight of a vehicle and send electronic signals to the traffic light so that it can change, if appropriate.

Growing in their importance in IVHS are positioning technologies that tell drivers not only where they are but also what's approaching them. Radar, the standard-bearer of navigation systems, plays a key role in collision warning systems currently under development. Vehicles can be equipped with radar systems that emit radio waves, and if bounced back, indicate the presence and approximate position of potential obstacles. The information is routed to alarm systems that alert drivers to impending collisions. The GLOBAL POSITIONING SYSTEM (GPS) also is perceived as an important positioning technology in IVHS. Engineers predict that properly equipped vehicles can be tracked by satellites and the information processed by computers. Such tracking would help owners of commercial vehicles to direct their fleets more efficiently and would assist traffic managers in pinpointing congested spots on roadways. Numerous technologies continue to be developed under the IVHS umbrella, and some engineers predict that their integration will emerge as the most technologically significant aspect of the concept.

Historical Development

When Henry Ford first unveiled a working automobile in 1896, he probably did not conceive of an intelligent highway. Yet the makings for one were in the works less than 14 years after his invention. In 1910, a handful of drivers were using the Jones Live Map, a primitive navigation system attached to steering wheels and based on dead reckoning, a method of judging distance by measuring the amount of time passed from a fixed point to another. That method is error prone, however, and the Jones Live Map never prospered.

In 1939, at the World's Fair in New York City, General Motors featured a primitive automated highway in which cars were controlled by radio waves. The company continued its research, and in the 1950s and 1960s began installing analog vacuum-tube electronics in vehicles. These and other research projects were steppingstones to more sophisticated systems on national levels.

Japan developed the first roadway guidance system during the early 1970s. From there, the country developed its main research program, called VICS (vehicle information and communications systems). By 1993, Japan had put 200,000 vehicles on the road with onboard electronic navigation equipment. In 1996, Japanese researchers demonstrated an automated highway near Komoro City using magnets, video cameras, and radio waves.

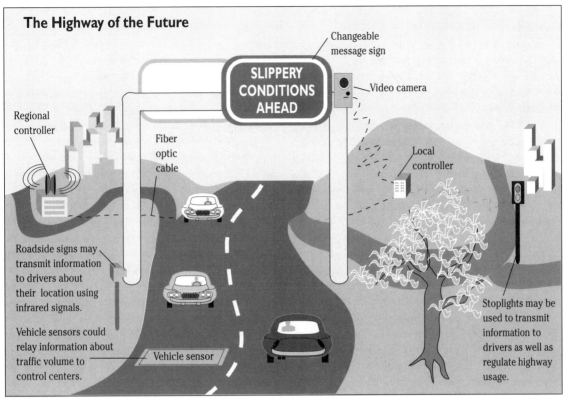

The Highway of the Future

Changeable message sign

SLIPPERY CONDITIONS AHEAD

Video camera

Regional controller

Fiber optic cable

Local controller

Roadside signs may transmit information to drivers about their location using infrared signals.

Vehicle sensors could relay information about traffic volume to control centers.

Vehicle sensor

Stoplights may be used to transmit information to drivers as well as regulate highway usage.

Adapted with permission from Deborah Perugi.

Meanwhile, European researchers were developing their own sophisticated IVHS systems. In 1986, PROMETHEUS (Program for European Traffic with Highest Efficiency and Unprecedented Safety) was initiated by car manufacturers in Europe under the aegis of the European Community (EC). With a budget of $875 million, the program developed onboard computers that told drivers which routes to take and control systems that measured distances between vehicles, applying brakes or turning steering wheels to avoid collisions. Today, the European Union (formerly the EC) continues to develop other major IVHS research programs.

In Hong Kong, engineers unveiled an electronic road pricing project in 1983 that used transponders at tollbooths to register tags in passing cars electronically. The data were sent via telephone lines to control centers that identified the car and cost of the toll. Invoices were then sent to drivers on a monthly basis. The program formed the basis of the current electronic toll debiting system being established in several states and countries (see SMART CARDS).

In 1988, U.S. transportation experts formed a group called Mobility 2000 to reevaluate highways in the United States. The group later changed its name to the Intelligent Vehicle Highway Society of America (IVHS America) and grew to include more than 1000 members. The group was designated as a federal advisory committee and was charged with incorporating intelligent vehicle and highway systems throughout the United States.

In 1991, the U.S. Congress passed the Intermodal Surface Transportation Efficiency Act (ISTEA), which allocated $660 million over six years to be spent on IVHS research. The act bolstered IVHS America significantly, and in 1997 the most visible and well-publicized IVHS innovation in the country was demonstrated on 7.6 miles of Interstate 15 near San Diego, California. It was a prototype of an automated highway that was developed by the National Automated Highway System Consortium (NAHSC), which included private companies, federal agencies, and universities.

The demonstration consisted of several tests. In one, two buses and three cars changed lanes using side- and rear-looking sensors. In a related demonstration, the lead vehicle transmitted to other vehicles that an obstacle in the road had been detected, so they could avoid it. Another showed platooning, a group of cars spaced automatically from one another and traveling at the same speed; when one of the cars left the line, the other cars adjusted the spacing accordingly. In 1998, however, the Transportation Department drastically cut NAHSC's funding to $2 million, from a budget of $54 million in 1994. Since then, the consortium has reverted to separate groups that conduct their own research.

Uses, Effects, and Limitations

Engineers have grouped IVHS technologies into five general categories that reflect their uses: advanced traffic management systems, advanced travel information systems, advanced vehicle control systems, commercial vehicle operations, and advanced public transportation systems. At its most fundamental level, IVHS begins with *advanced traffic management systems* (ATMS). These systems seek to relieve congested highways by offering real-time data on gridlock and accidents and by controlling drivers' access to roadways. Advanced systems transmit the number of vehicles that pass specific points to computer stations along roadways. The computers then relay the data to other computers in traffic centers, which are staffed by personnel who monitor the data. Personnel can also watch the traffic through video cameras mounted along roads, and in the case of accidents, send emergency vehicles.

Advanced travel information systems build on ATMS and help drivers find destinations and avoid congested roadways. These systems can integrate technologies such as the global positioning system, onboard computers, and databases to track vehicles. *Advanced vehicle control systems* (AVCS) focus on safety and efficiency of roadway use. AVCS encompasses collision warning technologies and seeks to increase the number of vehicles that can operate on existing roadways. One way to improve highway efficiency would be an automated highway that would virtually allow cars to drive themselves.

IVHS technologies could be employed in *commercial vehicle operations* to track fleets of buses, trucks, vans, or taxis. Using positioning systems and onboard navigation, dispatchers can monitor entire groups of vehicles and track where they are in their routes. Two-way radios can allow dedicated communication between drivers and dispatchers; dispatchers may also be able to transmit instructions to onboard computers as another means of communication.

Advanced public transportation systems are meant to increase the efficiency of mass transit. Consumers can purchase smart cards for use on buses and trains, which debit fare amounts from a prepaid card. Automatic toll collection is another application in this area.

Issues and Debate

Critics of IVHS have pointed out several potential problems with the concept, not the least of which are technological slips. A system based largely on the successful integration and interfacing of many technologies must ensure that individual components do not fail. They must be maintained and checked regularly, with particular attention to those that are exposed to climatic changes. Another obvious potential flaw in IVHS systems is the possible diversion of many cars onto the same roads. A system that avoids congestion on a highway only to clog the surrounding roads is problematic, and researchers will need to address how to avoid sending too many cars to the same roadways.

Ironically, an IVHS could mitigate some of its benefits because it allows more vehicles to travel on roadways.

IVHS advocates say that the technologies could increase highway capacity three- to sevenfold, while critics point out that putting more vehicles on the road will exacerbate the very congestion that an IVHS is meant to relieve. Urban officials have expressed concern that city roads and parking spaces will be overwhelmed by an increase in vehicles. Environmentalists have questioned whether the advantage of fuel efficiency with an IVHS will be nullified by the presence of more cars on the road.

For consumers, an IVHS means buying vehicles that come equipped with the technologies to coordinate with those of highways. Based on the cost of current prototype cars, that means adding an extra $10,000 to a car's price tag. Market analysts predict that the extra cost will decrease over time, but no one is certain by how much. Advocates for lower-income people have worried that an IVHS could create two classes of drivers, only one of which has the advantages of a technologically enhanced highway system. Consumers will not be the only ones who will need to pay extra money. State and local agencies will be expected to fund the maintenance and repair costs for elaborate systems that smaller communities may not be able to afford. That cost, in turn, may be passed on to taxpayers.

Critics also point to privacy issues as a major drawback of IVHS. They find the thought of centralized computer systems that monitor the exact locations and destinations of citizens worrisome and charge that some smart highway technologies are too invasive. Finally, accidents, the bane of traditional highway systems, remain an issue with an IVHS. Should drivers be held responsible for accidents when they are not actually driving the cars? In an automated highway scenario, critics say, the potential for disaster increases. With so many cars traveling in tight formation, a system that goes awry could cause considerable damage.

—Christina Roache

RELATED TOPICS
Computer Modeling, Global Positioning System, Smart Trains

BIBLIOGRAPHY AND FURTHER RESEARCH

BOOKS
Catling, Ian, ed. *Advanced Technology for Road Transport: IVHS and ATT.* Norwood, Mass.: Artech House, 1994.
Drane, Christopher R. *Positioning Systems in Intelligent Transportation Systems.* Norwood, Mass.: Artech House, 1998.
Parkes, Andrew M. and Stig Franzen, eds. *Driving Future Vehicles.* Philadelphia: Taylor & Francis, 1993.
Pauwelussen, J. P., and H. B. Pacejka, eds. *Smart Vehicles.* Lisse, The Netherlands: Swets & Zeitlinger, 1995.
Whelan, Richard. *Smart Highways: Smart Cars.* Norwood, Mass.: Artech House, 1995.

PERIODICALS
Amato, Ivan. "Thumbs-Sideways on Smart Cars and Roads." *Science News,* February 4, 1989, 79.

Anselmo, Joseph C. "'Smart Highway' Business Attracts Aerospace Firms." *Aviation Week and Space Technology,* January 31,1994, 56–57.

Ben-Akiva, Moshe, et al. "The Case for Smart Highways." *Technology Review,* July 1992, 40.

Candler, Julie. "Smart Cars, Smart Roads." *Nation's Business,* February 1996, 31–32.

Cogan, Ron. "Drive by Wire: Testing an Automated Highway, We Buckle Up in a Buick That Drives Itself." *Popular Science,* September 1997, 74–75.

Del Valle, Christina. "Smart Highways, Foolish Choices?" *Business Week,* November 28, 1994, 143–144.

Flam, Faye. "Military Electronics Follows a Rougher Road." *Science,* February 4, 1994, 623–624.

Gibbs, W. Wayt. "Not So Fast." *Scientific American,* September 1997, 42.

Hong, Patrick. "Automated Highway System." *Road & Track,* February 1998, 107–109.

Hughes, David. "Aerospace Electronics May Guide Smart Cars." *Aviation Week and Space Technology,* November 3, 1993, 63–64.

Jochem, Todd, and Dean Pemerleau. "Life in the Fast Lane: The Evolution of an Adaptive Vehicle Control System." *AI Magazine,* Summer 1996, 11–50.

Lowe, Marcia D. "Road to Nowhere: A Proposed Technological Cure for Smog, Gridlock, and Traffic Accidents Actually May Exacerbate the Problems It Is Supposed to Solve—Yet Within It Lies a Golden Opportunity That the Experts Have Overlooked." *World Watch,* May–June 1993, 27–34.

Rillings, James H. "Automated Highways: Cars That Drive Themselves in Tight Formation Might Alleviate the Congestion Now Plaguing Urban Freeways." *Scientific American,* October 1997, 80–85.

Sheldrick, Michael G. "Driving While Automated: Planning Smart Highways for Tomorrow's Smart Cars." *Scientifc American,* July 1990, 87–88.

Siuru, William. "Automated Highway Systems." *Popular Electronics,* December 1998, 33–35.

"Smart Highways Face Roadblocks." *Futurist,* January–February 1995, 46.

Walton, Chris. "First Drive: Buick's 'No-Hands' Future Car." *Motor Trend,* July 1997, 154.

Wilson, Jim. "A Real Smart Road." *Popular Mechanics,* February 1998, 15.

Wu, Corrina. "Look, Ma, No Hands!" *Science News,* September 13, 1997, 168–169.

INTERNET RESOURCES

Access Intelligent Transportation Systems
 http://www.itsa.org
Bureau of Transportation Statistics
 http://www.bts.gov
Federal Highway Administration
 http://www.fhwa.dot.gov
Federal Transit Administration
 http://www.fta.dot.gov
Human Factors in the Automated Highway System: Transferring Control to the Driver
 http://www.fhwa.dot.gov/aard/hum.html
"Look Mom, No Hands," by Judi Chew
 http://www.wfl.fhwa.dot.gov/td/newsltrs/Nov97new.pdf
Moving Safely Across America
 http://www.fhwa.dot.gov/utdiv/programs/saftymov.htm
National Transportation Library
 http://www.bts.gov/ntl
Volpe National Transportation Systems Center
 http://www.volpe.dot.gov

SMART HOMES

A variety of technologies and systems are available for converting an ordinary home into a smart home, in which almost every appliance is controlled by a central computer system. In a smart home, lights can be turned on and off automatically, following people from room to room; windows can open themselves when the sun comes out and close when it starts to rain; plants can be watered while the occupants are on vacation; and the fire or police department can be telephoned automatically in an emergency.

Smart homes seem like something out of a science fiction movie, but the technology has evolved out of domestic appliances developed over the last 100 years. The manufacturers of smart-home systems claim that they save time, energy, and money, but not everyone is convinced of their merits; automated homes can be very expensive and could cause chaos for their owners if they go wrong.

Scientific and Technological Description

A smart home consists of a number of domestic appliances, which are operated automatically by a central control system (see figure). Electrical appliances are most often powered from a 110-volt (V) domestic electricity supply (240 V in the United Kingdom), and in smart-home systems this standard household electrical wiring doubles up as a kind of computer network for linking the appliances to the control system without the need for additional wires.

Each appliance is plugged into a small control unit, which in turn is plugged into the standard electrical outlet. The central control system, which may be anything from a simple control box to a personal computer (PC) dedicated to the task of home automation, superimposes low-voltage, high-frequency control signals on top of the standard high-voltage, low-frequency household current. These signals are decoded by the small control units attached to each appliance. Each signal combines two pieces of information: the address (code number) of the appliance to which the signal relates and the message to be sent to the appliance (such as "turn on" or "turn off"). In the oldest and best-known automation system, X-10, up to 256 appliances (or groups of appliances) can be individually controlled.

A sophisticated PC-based control unit can work in two ways. First, it can turn appliances on and off according to a simple calendar program, which is set using the PC. Thus, central heating may be set to turn on and off at the same time each weekday, but at different times at the weekend, and for longer periods in winter than in fall. If the occupants expect to be away from home for a long period in summer, a lawn sprinkler system might be set to switch on for a couple of hours every couple of days. Apart from these calendar-based commands, the control unit can also accept immediate commands in a variety of ways. For example, the spoken command "Lights on!" can be used to turn on the lights at that moment, irrespective of the preprogrammed setting or time of day (see LANGUAGE RECOGNITION SOFTWARE).

Second, the control unit can be programmed to respond to other kinds of directly sensed input. Home automation systems usually contain a variety of fire, security, and safety detectors. Sensors placed around the house can detect smoke, upstairs windows being opened by small children, pressure on floorboards caused by intruders, and so on, and send emergency signals back to the control unit. The occupants have to program the system to take particular action when such an emergency occurs; this might include setting off an alarm, dialing 911, or telephoning the occupants at work with a prerecorded emergency message. The control unit can also be operated or reprogrammed by phone, which means that the occupants can call home to turn on the oven or run the hot tub so that it is ready for when they return from work.

Historical Development

Home automation dates back to 1879, the year when Thomas Edison developed the first electric light in his laboratory in Menlo Park, California. By 1893, a laborsaving electrical house of the future was being shown off at the Chicago Electrical Fair, and today's familiar electric appliances appeared soon afterward: electric tea kettles, also in 1893; washing machines in 1907; dishwashers in 1912; toasters in 1913; and electric razors in 1928. Early laborsaving appliances were sold primarily to create a demand

for electricity, and beginning in the 1920s were marketed as an acceptable way to a comfortable standard of living without the need for servants. Manufacturers sometimes attempted to induce guilt among housewives to sell their products; a 1930 advertisement for vacuum cleaners by the Pennsylvania Power and Light Company ran: "Got a baby? Want to keep it? Every day is dangerous to a woman who has no modern tools to work with. Own a vacuum cleaner! Take the dirt out of the house." By 1941, 80 percent of U.S. homes were connected to a central power grid for their electricity, 79 percent had an electric iron, and 47 percent were using an electric vacuum.

Americans became fiercely proud of their automated homes. In July 1959, then-U.S. vice president Richard Nixon and Soviet premier Nikita Khrushchev met at the American National Exposition in Moscow. In a high-tech mock-up of a laborsaving kitchen, they debated the relative merits of capitalism and communism, each arguing that their political system was best placed to deliver affluence and home comforts. Despite these promised benefits, the average time devoted to housework in U.S. homes remained constant between the 1920s and the 1960s. Only with the invention of the microprocessor (the self-contained computer on a single chip of silicon) in 1971 did true home automation become possible. During the 1970s and 1980s, microprocessor control systems became small and cheap enough to incorporate into appliances such as microwave ovens and domestic hot-water systems.

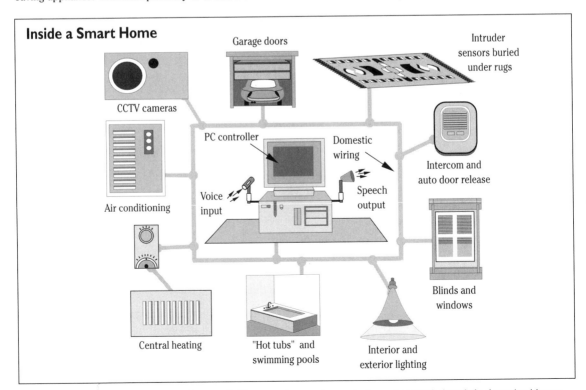

Inside a Smart Home

Garage doors

Intruder sensors buried under rugs

CCTV cameras

PC controller

Domestic wiring

Intercom and auto door release

Voice input

Speech output

Air conditioning

Central heating

"Hot tubs" and swimming pools

Interior and exterior lighting

Blinds and windows

In a smart home, almost any appliance can be controlled by a central personal computer by sending signals through the domestic wiring system. The PC can accept commands (such as "Lights on!") by voice, and can issue spoken alerts and warnings (such as "Intruder detected!").

Xanadu, Bill Gates' home in Medina, Washington, was designed entirely around smart-home technologies (Corbis/Anthony Bolante/Reuters).

Smart homes (by then the general term used to refer to home automation technologies and systems) became even smarter in the 1990s. In 1991, Matsushita Electronics developed the first voice-controlled VCR, but it could understand only Japanese. The same year, computer scientist and psychologist Alex Pentland and his colleagues at Massachusetts Institute of Technology (MIT) constructed a smart room that could locate its occupants and anticipate their needs by sensing their gestures and spoken commands and comparing them with past patterns of behavior. In 1997, the Electrolux company demonstrated the first radar-guided, robotic vacuum cleaner, which it claimed picked up 97 percent of household dirt compared to a typical human-operated machine, which collects only 75 percent. However, the machine took an hour to clean just two rooms.

What of the future? In March 1999, scientists from MIT and engineers from Motorola announced a $5 million partnership to develop smart-home technologies. At the launch of the Motorola–MIT DigitalDNA laboratory (so named because its products will be lifelike machines), product designers heralded a future where machines could talk to one another, explain themselves, and even get to know their users' likes and dislikes and adjust their behavior accordingly. Neil Gershenfeld of MIT has speculated that today's appliances will evolve into Things That Think (TTTs), such as washing machines that can dial the supermarket to order their own soap refills.

Uses, Effects, and Limitations

The purpose of smart-home technologies is to increase convenience, safety, and security and to save energy, but there is almost no limit to the number of ways in which the technologies can do this. Automatic lighting follows the occupants of a smart home around and can turn on automatically in an emergency. Air conditioning and heating systems can be regulated according to outside temperature and wind speed. Windows can be programmed to open when the temperature exceeds a certain level, and to close if rain is detected. Domestic chores can be tackled by appliances

such as automated lawn mowers that memorize the areas they need to cut and smart vacuum cleaners that use radar guidance to avoid obstacles. Home automation systems sometimes include advanced entertainment, including voice-controlled audio systems, which can track the occupants around the house, play the music they are listening to in whichever room they are in, and respond to spoken commands such as "Turn music down" or "Change CD."

For disabled or elderly people, home automation can dramatically increase independence and convenience. A person confined to a bed or wheelchair can use a voice-controlled computer system to turn lights on and off, open and close windows, check the identity of visitors using a speakerphone and open the door automatically, water plants, dial telephone numbers, and carry out many other jobs that they might otherwise have to rely on another person to do for them.

Simple electrical appliances are not the only things in the home that have been automated. The Sony Corporation has developed an electronic pet dog called AIBO (a Japanese word that means *artificial intelligence robot*), which retails for $2500 and is designed to have "the familiar lovable shape of small four-legged animals." It comes in two versions: the Emotion model, which switches between happiness, sadness, anger, and surprise; and the Instinct model, which has preprogrammed love, search, movement, and recharging instincts (see ROBOTICS). Unlike a real pet, it needs no house training, dog food, or trips to the vet.

The benefits of home-automation systems include savings in time, energy, and effort; better safety and security; and increased independence for disabled and elderly people. The biggest drawback is cost: A typical system may cost $10,000; Xanadu, the home built by Microsoft chairman and CEO Bill Gates in Seattle, was designed entirely around smart-home technologies and cost a total of $53,392,200. Other drawbacks include the possibility of a malfunction in the control system that would cause chaos throughout the house, and the inability of a programmed system to react sensibly to unexpected events (most people would not call the fire department if they burned the toast, but a smart house might not know any better).

Issues and Debate

Smart homes that plunge their owners into chaos when they go wrong have been a rich source of humor in popular culture. French comedian Jacques Tati's 1958 film *Mon oncle (My Uncle)* ridiculed the life of a wealthy industrialist and his wife whose smart home was continually developing hilarious faults. U.S. comedian Woody Allen's 1973 film *Sleeper* painted a bleak picture of a mechanized future in which "domestic-service menials" (robots with artificial personalities) do the chores and mechanical dogs provide companionship. Less than 30 years later, wealthy industrialist Bill Gates has constructed a smart home, and Sony has developed mechanical dogs.

The potential benefits of home automation systems are persuasive. Despite their cost, they can save around $600 per year by saving up to 40 percent of the energy used in a conventional home; advanced security systems alone can pay for themselves by preventing a single burglary. Smart homes save effort as well as time. Why should people cut lawns or vacuum carpets when computerized machines can do these jobs just as well? Home automation systems can also save lives, for example, by allowing parents to watch small children using camera systems mounted throughout the house and by setting off sprinkler systems in the event of fire.

Critics argue that these devices are gimmicks and question whether anyone is really so busy that they cannot open their own windows. Physicians have criticized even TV remote controls for removing a simple form of exercise from people's home lives. If domestic chores provide the only form of daily exercise for many people, smart homes could make people lazy, obese, and unhealthy. On the other hand, smart homes could help elderly people to enjoy their retirement in greater comfort and dignity, and could give disabled people much greater independence.

Early domestic appliances were advertised using images of smiling, apron-wearing women performing a variety of domestic chores. The idea was usually to encourage men to buy appliances for their housewives rather than to encourage men to do the housework themselves. Smart homes, which offer the prospect of complete automation, may finally solve the problem of who does the housework, by shifting the load entirely onto machines.

—*Chris Woodford*

RELATED TOPICS

Artificial Intelligence, Language Recognition Software, Microwave Communication, Robotics

BIBLIOGRAPHY AND FURTHER RESEARCH

BOOKS

Allen, Edward. *How Buildings Work: The Natural Order of Architecture*. New York: Oxford University Press, 1980.

Briere, Daniel, and Pat Hurley. *Smart Homes for Dummies*. Foster City, Calif.: IDG Books, 1999.

Carrow, Robert, and Nick Brown. *Turn on the Lights—From Bed! Electronic Inventions, Contraptions, and Gadgets Kids Can Build*. New York: McGraw-Hill, 1996.

Chant, Colin, ed. *Science, Technology, and Everyday Life, 1870–1950*. New York: Routledge, 1989.

Edwards, Alister, ed. *Extra-ordinary Human–Computer Interaction: Interfaces for Users with Disabilities*. New York: Cambridge University Press, 1995.

Gates, Bill. *The Road Ahead*. New York: Viking Penguin, 1996.

Gershenfeld, Neil. *When Things Start to Think*. New York: Henry Holt, 1999.

Negroponte, Nicholas. *Being Digital*. New York: Alfred A. Knopf, 1995.

Rosenberg, Paul. *The Complete Electronic House*. Upper Saddle River, N.J.: Prentice Hall, 1990.

PERIODICALS

Holmstrom, David. "House with an IQ." *Christian Science Monitor*, November 18, 1998, 11.

Pentland, Alex. "Smart Rooms." *Scientific American*, April 1996, 54.

Phillips, William. "Just Call HAL." *Popular Science*, June 1998, 36.

Phillips, William. "Smart Windows Get Smarter." *Popular Science*, December 1998, 40.

Stover, Dawn. "Reengineering the American Home." *Popular Science*, May 1997, 62.

INTERNET RESOURCES

General Resources

45 Practical Applications for Home Automation
    http://www.alhb.com/examples.html

"AIBO: The Freedom Fighter?" Rachel Ross, MSNBC News
    http://www.msnbc.com/news/274532.asp

Electronic House Magazine
    http://www.electronichouse.com

Home Automation Index
    http://www.infinet.com/~dhoehnen/ha/list.html

Home Toys: The Home Automation Library: Home Automation Standards
    http://www.hometoys.com/standards.htm

Popular Home Automation Magazine
    http://www.pophome.com/index.shtml

U.S. News Online: Bill Gates' House
    http://www.usnews.com/usnews/nycu/tech/billgate/gates.htm

Products, Retailers, and Manufacturers

Automated Living for Home and Business
    http://www.alhb.com

Friendly Machines
    http://www.friendlymachines.com/main.html

Home Plug and Play
    http://www.cebus.org/

Honeywell Inc. Product Information
    http://www.honeywell.com/products/

IBM Home Director
    http://www.pc.ibm.com/us/homedirector/homeowner.html

Multimedia Designs
    http://www.multimediadesigns.com

Sony AIBO Home Page
    http://search.sony.co.jp/~backup/www.world.sony.com/robot/top.html

SMART MATERIALS

Smart materials change their physical characteristics in response to some stimulus. For example, shape-changing piezoceramics—ceramic materials whose molecules realign in the presence of an applied electrical voltage—may bend or get longer when electrified. Light-sensing glass may darken when exposed to strong light. Shape memory alloys (SMAs) may return to a previous shape when heated.

Devices made from smart materials typically include sensors to recognize some physical change (vibration in a building or high temperature in an engine, for example), a control device (such as a computer) to initiate a signal in response (usually an electric current), and actuators that respond to the signal by initiating some physical change (perhaps sliding weights around in the building to counteract the vibration, or turning on the car engine's cooling fan). Smart materials may change the way that buildings and other physical structures are designed, and may be instrumental in creating a world in which objects respond and adapt to their surroundings.

Scientific and Technological Description

The term *smart materials* refers primarily to materials that change in response to some stimulus. In a broader sense, it refers to compounds, structures, or devices that employ such materials to monitor and control processes, to balance, strengthen, or stabilize buildings, airplanes, and other structures, or to make machines and similar devices lighter, stronger, and more efficient. Smart materials function as sensors and actuators. *Sensors* respond to environmental factors such as temperature or pressure by changing in some way that can be detected and used as a signal. *Actuators* respond to a signal by changing length, electrical resistance, or some other property to do useful work.

Types of Smart Materials

Chromogenic materials are smart materials that vary in transparency according to the degree of illumination, temperature, or electric voltage to which they are exposed. *Liquid crystals* are an example of chromogenic materials—their crystalline structures realign in the presence of an electric voltage and thereby change from transparency to opacity. The term *liquid crystal* does not name a type of material but a phase of many materials. The usual phases of matter are liquid, solid, and gas. The liquid crystal phase occurs in the temperature range between liquid and solid. Liquid crystal displays (LCDs) are found on many portable electronic devices, such as laptop computers.

Electrorheological fluids and solids alter their stiffness and density in the presence of an electrical field. These are liquids at room temperature, but when an electrical current is applied, they gradually become stiffer and less fluid until they actually solidify. In an electric automobile clutch, for example, if the motor spins a wheel immersed in an electrorheological fluid, with a parallel output wheel connected to the driveshaft, until voltage is applied, the motor runs but the driveshaft doesn't turn. As voltage is applied, the fluid becomes more viscous and starts dragging the driveshaft around. Although the electrorheological effect is not understood completely, it is known to occur primarily in colloidal dispersions (solid particles suspended, but not dissolved, in a fluid), where the diameter of the solid and electrically sensitive particles is much greater than the diameter of the electrically neutral molecules of the fluid in which the particles are suspended. Evidently, the solids "grip" each other more effectively when the electric field causes them to align.

Piezoelectrics generate electricity as a result of changing shape, and conversely, change shape in the presence of an electric current. Manufactured piezoelectric ceramic materials are the best known examples, although some

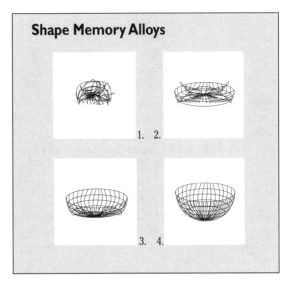

Shape Memory Alloys

1. 2.

3. 4.

Shape-memory alloys, such as Nitinol, are easily formed into any shape at a low temperature, but when heated beyond a certain temperature, return to a predetermined parent shape. Here, an antenna is shown returning to its parent shape. (Adapted with permission from Lockheed Martin Tactical Defense Systems, Akron.)

piezoelectric crystals do occur naturally (e.g., quartz, zinc sulfide, cadmium sulfide). When a piezoelectric crystal is connected to a source of electricity, the charged molecules in the crystal all line up. Since the molecules are also longer in one orientation than in the other, when they are all oriented in the same direction, the solid expands along that axis. Conversely, when no electricity is applied, the molecules are distributed randomly. If the material bends, the molecules on the side that is stretched align along their long dimension, and the molecules on the side that is compressed align along their shorter dimension. This alignment points the negatively charged ends of the molecules in one direction and the positively charged ends in the opposite direction. This generates a polarization that is electrically useful. For example, if the piezoelectric material is attached to a bridge girder, any bending of the girder will cause the piezoelectric material to generate an electric current that will send a signal that the bridge is flexing. In addition to ceramics and naturally occurring crystals such as quartz, polymers such as polyvinylidene fluorides and lead zirconate titanate exhibit piezoelectric characteristics.

Electrostrictives respond in approximately the same way as piezoelectrics, do, but these materials become "smart" only in the presence of a polarizing direct-current (dc) field. *Magnetostrictives* change shape or length in the presence of a magnetic field. These materials respond to magnetism as piezoelectrics respond to electricity: Asymmetric molecules of a magnetic material align themselves in the presence of a magnetic field and thus make the rod or actuator longer; conversely, when bent or twisted, magnetostrictives change their magnetic properties. The small security tags on some library books and retail goods contain magnetostrictives that act as switches, causing a signal to be emitted when they pass through the magnetic field of a security gate.

Fiber-optic sensors change the transparency, polarization, frequency, or arrival time of a light signal as a result of stretching or compressing the fiber. FIBER OPTICS can measure very small changes in very large dimensions, making them excellent candidates for detecting a building's shaking, bending, and stretching during high winds or earthquakes.

Shape memory alloys (SMAs) change from one shape to another when heated. These materials are formed into one shape (called the *parent shape*) at a high temperature, then another shape at a lower temperature. Like liquid crystals, SMAs exhibit phase changes not found in most materials. SMAs have two solid phases, called the *martensite* and *austenite* phases. When the material is formed into the parent shape and heated to about 500°C, the molecules enter the austenite phase and contract to the densest and most rigid alignment possible. As it cools below the transition temperature, the molecules loosen their grip and reenter the normal, martensite phase and can be shaped into a different configuration. When reheated, the molecules again seek the densest configuration and thus resume the austenite phase and the parent shape. A final type of smart material, *thermoplastics,* behave like SMAs but are made of plastic POLYMERS rather than metal alloys.

Historical Development

The field of smart material research is comparatively new and the technology is not yet in widespread use. Some of the phenomena involved have been recognized for some time, but only recently applied to design and manufacture. The piezoelectric effect, for example, was first described in 1880 by Pierre and Jacques Curie. Its first practical application was in the crystal radio sets of the 1940s. These radios were based on Greenleaf Whittier Pickard's turn-of-the-century discovery that galena, a naturally occurring crystalline mineral, would vibrate in the presence of radio waves. The vibration could be physically detected by touching the crystal with a metal point, then amplifying the vibration through headphones. In the early days of radio, cheap crystal radio sets were very popular in the unelectrified rural parts of the United States because they needed no batteries or other external power source—the piezoelectric vibration of the crystal in response to the broadcast power of the radio system was the only power used.

The phenomenon underlying liquid crystals was first noticed by Friedrich Reinitzer, an Austrian botanist, in 1888, in cholesteryl benzoate. One of the earliest commercial applications of liquid crystals was in the "Mood Ring" of the 1970s. The liquid crystal "stone" of the ring changed color in response to changes in skin temperature. A more refined and sensitive liquid crystal now employs the same effect in medicine as a color-changing "mapping" thermometer that

locates tumors by showing the variation in temperature over the surface of the skin.

Photochromic substances were first developed in 1964 by Corning Glass in a pair of sunglasses. The "shades" adjusted to light intensity automatically so that the wearer received approximately the same brightness through the lenses at all times. Similar auto-darkening materials have been applied to windows in automobiles, buildings, and greenhouses. In 1998, Gentex was selling a million self-regulating rearview automobile mirrors per year.

The first observation of the shape memory phenomenon was by L. C. Chang and T. A. Read in 1932, but it did not receive serious attention from researchers and engineers until the development of today's most widely used SMA, Nitinol ("Nickel Titanium Naval Ordnance Laboratory"). Nitinol is so named because its shape memory character was discovered by accident at that lab in 1961. William J. Buehler had created the alloy and was handing it around at a laboratory management meeting. The sample strip was bent many times; then David S. Muzzey heated it with his pipe lighter and was amazed to see it return to an earlier shape. Today, an entire industry has grown out of that pipe lighter.

Uses, Effects, and Limitations

In general, smart materials hold out the promise of making many buildings, machines, appliances, personal accessories, and clothing more efficient, useful, safe, and affordable. Photochromic substances are useful in sunglasses and rearview mirrors. Electrochromics can be used as glazing materials, which can be regulated to, for example, admit more sunlight on a bright but cold winter day. These materials may save billions of dollars in heating and air-conditioning costs.

One very important use of smart materials is in architecture, where they may make it possible to design structures that can adjust to natural forces such as winds and earthquakes. These buildings could be built of lighter and less expensive materials than those used today while actually increasing safety and damage resistance. Smart materials can also improve structure maintenance. Rust and other corrosion is the greatest threat to a structure such as a bridge. A smart materials maintenance approach could have the bridge signal its need for painting before it began to rust, perhaps by changing color in the presence of air much more quickly than rusting would begin so that a robot could see the color change and paint over the section in response. Or the materials of the bridge might even repair or protect themselves, by releasing additional protectants when exposed to air.

In addition, if new structures can be built of materials that actively adjust to, absorb, and dissipate unwanted sound energy, the noise level of many different environments can be reduced significantly. The damping effect of smart materials is already available in skis that sense unwanted vibrations and damp them out to achieve the best possible contact between the ski and the surface of the snow. In airplanes and automobiles, new materials in motor mounts, bearings, driveshafts, and other vibrating components may eliminate much of the sound-causing vibration rather than merely damping it.

Indeed, smart materials can both create and eliminate sound. Piezoelectric crystals that transform electricity into sound vibrations are used in singing greeting cards and neckties, flat speakers for stereo systems, and active sonar (sound navigation and ranging) systems in naval vessels and submarines. The helicopter and airplane may also benefit from other uses of smart materials. An airplane wing could be made to reshape itself directly, like a bird's wing, rather than requiring mechanically controlled surfaces. The helicopter's rotor mechanism might be replaced entirely by piezoelectric effects that reshape the rotor as it rotates. Another major area of smart material research and application is miniaturization, especially in biomedical appliances. A catheter, for example, can be guided to a tumor by using piezoelectric segments to change shape at bends and junctions as it passes up the vein.

Shape memory alloys are used as actuators, *superelastics,* and *retainers.* The force exerted by the material as it tries to return to the parent shape can be considerable, so SMAs are suitable for use in valves, motors, robot arm and leg joints, and similar applications. As actuators, SMAs are particularly well suited for moving things short distances in response to an electric current. They are used in consumer products such as computer disk ejectors, self-powered hot-air exhaust fans, antiscalding shower valves, and coffeepot regulators. In robotics, a *muscle wire* that contracts when heated by an electric current can move an arm, leg, or finger repeatedly An experiment conducted as part of NASA's *Mars Pathfinder/Sojourner* mission in 1997 used a muscle wire to cover and uncover a sensor on command from Earth. Superelastic SMAs are 10 times as elastic as steel. They make excellent dental retainer wires, bra underwires, and cellular telephone antennas. This may be the future of smart materials—to replace traditional complicated mechanisms with simpler, cheaper, and more effective solutions in everyday life. Smart materials can also be integrated with silicon chips to form smart microelectromechanical systems (MEMS), which provide new, smaller, and more efficient sensors. Disposable blood pressure sensors using MEMS are already on sale.

Issues and Debate

Each smart material comes with its own environmental considerations. Manufacturing piezoelectrics from lead compounds, for example, raises the many dangers and issues involved in working with lead, longtime exposure to which is toxic to human beings. In addition, smart materials make possible some things that many people find morally offensive. The U.S. Army, for example, is developing a smart bullet that uses piezoelectric actuators to steer a bullet to a target illuminated by a laser beam, just as smart bombs are already guided to precision targets (see SMART WEAPONS).

Such a system might increase the effectiveness and efficiency of an army tremendously, but effectiveness and efficiency are not virtues if employed for morally undesirable ends, and some people object to creating weapons that give people virtually no chance against them. Smart material research will no doubt be applied to build "better" weapons, with all the moral consequences that follow.

In medicine, it now seems inevitable that someone will develop an artificial heart. But even this would not be an unmixed blessing, as such an organ would be tremendously expensive and would not be available to everyone. Critics charge that such a development would help bring about longer, healthier lives—but only for the wealthy few—by using research and engineering funds that could have been directed toward improving the health and welfare of the general population.

More generally, however, smart materials pose the philosophical threat of blurring the line between the natural and the artificial in two ways: by creating artificial devices that do what humans have in the past relied on nature to do, and simultaneously, by making artificial devices more nearly "organic" in their structure and operations. Robots could, perhaps, incorporate smart materials and systems that would respond to stimuli with the same combination of instinctive and learned behaviors that humans employ rather than the preprogrammed response of today's simple machines. Then we might develop a robotic rooster, to signal dawn's arrival, or a robotic dog, to give companionship, or ultimately, a robotic person to be our friend and helper.

—Roy Weatherford

RELATED TOPICS
Artificial Organs, Composite Materials, Fiber Optics, Robotics, Surgical Robotics

BIBLIOGRAPHY AND FURTHER RESEARCH
BOOKS
Culshap, Brian. *Smart Structures and Materials*. Norwood, Mass.: Artech House, 1996.
Otsuka, K., and C. M. Wayman, eds. *Shape Memory Materials*. Cambridge: Cambridge University Press, 1998.

PERIODICALS
Berardelli, Phil. "Smart Materials Let Planes and Buildings Morph." *Insight on the News,* March 31, 1997.
"Five Easy Pieces: A Smart Material Could Make Now-Complex Helicopter Rotors Simpler—and Safer." *Popular Science,* September 1997.
Flinn, Edward D. "Shaping Quieter Rotors with Smart Materials." *Aerospace America,* July 1998, 24.
Gibbs, W. Wayt. "Scientific American Explorations: Smart Materials." *Scientific American,* May 1996.
Noor, Ahmed K., Samuel L. Venneri, Donald B. Paul, and James C. I. Chang. "New Structures for New Aerospace Systems." *Aerospace America,* November 1997.
Santo, Brian. "Smart Matter Program Embeds Intelligence by Combining Sensing, Actuation, Computation: Xerox Builds on Sensor Theory for Smart Materials." *Electronic Engineering Times,* March 23, 1998.

INTERNET RESOURCES

"Emerging Smart Materials Systems," by Harold B. Strock
http://world.std.com/~hbstrock/sta/exec.html

Global Link on Active Materials and Smart Structures
http://www.rpi.edu/~huangl2/smart.html

"Smart" Materials for Control and Automation
http://www.design-engine.com/d-engine/smartmaterials.html

Smart Materials/Structures Home Page
http://bass.gmu.edu/~nmutlu/smart_mat.html

Smart Materials: The Copernicus Programme of the European Commission
http://smart-www.ae.ic.ac.uk/

Tech Monitoring: Smart Materials
http://future.sri.com/TM/aboutSM.html

SMART TRAINS: COMMUNICATIONS- AND SATELLITE-BASED CONTROL

Communications-based train control (CBTC) is the most general of several acronyms that describe a central concept—integrating advanced communications technologies such as computers and satellites into railroad infrastructures. Advocates of CBTC espouse many goals, the most prominent of which are to increase safety, pinpoint the location of trains, allow more trains on the same tracks, and lower operational costs.

Aspects of CBTC have been demonstrated in some current rail systems, but the concept continues to be developed by engineers who envision an encompassing network of trains, positioning systems, and computer controls that speed people and cargo with unprecedented safety and efficiency. CBTC applies to many railroad systems, including conventional passenger and freight trains, high-speed rail, locomotives, and subways. Multiple government-sponsored and private industry projects are under way, but the implementation of a national CBTC system remains questionable and comes with a multi-billion-dollar price tag.

Scientific and Technological Description

Related to CBTC are the terms *positive train control, positive train separation,* and *advanced automated train control.* All are variations on a similar theme, and they are rooted in a conventional railroad concept—that of automated train control systems. These systems consist largely of transistors installed under trains that absorb electrical signals from the rails and transmit them to computers within the trains' control center. The computers monitor the trains' speed and apply the brakes automatically if engineers should fail to do so.

Several projects seek to build on automated train control systems. One CBTC system, the PINPOINT Locomotive Tracks System, developed by General Electric and the Harris Corporation, allows railroad operators to locate trains to within 328 feet (ft.) of their actual location anywhere in the world. This is done through the use of satellites that track the trains and send their locations to centralized computer stations. The satellites can also transmit other pertinent information, such as how much fuel is available on the train, how fast the train is traveling, and how far it needs to go to reach its destination.

Also using satellites as a basis for technology is a *positive train control* (PTC) program being developed by Union Pacific railroad. PTC focuses on the safety aspects of CBTC, such as the automatic application of brakes at excessive speeds. The Union Pacific model uses the nationwide differential GLOBAL POSITIONING SYSTEM (GPS), which consists of more than 23 satellites deployed by the Defense Department and now also used by private parties to pinpoint the locations of objects on Earth. The satellites track each train and a digital radio network connects the trains to a control center. Onboard computers transmit the speeds of the trains to computers in the control center. The control center computers maintain a real-time database of signals and accessibility of tracks so that they can stop trains if they are approaching a closed track, are ignoring a red signal, or are traveling too quickly. The computer also can be linked to signals at highway–rail intersections so that drivers of cars at railroad crossings can be alerted well in advance of a train's approach.

One CBTC-related concept chooses radio technology over satellites. The *advanced automated train control* (AATC) system was developed for San Francisco's Bay Area Rapid Transit (BART) District. In the AAT C system, transmitters and receivers are installed on trains and along the track at widely spaced intervals. These radios communicate information to and from control stations. Computers in control rooms monitor the BART track in sections, or control zones. Distributed radios form a network along the tracks, and trains are assigned time slots during which radio transmissions are permitted; this timing allows up to 20 trains to be tracked within each control zone. The position of trains is determined by measuring how long it takes for a radio transmission to go from one transmitter to a receiver. Researchers report that the AATC technology enables all trains in the system to be located within an accuracy of about 15 feet, a great improvement over most systems' existing ability to locate trains within about 500 feet.

Historical Development

The Industrial Revolution, beginning in the mid-18th century, produced an influential railroad industry, but its safety and communications systems were crude. At switching yards, engineers were given flags to thrust out of their trains to indicate that they had the right-of-way. Brakes were applied via blocks of iron pressed against the turning wheels; that system has survived today in large part because the blocks are inexpensive and easy to replace.

As technologies progressed and railroads grew, engineers sought improved efficiency, and in 1959, in

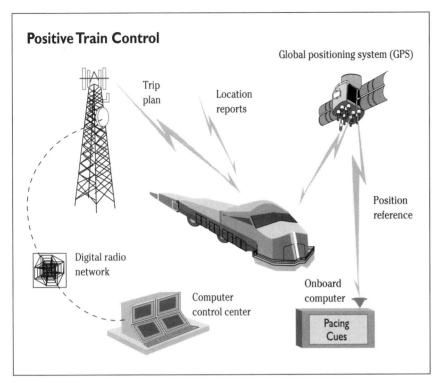

Positive Train Control

Trip plan

Location reports

Global positioning system (GPS)

Position reference

Digital radio network

Computer control center

Onboard computer

Pacing Cues

In the Positive Train Control system shown here, a computer control center keeps track of many trains' locations and speeds, and sends out instructions to individual trains as to how they should alter their schedule or speed. Onboard systems display the instructions to the train crew, as well as gather location data from satellites and transmit it back to the control center. (Adapted from GE Harris Railway Electronics, Whitepaper, Fig.1.0-1.)

Stockholm, Sweden, a train system called the Automatic Pilot was unveiled that practically drove itself. In addition to increasing efficiency, engineers were also pursuing safety issues. In the United States, a fatal rail accident in Darien, Connecticut, in 1969 prompted the National Transportation Safety Board (NTSB) to encourage a *positive train separation system* on a national level for the first time. In the 1970s, the NTSB issued a series of advisements to railroad operators on how to keep trains apart safely. Research continued into the feasibility of various ways of improving railroad safety, and by 1980, railroad and worker-related accidents were down by 79 percent.

In 1992, the U.S. Congress enacted the Rail Safety Enforcement and Review Act, which instructed the Federal Railroad Administration (FRA) to seek new safety initiatives. Two years later, FRA representatives reported to Congress that new systems that would help trains avoid collision could be installed on high-risk corridors by the year 2000, but they balked at making the systems legally mandatory. By 1995, the FRA was becoming increasingly interested in integrating GPS with railroads, and in June, it published *Differential GPS: An Aid to Positive Train Control*. In 1996, the FRA converted 22 ground-wave towers into differential GPS stations; these 300-ft. towers had been used by the U.S. Air Force as radio transmitters for an emergency network system.

Satellites continued to emerge as a focus for railroad researchers in the 1990s. In 1993, a train collision in Kelso, Washington, that killed five people prompted the Union Pacific and Burlington Northern–Santa Fe railroads in 1995 to test a satellite tracking system for 845 miles of rail stretching from Oregon to Washington. The Kelso accident preceded two major rail disasters in 1996. Within weeks of each other, two engineers on commuter trains failed to heed a signal to slow down and collided with other trains, spurring the NTSB to ask Congress for a timetable for implementation of advanced train controls on a national level. The FRA also demonstrated PTC for the first time in 1996; the agency used 20 miles of rail between Kalamazoo and Niles, Michigan.

The biggest boost for PTC so far came in 1998, when the FRA agreed with the Association of American Railroads (AAR) to build a PTC project in Illinois. The project should serve as a prototype for other PTC systems. The AAR committed $20 million to a partnership with the Illinois Transportation Department, and the FRA allotted a further $3.8 million. The project will support high-speed passenger service as well as freight trains by using global positioning locators and computer-controlled advanced radio systems.

Other countries, including Japan, France, and Germany, are also exploring CBTC technologies. These countries feature railway systems that are far more advanced than those in the United States. They feature high-speed and related magnetic levitation trains (see HIGH-SPEED AND MAGLEV RAILWAYS) that already use some principles of CBTC, and in these countries as in the United States, private companies and government officials are looking to satellites to improve safety and efficiency.

Uses, Effects, and Limitations

There are 180,000 miles of railroad track in the United States on which 19,000 locomotives and 1.2 million railcars travel each year. CBTC is meant to help monitor these trains, put more of them on tracks, and increase their speed, efficiency, and safety. Of these uses, perhaps safety is discussed most often at national levels. Although most rail accidents result from cars trying to beat trains at crossings and not from train-to-train collisions, railroad experts have estimated that fatalities occur most often from the latter. Moreover, 80 percent of crashes are caused by human error. CBTC and, specifically, PTC would diminish the human role in train operation by controlling train movements automatically when necessary.

CBTC's safety features may appeal to consumers, but the technology's promise to put more trains on the same track offers a strong incentive to railroad operators. Building new rails is expensive and often impractical, so industrialists tend to make do with what exists. To increase efficiency, railroad owners traditionally have looked to technological developments within locomotives and trains themselves, and they put an average of 600 to 800 new locomotives on the tracks each year. But CBTC shifts the emphasis away from new locomotives and to a systemwide network that coordinates many components to increase rail use. Because satellites can track the trains and computers can regulate speed and braking, less space is needed between trains and more cars can fit on one track.

Studies at the Draper Labs in Cambridge, Massachusetts, in the late 1990s indicated that PTC could increase track capacity by 15 to 20 percent. AATC, for example, provided BART with the ability to operate twice as many trains on existing tracks. The program's developers have increased the capacity of one line by two-thirds, allowing 30 trains to run each hour rather than 16. Without AATC, advocates said, the increase would have required building a second tube and set of tracks through San Francisco at an estimated cost approaching $4 billion.

CBTC also offers potential business benefits. Owners of trains can track their fleets and issue further instructions to engineers through two-way communications systems. CBTC is less wearing on equipment and reduces job stress for dispatchers, say advocates, because computer-controlled instructions are more precise than those given by humans. For consumers, CBTC could mean more choice among trains that arrive and depart on time. What remains unclear is how much, if any, of the cost of a CBTC overhaul would be passed on to consumers.

Issues and Debate

There are so many facets to CBTC that interoperability has emerged as the top issue in the field. With so many groups researching various aspects of CBTC, some experts fear that competing technologies will be created that are incompatible with one another. A state may choose a PTC system that does not work with a neighboring state's system, for example. Aware of that potential pitfall, the FRA is working to ensure that technologies in development will be compatible by creating industry standards. Also, in March 1998, the agency announced that it would launch a four-year research project into PTC on 123 miles of rail between Springfield and

San Francisco's Bay Area Rapid Transit (BART) system has implemented an Advanced Automated Train Control (AATC) system that uses transmitters and receivers installed on trains and along tracks to communicate information to and from control stations. The technology enables all trains to be located within about 15 feet (Corbis/Charles E. Rotkin).

Chicago, Illinois. The project aims to coordinate the FRA-sponsored Incremental Train Control System, Amtrak's Advanced Civil Speed Enforcement System, and other PTC systems to show how different PTC systems can operate on the same line. Conrail also has contracted Railroad Electronics Business of Rockwell to create a platform that can be used by a range of PTC systems.

Also being created to address coordination problems is the National Intelligent Transportation Infrastructure (NITI). The concept is described as a partnership among government bodies and private-sector organizations, and refers to all aspects of intelligent transportation systems (ITSs), including CBTC. The idea is to create a communication and information backbone that will allow ITS products and services to work together. The NITI will serve as a central warehouse of information on transportation and management systems so that data can be tracked across the country. It should also be a welcome source of data for use in transportation planning and statistical tracking.

Implementation of a national CBTC system would be expensive (a PTS system alone is estimated to cost between $1 billion and $3 billion), and it might not be equally effective on different kinds of lines. To address this potential problem, the FRA wants to identify specific corridors, or classes of corridors, for which advanced signal and train control systems would be most cost-effective. In 1995, the U.S. Transportation Department developed a model to analyze the potential risk reduction for specific railroad corridors if equipped specifically with PTC-like systems. The researchers decided that many factors had to be considered in any corridor before PTC application is employed, including historical accident data, the density and mix of freight and passenger trains operating in the corridor, the type of existing signal and train control systems, the number of tracks, track topography, population densities, passenger data, and freight types and densities.

But smaller transportation facilities may find installing a CBTC system to be less expensive than replacing existing technologies. PTC advocates reported that a system such as the one used in San Francisco can be installed at half the cost of traditional systems, and radio installation can be accomplished without interrupting operation of the existing train control equipment. Such a system also can be installed on existing tracks without the need to build new ones, and radio transmitter-based PTC is affordable because transmitters can be placed far apart, thus decreasing the amount of equipment needed.

Finally, the fact that CBTC can put more trains on a single track may create problems as well as benefits for railroad operators. Railroad owners may see that costs saved on repair are instead consumed by the need to maintain a more active fleet. Also, the use of more trains demands the employment of more people or the extension of their work responsibilities. Although more jobs may help the economy, they also may drain the coffers of railroad owners with limited budgets.

—*Christina Roache*

RELATED TOPICS

Global Positioning System, High-Speed and Maglev Railways, Smart Highways

BIBLIOGRAPHY AND FURTHER RESEARCH

BOOKS

Andrews, H. *Railway Traction: The Principles of Mechanical and Electrical Railway Traction.* New York: Elsevier Science Publishers, 1986.

Automatics Railways: A Convention Arranged by the Railway Engineering Group, September 23–25, 1964. London: Institution of Mechanical Engineers, 1965.

Leach, Maurice, ed. *Railway Control Systems.* London: A&C Black, 1991.

Rail Traction and Braking: Selected Papers from Railtech 96. London: Institution of Mechanical Engineers, 1996.

Railway Engineering, Systems, and Safety: Selected Papers from Railtech 96. London: Institution of Mechanical Engineers, 1996.

Railways in the Electronic Age. New York: Institute of Electrical Engineers, 1981.

PERIODICALS

"FRA to Conduct Positive Train Control Projects over Four Years." *Global Positioning and Navigation News,* March 11, 1998.

"Harmon + Hughes = Technological Landmark: A Giant Step for Wireless Train Control." *PR Newswire,* December 1, 1997.

Howe, Peter J. "Efforts Are Intensified for Safety on Rails: Satellite Surveillance among Systems Debated." *Boston Globe,* April 1, 1996, 1.

MacDonald, Neil. "DOT Starts Partnership Program for Transportation Initiatives." *Federal Technology Report,* November 5, 1998, 6.

"Railroads to Build in Safety Success with New Initiatives." *PR Newswire,* April 1, 1998.

INTERNET RESOURCES

Advanced Automatic Train Control
http://www.bts.gov/ntlDOCS/attc.html

Bureau of Transportation Statistics
http://www.bts.gov

Business Benefits of PTC
http://www.volpe.dot.gov/frarnd/PTC-Business-Benefits.htm.

Current Implementation of PTC in the United States
http://www.volpe.dot.gov/frarnd/PTC-RSAC-1.htm

Directory of World Wide Web Rail Sites
http://www.ribbonrail.com/NMRA/rrgov.html.NMRA

Federal Railroad Administration
http://www.fra.dot.gov

Federal Transit Administration
http://www.fta.dot.gov

Frequently Asked Questions about PTC
http://www.volpe.dot.gov/frarnd/PTC-faqs.html

Index to PTC Sites
http://www.volpe.dot.gov/frarnd/FRA-PTC.html

List of Related Sites
http://www.bts.gov/ntl/subjects/rail-gen.html

National Transit Geographical Information System
http://www.bts.gov/smart/cat/ntg.htm

National Transportation Library
http://www.bts.gov/ntl

Newsletter of the ITS Cooperation Deployment Network
http://www.nawgits.com/icdn

Text of FRA Administrator Jolene Molitoris' Letter on Positive Train Control to Class 1 Railroad CEOs.
http://www.volpe.dot.gov/frarnd/PTC-Administrator.html

Volpe National Transportation Systems Center
http://www.volpe.dot.gov

What Is PTC?
http://www.volpe.dot.gov/frarnd/PTC-Implementation-map.html

SMART WEAPONS

Smart weapons are the latest form of guided missile or bomb to be used by military forces. They differ from conventional missiles and bombs in that they have highly sophisticated control and guidance systems. They can be launched from land, sea, or air, depending on the type of smart weapon used, and their flight paths can be altered right up to the moment of impact with a target.

Elaborate claims have been made for the accuracy and efficiency of these costly weapons, and we speak of a surgical strike in describing the way that smart weapons can be used to destroy a target in a built-up area with little apparent damage to surrounding buildings. In practice, however, these claims do not always stand up to close scrutiny.

Scientific and Technological Description

Smart weapons are general-purpose bombs or missiles that have been fitted with additional components that allow the weapon to be steered, either manually or automatically, on to a target by means of a guidance system. The guidance system is mounted in the nose of the weapon and can be one of three types: electro-optical, laser, or infrared.

The electro-optical guidance system works by sending television pictures of the target area to a member of the aircrew who released the weapon. Using these pictures, the aircrew can choose to lock the weapon onto the target at the appropriate moment or to actually guide the weapon directly onto the target, maintaining control up to the moment of impact.

The essential components of a laser-guided missile are the computer-control group (the brains of the weapon), small, adjustable wings attached to the front of the missile to provide directional control, and a wing assembly (another small set of wings) at the rear of the missile to provide lift during flight. The flight path of a laser-guided weapon, when released from an aircraft, divides into three phases: ballistic, transitional, and terminal guidance. During the ballistic phase, which occurs just after the weapon is released, the missile follows the trajectory of the releasing aircraft. The transitional phase begins as soon as the target has been recognized. This point, known as *target acquisition,* marks the moment at which the weapon aligns its direction of flight to the target.

Terminal guidance begins as the missile closes in for the kill. At this point, laser light reflected off the target is picked up by a photosensitive detector on the missile. This triggers the computer-control group to make any necessary in-flight adjustments to ensure that the missile stays locked on to the reflected laser light. From this point onward the missiles flies down the path of the reflected beam of laser light until it explodes on impact with the target. The original beam of laser light, which was reflected off the target, is provided by a laser designator. This can be operated from the aircraft that fired the missile, another aircraft, or by a special forces unit on the ground.

Laser designators, and the photosensitive detectors aboard the laser-guided missile, operate on a pulse-coding system. Each pulse code is based on a simple three-or-four-digit pulse repetition frequency. By setting the same code for the designator and the missile's onboard detector, it is possible to ensure that the missile will follow only those laser beams sent out by the appropriate designator. This allows multiple simultaneous attacks to be made on a target from aircraft dropping weapons set to different pulse codes.

Smart weapons fitted with a guidance and control system that responds to infrared radiation seek out heat sources such as engines, generators, and even air-conditioning vents. Another category of smart weapon is the *cruise missile.* This is essentially a pilotless aircraft that can be either programmed to fly directly to its target or guided to it from the firing base. The Tomahawk cruise missile is a classic example of this type of weapon. It flies at over 500 miles per hour, remaining close enough to the ground to evade enemy radar. In programmed flight, constant checks on the supposed location of the missile are made using satellite-based global positioning technology, and any necessary corrections are carried out by the onboard systems (see GLOBAL POSITIONING SYSTEM).

Historical Development

Smart weapons could not have been built before the 1950s. Post–World War II advances in the fields of electronics and computing, combined with improvements in our understanding of aerodynamics and propulsion, came together to make these weapons a practical proposition. Smart weapons can be distinguished from conventional weapons by virtue of their guidance systems, and it is the extraordinary extent to which the path of the weapon through the air during flight can be controlled that defines it as a smart weapon.

The first weapon of this type was developed and adopted by the United States in the late 1950s. Called the AGM-12 (Aerial Guided Munition-12), it was a rocket-powered weapon fitted with a visual tracking system and a radio-controlled guidance system. The first missile fitted with an infrared tracking system was the AGM-45, also known as the Shrike, which was used in Vietnam against enemy radar

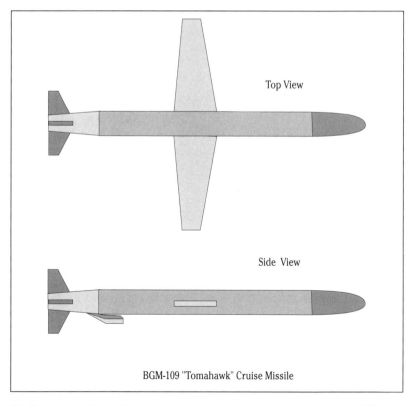

Top View

Side View

BGM-109 "Tomahawk" Cruise Missile

The Tomahawk cruise missile is a computer-guided smart weapon that travels at about 550 miles per hour. The missile is guided by onboard systems, including the global positioning system (GPS) to within 30 feet of its target. (Adapted with permission from Softwar.)

sites. The 1960s and 1970s also saw the extensive use of missiles fitted with electro-optical devices. Commonly called *Walleyes,* these were unpowered bombs that were coaxed on to their targets with the aid of television guidance.

The first of the modern smart weapons made their appearance during the U.S.-led attacks on Baghdad at the start of the Persian Gulf conflict in early 1991. Journalists based in the city reported seeing missiles fly past their hotel windows that were apparently following a street map of the city. Many claims were made for the accuracy of these weapons, and it was during this period that the term *surgical strike* was first coined to describe the way in which smart bombs were used in built-up areas against specific targets with the minimum of accidental damage to surrounding buildings. Unfortunately, any computer-based guidance system is only as good as the information with which it is programmed. According to the U.S. Air Force's official five-volume *Gulf War Air Power Survey,* of the 1460 air strikes carried out against Iraq's Scud missile batteries, not a single one reached its target.

At the present time, much of the work on advancing smart weapons technology is concentrated on increasing the accuracy of these weapons. A measure of the success of this work was seen during the 1999 conflict over Kosovo, where improvements to the NATO forces' military satellite global positioning system meant that cruise missiles could be delivered on target to an accuracy within less than 4 inches.

Uses, Effects, and Limitations

The principal advantage of smart weapons over conventional weapons is that they allow accurate strikes to be made against targets with little or no risk to the person who fired the weapon. This does not, however, mean that these weapons are without flaws. With the exception of sophisticated guidance and control systems, smart weapons are very similar to ordinary bombs and will, on occasion, fail to explode. That said, it is far more likely that a smart weapon will not perform according to plan as a result of a fault related to its guidance or control systems.

Mechanical failure of the type that can affect everything from automobiles to answering machines can result in a weapon flying off target. Simple human error can also result in the laser designator being directed against the wrong target. A tragic example of this kind of error occurred during the NATO bombing of Belgrade in 1999, when the Chinese embassy was hit from three different sides by smart weapons that had been directed with great accuracy against the wrong target.

Laser-guided weapons can also be sent off target by the scattering effects of rain, dust, or smoke particles, all of which refract the laser light reflected from the target and confuse the missile's guidance system. Fog and low clouds also reduce the chances of the weapon reaching its intended destination by obscuring the view of the person whose task it is to find the target with the laser designator. This reduces the amount of guidance time available for the missile and therefore decreases the probability that the target will be hit.

The phrase *one target, one bomb* is said to describe the way in which smart weapons are used. In practice, however, this is seldom the case, and it is often necessary to use several weapons against a target to ensure that it has been destroyed. During Operation Desert Storm, the U.S.-led coalition that bombed Iraq in the Persian Gulf crisis of 1991, for example, an average of four smart weapons were used against each designated target.

Issues and Debate

The notion that a war can be won by means of surgical strikes—characterized by minimal damage and loss of life—is an attractive one. It appeals to politicians, who understand that public support for a conflict is often closely related to the number of casualties suffered. Surgical strikes also appeal to military personnel, who would otherwise be expected to risk their lives in the course of the conflict. During the conflict over Kosovo, not a single NATO Alliance life was lost to enemy fire. This was due mainly to a policy of high-altitude bombing and the use of smart weapons technology. Concerns have been raised over the reliance on such tactics, however, with some people claiming that war without risk to one side makes conflict more likely.

Another area of concern is the price of smart weapons technology. Although smart weapons represented only 8 percent of the total munitions dropped on Iraq during the Persian Gulf conflict, they accounted for 84 percent of the total munitions costs. This is not surprising given that every Tomahawk cruise missile used during the conflict cost the American taxpayer $1,000,000. Technological improvements in these weapons since that time have served only to increase these costs.

Military leaders justify the expenditure on these weapons by emphasizing the improvements in efficiency that smart weapons offer over conventional technology. They point out, for example, that during World War II it may have taken up to 900 bombs to destroy an aircraft hangar that can now be destroyed using a single smart weapon. Regardless of taxpayers' feelings about paying such high prices for smart weapons, these guided bombs represent the state of the art in modern weaponry. That being the case, citizens will probably see their nations buy more and better smart weapons in coming years, not less.

—Mike Flynn

RELATED TOPICS
Global Positioning System, Lasers

BIBLIOGRAPHY AND FURTHER RESEARCH
BOOKS
Carus, W. Seth, Center for Strategic and International Studies. *Ballistic Missiles in Modern Conflict.* Westport, Conn.: Praeger Publishers, 1991.

Dunnigan, James F. *How to Make War: A Comprehensive Guide to Modern Warfare for the Post-Cold War Era.* New York: Quill, 1993.

Laur, Timothy M., et al. *Encyclopedia of Modern U.S. Military Weapons.* New York: Berkley Publishing Group, 1998.

Oliver, David, and Barton Strong. *Smart Weapons: Top Secret History of Remote Controlled Weapons*, int. Kenneth Israel. New York: Welcome Rain, 1998.

PERIODICALS

"Menacing Subs, Smart Torpedoes" (editorial). *New Scientist,* October 5, 1996, 11.

Morrison, Phillip, Kosta Tsipis, and Jerome Wiesner. "The Future of American Defense." *Scientific American,* February 1994, 20–27.

Zorpette, Glenn. "A Day at the Armageddon Factory." *Scientific American,* September 1996, 20.

INTERNET RESOURCES

"Crisis in the Gulf: Forces and Firepower," BBC News Online
http://news1.thdo.bbc.co.uk/low/english/events/crisis_in_the _gulf/forces_and_firepower/newsid_55000/55686.stm

History and Development of Autonomous Weapons
http://www-cse.stanford.edu/class/cs201/current/Projects/ autonomous-weapons/html/history.html

Laser Guided Bombs
http://www.fas.org/man/dod-101/sys/smart/lgb.htm

"Smoked Out: Smart Bombs," *Newsweek*
http://www.newsweek.com/nw-srv/issue/14_99a/tnw/today/ps/

"U.S. Bombing: The Myth of Surgical Bombing in the Gulf War," by Paul Walker
http://deoxy.org/wc/wc-myth.htm

"U.S Bombs Not Much 'Smarter,'" by Fred Kaplan, *Boston Globe*
http://www.fas.org/news/iraq/1998/02/20/us_bombs_not_much _smarter_.htm

SOLAR ENERGY

Solar energy technologies harness the sun's radiation and convert it to usable forms of energy. Solar thermal and photovoltaic systems are the two main variations of solar energy technology. The basic design of a solar thermal system involves the collection of solar radiation and the transfer of heat to a fluid, which can be either water or another fluid with a lower boiling point. This heat-transfer fluid is then used to heat a large amount of water, creating steam to drive an engine. In addition, this water can be used as the hot water for a building, or it can heat air to be distributed throughout the structure. Photovoltaic panels, on the other hand, collect sunlight and convert it directly to electricity that can be used in a manner similar to conventional electricity.

Solar energy research began in the middle to late 19th century, and most current designs for solar thermal systems were developed during that time. Solar energy is now a versatile technology, powering calculators, heating homes, and driving turbine generators. However, although the U.S. government has taken steps to encourage solar energy use, conventional energy sources are still more cost-efficient for most consumers.

Scientific and Technological Description

Solar Thermal Systems

There are two principal types of solar thermal systems. The first, the *flat-plate solar collector,* contains a transparent cover and a dark metal absorber at the bottom. A number of plates can be connected together with tubes on the roof of a building. A heat-transfer fluid moves through the tubes, picking up heat from the absorbers. A pump then moves the water to a storage tank. A heat exchanger transfers the heat from the water either to the air, to be distributed throughout the building, or to a hot-water heater.

The second type of solar thermal technology is the *solar concentrating collector system.* There are three design variations that use a reflection device to concentrate solar rays. The first is the *parabolic dish system,* in which dish-shaped mirrors reflect solar rays onto a receiver device situated above the dish. The mirrors must track the sun so that they reflect the greatest amount of sunlight throughout the day. The second type, the *central receiver system,* involves a field of thousands of mirrors (called *heliostats*) that surround a tower containing a heat-transfer fluid. These mirrors must track the sun as well, so that they are midway between the tower and the sun. The *parabolic trough* is the third system. A barrel-shaped trough (similar to an oil drum cut in half lengthwise) reflects solar radiation along a line rather than a point (see figure). This heats the heat-transfer fluid in a tube at the focal line of the parabola. The trough must follow the sun in only one direction, east to west, as opposed to the other systems, which must rotate in an arclike fashion around a central receiver. In general, the heat from the fluid in each of these systems is used to heat water, creating steam to power a turbine generator.

Photovoltaic Cells

Solar radiation can also be converted directly into electrical energy through the use of a photovoltaic cell. This is a solid-state device made of silicon. When exposed to light, the elec-

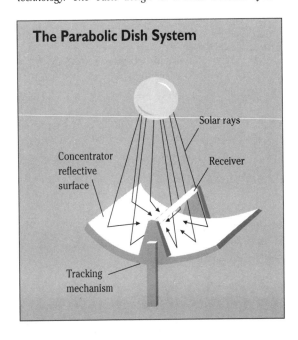

The Parabolic Dish System

Solar rays

Concentrator reflective surface

Receiver

Tracking mechanism

trons in the cell are excited by sunlight and move through the silicon. This movement creates direct current electricity, the type found in batteries. This must be converted to alternating-current electricity if it is to power household appliances.

Historical Development

The development of solar energy technology began in the 19th century and has taken two different paths, one for solar thermal systems and one for photovoltaics. The first instance of a thermal system converting solar energy to mechanical power was recorded in 1860. Auguste Mouchout, a French mathematics instructor concerned about his country's dependence on coal, began working on a solar-powered motor. His basic design involved a reflector dish with slanted edges that focused the rays of the Sun on a glass-enclosed cauldron of water. In 1865, he was able to run a steam engine with solar energy. By 1878, he used a solar-powered engine to power a refrigeration device. Eventually, he was able to generate ½-horsepower of energy.

In 1878, William Adams, an Englishman in Bombay, India, improved on Mouchout's design. He set up flat, silvered mirrors in a semicircular rack that could be rotated to reflect the Sun most intensely. The mirrors projected the solar rays onto an enclosed, stationary boiler, creating steam to run an engine. He ultimately ran a 2½-horsepower engine. Many modern solar plants follow Adams's original design of flat or slightly curved mirrors arranged in a semicircular track that is rotated to follow the Sun.

Charles Tellier, a Frenchman considered the father of refrigeration, worked on solar energy devices during the 1880s. He placed a solar collector on his roof, using ammonia as a working fluid for heat transfer rather than water, due to its lower boiling point. His solar collector consisted of an insulated bottom with a top of glass. This was filled with ammonia, which, when heated by the Sun, formed pressurized ammonia gas that could power a water pump. During daylight hours, his device could pump 300 gallons per minute.

Also in the 1880s, John Ericsson, the designer of the Civil War warship the *Monitor,* worked on a parabolic trough system. This mirrored trough focused solar rays in a line, eliminating the need for much of the complex tracking machinery in Adams's system. This method is also commonly used today because of its compromise between efficiency and ease of operation.

The next major advance in solar energy came in 1904, when inventor Henry E. Willsie began to develop a solar-powered motor that could operate at night. He used flat-plate collectors to heat hundreds of gallons of water that he kept heated all night in an insulated basin, powering an engine even after the Sun went down. Willsie began a company, but it failed, and much of solar energy research began to fade as World War I approached. Coal and oil producers acquired stable supplies and markets, and no desire for further research into solar thermal systems emerged until the 1970s. Solar power plants now exist that use solar concentrating systems to create large amounts of electricity for general use.

The history of photovoltaic technology began in 1839, when Edmund Becquerel, a French physicist, first noted the photovoltaic effect. He realized that an electric voltage occurred when he exposed two electrodes in a weak conducting solution to sunlight. However, the effect was mostly a curiosity and its potential was not fully developed for many years. The next large development in photovoltaics occurred in 1953, when researchers developed the ability to produce highly pure crystalline silicon. In 1954, Bell Telephone Laboratories used this technology to create a photovoltaic cell that could convert 4 percent of the incoming sunlight into electricity. The U.S. space program incorporated photovoltaic technology in 1958. Because photovoltaic panels are a relatively light and renewable medium for energy, they were used to power a small radio on a U.S. space satellite, and photovoltaic technology is still used in the space program. However, because of their high production cost, photovoltaics were not considered an efficient medium for general solar energy use.

In 1973, the Arab oil-producing nations of the Middle East embargoed oil shipments to the United States, protesting U.S. support of Israel. This set off an energy crisis, causing gas shortages and increases in energy prices, and thus reawakening the desire for an alternative to U.S. dependence on oil. Research in solar energy, both solar thermal and photovoltaics, began again. However, much of the new research was either redundant or a refinement of older technologies. Nonetheless, new methods of creating photovoltaic panels were developed, making them more cost-efficient and available for use by the public. Today, small devices such as calculators and watches are powered by solar power with *amorphous silicon,* a thin layer of photovoltaic material that is cheaper to produce than the most prevalent type, *crystalline silicon.*

Uses, Effects, and Limitations

Energy from the Sun can be used in varied ways. Flat-plate collectors can heat air and water for buildings, although

In Soldiers Grove, in southwest Wisconsin, all of the town center buildings are energy-efficient and solar-heated. When it was completed in 1983, Soldiers Grove's Main Street became the first business district of its kind in the nation (Corbis/Richard Hamilton Smith).

they are mainly limited to these tasks. Once installed, they require only minimal maintenance and have few moving parts. This technology is currently used in 1 million buildings in the United States. Solar-concentrating systems can generate utility-scale electricity, and the parabolic trough is the most commercially used form. Central receivers are also capable of generating a large amount of electricity, but they are expensive and damage prone. Parabolic-dish systems are efficient generators of electricity but are expensive and difficult to maintain.

Photovoltaics are a more versatile form of solar power. Small photovoltaic panels power calculators and watches. Larger photovoltaic systems run water pumps, electric fences, and radio relay stations. Highway emergency telephones use photovoltaic panels for power. This technology is also well suited for powering isolated cabins, when running power lines would be more expensive than building a structure that uses solar energy. Photovoltaics are also useful in developing nations in which much of the population has little access to conventional power. Here, solar technology provides electricity for the refrigeration of medicines and the irrigation of crops. The U.S. space program also uses photovoltaic panels in the design of satellites and space shuttles. Photovoltaic panels can also generate utility-scale electricity, although they are less commonly used than are solar thermal systems.

A primary advantage of photovoltaic panels is their simplicity of design. Like flat-plate collectors, they are mechanically simple and have no moving parts. They create direct-current electricity that allows for simple battery storage and have a life of 20 to 30 years. In addition, they create no noise or exhaust. However, at present, using photovoltaic technology is often more expensive than using conventional power. The cost of setting up a home to use it is equal to buying 20 years' worth of fuel up front. Inverters must also be used to change the direct current to alternating current. Furthermore, some of the materials used in the manufacture of photovoltaic materials are toxic.

There are a few drawbacks to solar energy use in general. The amount of solar energy varies according to the time of day, the time of year, the region of the world, and variations in weather. Therefore, the size and type of the appropriate solar energy system will vary by region. In addition, these systems must provide energy storage or a backup energy system when the Sun cannot meet energy needs.

Issues and Debate

Steam power plants account for most of the electricity produced in the United States. The burning of oil, coal, or natural gas or the heat from nuclear reactions turns water to steam that drives turbine generators, producing electricity.

Yet all of these methods have their shortcomings. Fossil fuels are depleted at a rate 100,000 times greater than they are produced. Emissions from the burning of fossil fuels contribute to acid rain, global warming (the gradual increase in the Earth's temperature over time due to a buildup of greenhouse gases in the atmosphere), and low air quality. Nuclear energy does not cause air pollutants, but it results in radioactive wastes.

Solar energy can also produce steam to run power plants, but it avoids all of these drawbacks. It has a minimal effect on the environment, is renewable (does not draw on a finite resource), and is versatile in its application. It can be used individually or on a large scale. An increased use of solar energy would also allow for less dependence on foreign oil. Yet for most energy consumers, conventional energy is still a less expensive choice. That may be changing, however. In 1997, U.S. president Bill Clinton announced the Million Solar Roofs Initiative, tax incentives for homes, schools, and businesses that install photovoltaic panels on their buildings. Clinton hopes to have photovoltaic panels on 1 million buildings by the year 2010. If accomplished, this would reduce carbon dioxide emissions by a level equal to that produced by 850,000 cars annually. In addition, a number of states deregulated their electricity industries in 1998, allowing independent companies to compete in offering energy to consumers. This may open the door for more solar power plants to enter the utility market. People in states such as California and Massachusetts already have the opportunity to buy energy from solar plants. Nonetheless, because of the stability and established cost structure of current energy markets, the government may have to take a more active role if it wants to encourage the use of solar energy.

—Paul Candon

RELATED TOPICS
Hydroelectric Power, Nuclear Energy, Wave and Ocean Energy, Wind Energy

BIBLIOGRAPHY AND FURTHER RESEARCH

BOOKS

Beattie, Donald A., ed. *History and Overview of Solar Heat Technologies.* Cambridge, Mass.: MIT Press, 1997.

Fujii, Iwane. *From Solar Energy to Mechanical Power.* New York: Harwood Academic Publishers, 1990.

Nansen, Ralph. *Sun Power: The Global Solution for the Coming Energy Crisis.* Seattle, Wash.: Ocean Press, 1995.

PERIODICALS

Bownell, Bruce. "The State of Solar Power: 1998." *Mother Earth News,* October–November 1997, 50.

Fitzgerald, Mark C. "Renewable Energy Today and Tomorrow." *The World & I,* March 1999, 150.

Hasek, Glenn. "Solar Shines Brighter." *Industry Week,* April 20, 1998, 24.

Hayhurst, Chris. "Solar Products Offer a Clean Alternative." *E, The Environmental Magazine,* January–February 1999, 48.

Hill, Roger. "Commercializing Photovoltaic Technology." *Mechanical Engineering,* August 1994, 80.

Mancini, Thomas R., James M. Chavez, and Gregory J. Kolb. "Solar Thermal Power Today and Tomorrow." *Mechanical Engineering,* August 1994, 74.

Siuru, Bill. "What's New in Solar Power." *Electronics Now,* August 1998, 53.

Smith, Charles. "Revisiting Solar Power's Past." *Technology Review,* July 1995, 38.

INTERNET RESOURCES

American Solar Energy Society
http://www.ases.org/

North Carolina Solar Center
http://www.ncsc.ncsu.edu/11overvw.htm

Photovoltaic Power Resource Site
http://www.pvpower.com/

Solar Energy Network
http://www.solarenergy.net/tsen/info.html

SOLAR POWER SATELLITES

The solar power satellite (SPS) is a potential future technology that would use the unlimited, constantly available supply of sunlight found in outer space to generate electricity on Earth. Very large orbiting satellites would collect SOLAR ENERGY, process it, and beam it to selected sites on Earth, where it would be converted to electricity.

The SPS represents an innovative, technically feasible alternative energy concept that can be used to satisfy the world's energy demands in the mid-21st century. Before STS technology can be utilized, however, an enormous investment in space technology will have to be made, and many environmental, political, and economic issues will need to be resolved.

Scientific and Technological Description

Since the early 1960s, most Earth-orbiting spacecraft have taken advantage of sunlight to generate electricity for their onboard needs. The solar power satellite carries this successful solar-to-electric energy transformation a giant step forward. The solar power satellite is a very large space structure (roughly 5 by 5 kilometers [km] in size) that is constructed and operated in Earth orbit. It takes advantage of the nearly continuous availability of sunlight to provide useful energy to Earth. At Earth's average distance from the Sun (a distance defined as 1 astronomical unit), the energy content of raw sunlight is approximately 1370 watts per square meter. A simple calculation for a representative 5- by 5-km satellite platform indicates that about 34 billion watts (i.e., 34 gigawatts [GW]) of raw sunlight is available for harvesting by just one satellite.

The basic concept behind the SPS is quite simple. Each SPS unit is placed in geosynchronous (GEO) orbit above the Earth's equator, which means that it revolves around Earth once per day, maintaining the same position relative to the

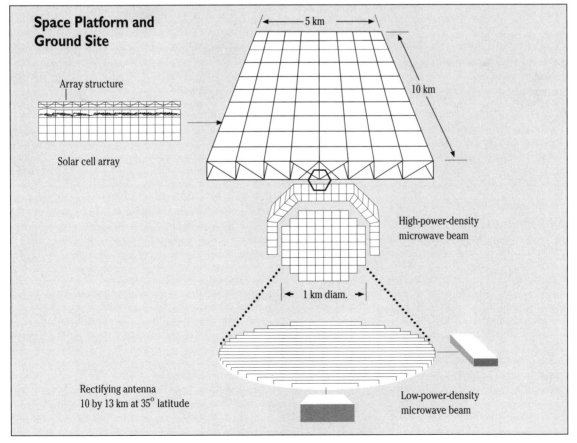

Adapted from an original illustration courtesy of NASA and the Energy Department.

Earth's surface, at an altitude of about 35,900 km. In such a GEO orbit, an SPS would experience sunlight more than 99 percent of the time. These large orbiting space structures can gather the incoming sunlight for use on Earth in one of three general ways: microwave transmission, laser transmission, or mirror transmission.

Microwave Transmission

In the microwave transmission technique (see figure), solar radiation is collected by the orbiting satellite and transformed into electricity, which is immediately converted into radio-frequency (RF) or microwave power and then beamed to a ground site on Earth. The sunlight-to-electricity conversion can be accomplished either directly, using photovoltaic solar cells, or indirectly, using a solar-thermal heat engine to drive appropriate turbogenerator equipment. The heat engine is a device, operating in a thermodynamic cycle, that converts a portion of an incoming quantity of heat into more useful mechanical work. Microwave transmitting tubes (called *klystrons*) convert the solar-generated electrical current to microwave energy, typically at a frequency of 2.45 gigahertz (GHz). This microwave energy is beamed precisely to a large receiving antenna site on Earth, where it is converted into electricity.

The process of beaming energy using the RF portion of the electromagnetic spectrum is called *wireless power transmission* (WPT). At the ground site, the incoming microwave energy is collected by a *rectenna* (a receiving antenna) and converted into direct-current (dc) electricity. The dc electricity can be inverted to create alternating current (ac) and then "stepped up" to high voltage for transmission into and use within the terrestrial power grid. For a 5-GW-class SPS, the rectenna site would cover a ground area of about 100 square kilometers (km^2) and would also require an exclusion area around the site of an additional 70 km^2 to protect human and animal life from continuous exposure to low-level microwave radiation.

Laser Transmission

In the laser transmission technique, incoming solar radiation is converted into infrared laser radiation, which is then beamed down to a special receiving facility on Earth for the production of electricity. LASERS represent an alternative to microwaves for the transmission of power over long distances. Compared with microwaves, lasers have a much smaller beam diameter. Light from an infrared laser can be transmitted and received by apertures more than 100 times smaller in diameter than that of a comparable microwave-beam system. This reduces both the size of the satellite portion of the SPS and the area of the companion ground site.

Mirror Transmission

In the mirror transmission technique, very large orbiting mirrors are used to reflect raw (unconverted) sunlight directly to terrestrial solar energy conversion facilities that would operate 24 hours a day. In one approach, about 900 very large orbiting mirrors, each 50 km^2 in area, would be used to create a global power system capable of generating 810 GW of electricity. The space segment of the SPS mirror concept is quite simple. However, a major disadvantage is that it requires an extremely large contiguous land area for the ground site and illuminates much of the night sky within a 150-km radius of the center of the site.

Historical Development

The SPS concept is generally credited to Peter Glaser, who proposed it in a 1968 technical paper. Because of its potential to relieve long-term national and global energy shortages, the concept was studied extensively in the 1970s. A series of NASA-sponsored studies embraced a sweeping vision of space technology development and suggested that giant SPS units could be constructed using extraterrestrial materials gathered from the Moon- and Earth-crossing asteroids. Within this visionary scenario, manufacturing and construction activities would be accomplished by thousands of space workers, living with their families in large, permanent space settlements established at selected locations in *cislunar space,* the region of outer space between the Earth and the Moon.

In 1980, the U.S. Department of Energy and NASA defined an SPS reference system to serve as the basis for conducting initial technical, environmental, and societal assessments of the technology. This SPS reference system represents one potentially plausible approach for achieving SPS concept goals. The proposed microwave transmission configuration would provide 5 GW of electric power at the terrestrial grid interface. In that scenario, 60 SPS units would be placed in GEO orbit and provide about 300 GW of electric power for use on Earth. It was optimistically estimated that it would take only about six months to construct each SPS unit once an extensive space technology infrastructure had been developed. This infrastructure included reusable launch vehicles capable of placing massive payloads into low Earth orbit at very low cost. Despite the exciting visions of large human settlements in space, permanent lunar mining bases, and harvesting the Sun's limitless energy to provide plentiful "nonpolluting" energy to a developing world, the original SPS concept remained just that. No single government or consortium of nations committed the long-term resources needed to develop the necessary technology infrastructure.

As the 20th century came to a close, an examination of expected global energy needs in the post-2020 period and concerns about rising levels of carbon dioxide (CO_2) levels in the atmosphere encouraged scientists to revisit the SPS concept. The major purpose of this "new-look" effort, conducted by NASA in the 1990s, was to determine whether a contemporary solar power satellite system could be defined that would deliver energy into terrestrial electrical power grids at prices equal to or less than the cost of ground alternatives. The new-look SPS would have to be a technically feasible system that could be developed at a fraction of the capital investment that had been projected for the SPS reference system of the 1970s.

In response to the NASA-sponsored effort, several interesting concepts emerged for a contemporary SPS. Building on the basic SPS concept, these designs generally emphasize the following: use of advanced energy-beaming and power conversion technologies, modularity (making the solar power conversion units on the satellite modular means constructing them in smaller discrete units versus one large assembly, a method that allows the SPS to be sized to meet a wide variety of future energy needs), extensive use of robotic systems for assembly, and minimal use of human space workers. In addition, more modest power output goals (typically, several hundred megawatts) for each orbiting platform have been established.

Uses, Effects, and Limitations

It is perhaps too early to validate the SPS concept (original or new look) or to dismiss it totally. What can be said is that the controlled beaming of solar energy to Earth (either as raw concentrated sunlight or as converted microwaves or laser radiation) represents an interesting energy alternative. However, any SPS concept also involves potential impacts on Earth. Some of these environmental impacts are comparable in type and magnitude to those arising from other large-scale terrestrial energy technologies, while other impacts are unique to the SPS concept. Some of these "SPS unique" environmental and health impacts are potential adverse effects on Earth's upper atmosphere from launch vehicle effluents and from energy beaming (e.g., microwave heating of the ionosphere); potential hazards to terrestrial life from nonionizing radiation (microwave or infrared laser); electromagnetic interference with other spacecraft, aircraft, terrestrial communications, and astronomy; and potential hazards to space workers.

A major concern with *wireless power transmission* (WPT), energy beaming that uses microwave radiation, is the possibility that humans, animals, and plants will be exposed to higher power densities than they currently experience in daily life, levels that may be dangerous. Inside a typical microwave oven the power density is about 1000 milliwatts per square centimeter (mW/cm²), while just outside the same microwave the power density is less than 10 mW/cm². Microwave radiation is nonionizing radiation, so it is simply too weak to rip electrons off atoms and molecules. However, when microwave beam power density levels exceed 100 mW/cm², heating effects can occur that are of biological concern. If microwave

transmission STS technology were pursued, precautions would have to be taken to exclude living things from regions where the microwave power density exceeds safe levels for continuous exposure. Although prolonged exposure to microwave radiation at the planned SPS power density may not prove directly harmful to plants and animals, the effects of chronic RF exposure must be studied. In addition, the general accumulation of migrating wildlife at warm locations may create unusual operating limitations at receiver sites.

The laser transmission technique suffers from three important disadvantages. First, infrared radiation is subject to severe degradation or absorption by clouds in the Earth's atmosphere, reducing in an unpredictable manner the amount of energy that can reach the ground site at any particular time. Second, high-powered continuous-wave (CW) lasers have relatively low overall power conversion efficiencies. Typically less than 25 percent of the collected solar energy can be converted into energy that is transmitted in the laser's beam. Third, the incoming beam intensity represents a real health and safety hazard, requiring the use of high perimeter walls to isolate the ground site. The raw sunlight conveyed via the mirror transmission technique, as with the laser-transmission technique, would be subject to reduction or elimination by cloud cover. The foregoing limitations will have to be resolved favorably if the SPS concept is to go forward.

Issues and Debate

In addition to the more technical difficulties involved in implementing the SPS concept, a number of international issues and economic concerns arise as well. Supporters of the SPS concept consider it a promising energy option, yet opponents see the concept as being of high risk. SPS backers believe the system will be cost competitive in the future energy market. As a result of NASA's studies in the mid-1990s, a "sun tower" concept emerged that sought to reduce the cost of the system while broadening its energy market flexibility. According to a 1997 NASA report, a 250-megawatt sun tower platform would cost between $8 billion and $15 billion. This projected cost is 30 times less than the investment that would have been required for the 1979 reference SPS. But SPS opponents argue that even such new system concepts are unlikely to be competitive without extensive government subsidies. Proponents of SPS argue that the system is potentially less harsh to the environment than are other energy options, such as fossil fuels, which release CO_2 into the atmosphere when burned. Opponents argue that the risks that SPS poses to humans and the environment are too great, certainly greater than those associated with more traditional terrestrial solar technologies.

Wireless power transmission will continue to raise general concern over the chronic exposure of animals (including people) and plants to higher levels of RF radiation than may be safe. Today's RF exposure standards suggest that no health hazards have been found at the microwave exposure limits now established. The current scientific consensus is that there is no known mechanism for *mutagenesis* (biological mutation) or any other nonthermal effect of microwaves on biological systems. There is also a lingering concern about the military potential and implications of large space platforms that can beam energy to points on Earth. An intense microwave beam, for example, could be used to interfere with an opponent's radio communications over a broad area, and a more focused laser beam could be used to attack targets in space or on the ground.

All these issues and concerns must be addressed. However, as the demand for energy stretches traditional energy resources and environmental burdens to their natural limits, the solar power satellite concept will continue to be revisited. Breakthroughs in space technology, especially dramatic reductions in the cost of launching payloads into space from Earth and the availability of more sophisticated robot devices to assist in in-orbit construction, may help overcome many barriers and shift the SPS debate in the direction of implementation and action.

—Joseph A. Angelo, Jr.

RELATED TOPICS

Lasers, Reusable Launch Vehicles , Satellite Technology, Solar Energy, Space-Based Materials Processing, Space Shuttle

BIBLIOGRAPHY AND FURTHER RESEARCH

BOOKS

Angelo, Joseph A., Jr. *The Dictionary of Space Technology,* 2nd ed. New York: Facts On File, 1999.

Glaser, Peter E., F. P. Davidson, and K. Csigi, eds. *Solar Power Satellites: A Space Energy System for Earth.* Chichester, West Sussex, England: John Wiley, 1998.

Johnson, Richard D., ed. *Space Settlements: A Design Study.* NASA SP-413. Washington, D.C.: U.S. Government Printing Office, 1977.

U.S. Congress (Office of Technology Assessment). *Solar Power Satellites.* Washington, D.C.: U.S. Government Printing Office, 1981.

PERIODICALS

Glaser, Peter E. "The Future of Power from the Sun." IEEE Publication 68C-21-Energy. Presented at the Intersociety Energy Conversion Engineering Conference (IECEC), 1968, 98.

Glaser, Peter E. "An Overview of International Activities on Power from Space." Presented at the International Conference on Macro Engineering Development, MIT, Cambridge, Mass., December 5, 1997.

Glaser, Peter E. "Power from the Sun: Its Future." *Science,* November 22, 1968, 857.

Mankins, J. C. "Space Solar Power: A Fresh Look." AIAA 95-3653. Presented at the 1995 AIAA Space Programs and Technologies Conference, Huntsville, Ala., September 1995.

INTERNET RESOURCES

Space Studies Institute
   http://www.ssi.org/

SUNSAT Energy Council
   http://www.tier.net/sunsat/index.htm

SPACE ALERTING AND DEFENSE SYSTEM

A space alerting and defense system is a future system that would function as a "planetary life and health insurance policy" against large asteroids or comets whose trajectories take them on a collision course with Earth. The system consists of two major components: a surveillance function and a mitigation function. The surveillance function uses optical and radar tracking systems to monitor space continuously for threatening near-Earth objects (NEOs). Should a large impactor be detected, defensive operations would be determined by the space technologies and the warning time available. With sufficient time, the impactor might be deflected or be disrupted sufficiently to save Earth.

Although advanced surveillance technologies are available, the proposed spectrum of defensive techniques requires more research, testing, and development. International cooperation is also essential because of the legal and political issues involved in the deployment of powerful nuclear explosives in space.

Scientific and Technological Description

Earth-approaching asteroids and comets are often collectively called near-Earth objects (NEOs). Over millennia, these wandering space objects can pose a significant future threat to our planet. Although the annual probability of Earth being struck by a large asteroid or comet is quite small, the consequences of such a collision would be catastrophic on a regional or even global scale. In fact, the collision between Earth and a large asteroid or comet could definitely become an extinction level event (ELE), because the global consequences of this incredibly energetic event would make our planet uninhabitable for most, if not all species.

Scientists consider an object with a diameter greater than 1 kilometer (km) a "large" impactor. Studies suggest that the minimum mass of an Earth impactor needed to produce an ELE is several tens of billions of tons. High-speed (more than 10 km per second) impact of an object this large would result in a ground burst explosion with an energy yield equivalent to about 1 million million megatons (10^6 MT) of the chemical high explosive trinitrotoluene (TNT). (The nuclear blast that destroyed the Japanese city of Hiroshima at the end of World War II had an equivalent explosive yield of between 15 and 20 kilotons [kT] of TNT.) The corresponding threshold diameter for extinction-level NEOs is between 1 and 2 km. Smaller impactors (from hundreds of meters down to tens of meters in diameter) can cause severe regional or local damage, but are not considered to pose a global threat.

The impact of a 10- to 15-km-diameter asteroid would most certainly cause massive extinctions. Locally, there would be intense shock-wave heating and fires, tremendous earthquakes, giant tidal waves (if the impactor hit in a watery

Space Alerting and Defense System

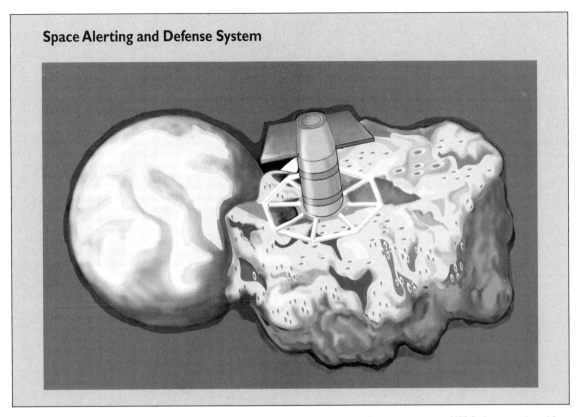

Above, a future drilling operation is shown on a small near-Earth object as part of an asteroid deflection experiment, c. 2025. Such an operation might be charged with simply breaking up the asteroid, or it could also mine materials from the rock. (Adapted from an original illustration courtesy of NASA.)

region), hurricane-level winds, and hundreds of billions of tons of debris thrown into the air. As it spread across the planet, this giant cloud of debris would cause months of darkness and much cooler temperatures. In addition to causing millions of immediate casualties (depending on where the impactor hit), global food crops would be destroyed.

Potential NEO Threats

Earth resides in a swarm of asteroids and comets that can, and do, impact its surface. An asteroid is a small, rocky body without atmosphere that orbits the Sun independent of a planet. Known asteroids range in size from about 1000 km in diameter down to "pebbles" that are just a few centimeters in diameter. The vast majority of asteroids have orbits that congregate in the main asteroid belt, a doughnut-shaped region of heliocentric space located between the orbits of Mars and Jupiter. There are also several "families" of Earth-approaching asteroids that represent potentially threatening NEOs.

About 2000 large NEOs are thought to reside in the inner Solar System. However, fewer than 200 have been detected so far. Over millions of years, between 25 and 50 percent of these NEOs will eventually impact Earth. Fortunately, the average time interval between large-object impacts is long, typically more than 100,000 years. These currently undetected large NEOs constitute the major cosmic threat to our planet. The

other NEO of concern is the comet, a dirty ice "rock" orbiting the Solar System. Comets represent the "cosmic leftovers" from when the Solar System formed over 4 billion years ago. Most comets reside far from the Sun in a frigid spherical cloud called the *Oort Cloud*. The *Kuiper belt* is another, closer region containing cometary nuclei. As a comet approaches the Sun from the frigid regions of deep space, the Sun's radiation causes its frozen surface materials to vaporize. The resulting vapors form an atmosphere or *coma* around the solid nucleus and a long luminous tail.

Once a comet approaches the planetary regions of the Solar System, it is subject to the gravitational influences of the major planets, especially Jupiter. A comet can achieve a quasistable orbit within the Solar System, showing up on a predictable basis, as does Halley's Comet. On occasion a comet can collide with a planet or even crash into the Sun. Astronomers classify comets as either *long-period comets* (which have orbital periods around the Sun in excess of 200 years) or as *short-period* or *periodic comets* (which have periods of less than 200 years). Planetary defense studies often restrict the term *short-period comet* to comets with periods of less than 20 years.

The short-period comet represents only about 1 percent of the NEO threat, while the long-period comet represents the second most important hazard, after asteroids. Although their numbers amount to only a few percent of the

NEO impacts our planet has experienced over geologic time, long-period comets appear from deep space with little warning (usually less than a year or so) and approach Earth with greater speeds and therefore higher kinetic energy in proportion to their mass. It is now estimated that as many as 25 percent of the impactors that have hit Earth with energies in excess of 100,000 MT have been long-period comets.

Mitigation Techniques

The surveillance portion of the planetary defense system must detect and identify all possible impactors in sufficient time to deploy one or more mitigation techniques. Early warning is the ultimate key to success. Both ground and space-based surveillance assets can be used to provide warning. In the 21st century, optical astrometry (precise measurement of the location of celestial objects) will be used to search for, identify, and confirm the precise trajectory of any threatening near-Earth object (asteroid or short-period comet). Radar astrometry will then provide detailed range and velocity data for the object. The surveillance effort should provide sufficient warning time (perhaps on the order of a decade or more) for about 90 percent of the potential impactors detected.

The remaining 10 percent of the impactor threat comes from long-period comets. The surveillance effort (ground- and space-based) will probably be able to provide only about a year or so of warning under the most favorable detection circumstances. To deal effectively with all threats, proponents maintain that the interception/mitigation effort should be deployed in space and ready to operate. Mitigation techniques depend significantly on the amount of warning time and fall into two broad categories: techniques that destroy the threatening object and those that deflect it. Within today's technology horizon, nuclear explosives appear to be the tool of choice in deflecting or disrupting a large impactor. Other techniques may become feasible as the space technology infrastructure matures in the 21st century. For example, some scientists have suggested using advanced propulsion systems to "capture" the threatening object and place it in a safe, accessible orbit in the Earth–Moon system.

With sufficient warning and reaction time, advanced space technologies can be used to nudge the threatening object into a harmless (or even useful) orbit. Focused-nuclear detonations, high-thrust nuclear thermal propulsion rockets, low (but continuous)-thrust nuclear or solar-electric propulsion vehicles, and even mass-driver propulsion systems (which use pieces of the object as reaction mass) have been suggested as possible techniques to deflect the impactor by changing its velocity. Interceptions far from Earth, made feasible by early warning, are much more desirable than interceptions near the Earth–Moon system.

Should deflection prove unsuccessful or impractical due to the time available, physical destruction or fragmentation of the approaching impactor would be required. Multiple-megaton-yield nuclear detonations, including explosive devices placed deep within the object at critical fracture locations, could be used to shatter it into smaller, less dangerous projectiles. With sufficient time, a smaller asteroid (perhaps 100 meters [m] or so in diameter) might be maneuvered into the path of the larger impactor, causing a collision that shatters both objects at a safe distance from the Earth–Moon system.

Today, scientists are carefully examining a variety of intriguing options—both nuclear and nonnuclear—as to how to nudge or blast a threatening NEO. Future scientific missions to asteroids and comets might include practice deflection or demolition experiments, as early field demonstration tests for the planetary defense system.

Historical Development

Many scientists suspect, on the basis of growing physical evidence, that a large asteroid, about 10 km or more in diameter, struck ancient Earth some 65 million years ago in the Gulf of Mexico near the Yucatan region. From archaeological and geological records it is clear that a tremendous catastrophe occurred at that time. In the ancient cataclysm, about 70 percent of all species then living on Earth, including the dinosaurs, disappeared within a very short period.

Meteor Crater in Arizona is one of Earth's youngest impact craters. It is believed to have been formed about 50,000 years ago by an iron mass (about 50 to 100 m in diameter) traveling in excess of 11 km per second and releasing between 10 to 20 MT (of TNT equivalent) energy. The well-preserved bowl-shaped crater near Flagstaff is approximately 1 km across and 200 m deep.

The Tunguska event occurred in 1908 over a remote region of Siberia. Scientists think that it is the result of a small (50- to 100-meter) comet (or possibly asteroid) that entered Earth's atmosphere and exploded at about 8 km altitude with an equivalent yield of between 10 and 20 MT (of TNT). Trees in the Siberian forest were knocked to the ground and fires ignited out to distances of about 20 km from the end point of the fireball trajectory.

In July 1994, chunks of Comet Shoemaker–Levy 9 that were 20 km or so in diameter plowed into the southern hemisphere of Jupiter. These cometary fragments collectively deposited about 40 million megatons of TNT-equivalent explosive energy. Plumes of hot, dark material rose higher than 1000 km above the planet's visible cloud top through holes punched in the Jovian atmosphere by the exploding comet fragments.

In the 1990s, at the request of the U.S. Congress, NASA, the Defense Department, and the Energy Department have conducted several studies on the planetary risk posed by NEOs and on ways to save the planet from future impactors. The studies identified the NEO risk as real (although not immediate) and suggested the development of a planetary defense effort based on surveillance to provide early warning and interception and deflection techniques, centering around use of advanced space technology and powerful nuclear explosives. This space alerting and defense system would be international in nature.

At present, NASA is sponsoring modest asteroid and comet search efforts. For example, there is the Near Earth Asteroid Tracking (NEAT) program that since the mid-1990s has been conducted in collaboration with the U.S. Air Force and Livermore National Laboratory, with instruments on Haleakala in Maui, Hawaii. This program uses modern electronic detectors and automated detection algorithms to identify moving NEOs against the background of star fields. Work on various interception/deflection systems remains primarily at the conceptual level.

Astrometry is the branch of science that involves the very precise measurement of the motion and position of celestial objects. Optical telescopes and radar systems are used in astrometry and are often coupled with low-light-level television and computer systems to achieve highly automated object identification. Civilian astronomers are using such equipment to detect and identify interesting new objects in the Solar System, including NEOs. Military personnel use similar instrumentation to perform space object surveillance in support of national defense. For example, the U.S.–Canadian North American Aerospace Defense Command (NORAD) has a space surveillance mission that includes responsibility for detecting, tracking, identifying, cataloging, and maintaining the status of human-made objects in space, especially those in orbit around Earth. In support of NORAD, the U.S. Space Command's Space Surveillance Center, located within Cheyenne Mountain in Colorado, is directly responsible for keeping track of everything in Earth orbit.

The United States, former USSR, and several other nations have developed extremely powerful, compact-sized nuclear explosives in the megaton-yield range. These thermonuclear explosive devices now represent the basic technical tool for a planetary defense system in the early 21st century.

Uses, Effects, and Limitations

The primary use of the space alerting and defense system is to protect Earth from all future large NEO threats. At present, no asteroid or comet is known to be on a collision course with Earth. The chances of a collision within the 21st century with an object 1 km in diameter or larger are very small (estimated at about 1 in 10,000), but such a collision is statistically possible. With sufficient warning, interception/deflection systems can be developed and deployed. However, the deflection task is not easy, and success depends on good surveillance information, long-term warning, a thorough understanding of the properties of the threatening object, and powerful nuclear systems (explosives and propulsive) to achieve the desired deflection or disruption.

Robotic space systems that have the ability to rendezvous with and land on an asteroid or comet are useful in scientific exploration as well as planetary defense (see ROBOTIC SPACE EXPLORATION). Powerful nuclear systems that can nudge an asteroid or comet and change its path can be developed; but testing and operational deployment in space will require political and legal decisions as well as technical demonstrations.

Issues and Debate

Given an immediate impactor threat, no reasonable person would debate the need for a system to protect our planet. However, in the absence of an immediate threat, budget-minded bureaucrats and legislators keep asking how soon we really need such a system. Advocates claim that it should be built as soon as possible so that the total NEO threat can be evaluated. Opponents look at the statistical threat data and view as premature the expenditure of resources on such long-term concerns. Another major issue centers around the global, international nature of the problem. Should one nation fund and operate the system unilaterally, or should the responsibility and costs be shared by all nations through an international organization?

The "sky-is-falling" issue of premature notices, false alarms, and public panic is also of concern. Competent authorities (technical and political, national and international) need to review NEO surveillance data. Data that reveal a potential threat should be subjected to rigorous technical review and confirmation procedures before a worldwide threat announcement is released. Otherwise, needless panic could spread uncontrollably around the world at the speed of electronic communication.

The use of nuclear explosives in space raises other concerns. By current international law, no nation can deploy or test nuclear explosive devices in outer space. However, the majority of the proposed defensive techniques require the use of massive nuclear explosions. Should these nuclear explosives be deployed in space and made available for planetary defense purposes? If yes, under whose control will such devices remain?

—Joseph A. Angelo, Jr.

RELATED TOPICS
International Space Station, Reusable Launch Vehicles, Robotic Space Exploration, Space Shuttle

BIBLIOGRAPHY AND FURTHER RESEARCH

BOOKS
Angelo, Joseph A., Jr. The *Dictionary of Space Technology*, 2nd ed. New York: Facts On File, 1999.
Desonie, Dana, Carolyn Shoemaker, and David Levy. *Cosmic Collisions*. New York: Henry Holt, 1996.
Gehrels, Tom, et al., eds. *Hazards Due to Comets and Asteroids* (Space Science Series). Tucson, Arizona: University of Arizona Press, 1995.
Lewis, John S. *Rain of Iron and Ice: The Very Real Threat of Comet and Asteroid Bombardment*. Reading, Mass.: Addison-Wesley, 1996.

Steel, Duncan, and Arthur Charles Clarke. *Rogue Asteroids and Doomsday Comets: The Search for the Million Megaton Menace That Threatens Life on Earth.* New York: Wiley, 1997.

PERIODICALS AND REPORTS

Bilyeu-Gordon, Bonnie. "That Asteroid Caper." *Astronomy,* July 1998, 6.

Canavan, G. H., D. G. Rather, and J. Solem, eds. *Near-Earth-Object Interception Workshop.* Report LANL 12476-C. Los Alamos, N.Mex.: Los Alamos National Laboratory, 1993.

Cowen, Ron. "Doomsday Asteroids." *Science News,* February 5, 1994, 90.

Morrison, D., ed. The Spaceguard Survey: *Report of the NASA International Near-Earth-Object Detection Workshop.* NASA Publication, Washington, D.C.: National Aeronautics and Space Administration, 1992.

INTERNET RESOURCES

"Asteroid and Comet Impact Hazard," NASA/Ames
    http://impact.arc.nasa.gov/

"Comet Shoemaker–Levy Encounters Jupiter," Space Telescope Science Institute
    http://www.stsci.edu/EPA/Comet

Minor Planet Center at Harvard
    http://cfa-www.harvard.edu/iau/mpc.html

Near Earth Asteroid Tracking (NEAT)
    http://huey.jpl.nasa.gov/~spravdo/neat.html

NEO Program in NASA
    http://neo.jpl.nasa.gov/

SPACE-BASED MATERIALS PROCESSING

The low-gravity environment provided by spacecraft and satellites in Earth orbit provides unique opportunities for manufacturing materials and understanding the processes by which they form. Without the distortions caused by gravity, it is possible to create extremely pure and uniform structures in materials such as metals, ceramics, and composites.

Despite this technology's failed initial promises of factories in space and of new wonder materials revolutionizing everyday life, space-based processing of materials is a promising young field, one that will become only more important in the future.

Scientific and Technological Description

Any space vehicle experiences effective *zero gravity* at its exact center of mass, but most of the vehicle and its contents will be at some distance from the center of mass. Effectively, each object in, and part of, the vehicle is in its own orbit, and, because they are at different distances from Earth, each will experience a slightly different pull of gravity. The difference in gravity experienced by an object on a spacecraft and by the spacecraft itself is typically thousandths or even just millionths of the 1g experienced on Earth. This vastly reduced gravity, known technically as *microgravity,* can be used for the study and manufacture of materials outside the constant gravitational pull experienced on Earth. In addition, a spacecraft orbits beyond the bulk of Earth's atmosphere, and the conditions of near-per-

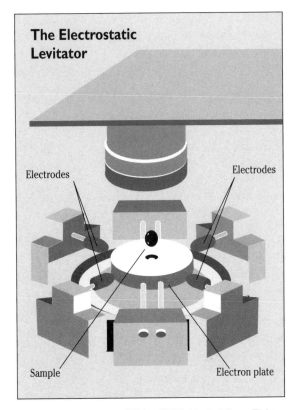

The Electrostatic Levitator

Electrodes   Electrodes

Sample   Electron plate

The Electrostatic Levitator (ESL) at NASA's Marshall Space Flight Center is a ground-based tool that uses static electricity to suspend metal and alloy samples in a vacuum. The ESL allows scientists to replicate space conditions for experiments, yet avoid the expense of getting into space. The ESL vacuum chamber contains a pair of electron plates and four electrodes that position the sample being processed. A side-view of the ESL shows a 3 mm drop of nickel-zirconium hovering between the two electrically charged plates.

fect vacuum that are available there can also be useful for creating materials.

Modern manufacturing techniques allow complete control of almost every factor that might have an influence on the end result of the process. However, gravity is one factor that cannot simply be "switched off." This places constraints on the manufacturing process itself, effectively compromising the quality of the finished materials. Even the use of containing vessels for handling materials is a necessity caused by the presence of gravity.

Many high-performance materials used today, ranging from advanced metal alloys to silicon chips, pass through a molten stage in their manufacture. In this phase they can experience two of the most important phenomena caused by gravity, sedimentation and convection. *Sedimentation* is the simple phenomenon that causes a mixture of particles with different weights or densities to settle into layers, with the heaviest particles at the bottom. *Convection* is the formation of currents as warmer materials expand, becoming less dense and moving up through their colder surroundings. Often, convection also acts to separate different types of particle in a mixture because one material heats up more rapidly than another.

In both cases, the end result is that different materials concentrate in different regions of the mixture and are trapped there when the molten mixture solidifies. This less-than-perfect mixing will usually have an effect on the performance of the finished material. Experiments have shown that materials created in microgravity are far less prone to these problems, although there are other forms of convection that cannot be suppressed by the absence of gravity.

Another important factor in material formation that is affected by gravity and the need for containment of reacting materials is *nucleation,* the formation of the initial nuclei of particles around which a material crystallizes and solidifies. In the case of alloys, for example, nuclei form and then grow into long, thin *dendrites*. These delicate threads tend to snap off under gravity and be carried by convection through the rest of the solidifying metal, disrupting the crystallization process and creating faults. In addition, the boundary with a container can act as a formation and anchor point for nuclei, and material from the container itself can contaminate the substance forming within it.

Nucleation is also important in the formation of crystals, which grow by the symmetrical addition of new molecules around a central nucleus. On Earth, gravity places an upper limit on the sizes of crystals: Above a certain mass, they begin to fracture and shear under their own weight. In addition, convection and sedimentation in the liquid *melt,* from which the crystal forms, can lead to the addition of molecules in a nonsymmetrical fashion. These phenomena can be avoided by manufacturing in microgravity, which allows the formation of larger crystals more rapidly.

Historical Development

The possible benefits of manufacturing in zero gravity were recognized in the early days of the U.S. space program and were often used as a long-term justification for the huge amounts of money poured into space exploration in the 1960s. However, early manned spaceflights were concerned primarily with simply getting people into space and investigating the conditions there, while space SATELLITE TECHNOLOGY was not advanced enough to support complex automated experiments.

In the early 1970s, as the U.S. space race with the Soviet Union came to an end, both nations turned their attention toward the concrete benefits to be had from exploitation of space. The first major experiments in microgravity were carried out aboard the U.S. *Skylab* space station (1973–1974). These early experiments often involved the study of phenomena that had already been encountered by astronauts in orbit—for instance, the way in which liquids in microgravity are pulled together by their surface tension to form spherical globules. Another significant phenomenon was the fact that gas bubbles liberated in a liquid (e.g., during boiling or a chemical reaction) did not automatically rise to the surface. Instead, they could carry on growing indefinitely. Other experiments carried out on *Skylab* included the growing of crystals of germanium selenide. Scientists had expected the crystals to be unusually symmetrical and flawless, but were surprised by the increased rate at which they grew and the much larger sizes achieved in microgravity.

The astronauts on board the joint U.S.–Soviet *Apollo-Soyuz* mission of 1975 investigated the behavior of other phenomena in space, including the important area of electrophoresis. This is a filtration method for separating a mixture of liquids using an electric field and is often used in the refinement of medicines. The *Apollo–Soyuz* results demonstrated that electrophoresis worked far better in the absence of gravity and sparked renewed enthusiasm for the idea of manufacturing medicines in space.

Although some satellite experiments were carried out during the hiatus in U.S. manned spaceflight of the late 1970s, the Soviet Union now took the lead in the development of microgravity processing technology. Experiments carried out aboard the Soviet space stations *Salyuts 5* and *6* involved smelting metals and manufacturing alloys in space, creating new types of glass, and growing crystals. Another particularly significant experiment involved manufacturing semiconductors by "doping" molten silicon with impurities, a process that lowers the silicon's electrical resistance and allows it to be turned into useful electronic components. Microgravity allowed the doping molecules to be dispersed evenly throughout the entire material, improving the semiconductor's efficiency. This success led to the inclusion of a computer-controlled furnace on board *Salyut 7.*

The final Soviet space station, *Mir,* which began operating in 1986, continued the commitment to materials science, with the attachment of a purpose-built laboratory module, Kristall, in 1990. Among the equipment carried on board this module were furnaces for semiconductor refinement and alloy manufacture, as well as vats for the growth of proteins and other crystals, and X-ray equipment which enabled materials to be analyzed while in space.

NASA returned to the development of microgravity with the *Spacelab* module, carried aboard the SPACE SHUTTLE. Large amounts of time and money were poured into the development of a microgravity electrophoresis system, with the intention of large-scale processing of some extremely hard-to-refine pharmaceuticals in space. However, delays caused by the increasing expense of the space shuttle program and the *Challenger* disaster in 1986 (in which seven astronauts were tragically killed) meant that the use of electrophoresis was overtaken by ground-based manufacturing methods, using sciences such as GENETIC ENGINEERING to make pharmaceuticals.

Factory satellites for the manufacturing of alloys and semiconductors are still only a distant prospect, and continued delays to the INTERNATIONAL SPACE STATION led NASA to rethink its microgravity research program in the late 1980s. Today, the focus of research is aimed at understanding the various factors that affect the formation of materials, factors that are often masked by gravity on Earth. Improved models of these delicate processes are helping to better manufacturing techniques on Earth, even if routine large-scale manufacturing in space is still some way off.

Uses, Effects, and Limitations

Space-based manufacturing has a wide range of applications. As outlined above, the manufacture of advanced alloys and semiconductors is one important area of research. Crystal growth studies have wide-ranging implications for the manufacture of pure and effective pharmaceuticals, but there are many other important fields of study. One such area is the purification of material samples. This is often done by a technique called *zone refining*, in which a small region of a bar or ingot of material (such as a metal or silicon) is melted, dissolving the impurities into a molten form. By careful application of heat, the molten zone can be moved along the bar, carrying the impurities with it and allowing purified crystals to freeze behind it. Repeating this process many times produces high-purity samples, but some impurities always remain behind, settling under gravity rather than moving through the *float zone,* the name given the moving molten zone. In orbit it is possible to create a much longer float zone, and without the effects of gravity, all the impurities are caught up in it.

Other applications take advantage of the containerless processing possible in microgravity. Because containers can encourage the nucleation and crystallization of materials, they are an obstacle to the formation of certain materials that are created by *supercooling* liquids to below their freezing temperature while keeping them in the liquid state. If the environment is suitable and nucleation is prevented, it is possible for the molten material to solidify without crystallizing, forming a glass. On Earth the variety of glasses is severely restricted, but in space it is possible to create glasses of extremely unusual materials, such as metals.

However, manufacturing in space has disadvantages as well. Some of these are associated with difficulties in handling materials. New techniques have to be developed to deal with molten metals that form free-floating globules in midair: For example, the use of magnetic and electrical fields for containment of some materials has proved useful. Other problems are caused by the absence of gravitational effects that actually help to refine materials in the laboratory. As noted above, for example, bubbles do not rise out of liquids in microgravity, and sedimentation is often a very useful method of separating a desirable material from unwanted by-products.

The major limitation on the development of microgravity research, even as a method for understanding and improving Earthbound processes, is of course the expense of space travel. However, alternatives can be used for some types of work. NASA's microgravity research project maintains an electrostatic levitator (see figure) that uses static electricity to suspend metal and alloy samples in a vacuum. The samples can then be melted using LASERS and their properties studied with a variety of sophisticated instruments. In NASA's Drop Tube Facility, based around a platform previously used for the Saturn V Moon rockets, droplets of molten materials are dropped in at the top of a 105-meter tube and experience nearly 5 seconds of free fall—effectively, a gravity-free, containerless environment in which the freezing of materials can be studied.

Issues and Debate

Although space-based materials processing is still in its infancy, some of the materials it has created have already proved useful to society. At present, space agencies are concentrating on applying knowledge from space to improve processes on Earth, but the first space-made materials have already been sold. These were tiny spheres of latex manufactured in microgravity on the space shuttle in the late 1980s. Because these microspheres are extremely lightweight, and mere microns (thousandths of a millimeter) across, they have a variety of scientific uses. The costs of these space-manufactured components are astronomical, but as with many cutting-edge technologies, if government entities or private groups can find the courage and the resources to make an initial large investment, the potential rewards are massive, and the cost of the products is likely to fall rapidly as the market for them increases.

The most important development for microgravity processing in the immediate future will undoubtedly be the International Space Station (ISS). By releasing space research from the constraints of short space shuttle missions or the declining facilities of the Russian space stations, the ISS will allow long-term scientific studies of materials formation in space. However, the ISS is unlikely to become a space factory itself; if all goes well, it will act as a proving ground for technologies that can then be adapted for use on unmanned satellites. Only then will space-based processing of materials become a full-fledged industry.

—*Giles Sparrow*

RELATED TOPICS

Composite Materials, International Space Station, Lasers, Satellite Technology, Smart Materials, Space Shuttle

BIBLIOGRAPHY AND FURTHER RESEARCH

BOOKS

Malmejac, Y. *Fundamental Aspects of Materials Science in Space.* Elmsford, N.Y.: Pergamon Press, 1983.

National Research Council. *Future Materials Science Research on the International Space Station*.Washington, D.C.: National Academy Press, 1997.

PERIODICALS

Beardsley, Tim. "Science in the Sky." *Scientific American*, June 1996.

Beardsley, Tim. "The Way to Go in Space." *Scientific American*, Special Issue: The Future of Space Exploration, 1999.

INTERNET RESOURCES

Microgravity News and Research
http://www.microgravity.com/

NASA Microgravity Research Program
http://microgravity.nasa.gov/

SPACE-BASED REMOTE SENSING TOOLS

Space-based remote sensing tools are Earth-orbiting satellites that acquire a wide variety of data across the electromagnetic spectrum in support of national defense, scientific research, environmental monitoring, meteorology, and commercial activities. Similarly, REMOTE SENSING instruments placed on spacecraft that encounter or orbit another planetary body provide unique scientific data.

From the early 1960s, remote sensing instruments placed on civilian satellites greatly improved meteorology, provided new insights about the complexities of our home planet, and created a library of high-quality images of Earth's surface in support of scientific, environmental monitoring, and commercial applications. Remote sensing instruments carried on board interplanetary spacecraft have also enabled scientists to study "up close" all the planets in the Solar System except Pluto. However, the more sophisticated instruments that now

perform satellite-based remote sensing generate incredibly large quantities of data, creating a deluge that could easily overwhelm any human effort to fully exploit its contents. The timely and equitable distribution of modern satellite-derived data is another issue of great concern.

Scientific and Technological Description

Remote sensing can be defined as the examination of an object, phenomenon, or event without having the sensor in direct contact with the object being investigated. Information transfer from the object to the sensor is achieved by means of the electromagnetic spectrum. Modern sensors are designed to use different portions of the electromagnetic spectrum, not just the narrow band of visible light we see with our eyes.

All remote sensing systems (including those placed on spacecraft to observe Earth and other planets) can be divided into two general categories: passive sensors and active sensors. *Passive sensors* observe reflected sunlight or emissions characteristic of and produced by the object being studied (e.g., the thermal infrared signature of a forest fire or of a rocket's exhaust plume). *Active sensors* (e.g., an imaging radar system) provide their own illumination on the object and measure the reflected signals as they are returned by the target. Both passive and active remote sensing instruments are used to obtain images of an object or scene. Other types of instruments can measure the total amount of energy (within a certain portion of the electromagnetic spectrum) that occurs in the field of view of the sensor.

For many passive instruments, the Sun is the natural source of target illumination. Earth receives and is heated by electromagnetic radiation from the Sun. Some of this

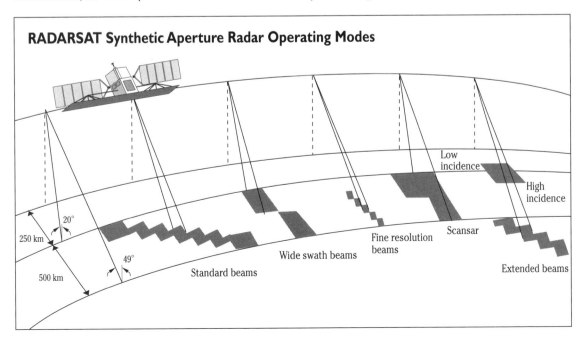

Operating modes for Canada's RADARSAT. (Adapted from an original illustration from the Canadian Space Agency.)

incoming radiation is reflected by the atmosphere, while most penetrates the atmosphere and is subsequently reradiated by gases within the atmosphere, clouds, and the planet's surface. Many types of passive sensors collect reflected sunlight or object-emitted radiation. These instruments include the imaging radiometer and the atmospheric sounder. An *imaging radiometer* detects an object's characteristic visible, near-infrared, thermal infrared, or ultraviolet radiation and then generates an image of the target or scene being viewed. The intervening atmosphere can block or absorb certain portions of the electromagnetic spectrum, so imaging radiometers are often restricted to viewing in selected wavelength bands, called *atmospheric windows*. The detector materials establish a sensor's response to a given type of the electromagnetic radiation. An *atmospheric sounder* collects the radiant energy (in the infrared or microwave portions of the spectrum) emitted by atmospheric constituents (such as water vapor or carbon dioxide). These remotely sensed data are then evaluated to infer temperature and humidity throughout the atmosphere.

Active sensors provide their own illumination (radiation) on the object or scene and then collect the radiation reflected back by the target. Active remote sensing instruments include imaging radars, scatterometers, radar altimeters, and lidars. An imaging radar system such as Canada's *Radarsat* is a very useful remote sensing tool that can penetrate through clouds and operate night or day independent of available sunlight. The imaging radar emits pulses of microwave radiation and then collects the microwaves reflected by the target to produce an image. The scatterometer emits microwave radiation and then senses the total amount of energy scattered back over a wide field of view; it is often used to measure surface wind speeds and direction. The radar altimeter emits a narrow pulse of microwave energy toward the surface and then accurately times the arrival of the return pulse reflected by the surface. This instrument provides a precise measurement of the space platform's altitude above the surface of a planet. Similarly, the lidar altimeter emits a narrow pulse of visible or infrared laser light toward the surface and then times the arrival of the light pulse reflected back from the surface. Remote sensing instruments like these are carried on Earth-orbiting satellites. They gather important data for a wide variety of scientific, military, environmental, and commercial applications.

Earth-orbiting space platforms provide a panoramic view of our planet's surface, its oceans, and atmosphere unhindered by political boundaries or natural barriers. Unlike aerial platforms, such as planes, balloons, and robot aircraft, satellites provide a vibration-free environment, greatly enhancing the quality of the data collected. Two basic factors influence satellite-based remote sensing: the choice of orbital parameters for the space platform and the region of the electromagnetic spectrum within which remote sensing instruments operate.

Depending on the orbital path selected, regions on Earth can be observed continuously or visited briefly on a repetitive basis. For example, a spacecraft can be made to "stand still" over a point on Earth's equator. Such a satellite, called a *geostationary satellite,* travels in synchronous orbit, which means that it takes as long to complete an orbit around Earth as it takes for Earth to complete one rotation about its own axis (about 24 hours). Certain types of military surveillance satellites and weather satellites take advantage of the geostationary orbit to monitor very large portions of the globe simultaneously. However, a satellite in geostationary orbit is at an altitude of 35,900 kilometers (km) above Earth's equator. Because of this altitude, the capability of certain remote sensing instruments can be significantly limited in spatial resolution. In general, the spatial resolution of an imaging instrument corresponds to the dimensions of the smallest object that can be seen in an image. For example, a 10-meter (m) resolution instrument would allow analysts to identify objects that were 10 m or more in size in each dimension, such as a ship, an office building, or a very large boulder.

Problems with spatial resolution can often be mitigated by making the instrumented platform orbit Earth at a much lower altitude. Another very useful orbit is the *polar orbit*. A remote sensing spacecraft in a polar orbit has an inclination (number of degrees that the orbit is inclined away from the equator) of about 90° and orbits Earth traveling alternately in north and south directions. A polar satellite eventually passes over the entire globe because Earth is rotating from west to east beneath it. The polar orbit is used extensively by military photo reconnaissance satellites as well as civilian environmental monitoring satellites such as NASA's Landsat 7 and Canada's Radarsat.

Remote sensing instruments on polar orbiting platforms can more easily achieve high spatial resolution measurements (because of the low altitude), but the observation time over a particular spot on Earth's surface is quite limited during each orbital pass. The revisit time (typically, several days) to a particular location is also determined by the orbital parameters selected for the satellite. If more frequent site visits are needed, several polar orbiting satellites (of the same type) can be deployed.

Historical Development

The Cold War and the nuclear arms race between the United States and the former Soviet Union stimulated the development of photo reconnaissance and missile surveillance satellites. These often "supersecret" military satellites provided and continue to provide national leaders with valuable early warning and intelligence. For example, following the surprise launch of the Soviet *Sputnik 1* satellite (on October 4, 1957), U.S. president Dwight D. Eisenhower approved a spy satellite program that would answer important questions about Soviet missile and nuclear arms capability and replace risky U-2 flights over restricted Soviet territory. Following Eisenhower's direction, in early 1958 the top secret *Corona* project was developed, a system that used photographic film to take high-resolution pictures from space and would then "deorbit" a film canister to be recovered in midair by an Air Force C-119 aircraft over the Pacific Ocean. The first successful recovery of a *Corona* capsule occurred in August 1960 (during the *Corona XIV* mission) and the age of satellite-based remote sensing for military purposes began. The film capsule retrieved from *Corona XIV* was processed immediately and exploited rapidly by the intelligence community. This one film capsule provided more coverage of the Soviet Union than that of all previous U-2 flights combined.

The 145th and final *Corona* launch occurred on May 15, 1972. During this program the quality of the remotely sensed photo reconnaissance imagery improved from an initial ground resolution of 7.6 to 12.2 m to a final capability of about 1.82-m resolution. (The smaller the spatial resolution, the more information provided about objects in a scene and the smaller the size of the objects that can be identified clearly.) Early imagery collections were driven by the need of the United States to confirm suspected Soviet strategic arms capabilities. *Corona* satellites also provided photographic coverage that was used to produce maps and charts for the Defense Department and other U.S. government agencies.

In 1995, a presidential executive order declassified more than 800,000 images collected by early U.S. photo reconnaissance systems from 1960 to 1972. The historic, declassified imagery is now being released to assist environmental scientists in studies of global change. Today, the National Reconnaissance Office (NRO), a Defense Department agency, meets the U.S. government's needs through spaceborne reconnaissance.

Military remote sensing satellites have also been placed in geostationary orbits to monitor large regions of the globe continuously. For example, in 1966 the U.S. Air Force started development of a missile warning system now known as the *Defense Support Program* (DSP). Orbiting at an altitude of 35,900 km above the equator, these surveillance satellites use infrared detectors to scan Earth's surface continuously for the hot exhaust plume signatures characteristic of ballistic missile launches. Since 1970, DSP has been the mainstay of the U.S. early warning program. Providing 24-hour worldwide surveillance, these remote sensing sentinels stand ready to alert national authorities about any large-scale intercontinental ballistic missile (ICBM) attack.

Satellite-based remote sensing of Earth is not limited just to military applications. The first civilian application also occurred in the early 1960s and helped improve the understanding and prediction of weather. Today, the United States and many other nations depend on two basic types of weather satellite: polar orbiting and geostationary.

Data from these systems have revolutionized meteorology and have also established the technical basis for global climate change research. More sophisticated environmental satellites, such as NASA's *Earth Observing System,* will greatly expand our understanding of the state of the atmosphere, land, and oceans, as well as of their interactions with solar radiation and one another. (The polar-orbiting *EOS AM-1* satellite—also called *Terra*—is scheduled for launch in late 1999.)

The concept of a civilian Earth resources satellite was developed by the Interior Department in the mid-1960s. NASA then embarked on an initiative to develop and launch the first Earth monitoring satellite, a system (eventually called *Landsat*) that used remote sensing technology (here, MULTISPECTRAL IMAGING) to meet the needs of resource managers and environmental scientists. *Landsat-1* (originally called *ERTS-A*) was launched on July 23, 1972. On April 15, 1999, NASA launched *Landsat-7,* carrying an advanced imaging system. It continues the *Landsat* tradition of providing calibrated remote sensing imagery data to a broad user community.

In November 1995, Canada's *Radarsat* was placed into polar orbit from Vandenberg Air Force Base in California. Its primary remote sensing instrument is an advanced *synthetic aperture radar* (SAR) that produces high-resolution images of Earth's surface despite clouds and darkness. *Radarsat's* SAR is an active sensor that transmits and receives microwave pulses. In an SAR system, the motion of the spacecraft is an integral part of the process of sending, receiving, and processing thousands of pulses of microwave energy per second to create useful images of Earth's surface, even when the surface cannot be observed by optical instruments.

Uses, Effects, and Limitations

Remote sensing of Earth from space provides government, military, scientific, industrial, and individual users with large quantities of high-quality data for a variety of important tasks. In many cases the application of these data is limited by the user's ability to process and fully exploit the entire quantity available. Space scientists experience a similar situation, as sophisticated robotic spacecraft provide large streams of interesting data from other bodies in the Solar System.

It is not unreasonable to suggest that remote sensing by military satellites has helped the U.S. government (and other governments) monitor treaties, reduce suspicions and distrust, and generally keep the world "sane" during the tension-filled Cold War. With satellite data from the *Corona* program, for example, the United States had the space-based "eyes" it needed to gather crucial information about the Soviet Union's capabilities. Advanced military remote sensing satellites will continue to serve as silent sentinels to support the ever-growing information needs of civilian and military leaders as the world confronts emerging new threats from rogue nations and terrorist groups.

Remote sensing from space is also a technical key to sustainable growth and intelligent stewardship of planet Earth, its resources, and the intricately interwoven biosphere. It enables many critical activities, including the monitoring of the oceans; the global monitoring of clouds, rainfall, atmospheric temperature, and other key meteorological parameters; regional environmental monitoring and change detection; and the efficient identification of unusual surface features, especially with multispectral and radar imagery. All of these observations can be performed from space unhindered by political boundaries or natural barriers. Properly calibrated and processed, these data comprise a highly reliable information source with which to conduct science, manage natural resources, or make business decisions.

The current high cost of accessing space (between $5000 and $20,000 per kilogram in low Earth orbit) and the need for robust, reliable satellite systems, makes an Earth-orbiting remote sensing platform a very expensive undertaking. Earth's atmosphere limits sensors to certain wavelength regions, and the physics of remote sensing limits the spatial and spectral response of certain detectors. But perhaps the biggest limit facing those who wish to use remote sensing from space is the lack of the ability to process the large volume of data generated by contemporary sensors efficiently and inexpensively. Efficient processing of vast data streams is critical to the successful use of these data by human analysts. A process called *data fusion* is often undertaken in this regard. In this process, simultaneous data streams for a specific target or scene acquired by different types of remote sensing instruments are blended and carefully analyzed. If done properly, this integrated data evaluation can reveal subtle changes and unusual phenomena that are not readily apparent in data from a single instrument. However, if not properly planned or performed, data fusion can be ineffective, expensive, and time consuming. This is the real challenge for those who wish to take full advantage of the incredible power of satellite-based remote sensing.

Issues and Debate

Perhaps the most significant issue that continues to be raised regarding satellite-based remote sensing involves costs versus benefits. This applies alike to military, scientific, and commercial space platforms. Satellite systems are very expensive. Therefore, during the development of any major space platform, the sponsors of individual instruments need to demonstrate that their remotely sensed data are of special value in achieving the overall mission objectives of the platform. In some cases this gets into a technical debate over such issues as choice of wavelength, ground spot size (spatial resolution), and coverage. In other cases, geopolitics and competing goals begin to temper the debate. Is it more important to measure the depletion of tropical rain forests or the surface temperatures of oceans?

There is also a continuing debate concerning governmental regulation of satellite-derived imagery and the quality of the data that can be publicly offered by commercial companies entering the Earth-observing business. For example, national security leaders are concerned that power asymmetries might arise quickly if a rogue nation suddenly acquired high-quality commercial Earth-observation data that revealed unsuspected weaknesses in the military forces of neighboring states.

From the very beginning of the space age, observing Earth's surface from space has been regarded by the international space law community as an acceptable, legal act open to any nation or group that has the resources and technology to deploy appropriate spacecraft. Nevertheless, privacy—personal and national—will continue to be an issue in the 21st century. As the quality of the publicly or commercially available remotely sensed data continues to improve, nations and individuals must become accustomed to "living on a glass globe."

—*Joseph A. Angelo, Jr.*

RELATED TOPICS

Computer Modeling, Multispectral Imaging, Remote Sensing, Robotic Space Exploration, Satellite Technology

BIBLIOGRAPHY AND FURTHER RESEARCH

BOOKS

Ginsberg, Irving W. and Joseph A. Angelo, Jr., eds. *Earth Observations and Global Change Decision Making, 1989: A National Partnership.* Melbourne, Fl.: Krieger Publishing, 1990.

Peebles, Curtis L. *The Corona Project: America's First Spy Satellites.* Annapolis, Md.: Naval Institute Press, 1997.

Rees, Gareth. *The Remote Sensing Data Book.* New York: Cambridge University Press, 1999.

Sabins, Floyd F., Jr. *Remote Sensing: Principles and Interpretation,* 3rd ed. New York: W.H. Freeman, 1996.

Short, Nicholas M., Paul Lowman, Stanley Freden, and William Finch. *Mission to Earth: Landsat Views the World.* NASA SP-360. Washington, D.C.: U.S. Government Printing Office, 1976.

PERIODICALS

NASA/Goddard Spaceflight Center. "NASA's Earth Observing System: EOS AM-1." NP-1998-03-018-GSFC, 1998.

Prins, Eric. "Remote Sensing Data for Biodiversity Management in Sahelian Africa." *Earth Observation Magazine* (EOM), August 1997, 18.

Thompson, M. D. and J. B. Mercer. "Digital Terrain Models from RADARSAT." *Earth Observation Magazine* (EOM), March 1996, 22.

INTERNET RESOURCES

AF Space Command Home Page
http://www.spacecom.af.mil/hqafspc/
Commercial Satellite Imagery Company
http://www.spaceimaging.com
Earth Observation Magazine Online
http://www.eomonline.com/
EROS Center Home Page, UGSG
http://edcwww.cr.usgs.gov/eros-home.html
Landsat History
http://geo.arc.nasa.gov/
Los Angeles AFB Home Page—Military Satellites
http://www.laafb.af.mil/

National Reconnaissance Office Home Page
   http://www.nro.odci.gov/
Remote Sensing Tutorial
   http://rst.gsfc.nasa.gov/
USGS—"Earthshots"—Images of Environ Change
   http://www.usgs.gov/Earthshots

SPACE SHUTTLE

The Space Shuttle is a reusable, delta-winged aerospace vehicle that is launched into space like a rocket but returns to Earth by gliding through the atmosphere and landing on a runway similar to that used by an airplane. It is the prime element of the U.S. Space Transportation System. In its first two decades of operation, the Shuttle has performed almost 100 human-crewed missions in Earth orbit.

However, the more than $100 billion system is also overshadowed by the tragic loss of the Challenger vehicle and its seven-person crew in 1986, a chronic inability to meet anticipated reductions in the cost of delivering payloads to low Earth orbit, and shrinking operational budgets for an aging fleet of vehicles. Until a fully reusable launch vehicle is available, this shuttle fleet will be pressed into tightly scheduled service over the next decade, primarily as the "space taxi" for the construction and initial operation of the INTERNATIONAL SPACE STATION.

Scientific and Technological Description

Unlike all previous expendable ("throwaway") rockets, all the major components of the Space Shuttle system are reused, with the exception of a large external tank that is discarded during each mission.

Major Shuttle Components

The Shuttle flight vehicle has three main components: the winged-orbiter vehicle; the giant external tank, which feeds cold liquid hydrogen and liquid oxygen propellants to the three Space Shuttle main engines; and two large solid rocket boosters. The orbiter vehicle contains a pressurized crew compartment, a huge cargo bay (18.3 meters [m] long and 4.57 m in diameter), and three main liquid propellant engines mounted on its aft end. The delta-winged vehicle is 37 m long, 17 m high, and has a wingspan of 24 m. Each of the orbiter's three main engines uses very cold (cryogenic) propellants to generate a thrust of about 1.7 million newtons at sea level. The liquid hydrogen/liquid oxygen-fueled engine is a reusable high-performance rocket engine. Each orbiter also has two smaller orbital maneuvering system (OMS) engines that operate only in space. The OMS engines are located in external pods on each side of the aft fuselage and provide thrust for orbit insertion, orbit change, orbit transfer, rendezvous operations, and de-orbit.

The huge external tank is 47 m long and 8.4 m in diameter. The two inner propellant tanks contain a maximum of about 1.46 million liters of liquid hydrogen (LH_2) and 0.54 million liters of liquid oxygen (LO_2). The external tank is the only major component of the Shuttle flight vehicle to be expended during each launch.

Each solid rocket booster (SRB) is 45.4 m high and 3.7 m in diameter. Each booster produces a thrust of about 13.8 million newtons for the first few seconds after ignition. The thrust then gradually declines for the remainder of its 2-minute burn. This tapered thrust design prevents overstressing of the Shuttle flight vehicle. When a pair of SRBs burn together with the three main liquid-propellant engines, the Shuttle vehicle generates a total thrust of about 32.5 million newtons at liftoff.

Typical Mission Profile

Shuttle vehicles are launched from either Pad 39A or 39B at NASA's Kennedy Space Center in Florida. A Shuttle can carry up to about 22,700 kilograms of payload into low Earth orbit (LEO), located at an altitude of 125 miles (200 kilometers [km]) or more above Earth. Each mission has its own unique set of requirements, but a typical mission involves the following sequence of events and activities. A few seconds before the final commitment to launch, the three main liquid-fueled engines (the Space Shuttle main engines [SSMEs]) are ignited and brought up to full power. If any problems or anomalies are detected, they are shut down automatically and the two SRBs are never ignited. If the SSMEs are functioning well, a commitment to launch is made and the giant SRBs are ignited.

The SRBs generally burn until the Shuttle reaches an altitude of about 45 km. They then separate and fall back into the Atlantic Ocean to be retrieved, refurbished, and prepared for another flight. After the SRBs have been jettisoned, the orbiter's three main engines continue to burn for another 6 minutes before they too are shut down. At this point in the flight, the external tank is jettisoned and falls back to Earth (see figure). The orbiter's OMS engines then provide the final thrust necessary to achieve the desired orbit. While in space, the OMS engines can be fired to raise or adjust the vehicle's orbit to satisfy the needs of a particular mission.

After performing its task—deploying a satellite, operating onboard scientific instruments, making observations of Earth or the heavens, performing in-orbit satellite repair, or accomplishing space station assembly tasks—the orbiter vehicle and its crew are ready to return to Earth. A Shuttle flight can last from a few days to more than a week. Since the orbiter operates only in low Earth orbit, a Shuttle-deployed satellite that needs to be placed in a higher orbit often has an attached upper-stage chemical rocket propulsion unit. Once at a safe distance from the orbiter, the upper stage ignites and delivers the payload to its higher altitude or operational orbit, or places it on an interplanetary trajectory.

To return to Earth, the orbiter reverse-fires its OMS engines, thereby reducing its orbital velocity and enabling the vehicle to reenter the upper regions of Earth's atmos-

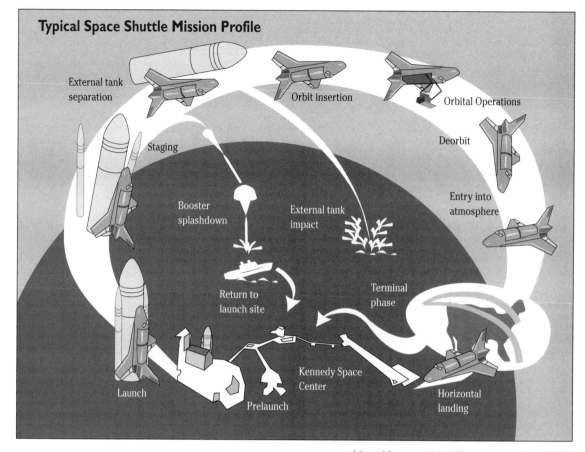

Typical Space Shuttle Mission Profile

External tank separation

Orbit insertion

Orbital Operations

Deorbit

Staging

Booster splashdown

External tank impact

Entry into atmosphere

Return to launch site

Terminal phase

Kennedy Space Center

Horizontal landing

Launch

Prelaunch

Adapted from an original illustration courtesy of NASA.

phere. Then, operating like a giant, unpowered glider, the Shuttle can maneuver to the right or left of its entry path as much as 2000 km. Special insulation serves as the orbiter's heat shield, protecting it from burning up as it reenters the atmosphere. After gliding down through the atmosphere in a series of energy-dissipating maneuvers, the orbiter touches down like an airplane on a runway at the Kennedy Space Center in Florida (the primary landing site) or at Edwards Air Force Base in California (the alternative landing site in case of bad weather).

Historical Development

The concept of a rocket-boosted glider can be traced back to work in Germany during the 1920s and 1930s by a variety of technical societies that were fostering rocketry and interplanetary space travel. In 1958, the U.S. Air Force began development of a hypersonic boost-glide vehicle called the X-20 or Dyna Soar. Although canceled in 1963 before it flew, the Dyna Soar program was the technical ancestor to a wide variety of Air Force- and NASA-sponsored lifting body vehicles, which in turn supported evolution of the Shuttle.

From its establishment in 1958, NASA studied aspects of REUSABLE LAUNCH VEHICLES. Beginning in 1963, NASA joined the Air Force in research toward the development of the

Aerospaceplane, a human-crewed vehicle that could go into orbit and return, taking off and landing horizontally.

In 1969, an Air Force–NASA Space Task Force Group was asked to help define post-*Apollo* era space objectives for NASA's *Apollo* Program (1968–1972), which fulfilled President Kennedy's national goal of achieving a successful human lunar landing and safe return to Earth by the end of the 1960s. This group, with vision toward the 21st century, recommended the development of an economical, reusable space transportation system capable of supporting a space station in LEO. However, despite the *Apollo* program's amazing triumph of landing the first humans on the Moon, the entire post-*Apollo* space program recommended by the task force group was distilled by budget pressures to a single new program—a recoverable, reusable new launch vehicle called the Space Shuttle.

Between 1969 and 1971, the aerospace industry studied a reusable space transportation system. These studies were heavily influenced by continued budget pressures, which meant that an early design for a two-stage fully reusable liquid-propelled vehicle eventually became a delta-winged aircraftlike vehicle mounted on a large expendable external tank and supplemented in thrust by two large reusable solid rocket boosters. This austerity-driven final design, which fell short of the original vision of a completely reusable launch vehicle, became the Space Shuttle.

On January 5, 1972, President Nixon announced that the United States would develop the Space Shuttle. Each Shuttle would have a lifetime of 100 space missions and be capable of carrying up to 29,500 kilograms of payload. The first crewed operational flight was scheduled for March 1979. Optimistically, NASA projected 60 or 70 flights a year by the mid-1980s. However, NASA now had to compensate for the absence of a "prime purpose" for the Shuttle (i.e., there was no planned space station to and from which the crew and cargo would go). NASA quickly suggested that the Shuttle would usher in a new era of space flight in which a wide variety of people could now have the opportunity to fly in space. The Shuttle was identified as a vehicle that could carry satellites to orbit, repair them there, or even return them to Earth for refurbishment. In the absence of a space station, the Shuttle would also carry a special research facility, called *Spacelab,* in its cargo bay. Built by the European Space Agency, the pressurized habitat configuration of *Spacelab* would permit up to four scientists to perform experiments while in Earth orbit.

A new era in human space flight opened on April 12, 1981, when the space shuttle *Columbia,* piloted by astronauts John W. Young and Robert L. Crippen, blasted off from the Kennedy Space Center. This orbital flight test mission, called STS-1, was planned as a short duration flight with the major objectives of safe ascent, orbital flight, and landing. After a nearly flawless space voyage, Young and Crippen made a perfect landing at Edwards Air Force Base in California.

Uses, Effects, and Limitations

Despite the tragedy of the *Challenger* (51-L) mission, the Space Shuttle program represents a very productive and very important period in the overall U.S. human space flight program. The in-orbit repair and servicing missions to the Hubble Space Telescope (1993 and 1997) have improved the value and extended the mission lifetime of this unique scientific instrument. The *Spacelab* missions collectively have returned a wealth of scientific and engineering data over a wide range of technical disciplines. The successful deployment of many scientific satellites on interplanetary trajectories (such as NASA's *Galileo* to Jupiter and *Magellan* to Venus) have opened up new frontiers in space science. Finally, the rendezvous and docking operations with the Russian *Mir* space station represented the first phase of the International Space Station program.

NASA's space-faring fleet of orbiters are named after pioneering sea vessels that established new frontiers in research and exploration on Earth. The operational fleet consists of *Columbia* (OV-102), *Discovery* (OV-103), *Atlantis* (OV-104), and *Endeavour* (OV-105). The *Challenger* (OV-99) was lost along with its crew in a tragic launch accident that occurred on January 28, 1986. The *Challenger* lifted off from the Kennedy Space Center on its final journey at 11:38 A.M.

The Space Shuttle Discovery *at Kennedy Space Center in May 1998 (Courtesy of NASA).*

(local time) on January 28, 1986. A catastrophic explosion approximately 73 seconds after liftoff claimed crew and vehicle. The cause of this explosion was determined to be an O-ring failure in the right solid rocket booster.

The *Enterprise* (OV-101) was the first Space Shuttle orbiter manufactured. Originally to be named *Constitution,* a popular write-in campaign to the White House urged that this test vehicle be named *Enterprise,* after the starship in the popular TV show *Star Trek.* But *Enterprise* was never intended to fly in space. Instead, it was used in a series of approach and landing tests (ALTs) conducted in 1977 as a preparation for the space flight operations of the other orbiter vehicles.

Since the STS-1 mission, almost 100 Shuttle flights have been accomplished. All but one, the ill-fated *Challenger* mission, have produced pioneering results and varying degrees of technical success in many areas of space technology, life science, and astrophysics. In December 1998, the newest orbiter to join the fleet, the *Endeavour,* successfully accomplished a very historic mission, STS-88. During this mission, the Unity module (the U.S.-built Node 1 of the International Space Station) was delivered into orbit and joined to the Russian-built *Zarya* control module. (The module had previously been placed in orbit by an expendable Russian Proton rocket.) Using hours of carefully planned and performed extravehicular activity, Shuttle astronauts accomplished the first assembly activity for the International Space Station. Through 2004, over 40 Shuttle missions are planned to support the complete assembly and initial operation of the space station.

Issues and Debate

There is little debate that the Shuttle is an incredibly complex machine that has pushed the envelope of space technology in a number of important areas. It is also generally recognized that the Shuttle, because of its more gentle ascent acceleration environment, has made space flight available to a larger number of human beings. *G-force,* expressed in numbers of Gs, refers to the increased pull of the Earth's gravity as felt in flight. At 6g, a person who weighs 100 pounds on Earth would weigh 600 pounds in flight. Whereas early astronauts had to endure forces of 6g and more on liftoff, today's Space Shuttle astronauts endure fewer than 3gs.

Despite these important accomplishments, the "reusable" Shuttle has failed to reduce significantly the price of launching objects into space: It now appears to cost between $10,000 and $20,000 per kilogram to carry an item into space on the Shuttle. This cost is far above that envisioned in 1969 by the NASA–Air Force Space Task Force Group, which projected a totally reusable launch vehicle capable of hauling payloads into LEO for $100 per kilogram or less. Critics point out that over the Space Shuttle's first two decades of operation, the system has cost over $100 billion, even though there has been no prime mission for it to accomplish. In fact, only recently has space station assembly become part of the Shuttle program. Shuttle advocates tend to overlook the program's cost and point to the wide variety of missions accomplished by the Shuttle fleet. Opponents, many of whom are nonetheless strong space advocates, tend to regard the Shuttle as a huge money-sink that has absorbed fiscal resources from many other areas of space science and technology while actually providing very little return on the investment.

Another fundamental debate rages both inside and outside NASA: Why send people on high-risk missions, when machines can do the same job more cheaply? This debate extends across the entire spectrum of space exploration missions, and it is a difficult question to ignore. Why risk a $1 billion space vehicle and the lives of four to seven astronauts to launch a communications satellite that can be placed in space equally well by an expendable launch vehicle at a fraction of the cost? Advocates insist that such Shuttle missions are necessary to maintain crew capability and to justify the cost of the Shuttle fleet. Opponents say that Shuttle flights are unnecessary. They would prefer to see NASA money devoted to other space pursuits, such as ROBOTIC SPACE EXPLORATION and the development of a reusable manned launch vehicle. Those lines of research, they say, will ultimately allow NASA to reduce its costs and will help the U.S. aerospace industry to remain competitive in the international space marketplace. They also warn against the tragic technical, social, and political implications of another *Challenger*-like accident. Aerospace safety analysts now suggest that the risk probability of a major Shuttle launch accident is somewhere between 1 in 200 and 1 in 1000.

NASA planners are currently supporting the development of a new generation of totally reusable, crewed launch vehicles. Until these become available, however, the Space Shuttle fleet, with its capabilities, limitations, and issues, represents the currently available "space taxi" for construction of the International Space Station and a variety of other important space missions in low Earth orbit at the beginning of the 21st century.

—*Joseph A. Angelo, Jr.*

RELATED TOPICS
International Space Station, Reusable Launch Vehicles, Robotic Space Exploration, Satellite Technology, Space-Based Materials Processing

BIBLIOGRAPHY AND FURTHER RESEARCH
BOOKS
Angelo, Joseph A., Jr. *The Dictionary of Space Technology*, 2nd ed. New York: Facts On File, 1999.
Brown, Robert A., ed. *Endeavour Views the Earth*. New York: Cambridge University Press, 1996.
Cole, Michael D. *Challenger: America's Space Tragedy*. Springfield, N.J.: Enslow Publishers, 1995.
Joels, Kerry Mark, Gregory P. Kennedy, and David Larkin. *The Space Shuttle Operators Manual*. New York: Ballantine Books, 1988.
Neal, Valerie, Cathleen S. Lewis, and Frank H. Winter. *Spaceflight: A Smithsonian Guide*. Toronto, Ontario, Canada: Macmillan, 1995.
PERIODICALS
Boyer, William, Leonard David, and Theresa Foley. "Special Report: Building the International Space Station." *Final Frontier,* December 1994–January 1995, 20.
Kistler, Walter P. "Humanity's Future in Space." *The Futurist,* January 1999, 43.
Slakey, Francis, and Paul D. Spudis. "Robots vs. Humans: Who Should Explore Space?" *Scientific American,* Spring 1999, 24.
Warren, Michael. "$80 Billion Blunder? Is the Space Shuttle Worth the Cost?" *Final Frontier,* November–December 1995, 18.
INTERNET RESOURCES
NASA Homepage
http://www.nasa.gov/
NASA Kennedy Space Center: Upcoming Space Shuttle Launches
http://www-pao.ksc.nasa.gov/kscpao/schedule/schedule.htm
Upcoming Shuttle Launches
http://spacelink.nasa.gov/NASA.Projects/Human.Exploration..../
NASA Space Shuttle
http://spaceflight.nasa.gov/shuttle/reference/

SUPERCOMPUTERS

Supercomputers can solve problems that are too complex for conventional computers by processing greater volumes of information more quickly. Without supercomputers, weather forecasting, space research, and advanced product design would be more time consuming and less productive. However,

Two Kinds of Supercomputers

A traditional supercomputer is like a scaled-up version of a normal computer. Its processing unit may be made up of multiple processors linked together in parallel, connected to a massive memory by a bus known as a hyperchannel. Input and output are coordinated by a large mainframe computer. An alternative form of supercomputer can be created temporarily by linking PCs together over a network such as the Internet. Work is divided between the PCs using parallel processing techniques, and coordinated by a central mainframe computer linked to a web server.

supercomputers are extremely expensive, and although they are tackling ever more complex problems, it is not clear whether they are helping to solve the problem of information overload or are making it worse.

Scientific and Technological Description

Most supercomputers process information in the same way as it is processed by a personal computer (PC), although on a much larger scale. In any computer, data (information to be processed) are fed in via an input unit, stored in memory, and operated on by a central processing unit (CPU) according to a set of stored instructions known as a *program*. The results are then fed out via an output unit. In a PC, the input unit may be a keyboard or a mouse, the memory may be multiple megabytes of storage (equivalent to millions of characters of information) on silicon chips, the program could be a graphics package or word processor, and the output unit could be a monitor or printer. The central processor could be a single Intel Pentium chip, connected to the other components by cables, known collectively as a *bus,* that carry digital (numerical) information back and forth.

In a supercomputer, each of these components is scaled up. Large mainframe computers are used as the input units to feed in vast amounts of data. The CPU typically consists of multiple processors; 32 are used in one of the world's fastest supercomputers, the Cray T90TM, but no fewer than 64,000 were used in the Thinking Machines Corporation's Connection Machine CM-1, launched in 1983. Supercomputers also require enormous memories for storing their data during processing. A prototype Cray-3 machine had a memory of 8 gigabytes (8 billion bytes), compared to a typical PC, which can store 64 million bytes (megabytes). The bus in a supercomputer, called a *hyperchannel,* moves information at up to 100 megabytes per second. All of this means that supercomputers can work up to 5000 times faster than the fastest PCs. Thus the Cray Y-MP can work at a speed of 16 gigaflops (16 billion flops), where a flop is a measure of computing speed equal to one calculation performed in a second.

These speeds are achieved partly by making the distance between components in the processor as short as possible. It is this that gives Cray computers their distinctive C-shape. In the Cray-1 supercomputer, begun in 1976, there was some 60 miles (97 kilometers) of wiring, but no single wire was longer than 4 feet (ft.) [1.2 meters (m)]. The Cray-2 computer, announced in 1979, reduced the maximum wire length to 16 inches (in.) (41 centimeters [cm]). Several years later, the prototype Cray-3 had gotten this down to 3 in. (8 cm).

Packing so many components so tightly generates enormous amounts of heat and necessitates that supercomputers include ingenious cooling systems. PCs use small electric fans and heat-radiating fins attached to the CPUs known as *heat sinks*, but supercomputers are in another league even here. One 1960s supercomputer, the Control Data Corporation (CDC) 6600, circulated the refrigerator gas Freon around the processing unit. The Cray-2's circuits were immersed in baths of liquid coolant for the same reason.

Traditionally, computers carry out one operation at a time, which is known as serial processing. Some supercomputers use this approach, but increasingly they use a technique known as *parallel processing*. Typically, this involves using multiple CPUs to process more than one program instruction or piece of data at a time, which increases processing speed considerably (see PARALLEL COMPUTING).

Historical Development

Any computer built during the early 20th century might be described as a supercomputer. All were bigger and more complex than anything that had been seen before; all were capable of solving problems that had previously taken people months or years of intensive calculation. Thus the ENIAC, a 100-ft. (24-m)-long box containing 18,000 valves (the predecessors of transistors and silicon chips), built at the University of Pennsylvania between 1943 and 1946, might qualify as one of the world's first supercomputers. Certainly, it was one of the first machines to use parallel processing.

One person can be credited substantially with the historical development of supercomputers—U.S. electronic engineer Seymour Cray (1925–1996). In 1957, Cray co-founded the Control Data Corporation (CDC), where he designed the CDC 1604, one of the first computers to use transistors instead of valves. In 1963, Cray and his partner James Thornton designed the CDC 6600, which contained 350,000 transistors, used Freon cooling, and cost $7.5 million. It was the world's most powerful computer at that time.

Cray was largely unknown outside the computing world until 1972, when he left CDC to found Cray Research, Inc. During the next decade, the name *Cray* became synonymous with supercomputers. The Cray-1, the first example of which was installed at Los Alamos National Laboratory in 1976, is considered to be the world's first true supercomputer. It contained some 200,000 chips and could work at a speed of about 160 megaflops. The Cray-2, announced in 1979, had 256 times as much memory as the Cray-1. Later designs included the Cray X-MP (announced in 1982), which was 10 times faster than the Cray-1. A Cray-3 was designed,

In 1997, IBM's Deep Blue supercomputer beat chess Grand Master Gary Kasparov in a six-match tournament (Courtesy of IBM).

which would have had 16 processors and an 8-gigabyte memory, but it was never developed successfully.

Other manufacturers have also developed supercomputers, including IBM, Fujitsu, Hitachi, and Convex/Hewlett Packard. In 1983, Thinking Machines Corporation (TMC) began developing "massively parallel" supercomputers called Connection Machines, which contained up to 64,000 processors. Although the company attracted the brightest staff from around the world and developed three generations of Connection Machines, TMC suffered from a decline in the market and stopped making supercomputers in 1995.

Supercomputers have often caught the public's imagination. In 1968, a supercomputer known as HAL (whose initials were one letter off from those of IBM) became the star of Stanley Kubrick's film *2001: A Space Odyssey,* based on a story by Arthur C. Clarke. During the 1980s, supercomputers became formidable chess and checkers players, but it was not until 1997 that IBM's 288-processor chess-playing machine, Deep Blue, beat Grand Master Gary Kasparov in a six-match tournament by 3 ½ points to 2 ½. In 1997, Intel Corporation installed the world's fastest experimental supercomputer, the ASCI Red TFLOPS, at Sandia National Laboratory, in Albuquerque, New Mexico, by linking together over 9000 Pentium-Pro chips; it worked at a maximum rate of 1.8 teraflops (1.8 trillion flops). Currently using 9472 processors, it is still the world's fastest computer.

Uses, Effects, and Limitations

Supercomputers are the ultimate number crunchers, best known for their role in solving complex scientific problems involving vast amounts of data. These include modeling the world's climate and the phenomenon known as global warming (see COMPUTER MODELING), attempting to crack mathematical codes (see CRYPTOGRAPHY), and producing complex visual simulations of the real world (see VIRTUAL REALITY).

One of the first applications of supercomputers was in weather forecasting. Producing reliable predictions of the world's weather requires enormous amounts of data to be processed. The National Center for Atmospheric Research (NCAR) in Boulder, Colorado, and the European Center for Medium-Range Weather Forecasting in Reading, England, use powerful supercomputers to produce weather forecasts up to about two weeks in advance. The world's most powerful weather forecaster (and seventh most powerful computer) is a Silicon Graphics/Cray T3E-900 supercomputer with some 876 processors, installed at the U.K. meteorological office in Bracknell, England, in 1997.

The Human Genome Project is a worldwide collaboration between biochemical researchers who are attempting to understand the function of all the genes in human DNA (genetic material). The project uses parallel supercomputers to investigate human genes that are related to different diseases and functions in the body (see GENETIC TESTING). Enormous amounts of relevant data have been collected by researchers across the world, and supercomputers enable these data to be searched and processed in just a few seconds.

This type of collaboration is being used for other global projects, including geological models of seismic (earthquake) data and global climate models. Another type of collaboration enables PC users to connect their machines together over the Internet to produce temporary supercomputers. The University of California at Berkeley is home to a project known as the Search for Extra Terrestrial Intelligence (SETI), which uses a million ordinary Internet-linked PCs to analyze vast amounts of radio telescope data in an effort to determine whether there is intelligent life beyond Earth. But perhaps the best known example of this type of collaboration is the Great Internet Mersenne Prime Search (GIMPS), in which thousands of Internet-linked PCs work together to find the biggest prime number. Thanks to GIMPS, the record for this feat of supercomputing is held not by a Cray machine but by a single 350-megahertz IBM Aptiva desktop PC with a Pentium II processor, which found the 2-million-digit prime number $2^{6,972,593} - 1$ on June 30, 1999.

Supercomputers have demonstrated their ability to solve many complex problems, but they are based largely on mathematical models that approximate reality by making certain simplifying assumptions. Thus the original models of global warming, the gradual warming of the Earth's climate that many scientists believe is linked to increased levels of greenhouse gases in the atmosphere, took little account of how the oceans would respond to changes in the climate, because no one knew precisely how to model this. In other words, supercomputer-based models are still limited by human abilities to understand the world. Although supercomputers enable more complex models to be explored, these are still limited by the quality of the assumptions made at the outset.

Issues and Debate

By definition, supercomputers are more powerful computers; today's supercomputers are better than yesterday's. Yet their usefulness is still ultimately limited by what humans do with the information they receive from them. Weather predictions that once took days can now be carried out in seconds, but people still forget to carry umbrellas. Supercomputers can model global warming and predict its consequences, but people will continue to make unnecessary car trips, releasing more polluting emissions into the atmosphere, even when confronted with evidence of climate change. Supercomputers may be extremely powerful, but people will always be fallible. Once scientific data are generated, political bodies or appointees must decide how to use it.

In the widely read book *Future Shock,* published in 1970, U.S. journalist Alvin Toffler outlined a future in which people are overwhelmed by too much information and too much change. Commentators such as Toffler and U.S. architectural historian Lewis Mumford have pointed out the need to be aware of the moral dimension of technology, and even the need to regulate technological advance. It could be argued that supercomputers, which fuel the need for ever-more information and ever-more complex models of reality, are a powerful symptom of the future shock that Toffler envisioned. But ironically, supercomputers also provide the best means of tackling what Toffler termed *information overload.* Either way, computers do not necessarily make human beings wiser or better people or the world a better place. Supercomputers can perform staggering calculations; they enable cutting-edge research into genetics, meteorology, space research, and numerous other technologies. But they cannot make political decisions or moral judgments based on the results.

Many people have questioned the wisdom of handing over human responsibilities to computers; occasionally, people have gone even further in their opposition. In 1987, U.S. political activist Katya Komisaruk used a crowbar to haul out and destroy the processor chips from the $1.2 million NAVSTAR computer at Vandenberg Air Force Base in California, which controls the GLOBAL POSITIONING SYSTEM (GPS) satellite navigation system. She argued that the machine was being used to target missiles at the former Soviet Union and might lead to civilian deaths. But her "greater good" defense was thrown out by a judge, who noted that NAVSTAR was also used to help nonmilitary ships and airplanes navigate around the world, and Komisaruk was sentenced to five years in jail.

Despite moral opposition, some of the world's fastest supercomputers continue to be used for defense-related research. Because high-performance computers enable the development of weapons of mass destruction, their export from the United States to potentially hostile countries has been strictly controlled. But with the relaxation of those controls by the Clinton administration in 1995, the export of several powerful supercomputers to China and Russia, and the ever-increasing power of workstations (very powerful desktop computers whose export is not restricted), concern has been expressed in Congress that the restrictions are not tight enough.

Supercomputers may fall out of favor for reasons other than public mistrust and opposition to defense-related uses. The difference between a PC and a super-computer, although vast, narrows every day. PCs are now as powerful as large computers were 20 years ago, but far cheaper and easier to use. Parallel processing is increasingly being used on smaller machines. If PCs can be linked together using networks such as the Internet to make virtual supercomputers, perhaps the world will no longer need the real thing.

All of this is reflected in the turbulent state of the super-computer market, where development costs are high and there are relatively few buying customers to produce a worthwhile return on a massive investment. Kendall Square Research, a leading manufacturer, stopped making super-computers in 1994. Thinking Machines Corporation closed down its manufacturing in 1995. Even Cray Research was forced to merge with the workstation manufacturer Silicon Graphics in 1996. These factors, and increasing interest in technologies such as optical computers that work by processing light instead of electricity, may ultimately send today's supercomputers the way of the steam engine.

—*Chris Woodford*

RELATED TOPICS
Computer Modeling, Cryptography, Parallel Computing, Virtual Reality

BIBLIOGRAPHY AND FURTHER RESEARCH

BOOKS
Computer Modeling, Cryptography, Billings, Charlene. *Supercomputers: Shaping the Future*. New York: Facts On File, 1995.
Hsu, Jeffrey and Joseph Kusnan. *The Fifth Generation: The Future of Computer Technology*. Blue Ridge Summit, Pa.: Windcrest, imprint of TAB Books, 1989.
Kaufman, William and Larry Smarr. *Supercomputing and the Transformation of Science*. New York: Scientific American Library, 1993.
Kidder, Tracy. *The Soul of a New Machine*. New York: Avon Books, 1982.
Kuck, David. *High-Performance Computing: Challenges for Future Systems*. New York: Oxford University Press, 1996.
Stork, David, ed. *Hal's Legacy: 2001's Computer as Dream and Reality*. Cambridge, Mass.: MIT Press, 1997.
Toffler, Alvin. *Future Shock*. New York: Bantam Books, 1991.

PERIODICALS
Copeland, B. Jack, and Diane Proudfoot. "Alan Turing's Forgotten Ideas in Computer Science." *Scientific American*, April 1999, 77.
Corcoran, Elizabeth. "Calculating Reality." *Scientific American*, January 1991, 100.
Gibbs, W. Wayt. "Taking Computers to Task." *Scientific American*, July 1997, 64.
Lloyd, Seth. "Quantum-Mechanical Computers." *Scientific American*, October 1995, 44.
O'Malley, Chris. "Computing's Outer Limits." *Popular Science*, March 1998, 64.
Pournelle, Jerry. "Of Supercomputers, Sound Files, and Sugarscape." *BYTE*, June 1997, 143.

Sharp, Oliver. "The Grand Challenges: The Supercomputer Makers." *BYTE*, February 1995, 65.
Thompson, Tom. "The World's Fastest Computers." *BYTE*, January 1996, 44
Weingarten, Donald. "Quarks by Computer." *Scientific American*, February 1996, 104.

INTERNET RESOURCES
ASCI Red: The World's First TeraFlop UltraComputer at Sandia National Laboratory, Albuquerque
http://www.sandia.gov/ASCI/Red.htm
National Center for Supercomputing Applications
http://www.ncsa.uiuc.edu/ncsa.html
San Diego Supercomputer Center
http://www.sdsc.edu/
Silicon Graphics' Cray Supercomputers Web Site
http://www.cray.com
University of Mannheim's 500 Top Supercomputing Sites
http://www.top500.org
Participatory Internet Supercomputing Projects
GIMPS—The Great Internet Mersenne Prime Search
http://www.mersenne.org/prime.htm
SETI@home: Search for Extra Terrestrial Intelligence (SETI) at Home
http://www.setiathome.ssl.berkeley.edu

SUPERCONDUCTORS

Superconductors are materials that carry a charge of electricity with no resistance whatsoever. This means that an electrical charge fed into a ring of superconducting material will literally circle the ring at the same strength indefinitely. In addition, superconductors repel magnetic fields. Various materials can become superconductors under the proper conditions: extreme cold and an absence of strong magnetic fields. That is because superconductivity might be considered a state of matter, much as solidity and liquidity are states. Matter can pass from a superconducting state to a normal state, just as a solid can be melted into a liquid. Although many materials can become solid or liquid, fewer materials can become super-conducting. Materials that are known to superconduct are metals, metallic alloys, and certain ceramics. All of these materials also conduct electricity at normal temperatures.

Superconductors are currently used in some medical scanning devices, as well as supersensitive metal detectors, and they may one day be used to generate and transport power. The high cost of cooling materials down to superconducting temperatures, however, has so far limited their use.

Scientific and Technological Description

To understand how materials become superconductors at low temperatures, it is first essential to understand how they conduct electricity normally. All matter is made up of atoms, which in turn are made up of positively charged nuclei (containing protons and neutrons) surrounded by clouds of negatively charged electrons. In some matter, the electrons are tightly bound to the nucleus. But in other matter, especially metals, the electrons are not tightly bound. In this case, the atoms link up in such a way that the nuclei form a lattice and essentially share their electrons with each other. The electrons move around the lattice relatively freely.

An electric current is usually carried by a stream of electrons moving from one place to another. The strength of the current depends in part on how many electrons are in it. If a current is applied to atoms where the electrons are tightly bound to the nuclei, the electrons will stubbornly refuse to move, and the current will not be carried, or conducted. In matter in which the electrons can move freely, however, the electrons will move across the lattice when an electric charge is applied.

Even in metals such as silver and gold that are excellent conductors, however, the electrons do not move perfectly freely. A current passed through a silver wire will lose some of its strength by the time it reaches the end of the wire because some of the electrons do not complete the journey from one end to the other. This loss of strength is called *resistance*. Resistance has three causes. One is the inevitable impurities that are found in any material. Often, these impurities do not conduct electricity as well as does the base substance. The second cause is interactions between elec-

trons. Since electrons are negatively charged, they repel each other if they get close. These interactions can knock electrons out of the current stream, weakening the current. The third cause is interactions between the electrons and the lattice. Since the lattice is positively charged, it can pull electrons out of the current stream. Ironically, two of the three causes of resistance—interactions between electrons and interactions between the electrons and the lattice—are also the causes of *superconductivity,* the absence of resistance. That is because these interactions have a different effect when material is very, very cold.

Heat energy is very exciting on the subatomic level. It makes electrons move about faster and more chaotically, and it makes the lattice vibrate strongly. All of this increases the chances of the types of random collisions and interactions that knock electrons out of the current stream and cause resistance. But when the material is cooled to a superconducting state, everything is calmer. The electrons don't move around so much, and the lattice vibrates very minimally. Even a tiny electrical current added to this material will send an electron running across the lattice. As the negatively charged electron moves between two positively charged nuclei in the lattice, the electron attracts them both. The nuclei move slightly toward the electron, and more important, toward each other (see figure).

After the electron passes, the nuclei linger closely together for a short while, then move back into place. For the moments when they were close together, however, they created a spot in the lattice that had a relatively strong positive charge. This charge attracts a second electron, and it starts moving along in the wake of the first electron. Between the efficiency of this pair movement and the lack of chaotic, heat-induced motion that can knock the electrons off track, the electrons are able to move along indefinitely. A larger electrical charge gets millions of electrons moving cooperatively, and their movement is too strong for small impurities to affect it. The current is conducted with perfect efficiency.

Interaction between Electron Pairs and the Nuclei Lattice in a Superconductor

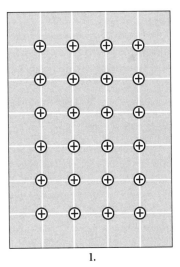

1.

1. A lattice of positive atoms

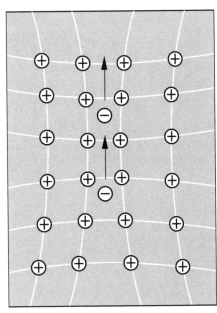

2.

2. As the first negatively charged electron in a pair moves between the positive atoms of a superconducting material, the electron attracts them, drawing them closer together. Once drawn together, the positively charged atoms more strongly attract the second electron in the pair. This effect allows a stream of electrons to move through a superconducting material with perfect efficiency.

The ease with which electrons move in superconducting material also explains why such material can resist a magnetic field. When normal matter is placed in a magnetic field, the field usually just goes right through the matter, perhaps with some distortions. But when a superconductor is placed in a magnetic field, the electrons in the surface layer of the superconductor move around to generate an opposing magnetic field, neutralizing the outside field so that the inner bulk of the superconductor is unaffected. (This powerful opposing magnetic field is why a magnet will levitate if placed over a superconductor.)

The fact that electrons go to such trouble demonstrates the truth of one of the basic theories of physics: Matter tries to stay at a low-energy state. When a magnetic field passes through a type of matter, the matter moves into a higher-energy state. But since superconductors have such mobile electrons, the electrons on the surface can move around enough to repel the outside field. The surface electrons are of course in a very high energy state, but on the whole, the superconductor is at a lower state than if the magnetic field flowed through it. If the magnetic field is very strong, however, it will put the superconductor at too high an energy state to keep repelling it. In that case, the superconducting material simply reverts to a normal state and lets the field flow through. Some superconducting material will essentially compromise, letting the magnetic field penetrate just a little.

Historical Development

Superconductivity was discovered in 1911 by the Dutch scientist Heike Kamerlingh Onnes, who was a major figure in the then cutting-edge science of supercooling. He cooled mercury to 4° K (-452°F) to see what unusual properties the material might have and discovered that it no longer demonstrated any resistance to electricity. Kamerlingh Onnes named the phenomenon *superconductivity,* but he had no idea why the mercury was acting in such an unusual fashion.

For decades, nobody else did either. Superconductivity was one of the great mysteries of physics, stumping such legendary figures as Albert Einstein. But scientists could still try to make new metals superconduct (one puzzle was that excellent conductors such as gold and silver were almost impossible to make into superconductors), and they could observe what superconductors did. In 1933, Austrian engineers Walther Meissner and R. Ochsenfeld realized that superconductors repelled magnetic fields as described above. The discovery of what is now called the *Meissner effect,* however, just added to the list of poorly understood properties of superconductors.

It was not until 1957 that a workable theory of superconductivity was proposed. John Bardeen, Leon N. Cooper, and John R. Schrieffer, working at the University of Illinois, proposed that interactions between electrons themselves and between electrons and the lattice—well known as causes of resistance—actually caused superconductivity as well.

The Bardeen–Cooper–Schrieffer theory (outlined above) explained just about every observable quirk of superconducting metals. For example, the puzzle over gold and silver suddenly was solved: Those metals are good conductors because their lattices and electrons barely interact, leaving the electrons free to flow. But since the electron–lattice interaction is essential to superconducting, silver and gold don't go into a superconducting state easily. The BCS theory became famous, and the team was awarded a Nobel Prize in Physics in 1972.

But while superconductor theory advanced, its practical applications were limited because of the high cost of cooling material down to a superconducting state. Then in 1986, K. Alexander Muller and J. Georg Bednorz, working in an IBM lab in Zurich, discovered a ceramic mixture that superconducted at the relatively high temperature of 35K (-397°F). The discovery was an exciting one (Muller and Bednorz won a Nobel prize the next year), in part because it was so unexpected. Conducting ceramics are very rare—in fact, many ceramics are used as insulators. Ceramics are also very different from the metals that had been used as superconductors in the past, so different, in fact, that neither BCS theory nor any other existing theory can explain exactly why or how ceramics superconduct.

One major advantage of ceramics over metals is that ceramics are improvable in a way that metals and even metal alloys are not. Ceramics can be made using many different methods and substances. Scientists quickly went to work formulating ceramics that would superconduct at even higher temperatures. Some of those ceramics now superconduct at temperatures as high as 125° K (-235°F), a temperature that can be achieved using inexpensive liquid nitrogen.

Uses, Effects, and Limitations

The primary limitation on superconductors is the fact that they must be kept extremely cold, as heat will cause them to return to a normal state. Devices that use superconductors always require bulky and expensive refrigeration systems to prevent the superconductors from warming. Nonetheless, superconductors have some very useful properties. One of these properties is, of course, that superconductors transmit electrical charges without resistance. Although still in the experimental stage, scientists have created superefficient power cables made of superconducting material. Such cables could be used to carry electricity from generators to distant homes and businesses without the waste of power that is caused by the electrical resistance of the cable. Some types of superconductors can function as extremely powerful magnets. Scientists are currently working on trains that would use superconducting magnets to float above the track and on motors and generators powered by superconducting magnets (see HIGH-SPEED AND MAGLEV RAILWAYS).

Superconducting magnets have also found a home in medicine—they form an integral part of MAGNETIC RESONANCE

IMAGING (MRI) machines. These powerful medical scanners enable doctors to see soft tissue that does not show up with conventional X-RAY IMAGING. Although MRI machines have helped countless patients, they are extremely expensive to use and to maintain because they use low-temperature superconductors that must be cooled with costly liquid helium. Scientists hope that high-temperature superconductors will eventually move from research laboratories into devices such as MRI machines.

Another useful property of superconductors is that two superconductors separated by an extremely thin piece of insulation will make an electric current. (Such separations are called *Josephson junctions,* after Brian Josephson of Great Britain, who predicted their behavior in 1962. Josephson won a Nobel Prize in Physics in 1973.) If both superconductors are kept cold and not exposed to any magnetic field, they will not make the current. But the tiniest bit of warmth or the tiniest increase in the magnetic field will turn the current on—and the current comes on all at once, not gradually, making the junction well suited for use as a switch.

Researchers say that such switches could be used for computer chips. Normal chips heat up because of their resistance to the electricity flowing through them, and if they are packed too closely together, they will literally melt a computer. But chips made of high-temperature superconductors would not heat up and could be packed very closely, making truly tiny but powerful computers possible, provided that the cooling mechanism to keep the superconductors at the right temperature could be made small enough. Since Josephson junctions are very sensitive to magnetic fields or heat, they are used to make extremely sensitive radiation detectors and magnetic field detectors. These are used to locate metal objects such as mines that cause distortions in the magnetic field.

Issues and Debate

Although superconducting technology has already proven useful in specialized fields such as medicine, it remains on the cusp of making a truly significant impact on the lives of most people. If cheap and portable superconductors can be made, they will have far-ranging implications for such fields as electricity production and transportation. Superconducting magnets could be used to power generators, which would make cheap and clean electricity. This electricity would be carried to homes with perfect efficiency by superconductor cables that would never heat up and melt during periods of heavy electricity use. Motors using superconducting magnets could be used to power cars, and high-speed mass train systems could become cheaper to build, lessening reliance on polluting fossil fuels.

But superconductors have already had a huge impact on the field of physics. Subatomic particles—electrons, protons, and neutrons—do not interact with each other in the same way that larger objects do. The "rules" that describe subatomic behavior, known as the *rules of quantum mechanics,* are quite different from the rules that describe the movement and interaction of larger objects. This difference is due primarily to the effect of heat: Heat excites and disorganizes subatomic particles, preventing groups of subatomic particles from working together and behaving in the same way that individual particles do. But supercooled, superconducting matter behaves very much like subatomic particles do. Physicists theorize that this is because the normally overwhelming effect of heat is absent, allowing quantum forces to come into play on a large scale. Quantum theorists such as Josephson have been able to predict accurately how superconductors will behave. This opens the exciting possibility that the behavior of superconductors may in turn play a large role in helping physicists better understand how matter operates on the subatomic level.

To an extent, this has already happened—the Bardeen–Cooper–Schrieffer theory, developed to explain superconductivity in metals, has had a substantial impact on particle physics. It has also helped cosmologists, who deal with nature on a much larger scale. Cosmologists explore the origins and basic structure of the universe, and they have looked at superconductivity research to help them understand how matter behaves in the extreme conditions found in space. The question of why some ceramics superconduct has already sparked a number of novel theories; there is little doubt that any answer found will challenge basic assumptions about how matter works and what it can be made to do.

—Mary Barr Sisson

RELATED TOPICS
Magnetic Resonance Imaging, Supercritical Fluids

BIBLIOGRAPHY AND FURTHER RESEARCH
BOOKS
Kresin, Vladimir Z., and Stuart A. Wolf. *Fundamentals of Superconductivity.* New York: Plenum Press, 1990.
Miller, Richard K. *Superconductors: Electronics and Computer Applications.* Lilburn, Ga.: Fairmont Press, 1990.
Poole, Charles P., Jr., et al. *Superconductivity.* San Diego, Calif.: Academic Press, 1995.
Ruggiero, Steven T., and David A. Rudman, eds. *Superconducting Devices.* San Diego, Calif.: Academic Press, 1990.

PERIODICALS
Baselmans, J. J. A., et al. "Reversing the Direction of the Supercurrent in a Controllable Josephson Junction." *Nature,* January 7, 1999, 43–45.
Cava, Robert J. "Superconductors Beyond 1–2–3." *Scientific American,* August 1990, 42–49.
Chu, C. W., et al. "Superconductivity Above 150 K in $HgBa_2Ca_2Cu_3O^{8+}$ at High Pressures." *Nature,* September 23, 1993, 323–325.
Clarke, John. "SQUIDs." *Scientific American,* August 1994, 46–53.
Kirtley, John R., and Chang C. Tsuei. "Probing High-Temperature Superconductivity." *Scientific American,* August 1996, 68–73.
Ouboter, Rudolf de Bruyn. "Heike Kamerlingh Onne's Discovery of Superconductivity." *Scientific American,* March 1997, 98–103.

Rodgers, Peter. "Theory Debate Gets Literary, and Ugly." *Science,* September 4, 1998, 1427.

INTERNET RESOURCES

CSI Reference Library
   http://www2.csn.net/~donsher/resources.html
Superconductivity for Electric Systems
   http://www.eren.doe.gov/superconductivity/
Superconductor Week
   http://www.superconductorweek.com
University of Maryland Center for Superconductivity Research
   http://www.csr.umd.edu/

SUPERCRITICAL FLUIDS

Supercritical fluids are used in the chemical, pharmaceutical, environmental, and food industries for separation and extraction of either valuable or contaminating substances from various materials. Supercritical fluids consist of matter at high temperature and pressure that has both gas-like and liquid-like physical properties. Supercritical fluids are better solvents than fluids in gas or liquid states. Supercritical carbon dioxide is the most commonly used fluid in supercritical fluid technology, and decaffeination of coffee is among the most commonly known applications of supercritical carbon dioxide. Supercritical fluid technology eliminates some of the drawbacks of traditional separation and extraction methods, such as the use and creation of toxic compounds.

Scientific and Technological Description

There are three states of matter visible in daily life: solid, liquid, and gas. In the case of water, we observe the three states often: Ice is water in solid form, a river flows with liquid water, and steam rising from a boiling kettle is gaseous water. Normally, we think of state changes as dependent on temperature only, because temperature is the most noticeable factor affecting state. But state changes also depend on pressure. Almost all of the state changes that we watch in our daily lives occur at the same pressure—the atmospheric pressure at Earth's surface—so we do not have much opportunity to notice how pressure affects state. Scientists, however, can manipulate pressure with laboratory equipment, and they can alter both the temperature and pressure of a fluid.

Changes in temperature and pressure cause a predictable pattern of state changes in most matter. Generally, at a constant pressure, matter changes from solid to liquid to gas with increasing temperature. At a constant temperature, matter changes from gas to liquid to solid with increasing pressure (at very low pressures and temperatures, the liquid state is skipped). At a certain high temperature and pressure, the distinction between liquid and gas breaks down; this combination of temperature and pressure is called the *critical point.* The value of the critical point is different for every substance. When both the temperature and pressure of a substance exceed its critical point, the sub-

stance becomes a supercritical fluid. Supercritical fluids retain many of the properties of gases, but they have much higher densities than gases, as do liquids. Density is a measure of the mass of a substance in a given volume. The combination of gaslike and liquidlike properties makes supercritical fluids good solvents, which means that they readily dissolve other substances.

Supercritical Fluid Chromatography

Supercritical fluid chromatography (SFC) is a technique in which the components of a mixture are separated. A supercritical fluid serves as the *mobile phase,* into which the mixture to be separated is dissolved. After the mixture has been dissolved, the mobile phase is poured over or through a *stationary phase,* which consists of a particular type, either liquid or solid material. The molecules that were part of the original mixture stick to particles within the stationary phase. The various components of the mixture behave differently with respect to the stationary-phase particles. The molecules of some components stick to the particles quickly and remain stuck for a long time, while others stick for short periods and then pass through. A scientist can thus separate the mixture's components by collecting them as they exit the stationary phase. Slight variations in chromatographic procedure exist; the precise technique depends on the properties of the mixture and the stationary-phase material.

Chromatography is a common technique in chemical analysis. In SFC, monitoring the time at which the various types of molecules exit the stationary phase allows scientists to gather useful information. In other types of chromatography, there are other possible indicators—each component might show up as a distinctly colored band in the stationary-phase material, for example.

Supercritical Fluid Extraction

In supercritical fluid extraction (SFE), supercritical fluids are used to remove particular compounds from mixtures. Supercritical fluid is poured through the mixture; as it passes through, it attracts and dissolves the target compound. Upon exit from the mixture, the supercritical fluid (carrying the target compound) is subjected to lowered pressure, which causes the target compound to fall out of solution. The compound is then collected, while the fluid is repressurized and sent through the mixture again. Various gases and liquids are heated and pressurized to become supercritical fluids for use in SFC and SFE, including carbon dioxide, ethylene, nitrous oxide, propane, water, methanol, and ammonia.

Historical Development

In 1822, French physicist Charles Caignard de la Tour noted the disappearance of the distinction between gas and liquid when he brought certain substances to a high temperature. His work represents the first documented observation of the critical point. English partners in chemistry, J. B. Hannay and J. Hogarth, took the second major step in the history of SFC and SFE when in 1879 they published their work on the solubility of supercritical ethanol (an alcohol). They reported that increasing pressure allowed more of a solid to be dissolved in ethanol, and that when the pressure was lowered, the dissolved material precipitated out of solution back into solid form. Their results were at odds with those predicted by theory, and they did not understand why the fluid behaved as it did. Little research on the topic took place during the next 20 years. In 1903, however, the technique of chromatography—using light petroleum as the mobile phase and chalk as the stationary phase—was developed by Russian botanist M. Tswett.

In the 1930s, various industrial researchers used fluids at pressures and temperatures in the vicinity of the critical point to separate the components of oil products, although with little understanding of the science of supercritical fluids. The major breakthrough in supercritical fluid technology occurred only in the 1960s, in the German laboratory of K. Zosel. Zosel investigated the solubility of numerous fluids at pressures and temperatures clearly above the critical point. His work was focused on separations and extractions of coal- and oil-based materials, but in 1970 he also reported the first decaffeination of coffee using supercritical carbon dioxide. Successes in SFC and SFE in the 1960s led to wide-scale research into supercritical fluid applications. In the 1980s, as the number of papers on supercritical fluids skyrocketed, techniques became sophisticated and supercritical theory well understood. In the 1990s, the push for environmentally friendly chemistry increased interest in supercritical fluids as solvents, because they are typically less toxic than traditional liquid solvents.

Uses, Effects, and Limitations

Supercritical fluid technology is the basis of numerous important separation and extraction processes. Supercritical fluids are used as solvents in the extraction of unwanted compounds, as in the removal of cholesterol from dairy products, of fat from meats, of caffeine from coffee and tea, and of toxic contaminants from waste materials. They are also used to extract valuable compounds, as in the removal and collection of perfumes, flavors, and pigments from plants, of vitamins and minerals from various products, and of oils from vegetables and fish. Various processes for the separation of fossil fuels into their components employ supercritical fluids.

Various environmental detoxification processes involving supercritical fluids are under investigation, as are the removal of nicotine from tobacco and the extraction of taxol (an anticancer drug) from the bark of the yew tree. Researchers are also interested in using supercritical fluids as the setting for certain chemical reactions, because some reactions (such as the hydrogenation of oils, the process by which hydrogen is added to oils, usually to improve their shelf life) occur more efficiently when in a supercritical environment. Scientists have begun to clean various types of equipment, including microchips and other electronic parts, with supercritical carbon dioxide. This application promises to curb the abundant use of toxic cleaning solvents such as chlorofluorocarbons (CFCs), which are harmful environmental contaminants. Carbon dioxide is nontoxic, readily available, nonflammable, and easy to bring to its supercritical state.

Supercritical fluids in general are more advantageous than traditional liquid and gas

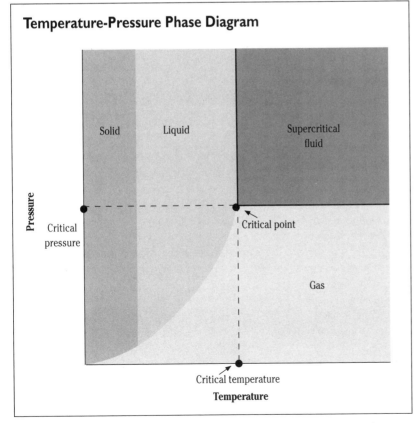

Temperature-Pressure Phase Diagram

Solid

Liquid

Supercritical fluid

Pressure

Critical pressure

Critical point

Gas

Critical temperature

Temperature

A substance becomes a supercritical fluid when its temperature and pressure both exceed the critical point.

solvents in a number of ways. Supercritical fluids yield faster extraction and exhibit higher selectivity for target extractives. The processes involved in SFE and SFC are relatively simple, reducing preparation and cleanup time. In SFE, the dissolved compound is more easily separated from the solvent, and the need for drying the compound after extraction is eliminated. (In most cases of traditional extraction, when the solvent is a liquid, the extracted compound must be dried.) Traces of toxic solvent rarely remain in the extract following SFE. The solvating power of supercritical fluids can be increased or decreased by mixing the supercritical fluid with other substances, called *modifiers,* giving the scientist close control over the separation or extraction.

There are a few factors that limit supercritical fluid technology. Compared to traditional extraction and separation procedures (in which liquid and gas solvents are used), supercritical fluid technologies involve more experimental factors that must be correctly implemented to produce the desired result. Pressure, temperature, choice of fluid, type and amount of modifier, and prevention of leaks and contamination are only some of the elements that must work together harmoniously to produce good results. Scientists are using experimental statistics to design computer programs to offset their difficulty in coordinating the elements of SFE and SFC.

Another problem that occurs in SFE is the extraction of unwanted compounds along with the desired compound. Supercritical fluids are such good solvents that they sometimes dissolve compounds that would not have been dissolved by a liquid solvent. Researchers are working on methods of cleaning the extract to wash away the unwanted material. The high cost of equipping a laboratory for SFE and SFC also limits opportunities to investigate the technology. Researchers expect that the cost will be lowered as less equipment is borrowed from other technologies and more equipment is built specifically for SFE and SFC.

Issues and Debate

The potential of supercritical fluids to reduce the use of toxic solvents has gained the technology many supporters. There is only one controversial issue related to supercritical fluids that has been publicly discussed: Some scientists have noted that SFE and SFC might use considerably more energy than their traditional counterparts. Bringing a substance to a high temperature and pressure and keeping it there for the duration of an experiment requires energy, which costs money and depletes resources. In the face of this criticism, advocates respond that carbon dioxide, the most commonly used supercritical fluid, has a low critical point relative to other substances. Carbon dioxide's critical temperature is $31°$ C and its critical pressure is 73 atmospheres (atm), 73 times the atmospheric pressure at Earth's surface. (By comparison, the critical point of ammonia is $132°C$ and 112 atm.) Critics contend that heating and pres-

surizing carbon dioxide still takes energy, and that any energy expenditure should be a consideration when choosing supercritical fluids as a method of chemical extraction or separation.

—*Tamara Schuyler*

RELATED TOPICS
Airplane Fuel Technology, Combinatorial Chemistry, Polymers, Superconductors

BIBLIOGRAPHY AND FURTHER RESEARCH

BOOKS
McHardy, John, and Samuel P. Sawan, eds. *Supercritical Fluid Cleaning: Fundamentals, Technology, and Applications.* Westwood, N.J.: Noyes Publications, 1998.
Taylor, Larry T. *Supercritical Fluid Extraction.* New York: Wiley, 1996.
Wenclawiak, B, ed. *Analysis with Supercritical Fluids: Extraction and Chromatography.* New York: Springer-Verlag, 1992.

PERIODICALS
King, Jerry, and Zhouyao Zhang. "SFE Starting to Make Headway in Food Analysis." *Research and Development,* October 1997, 46CC.
Lyons-Johnson, Dawn. "Enzyme Catalyst for Solventless Extractions." *Agricultural Research,* August 1997, 22.
Mermelstein, Neil H. "Supercriticial Fluid Extraction Gets Renewed Attention." *Food Technology,* January 1999, 78.
"Reducing Solvent in Analysis of Fat." *Resource,* November 1997, 5.
Wai, C. M. and Fred Hunt. "Chemical Reactions in Supercritical Carbon Dioxide." *Journal of Chemical Education,* December 1998, 1641.

INTERNET RESOURCES
Supercritical Fluids at Pacific Northwest National Laboratory http://www.pnl.gov/scrfluid/index.html
The Supercritical Times http://www.sc-times.com/

SURGICAL ROBOTICS

Surgical robotics refers to the use of human-controlled robotic equipment to perform surgery. A nascent but powerful technology, surgical robotics has been used in only a few operations but holds the potential to revolutionize surgical science. Robots offer the benefits of increased precision, greater dexterity, smaller incisions, and quicker recovery for the patient.

Although cost, insurance coverage, and government-approval issues will probably delay the widespread implementation of surgical robotics, many experts claim that robots will be central players in the operating rooms of the future.

Scientific and Technological Description

There are two main categories of robot. *Autonomous robots* are designed to operate automatically or with limited response to changing conditions, performing tasks without direct human intervention. An example is an industrial machine that assembles parts in a factory. Autonomous robots carry out some medical procedures, such as drilling holes in bones. But the revolution in surgical robotics concerns *telerobots,* robots that operate in a highly controlled manner almost exclusively in response to human directions.

A Telesurgical Workstation and Remote Surgery Site

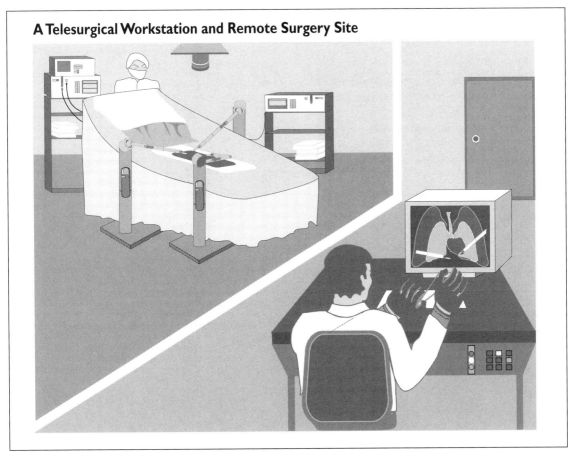

In telerobotic surgery, a surgeon may perform an operation from a remote location by inserting his hands into glove-like structures that are connected to a computer. The computer sends signals to a robot that works on the patient.

Nonmedical examples of telerobots include robots used to deactivate bombs and handle radioactive substances.

A few surgical telerobots have been built, and those are setting the standard for the field. Surgical robotics combines advanced computer technology and sophisticated telerobotics to yield powerful systems for the operating room. In the typical scenario for an operation using Robotics, the surgical telerobot has several jointed, flexible arms, at the ends of which are attached interchangeable Surgical Tools. The tools are similar in material and form to the tools that surgeons use in normal operations, although they may include tools of a smaller scale. The robot is positioned above the patient, suspended by strong, stable structures. Every moving part of the robot is wired to a high-powered computer system that is also connected to a console where the surgeon sits, either near the patient or in another room (see figure).

The surgeon is a skilled doctor trained in traditional surgical techniques. He or she watches the procedure on a three-dimensional video display; the camera can be moved (through directions given at the console) to the optimal viewing position as the surgery proceeds. Small cuts in the skin, or *ports,* are made at the appropriate locations in the

patient before surgery begins, and the robotic arms operate through these ports. The surgeon inserts her hands into glove-like structures, which are wired up to the computer system. Watching the three-dimensional display, she performs the procedure just as if she were operating directly on the patient. The system, which monitors the gloves more than 1000 times a second, senses each movement of the surgeon's fingers and hands and moves the robotic arms to mimic those movements exactly. The computer system gives *force feedback* to the surgeon: It detects the forces acting upon the robotic arms by surrounding tissues and causes the gloves to return those forces on the surgeon's hands. This assures that the surgeon uses the right amount of pressure to manipulate the tissues.

The video system magnifies the surgical site about ten-fold, changing a 2-millimeter artery into an image of a structure 2 centimeters in diameter. To accompany this magnification, the system has scaling capability: It translates the surgeon's hand movements into smaller-scale movements for the robotic arms. The system also filters out the natural tremors present in the hands of even the smoothest surgeons. The computer system is programmed so that if a system error occurs during the operation, the system alerts the surgeon, and in the absence of appropriate external action, it shuts itself down automatically.

Nanorobotics

NANOTECHNOLOGY takes its name from the nanometer, which is one-billionth of a meter. Nanodevices measure from a few to a few thousand nanometers in each of three dimensions. Engineers of medical robotics are investigating applications of nanotechnology in surgery, and a few prototypes of nanorobots have been built. One is an artificial red blood cell, nanorobot is made of 18 billion atoms, mostly carbon atoms arranged similarly to those in the lattice structure that makes up a diamond. The tiny robot can travel in a patient's bloodstream, carrying a cargo of oxygen, carbon dioxide, and potentially other compounds. It is capable of delivering hundreds of times as much oxygen to surrounding tissues as is a natural red blood cell, and engineers theorize that adding a large load of such nanorobots into a person's blood could enable the person to stay alive without breathing for several hours at a time. This might be useful for difficult surgical procedures that involve the respiratory system. Other types of nanorobots might travel through other bodily systems, delivering drugs or removing tiny tissue samples from hard-to-reach places.

Historical Development

The history of modern surgery dates back to the 16th century and the work of French medical scientist Ambroise Paré, who introduced humane improvements to the crude surgical techniques of the time. Major advances in surgery had to await the greater understanding of anatomy that arose during the 18th century, and finally in the mid-19th century, the introduction of anesthesia improved surgical conditions and made more procedures possible. Surgery took strides forward throughout the 20th century, in step with broader advances in medical knowledge and technology.

Robotics is a much younger technology. The first simple robots appeared in the 1950s, after the introduction of a telerobotic arm designed by Raymond Goertz for the Atomic Energy Commission for use in various industrial processes. The first industrial robot to serve on a production line, a model constructed by Unimate Robot Systems, was installed by General Motors. During the following two decades, robotics developed steadily, particularly for application in the automobile industry. Large robotic machines, capable of faster and more forceful manipulations than those humans can carry out, assembled car parts and spray-painted car bodies, lowering costs and churning out cars to meet increasing demand.

The joining of robotics and surgery began in the 1980s, as important developments in both fields led them toward each other. The first telerobots were built for handling hazardous material, such as toxic waste and radioactive substances, for deactivating bombs, and for creating special effects in the entertainment industry. In surgery, minimally invasive techniques were introduced. Minimally invasive surgery involves operating with long-handled tools through a few small cuts in a patient's skin, rather than the traditional surgical procedure, which involves opening a large surgical wound, pulling back the skin and outer tissues, and operating within a large open cavity. Minimally invasive surgery requires a viewing instrument that allows surgeons to see the tips of their tools inside the patient. The endoscope fills this role; it is a long, narrow, jointed rod with a tiny camera on the end. The camera end is inserted through a cut in the patient's skin, and the instrument is wired to a monitor, which displays the camera's internal view.

The benefits of minimally invasive surgery, including quicker recovery, less risk of infection, and better prognosis, were clear immediately, but its disadvantages stood out, too. The tools were difficult to manipulate, and the procedures were limited to simple operations, such as removal of the gallbladder. The medical community was anxious to extend the benefits of minimally invasive techniques to heart surgery and other common but complex operations. In the 1990s, several technology companies began to explore the possibility of building telerobots that could perform surgery within minimally invasive guidelines. The U.S. government also began to fund such research, as authorities saw the potential for surgeons to perform remote surgery on soldiers injured on the front lines of battle. In the mid-1990s, the first telerobotic surgery systems were built and tested, and the first telerobotic surgery was performed in Belgium in 1998. By 1999, several operations involving telerobots had taken place in France, Germany, Belgium, and California, at four of the leading sites of surgical robotics research.

Uses, Effects, and Limitations

As surgical robotics is a brand new technology, only a few applications have actually been realized. Operations performed by a telerobot have included heart valve repairs, heart bypass grafts (in which new sections of blood vessel are constructed to bypass blocked arteries), and removal of structures such as the gallbladder. Telerobots have also assisted in the positioning and placement of artificial hip joints. Robot engineers and physicians hope that someday surgical robotics will be used to introduce minimally invasive techniques to all types of surgery and to allow surgeries that are not currently possible. Plans are under way for telerobotic systems that will be capable of operating on delicate, sensitive structures such as the eyes, nose, throat, ears, spine, and brain.

The potential benefits of telerobotic surgery include all the benefits of minimally invasive methods. Undergoing fewer and smaller cuts than in traditional open-cavity surgery, patients under the robotic knife have a quicker recovery time and less risk of complications. Costs are also reduced, as fewer hospital personnel need to be present at a telerobotic procedure, and the patient's hospital stay is shortened. Surgeons controlling telerobotic surgery are far more comfortable than when standing hunched over a patient for hours during a traditional procedure. This increased comfort is likely to help surgeons perform to the best of their skill. Greater precision is possible in telerobotic surgery, due to magnified imaging, scaling of the surgeon's movements, and filtering of hand tremors.

There are two main obstacles in sight for the widescale implementation of surgical robotics. First, the cost of developing, building, and insuring the systems is considered by many to be exorbitant, and second, the process of obtaining federal approval for such a new and complex medical procedure could take more than two decades, according to some experts.

Issues and Debate

The potential benefits of surgical robotics are clearly understood by much of the medical community. But leaders of the medical community foresee problems in convincing the public that the procedure is safe. They imagine that many people will be scared of machines that move about without the directly visible control of a human being. They also think that insurance companies will be reluctant to cover surgical robotics, because of apprehension about a robot's potential to cause irreparable damage to a patient. Advocates of surgical robotics respond to this fear by assuring critics that the systems are actually safer than traditional surgical procedures, due to the microscaled movements of the robotic arms and the elimination of hand tremors. They argue that since the telerobot makes only the precise movements of its controlling surgeon and shuts down in the case of a system malfunction, there is little risk of problems other than human errors, which are equally possible in traditional surgery.

Advocates also point out that robotics may reduce surgical costs in the long run, by reducing the number of people who need to be present at a surgery, speeding up procedures, and limiting the financial drain of postsurgical hospital care. Critics respond that these cost-savers do not justify the great expense of making the technology widely available.

There is some debate about whether surgeons will unhappily view telerobots as a threat to their trade. Many surgeons take pride in their skills, and may be reluctant to give up their position at the patient's bedside during a lifesaving operation. Many observers believe, however, that although it may take many years, surgical robotics will arrive as a mainstream medical procedure in the future.

—*Tamara Schuyler*

RELATED TOPICS

Artificial Life, Artificial Tissue, Computer Modeling, Drug Delivery Systems, Nanotechnology, Organ Transplantation, Robotics, Surgical Tools, Telemedicine

BIBLIOGRAPHY AND FURTHER RESEARCH

BOOKS

Rutkow, Ira M. *Surgery: An Illustrated History*. St. Louis, Mo.: Mosby–Year Book, 1993.

Sayers, Craig. *Remote Control Robotics*. New York: Springer, 1998.

Timp, Gregory, ed. *Nanotechnology*. New York: AIP Press, 1999.

PERIODICALS

Paula, Greg. "Robotic Heart Surgery." *Mechanical Engineering,* June 1998, 8.

Skari, Tala. "The Cutting Edge: Heart Surgery Enters the Age of Robotics." *Life,* Fall 1998, 14.

"The Future of Medicine." *Futurist,* September–October 1997, 60.

Wu, Corinna. "Nanotech: Bigger Isn't Better." *Science News,* March 1, 1997, S14.

INTERNET RESOURCES

Medical Robotics at UC Berkeley
http://robotics.eecs.berkeley.edu/~mcenk/medical/#laparo

Nanotechnology
http://www.nano.xerox.com/nano/

Operation Robotics
http://www.macontelegraph.com/special/robot/robot.htm

The Heart of Microsurgery
http://www.memagazine.org/contents/current/features/microheart/microheart.html

SURGICAL TOOLS

Sophisticated medical devices such as LASERS, which are intense narrow beams of light, and gamma knives, which are streams of gamma radiation, have revolutionized surgery. Lasers are used to cut through tissue, remove tumors, and vaporize tissue in various eye and skin complications. Gamma knives are used to treat tumors and other abnormalities in the brain.

Some laser surgery carries risks, and gamma knives, although shown to be safe, are limited to treating small lesions. Both techniques offer surgery without blood and with quick recovery times. As laser and gamma knife surgery are new technologies, their efficacy is debated and their long-term effects are unknown.

Scientific and Technological Description

Laser Surgery

Laser is an acronym for "light amplification by the stimulated emission of radiation." Ordinary light contains radiation of many different wavelengths, but a laser consists of a narrow, intense beam of electromagnetic radiation of a single wavelength. The energy of a laser beam can be focused on an extremely narrow target, as small as one-thousandth of a millimeter (see *Lasers*). When a laser beam hits a patient's tissue, the water and pigments in the tissue absorb the laser's energy and convert it into heat. Consequently, the tissue's cells either vaporize (at high intensities), which is effective for cutting through tissue, or simply die (at lower intensities), which is effective for killing cancerous cells. A laser surgeon can control the intensity of the laser beam and choose lasers of various wavelengths to fit specific applications. Unlike simple surgical blades, laser beams actually cauterize (seal with heat) the blood vessels through which they cut, preventing blood loss at a surgical site.

In laser eye surgery, a laser beam is used to reshape the surface of the cornea, the transparent lens that focuses light toward the center of the eye. This procedure, technically known as *photorefractive keratectomy,* is the most common laser eye surgery performed. The procedure can correct both nearsightedness and astigmatism, an irregularity of the cornea resulting in blurred vision. In cosmetic surgeries, such as the removal of birthmarks, laser light is used to remove layers of skin until the blemish disappears. In hair removal, the laser damages the hair follicle, preventing the follicle from producing hair cells. In the removal of tattoos, the laser light simply eliminates the pigmentation in the tattoo dye.

Lasers are used for internal surgery as well. Long, thin instruments with lasers mounted on their ends can be inserted through the digestive tract or through skin incisions and into blood vessels. The laser is directed toward the target tissue and fired, while the surgeon views the event through a tiny camera also mounted on the end of an internally inserted device. This procedure is performed to remove tumors and other abnormal tissue in various parts of the body and to stimulate the generation of new blood vessels through the heart wall (for conditions such as hypercholestemia, in which a patient's heart is smothered by permanently clogged blood vessels).

Gamma Knife Surgery

Gamma radiation is high-energy electromagnetic radiation emitted during the spontaneous decay of the nuclei of some unstable elements, such as cobalt. In gamma knife surgery, a steel frame is screwed to a sedated patient's skull and fit into an immovable helmet (see figure). The patient's frame- and helmet-encased head is inserted into a machine that contains about 200 sources of radioactive cobalt-60. Once activated, the machine sends beams of gamma radiation into the patient's brain, all focused on a single location, typically a tumor or other abnormal growth. The gamma radiation, each beam of which would be independently harmless, collectively kills the target tissue. The gamma treatment itself is painless and lasts between 15 minutes and an hour. The tumor cells may begin to disappear immediately, or their complete removal may occur slowly over several years following gamma knife surgery.

Historical Development

The immediate precursor to the laser was the *maser* (microwave amplification by the stimulated emission of radiation), developed by U.S. physicist Charles Townes in 1954. Building upon ideas presented by Albert Einstein earlier in the century, Townes experimented with amplified light. He showed that microwave radiation could be amplified by causing the controlled emission of photons (particles of light) as molecules jump from high-energy states to lower-energy states. His maser was based on this type of amplification. Masers were used in atomic clocks and radio telescopes (see TELESCOPY), and Townes was awarded the 1964 Nobel Prize in Physics for this work.

In 1949, another U.S. physicist, Arthur Schawlow, had begun working with Townes. Schawlow and Townes experimented with applying maser-type amplification to other levels of the electromagnetic spectrum. Schawlow rigged a

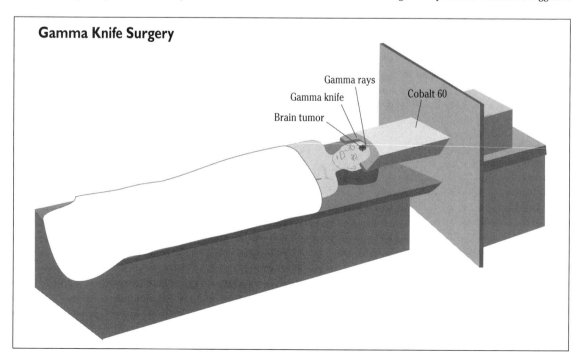

Gamma Knife Surgery

Gamma rays

Gamma knife

Cobalt 60

Brain tumor

long tunnel with mirrors at each end that helped concentrate and amplify the radiation, resulting in a beam of single-wavelength, high-intensity, narrow-focus light. In 1959 the two physicists filed an application for the theoretical description of this phenomenon, which they called an optical maser, and in 1960, Bell Telephone Laboratories (where Schawlow was employed) was awarded the patent. Schawlow earned a Nobel Prize in Physics in 1981 for his contributions to laser science.

The first working laser was constructed independently, also in 1960, by U.S. physicist Theodore Maiman. Lasers found numerous applications, the first of which was skin surgery; later examples include other surgical techniques, checkout scanners, compact disk (CD) technology (see CD-ROMS), and spectroscopy (the use of diffracted light to analyze the components of a chemical mixture or a distant galaxy). By 1999, several million laser surgeries had been performed.

New Zealand–British physicist Ernest Rutherford discovered gamma radiation in 1900, during the course of revolutionary experiments proving that the nuclei of some elements are not stable but decay over time, giving off radiation of three types: alpha, beta, and gamma. Investigations involving gamma radiation were a routine part of research in particle physics over the next half century, but it was not until the 1960s that gamma radiation entered surgery. Swedish scientists Lars Leskell and Borge Larsson recognized the potential of being able to reach structures inside the brain without cutting through the skull. In 1968, they developed a working model of the gamma knife. The technology took 20 years of improvement to be used routinely in medicine. Introduced in the United States in 1987, the gamma knife had served more than 50,000 patients worldwide by 1999.

Uses, Effects, and Limitations

Laser surgery is used for a myriad of complications, including correcting eyesight and blood vessel disorders, removing skin irregularities, performing cosmetic alterations, and disintegrating tumors and irregular tissue. In the case of sight problems such as nearsightedness and astigmatism, laser surgery is the least invasive surgical treatment currently available (although there are several others under investigation) and the only treatment for some disorders. Recovery from eye laser surgery is typically very rapid, and many patients return to normal activities the day following surgery. Laser eye surgery has not been approved for all eye problems, including farsightedness. Occasionally, the treatment solves one problem but makes a different one slightly worse. Also, children cannot undergo the procedure, because human eyes are not fully developed until adulthood and surgery might disrupt development.

The benefits of lasers as cutting tools and for internal surgery include the absence of blood loss and the reduction of scarring. The precise focus of a laser on target tissue causes less damage to healthy surrounding tissue than do conventional surgical techniques. Laser surgery circumvents the invasive methods of large-incision open-cavity surgery.

Cosmetic laser surgical applications include removal of wrinkles, acne scars, stretch marks, age spots, tattoos, varicose veins, spider veins, and hair. Such procedures have been touted as quick, effective, and less invasive than other face-lift techniques such as lipid removal and skin stretching. However, laser surgery carries risks. There is a greater risk of infection following laser surgery than with traditional face-lifts, because more raw skin is exposed on the face surface. These infections can cause prolonged discomfort and scarring. The healing process in some cosmetic laser surgery takes more than a week after at least two initial days of pain and swelling. Although this is comparable to recovery from treatments such as chemical peels, it is longer and more painful than recovery from a traditional face-lift (rhytidectomy), which involves little discomfort and typically looks normal within a week. Moreover, the long-term effects of cosmetic laser surgery are unknown.

Gamma knife surgery is used to treat brain tumors, abnormal and sometimes cancerous growths in the brain that are often fatal, and other irregular brain tissue. A disease of the brain called trigeminal neuralgia, which can cause severe facial pain and is sometimes untouchable by conventional surgery, can also be treated with gamma knives. Gamma knife surgery replaces one of the most invasive and complicated surgical procedures, brain surgery. Conventional brain surgery involves cutting open the skull, removing a piece of it, operating directly on brain tissue, and replacing the skull piece. The procedure lasts many hours, entails the risk of harming healthy brain tissue, and involves a long and uncomfortable recovery. In gamma knife surgery, the patient typically remains hospitalized for one night for observation and can return home the following day. Pain is minimized with a gamma knife because only a

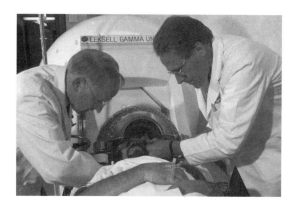

A medical team prepares a patient for gamma knife surgery in November 1991 by immobilizing his head in a metal frame. The procedure is noninvasive and allows gamma rays to be precisely aimed at a tumor (Corbis/Jim Sugar).

few incisions are made in the head to screw the steel frame to the skull, and the procedure is relatively quick.

Gamma knife surgery has lived up to its promise, yielding predominantly successful surgeries. In many cases, the problematic tissue stops growing and in others it actually disappears. However, the technology is limited to lesions smaller than 30 cubic centimeters. With larger tumors, there would be a danger of destroying normal brain cells because a larger dose of radiation would be required.

Issues and Debate

Controversy about the effectiveness and long-term safety of laser eye surgery has surrounded the technology since its approval by the U.S. Food and Drug Administration (FDA) in 1995. Eye doctors were split on whether to support the surgical technique, but those in favor collectively performed thousands of operations. Critics activated a public debate about laser eye surgery by publishing letters describing instances of serious complications and permanent eye damage. In 1997, the FDA, supported by the American Academy of Opthalmology, seized millions of dollars worth of laser equipment, citing unsafe procedures and unknown consequences. However, little documented evidence was uncovered to give credence to the harshest critics' concerns, and laser eye surgeons continued to treat patients.

Critics, including some opthalmologists, maintain that the safety of laser eye surgery is oversold. Although the statistics differ depending on the source (from only a few in several thousand to more than 10 percent), some patients do experience complications. A few lawsuits have been brought successfully against doctors for worsening patients' eyesight during laser surgery. Instances of overcorrection have occurred during surgery, as have instances of regression to nearsightedness following initial correction. Cases of decreased night vision following surgery have been documented, although this problem appears to have been largely solved. Advocates argue that most issues have been smoothed by improvements in the technology and the introduction of more sophisticated techniques.

So far no serious complications have been related to gamma knife surgery, although the long-term effects of the treatment have yet to be evaluated. Many medical facilities cannot afford the equipment to perform the technique, so it remains restricted to well-funded organizations.

—*Tamara Schuyler*

RELATED TOPICS
Lasers, Surgical Robotics

BIBLIOGRAPHY AND FURTHER RESEARCH

BOOKS
Alster, Tina S., and David B. Apfelberg, eds. *Cosmetic Laser Surgery.* New York: Wiley-Liss, 1996.
Caster, Andrew. *The Eye Laser Miracle.* New York: Random House, 1997.
Ganz, Jeremy C. *Gamma Knife Surgery.* New York: Springer, 1997.
Lunsford, L. D., D. Kondziolka, and J.C. Flickinger, eds. *Gamma Knife Brain Surgery.* New York: Karger, 1998.
Puliafito, Carmen A. *Laser Surgery and Medicine: Principles and Practice.* New York: Wiley-Liss, 1996.

PERIODICALS
Daugherty, Jane. "Lincoln Park Woman First in U.S. to Undergo New Laser Heart Surgery." *Detroit News,* September 7, 1997, B1.
Davis, Henry L. "Gamma Knife at Roswell Park Offers Bloodless Brain Surgery." *Buffalo News,* August 4, 1998, B10.
Mitchell, Peter. "Laser Surgery for Eye Defects—of Proven Use or Not?" *Lancet,* May 9, 1998, 1412.
Pinsky, Marilyn. "Treatment with Gamma Knife Is Delicate." *Syracuse Herald American,* October 25, 1998, AA3.
Westrup, Hugh. "Laser Eye Surgery Zaps Poor Vision." *Current Science,* November 14, 1997, 10.

INTERNET RESOURCES
Gamma Knife Overview
    http://www.gammaknife.uab.edu/overview/index.html
Gamma Knife Radiosurgery
    http://www.islandscene.com/health/1999/990407/head_zapper/index.html
Laser Eye Surgery
    http://www.prk.com/PRK_Glossary.html
OncoLink: Question: What Is a Gamma Knife?
    http://www.oncolink.com/specialty/med_phys/gamma.html
What Is Laser?
    http://www.asds-net.org/laser.html

TELEMEDICINE

Telemedicine refers to the use of telecommunications technologies to provide medical services. Resources such as e-mail, videoconferencing, and various digital transmission techniques offer communication solutions when physician and patient are in different locations.

Telemedicine services increased steadily during the 1990s and promise to expand medical care to remote areas and reduce medical costs in the long term. As a new technology, telemedicine is still plagued by legal questions related to insurance reimbursement, licensing, and liability, and ethical debates such as whether remote care constitutes a desirable form of medical attention.

Scientific and Technological Description

Telemedicine is a broad term, referring to various applications of telecommunications within the health care industry. A typical telemedicine scenario involves a telephone or video consultation between a physician at one location and a patient at another location (typically, there is a health care professional present with the patient as well). But telemedicine also includes events such as two physicians communicating over electronic mail about a patient, and theoretically, a surgeon performing an operation remotely through computerized robotic equipment (see SURGICAL ROBOTICS).

In telemedicine, information is relayed between patient and physician by way of electronic communication equipment such as telephones, fax machines, and computers. The information exchanged may consist of, for example, patients' questions, physicians' explanations, diagnosis and treatment advice, photographs of skin abnormalities, X-ray images, heartbeat audio, or images of brain scans. The information may travel over telephone cables, along cellular phone airwaves, or via satellite. Communication is enabled by voice and image transmission and digital video technology. Voice transmission occurs over the telephone and through videoconferencing, the latter of which involves the relay of complete audio and video coverage of a location. SCANNERS allow physicians to translate images, such as X-ray results, into digital format on a computer screen and send them to a distant physician for viewing. Other equipment that translates information into digital signals is emerging; for instance, digital stethoscopes can capture and transmit the sound of a heartbeat across telecommunication lines.

The Internet provides an international computer network through which written electronic communication and transmission of images and charts takes place. In telemedicine, communication is often encrypted (see CRYPTOGRAPHY) to protect messages from interception by a third party. Telemedicine communications may take place in *real time,* in which both patient and physician are present at communication terminals during the con-

sultation. Alternatively, communication occurs in a *store-and-forward* scenario: Information is sent from one party and is retrieved later by another. Telemedicine practitioners say that most telemedicine has been of the store-and-forward variety but that real-time consultations are expected to increase as communication technologies such as videoconferencing become more advanced and less expensive.

A few ambulances have been equipped with telemedicine systems. Where those systems are in place, emergency physicians at a hospital can view the scene inside an ambulance on its way to the hospital and give advice and verbal assistance to the ambulance staff. SURGICAL ROBOTICS, in which operations are performed by robotic equipment commanded by remote surgeons, is in the development stages. Many robotic surgeries have already taken place, but the surgeon in these cases has been in a nearby room. When the technology has proved itself sufficiently, remote surgeries may be called upon regularly.

Telecommunication technologies have begun to serve as effective medical training tools. Video and audio recordings of surgeries and other procedures are transmitted to medical classrooms so that students can learn from actual events without being present (where they would be a distraction to the physicians performing the procedure). In the

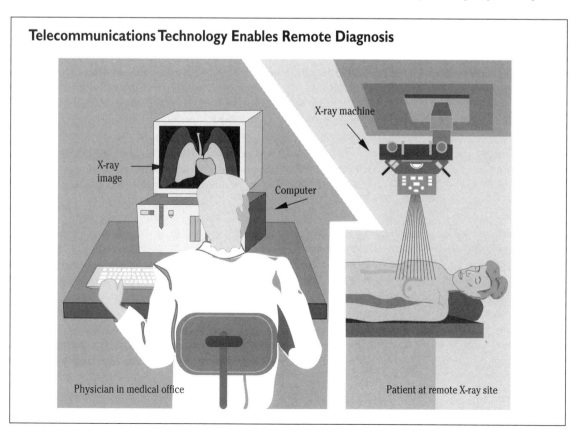

Telecommunications Technology Enables Remote Diagnosis

X-ray machine

X-ray image

Computer

Physician in medical office

Patient at remote X-ray site

In this use of telemedicine, the physician and patient are in different locations. The patient is having a chest X-ray, and the physician can view (usually at a later time) the X-ray image, which is sent electronically via computer.

future, it is hoped that VIRTUAL REALITY techniques will replace real cadavers in some medical training.

Historical Development

The emergence of telemedicine is rooted in the coupling of telecommunications technologies with modern medicine. Telemedicine began with the desire of physicians to increase communication capabilities within communities of health care providers. As technologies advanced, opportunities to provide remote diagnosis and treatment were seized. As the first mass communication tool, the telephone was also the first telemedicine instrument. In cities and suburbs, doctors' home visits were throughout the 1950s and 1960s slowly replaced by telephone calls in nonemergency situations. Medical hotlines, for answers to common questions related to poison control, substance abuse, cold and flu remedies, and infant care, emerged in the 1970s with the proliferation of telephone service. But modern telemedicine had to await the development of communication networks.

One of the seeds of telemedicine was a not-for-profit organization called Satel Life, started by physician Bernard Lown of Boston in 1989. In talking to doctors in nonindustrialized countries, Lown had learned that modern equipment and drugs were not their highest priority. Instead, they wanted more advanced means of communicating with each other and with the international medical community. Lown organized Satel Life, and in 1991 it funded the launch of several satellites intended to provide connectivity among physicians serving remote areas of nonindustrialized regions. The project was called HealthNet. The satellites were joined to ground stations and international telephone systems. Physicians began to exchange information through peer-reviewed online discussion groups and gained access to databases of newly emerging information about diseases, epidemics, and treatments.

The Internet became a public entity in the 1980s and the World Wide Web introduced vast numbers of people to computer network technology in the 1990s. As networks expanded and as more professional communities became connected, online medical information services arose. During the early 1990s, the first documented instances of remote medical consultations took place, and in 1993 there were an estimated 2000 such consultations. In 1995, the U.S. Justice Department initiated a pilot study to determine whether telemedicine would be effective in the prison system. The three-year experiment determined that prison telemedicine was indeed cost-effective, and it inspired further programs aimed at maximizing telemedicine's potential in jails.

In 1997, continuing the thread of Lown's HealthNet, the Internet was used as the basis for a project linking doctors and researchers studying malaria, a disease dangerously on the rise in Africa. Unrest and NATO intervention in Bosnia in the mid-1990s piqued the U.S. government's interest in funding research into telemedicine, particularly remote surgery,

for serving soldiers stationed overseas. In 1999, a survey taken by the Association of Telemedicine Service Providers and *Telemedicine Today* magazine found there to be at least 150 telemedicine programs operating in the United States. The survey's authors estimated that about 58,000 telemedicine consultations took place in 1998. Norway, a country where large numbers of people live in rural, hard-to-reach areas, also had a highly organized telemedicine system by the end of the 1990s.

Uses, Effects, and Limitations

The main use of telemedicine has been to serve people living in remote areas where there is a shortage of physicians. For nonemergency services, telemedicine has proven able to provide effective diagnosis and treatment and to reduce travel time and costs for both patients and doctors. Because it is heavily weighted toward clinical specialists (rather than general medicine) and the procurement of second opinions, telemedicine has entered the fields of mental health, cardiology, dermatology, and radiology (X-ray technologies).

In 1999, a telemedicine-based clinic was opened at a mall in suburban Philadelphia. The clinic operated on a walk-in basis, accepted no insurance claims, and charged $15 to $20 for a 5-minute videoconference consultation with a doctor working out of an office in a neighboring town. A registered nurse was present at the clinic with the patient and performed exams as needed. The clinic was intended to diagnose and offer prescriptions for simple ailments, such as viral and bacterial infections, colds, and rashes. The clinic's owners expected the business to serve as a template for similar clinics across the country, and their hope was that such telemedicine programs would decrease patients' and doctors' costs, serve more people in less time, and bypass the paperwork-laden system of insurance claims.

In general, the potential benefits of telemedicine are recognized as the expansion of health care access to more people and the sharing of health care knowledge more widely. The initial costs of purchasing telecommunications equipment and training medical professionals to use it are expected to give way to reduced costs due to less filling out of paperwork, image copying, transportation of paperwork, X-rays and other images, and mailing. In prisons, telemedicine promises to circumvent the security threats and expenses of transporting inmates out of jails to see medical specialists. Telemedicine has begun to allow patients with certain chronic ailments to stay at home rather than spend long periods of time at the hospital. Patients with high-risk conditions such as complicated pregnancies, acute diabetes, and risk of heart failure may be effectively monitored by doctors through digital equipment installed at the patients' homes. The comfort of home is likely to improve such patients' prospects in most cases.

Telemedicine carries risks as well as benefits. Liability is among the major concerns of telemedicine providers.

Technology failures may lead to faulty diagnoses or misunderstandings, and fear of malpractice suits prevents some doctors from choosing telemedicine. Also, legal issues related to telemedicine (such as licensing and insurance reimbursement) have not been fully clarified because telecommunications technologies themselves are still new to the law books and because remote consultations are even newer. Some physicians are reluctant to practice telemedicine until these issues are resolved.

Issues and Debate

Several legal questions pertaining to telemedicine are under debate. One question involves physician licensing. Some authorities view telemedicine as a virtual transportation of the patient to the location of the consulting physician; others think the situation is more appropriately seen as the virtual transportation of the doctor. Typically, a physician is licensed to practice medicine only in the state in which he or she works. In cases where telemedicine crosses state lines, the physician is unlikely to be licensed in the patient's state. By mid-1999, nearly 20 U.S. states had passed laws asserting that a physician must be licensed in the patient's state to provide remote care. Some physicians believe that this is unfair, as licensing takes time and money away from practicing medicine. Other questions of medical law also arise in cross-state telemedicine: Should the medical care follow the laws of the physician's state or those of the patient's state? In Norway and other European nations, medicine is organized under a single national socialist system, so issues of licensing and local laws are nonexistent.

Reimbursement by insurers is another problem in telemedicine. Private insurance companies do not always cover telemedicine consultations; others are simply undecided about whether to add telemedicine to their programs. The Balanced Budget Act of 1997, a U.S. law, put aside money for reimbursing telemedicine consultations in rural areas where there is a shortage of physicians. In 1999, Medicare—the U.S. government health care program for the elderly—began reimbursing for telemedicine in remote areas. But numerous states have laws mandating that store-and-forward telemedicine services, the most common telemedicine services, are not reimbursable.

Proponents of telemedicine acknowledge that there are issues that need attention before telemedicine can advance successfully. Laws must address not only licensing and reimbursement issues, but also the possibility of equipment failure, the allocation of public telecommunications lines to medical services, and the protection of patient privacy for electronically transmitted information. Telemedicine has the potential to alter the traditional relationship between patient and doctor, as it does not involve in-person contact between the two. Some critics see this lack of contact as a major disadvantage and posit that it may be a detriment to the health of the patient. Physical contact and the development of a relationship between physician and patient is believed by some people to play a large role in a person's recovery from illness or surgery. Advocates of telemedicine respond to this concern by noting that telemedicine is not intended to replace all contact between physician and patient, and that the cases in which personal contact is most important—for instance, life-threatening illnesses—are not candidates for the practice of telemedicine.

Some critics of telemedicine worry that it may turn out to increase the overall cost of health care. Many private health care providers are not able to afford the up-front costs of becoming equipped to practice telemedicine. State and federal governments may fill this gap, funding an infrastructure of national telemedicine in the interest of extending care to underserved rural areas. Such a system would mean that more services are available to more people, which might also mean that health care would become more expensive for everybody as a result of an increased number of insurance claims being filed. Advocates argue that telemedicine will eventually cut costs, however, by reducing paperwork and office visits.

—*Tamara Schuyler*

RELATED TOPICS
Cryptography, Internet and World Wide Web, Internet Protocol Telephony, Surgical Robotics

BIBLIOGRAPHY AND FURTHER RESEARCH

BOOKS
Berek, Britton, and Marilyn Canna. *Telemedicine on the Move: Health Care Heads Down the Information Superhighway*. Chicago: American Hospital Association, 1994.
Coiera, Enrico. *Guide to Medical Informatics, the Internet, and Telemedicine*. New York: Chapman & Hall, 1997.
Emery, Sherry. *Telemedicine in Hospitals: Issues in Implementation*. New York: Garland Publishing, 1998.
Reid, Jim. *A Telemedicine Primer: Understanding the Issues*. Billings, Mont.: Innovative Medical Communications, 1996.
Viegas, Steven F., and Kim Dunn, eds. *Telemedicine: Practicing in the Information Age*. Philadelphia: Lippincott-Raven, 1998.

PERIODICALS
Baldwin, Gary. "Attention Shoppers!" *American Medical News,* April 19, 1999, 21.
Fillion, Roger. "Sharing the Health: Technology Expands Long-Distance Care." *Denver Post,* April 5, 1999, C1.
Mitka, Mike. "Developing Countries Find Telemedicine Forges Links to More Care and Research." *Journal of the American Medical Association,* October 21, 1998, 1295.
Sandburg, Leslie A. "Telemedicine Continues to Wrestle Wicked Problems." *Health Management Technology,* February 1999, 133.
Strode, Steven W., Susan Gustke, and Ace Allen. "Technical and Clinical Progress in Telemedicine." *Journal of the American Medical Association,* March 24–March 31, 1999, 1066.

INTERNET RESOURCES
Department of Defense Telemedicine
  http://www.matmo.org/
Federal Telemedicine Gateway
  http://www.tmgateway.org/

Telehealth Magazine
    http://www.telemedmag.com/
Telemedicine Information Exchange
    http://tie.telemed.org/
Telemedicine Today Magazine
    http://www.telemedtoday.com/

TELEOPERATED OCEAN VEHICLES

The ocean depths are among the most dangerous environments on Earth but are also the largest unexplored and unexploited regions of the planet. People can venture to these depths only rarely, in advanced submersible vehicles.

But technology now also allows us to explore the oceans remotely, using robots called teleoperated ocean vehicles (TOVs). TOVs can not only reveal the natural wonders of the deep seas but can also carry out many practical tasks that would otherwise require skilled divers to risk their lives. They are already widely used in the salvage and oil exploration businesses, but the increased accessibility they allow also increases the harmful effects that humankind can have on Earth's oceans.

Scientific and Technological Description

The term *teleoperated vehicle* applies to a wide range of machines, often with completely different methods of operating. For instance, a large proportion of TOVs in operation around the world today are bottom-reliant or structurally reliant vehicles—they work by crawling along either the seafloor or undersea structures such as oil pipelines. Most vehicles are self-propelled, but a significant proportion are towed through the water (or along the seabed) behind a surface vessel. Most vehicles have a physical tether linking them back to the surface, but some are autonomous and capable of operating with a far greater degree of independence. However, the most familiar type of TOV is the free-swimming tethered vehicle. This is typically an instrument platform assembled around and within a roughly rectangular framework about 2 meters long, with its own propulsion system. The entire framework is linked to a support vessel by a cable that carries instructions and power down to it, and data and images back up to the surface.

Different types of TOV have different design requirements, depending on the conditions in which they operate. However, one of the most common demands is that the vehicle be heavy enough not to float on the surface while being light enough not to sink to the seabed. Ideally, once submerged, most TOVs should be capable of hovering at any depth without using power. This condition, known as *neutral buoyancy,* is achieved when the mass of the vehicle is equal to the mass of water it displaces. Normal submarines achieve neutral buoyancy easily because they are filled with air, but TOVs are designed to operate at depths of up to 6000 meters (20,000 feet), where any air-filled chambers would collapse under the enormous external pressures exerted by the sea. Therefore, to achieve neutral buoyancy, most TOVs carry several large tanks pumped full of foam. This foam is much lighter than water but is not easily compressible.

To move around underwater, most TOVs require a propulsion system. Usually, this is provided by traditional propellers, mounted behind the main vehicle framework. To move the vehicle up or down, additional propellers can be mounted around the vehicle, or alternatively, the main propellers may be allowed to tilt. Bottom-reliant vehicles often use wheels, caterpillar tracks (such as those seen on earthmovers), or an Archimedes screw (effectively, a screw-shaped propeller enclosed in a tube) to provide propulsion. Structurally reliant TOVs use wheels or hydraulic rams to push their way along a structure. The latter types of vehicle are often very different from free-swimming TOVs—bottom-reliant vehicles can resemble terrestrial digging machinery, while structurally reliant ones may take on a variety of shapes, such as rings, which fit around pipelines.

A TOV provides a platform to carry a wide range of devices into the sea. Probably the most widely used are cameras. High-resolution still cameras can provide valuable images of the seabed or marine life for scientific research, while television cameras allow a TOV's surroundings to be viewed in real time from the support vessel on the surface. For the cameras to function, of course, the TOV must be equipped with powerful lights. Television images from the TOV are relayed to the surface, where they appear on a screen on the operator's console. From here, a skilled operator uses a joystick and other controls to move the TOV. A variety of specialized tools can be attached to the vehicle for a wide range of uses (see *Uses, Effects, and Limitations*). These can also be controlled using joysticks or more sophisticated manipulators that allow the operator's movements to be reproduced exactly by the robot tools. Some TOVs function as part of hybrid systems, working in conjunction with a manned submersible. In these cases, the TOV is tethered to the submersible and controlled from there rather than from the surface.

Historical Development

The first TOVs were developed in the 1950s purely for scientific purposes. At the time, very little was known about the ocean depths, and the first deep-sea submersibles were only just being built. A remotely controlled instrument platform that could return pictures and even physical material from the deepest seafloors without endangering divers seemed an ideal alternative to building expensive pressurized vessels. At this time, electronic technology was in its infancy, so the earliest TOVs were drones, entirely dependent on control from the surface. This began to change in the 1960s, but the greatest spur to the development of more independent vehicles came in the 1970s, when world oil shortages suddenly made the exploitation of undersea oil reserves, such as those in Europe's North Sea, commercially feasible.

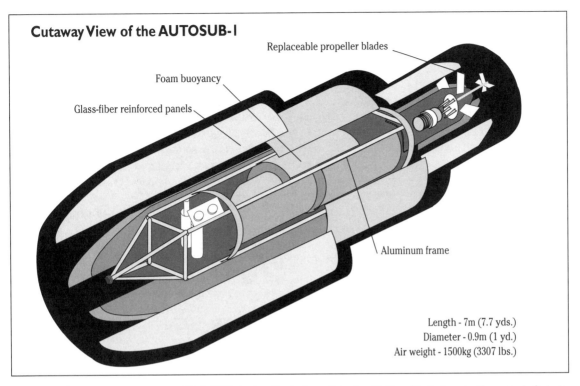

Cutaway View of the AUTOSUB-1

Replaceable propeller blades

Foam buoyancy

Glass-fiber reinforced panels

Aluminum frame

Length - 7m (7.7 yds.)
Diameter - 0.9m (1 yd.)
Air weight - 1500kg (3307 lbs.)

The Autosub-1 is an Autonomous Underwater Vehicle (AUV) developed by the Ocean Technology Division of the Southampton Oceanography Centre in the U.K. Designed to operate independently, the Autosub-1 is a robotic vehicle capable of carrying out a variety of scientific and exploratory tasks. The Autosub can gather physical, biological, chemical, and geophysical data. (Adapted with permission from Southampton Oceanography Centre, UK.)

Up until this time, most undersea industries had relied on specialist divers, who often spent long periods living in underwater saturation chambers (where they breathed an atmosphere similar to the mixture they required when diving). It was thought that TOVs would be too limited by the resolution of their television pictures and that their tools would not be versatile enough to meet a variety of undersea needs. One exception to this rule was in the field of undersea cable laying, where bottom-reliant TOVs were used to dig trenches, lay, and bury cable safely below the seabed.

In the late 1970s, as robotics became more sophisticated, engineers realized that TOVs could function effectively without the need for high-resolution imaging—after all, the vision of divers operating on the seafloor is also severely impaired. Instruments to provide TOVs with other "senses," however, had developed to a much higher level. For example, it was possible for instruments to "feel" for stress in pipelines, measure changing magnetic fields, and "smell" chemical compounds. These advances led to the rapid expansion of TOV use in the oil industry and in military applications. Whereas industrial TOVs tended to be designed for specific functions, the U.S. Navy funded a variety of multipurpose vehicles, many of which are used for civilian scientific as well as military purposes.

The rapid development of TOV technology from the early 1980s led directly to some of the most famous undersea vehicles, including the Woods Hole Oceanographic Institute's *Jason Jr.*, the TOV that swam through the wreck of the *Titanic* and returned the first pictures to a fascinated world in 1986. *Jason Jr.*, then a prototype robotic vehicle, was operated from the *Alvin*, a small manned submersible. Since the 1980s, computer programming and robotic technology have finally allowed the development of *autonomous underwater vehicles* (AUVs) with versatility approaching that of TOVs. AUVs also offer the advantage of being able to operate physically independent from their support vessel for days or even weeks.

Uses, Effects, and Limitations

TOVs originated as tools of scientific exploration, but they are now workhorses of many undersea industries. Some of their most important applications are listed below.

Search and retrieval. Free-floating TOVs and AUVs are frequently used in underwater exploration and salvage missions, ranging from the discovery of long-lost shipwrecks to the investigation of modern sea and air accidents. TOVs are able to remain underwater indefinitely, allowing salvage teams to conduct methodical searches of the seabed around a wreck or crash site over days or weeks. Often, an accurate map of a wreck's *debris field* is all the investigators need, but sometimes debris needs to be salvaged. TOVs can collect small items and return them to the surface, or pinpoint the positions of larger objects so they can be recovered later by divers or submersibles with greater lifting capabilities. TOVs have been used in this way on missions as diverse as the exploration of the *Titanic* and the investigation of the *Challenger* Space Shuttle explosion.

Science. The free-floating TOV was originally developed as a scientific tool, and it has revolutionized our understanding of the ocean floor. Instruments carried by TOVs fall into two categories: the geological and the biological. Studies of the seabed using magnetic instruments, for instance, have offered conclusive proof that the continental plates are slowly drifting apart, and rock samples returned to the surface for analysis have recently revealed the remains of an entire sunken continent in the Indian Ocean. TOVs have also enabled oceanographers to map the seafloor with new accuracy. Biological discoveries made by TOVs most famously include life found around deep undersea volcanoes. The strange and abundant organisms found in these hostile environments have led some biologists to speculate that the first life on Earth could have formed in these areas. As well as being used directly for scientific purposes, TOVs are also used for placing other scientific equipment underwater. The DUMAND telescope, for example, an instrument used to detect high-energy particles from outer space, was assembled on the Hawaii seabed using a TOV in conjunction with a manned submersible.

Industry. The advance of undersea industries has been the largest single influence on the development of TOVs.

The Jason Jr. *is a small submersible vehicle designed, built, and operated by the Woods Hole Oceanographic Institute that is remotely piloted from a small manned submersible, the* Alvin. Jason Jr. *and the* Alvin *were used to explore the wreck of the* Titanic *in 1986 (Corbis/Bettman).*

Today, they are used for a wide range of tasks, most of which are in the mundane area of day-to-day maintenance and safety checks on existing oil facilities. Structurally reliant TOVs have replaced divers in nearly all these operations, greatly reducing the risk to human lives and allowing skilled divers to be deployed only for the most difficult repair tasks. A TOV with a skilled operator on the surface can identify signs of structural weakness in, for example, oil pipelines. Often, sections of pipelines or other structures can also be removed and replaced remotely. Bottom-reliant TOVs are used in industry primarily for the laying and burial of cables and pipelines.

Military. Many of the tasks outlined above are also performed for military purposes, such as naval search and retrieval, surveying, and construction. Military funding has been another major encouragement to the building of TOVs and has led to the development of several new applications including surveillance and intelligence (e.g., placing instruments on the seabed to listen for submarine movements) and placement of mines.

Issues and Debate

Perhaps the most important issues attached to the rapid development of TOVs, and their use in the exploitation of the seas, are environmental ones. The oceans are a delicate and little-understood ecosystem, and the influence of undersea technology such as TOVs and submersibles thus far has been finely balanced between encouraging increased industrialization of the seabed and providing the scientific information necessary for oceanographers to monitor and perhaps regulate the changes that exploitation is causing (see DEEP-SEA VEHICLES).

Another interesting debate about TOVs, however, arises because they are probably the most successful example of a robot technology largely replacing human beings in the performance of a wide range of tasks (see ROBOTICS). In this specific case, TOVs offer the additional benefit that they do not put human divers at risk simply to perform routine procedures, but their overriding advantage to industry is undoubtedly their lower cost. There can be little doubt that if the economics were different, TOVs would not have supplanted divers, despite their safety benefits.

Of course, TOVs are not true robots—the diver below the surface is simply being replaced by an operator above it. Teleoperation is becoming increasingly widespread in a variety of other dangerous tasks, such as defusing bombs and disposing of nuclear waste. The advantage of teleoperation over true robotics is that it is much more versatile and less expensive—a human operator can respond to unexpected situations much more readily than a computer's complex programming (see SURGICAL ROBOTICS).

The new generation of AUVs are finally combining the versatility of TOVs with the independence of true robots. AUVs are able to operate on the seabed for long periods of time, carrying out complex preprogrammed instructions, and even adjust to changing situations within limits—a human operator is required only in emergencies. As this type of robotic technology becomes more commonplace in other fields of industry, it is bound to raise important questions about the future role of human beings in the workplace.

—*Giles Sparrow*

RELATED TOPICS
Deep-Sea Vehicles, Robotics, Space-Based Remote Sensing Tools, Surgical Robotics

BIBLIOGRAPHY AND FURTHER RESEARCH

BOOKS
Ballard, Robert D. *The Discovery of the Titanic.* Toronto: Madison Publishing, 1995.
Society for Underwater Exploration. *Advances in Underwater Inspection and Maintenance.* Boston: Graham & Trotman, 1990.
Society for Underwater Exploration. *Diverless and Deepwater Technology.* Boston: Graham & Trotman, 1989.

PERIODICALS
Ballard, Robert D. "High-Tech Search for Roman Shipwrecks." *National Geographic,* April 1998, 32–41.

INTERNET RESOURCES
Society for Underwater Exploration
    http://www.underwaterdiscovery.org/
Woods Hole Oceanographic Institute
    http://www.whoi.edu/

TELESCOPY

A telescope is a device that gathers radiation from distant objects and forms a magnified image. For centuries, telescopes were used only to study visible light, but in the past few decades, new types of telescopes have been developed that work with other types of radiation. Modern telescopes are not used just for creating images of distant objects, though—they can often split the radiation they collect into a spectrum, revealing the elements and molecules that make up distant planets, stars, and galaxies.

Throughout their history, telescopes have transformed our knowledge of the Universe and our place in it. They have not only shown us the unimaginable scale of the entire Universe, but also taught us lessons about our own environment through the study of nearby planets.

Scientific and Technological Description

A telescope performs two basic functions: It provides a larger light-collecting aperture than that of the human eye or a camera (thus providing brighter images), and it allows the images it forms to be magnified. The first telescopes used lenses to capture and bend (refract) light rays, and refracting telescopes are still popular as small instruments today. However, high-performance modern telescopes are always reflectors—they use a system of mirrors to capture light and bring it to a

focus. As well as producing higher-quality optical images than refracting telescopes, the principle of the reflecting telescope for visible light is easily transferred to telescopes for studying other types of radiation with different frequencies and wavelengths, such as infrared and ultraviolet.

The reflecting telescope is a simple instrument, because it is designed to focus light rays from distant objects only. These light rays will arrive at the telescope traveling in parallel and so can be focused by a single mirror ground into a precise concave shape called a *parabola.* Any light rays entering the telescope and striking this "primary" mirror will be reflected back into paths that converge at the telescope's focal point before spreading out again. As the light rays spread out beyond the focal point, they behave as though the object has been brought to the focal point of the telescope. A small lens, called the *eyepiece,* placed close to the focal point, can alter the paths of the diverging rays, allowing them to be viewed by an observer or an instrument. The angle at which the light rays diverge affects the apparent position of the object, and therefore its magnification.

This is the principle of the ideal telescope, but it neglects practical problems such as the position of the observer and the size of the telescope. In practice, at least one additional mirror is inserted into the optical system in front of the primary mirror. This "secondary" mirror intercepts the path of light from the primary, reflecting it and altering the position of the focal point. Primary mirrors can be ground accurately only with very shallow curves and long focal lengths, but the secondary mirror allows the length of the telescope to be reduced, and also moves the eyepiece position out of the path of incoming light rays.

There are several different ways to measure the effectiveness of a telescope. Magnification, although often quoted as a measure of strength on amateur instruments, is a fairly meaningless measurement. It is governed by the focal length of the eyepiece lens alone, and the ability to produce a highly magnified image is less significant than the telescope's quality, light grasp, and resolution. These last two properties are closely related—a telescope's light grasp is proportional to its aperture and is simply a measure of the amount of light it can collect compared to the human eye. Resolution is a measure of the telescope's ability to separate closely spaced objects or to distinguish fine details. The quest for greater light grasp and resolution has driven astronomers to develop telescopes that use huge mirrors and advanced technologies, but resolution, in particular, also has important effects when the reflecting telescope principle is applied to nonvisible radiations.

Historical Development

The principle of the telescope was supposedly discovered accidentally by the Dutch spectacle maker Hans Lippershey in about 1608. According to legend, children playing in

Lippershey's workshop one day lined up two lenses with the local church spire and were surprised to find that it appeared to come closer. Lippershey's telescope was the first *refractor.* Light rays from distant objects passed through the first lens, called the *objective,* and were bent inward, or refracted, to a focus. The second lens then acted as the telescope's eyepiece.

News of this discovery spread rapidly across Europe in the months that followed. Many astronomers began to make their own instruments, and this led to rapid growth in astronomical knowledge. Most famously, around 1610, Italian astronomer Galileo Galilei discovered that the Milky Way was made of separate stars, that the Sun had spots on it, and that Jupiter had four satellites orbiting it.

The telescope soon found uses outside astronomy, most notably in navigation at sea. One problem, however, was that the crossover of light rays in the tube resulted in an inverted image. Insertion of another lens between the focus and the eyepiece was found to correct the problem, but this additional lens dims the image, and so is used only on terrestrial telescopes today.

Although telescope objectives grew steadily throughout the 17th century, the larger, thicker lenses began to cause problems. Not only did thicker lenses absorb more light, they also produced *chromatic aberra-* *tion,* the formation of a series of colored fringes around the image (caused as a lens bends light of different colors and wavelengths by slightly different amounts).

The reflecting telescope, invented by Scottish mathematician James Gregory in 1663, and independently by Sir Isaac Newton in 1668, offered a way around these problems, and reflecting designs became the standard for large telescopes from the 18th century onward. Traditional telescopes grew steadily in size over the years, but in 1948 reached an upper practical limit with the huge 5-meter-diameter Hale Telescope on Palomar Mountain, California. Only recently have new designs allowed this size barrier to be broken.

The Earth's protective atmosphere prevented astronomers from investigating other wavelengths of radiation besides visible light until relatively recently. By shielding the Earth's surface from harmful high-energy radiations, the atmosphere is also blocking out valuable astronomical information. Only visible light and a small portion of the radio-wave spectrum are able to pass through the atmosphere and reach the surface.

The first radio waves from space were detected by U.S. astronomer Karl Jansky in 1932 using a long line of radio antennas. However, it was not until 1957 that Sir Bernard Lovell succeeded in applying the reflecting telescope principle to radio waves, collecting and focusing the waves from specific regions of the sky using a steerable dish at Jodrell Bank in England. Because radio wavelengths are so much longer than those of visible light, radio telescopes have to be much larger—huge, precisely engineered metal dishes that gather radio waves and reflect them to a detector horn at the focus point.

Other wavelengths of radiation from space have been discovered only since the beginning of the space age; early rocket flights carried radiation detectors that found infrared, ultraviolet waves, X-rays, and gamma rays. The reflecting telescope principle was again adapted for space-based telescopes such as the International Ultraviolet Explorer (1978), the Infrared Astronomical Satellite (1983), and EXOSAT

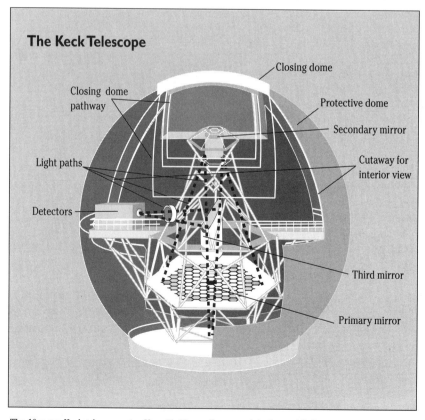

The Keck Telescope

Closing dome

Closing dome pathway

Protective dome

Secondary mirror

Light paths

Cutaway for interior view

Detectors

Third mirror

Primary mirror

The 10-meter Keck telescope atop Hawaii's Mauna Kea summit is the world's largest optical telescope. (Adapted from Durant et al., Encyclopedia of Science in Action, *Macmillan Publishers Ltd., 1995.)*

(1983), although these radiations frequently called for radical changes to the basic telescope design.

Uses, Effects, and Limitations

Detection and Analysis

The radiation collected by telescopes can be used in a variety of ways, usually by replacing an actual observer at the eyepiece with equipment that records and analyzes information from the telescope. The actual image produced by the telescope is often recorded using a camera. Until recently, this was done primarily with traditional photographic plates. Unlike the human eye, these can build up a brighter, more intense picture by long exposure to the image. The need for long exposures led to the development of powered telescope mountings that turn the telescope at the same rate as the Earth spins, so keeping objects fixed in the telescope's field of view.

In the past couple of decades, however, photographic plates have been replaced in most modern telescopes by electronic light detectors called charge-coupled devices (CCDs). A CCD is a silicon chip covered in millions of individual *pixels,* which collect electric charge each time they are struck by a photon of light. When a telescope image is projected onto them, the brightest portions of the image collect the greatest charge. Each individual pixel is then "read" by electrodes at the end of each row.

CCDs offer several advantages over traditional photography. Most important, they are far more sensitive, allowing astronomers to capture images of fainter objects. They can be manufactured to respond to radiation of different wavelengths, such as infrared and ultraviolet, and they also produce pictures in electronic format, suitable for processing and storage by computers. However, CCDs are always very small, so wide-angle images showing large regions of space are still often taken the traditional way. However, light from stars and planets carries far more information than simply their images—an instrument called a *spectrograph* can also reveal what they are made of. A spectrograph splits light collected by a telescope into a spectrum of different colors and wavelengths using a prism or a finely etched glass *diffraction grating,* which has the same effect.

When viewed in detail, this spectrum is usually covered by narrow dark and bright lines at specific wavelengths. Called *absorption* and *emission lines,* these form when atoms in, for example, a star's atmosphere or an interstellar gas cloud absorb or emit energy in the form of photons of radiation. The wavelengths of these photons are related directly to the internal structure of the atoms that emit them, so each element has its own unique series of lines. These lines serve as a "fingerprint" that reveals that the atom's light has passed through on its way to the telescope. Astronomers can use images from these telescopes to help them determine the composition of astronomical objects and their distance from Earth.

New Telescope Designs

Modern telescopes are used for studying many different types of radiation, from long-wavelength radio waves, through infrared and visible radiation, to ultraviolet, X-rays, and gamma rays, which have the shortest wavelengths and highest energies of all. Telescopes designed for these different radiations all use the same basic principles, but each has to overcome problems specific to the radiation it studies.

Even the largest radio telescopes, for example, are incapable of producing high-resolution images. However, images from several distant telescopes can be combined by computer, using a technique called *interferometry* to produce an image with the resolution (but not the light grasp) of a single telescope many kilometers across. Telescopes designed to study short-wavelength X-rays face a different set of problems. The high energies of those rays allow them to pass through many materials, including mirror glass. They can be focused only by ricocheting at a shallow angle off a carefully shaped metal surface, called a *grazing incidence reflector.* A series of these reflectors nested inside each other can capture and focus X-rays entering the telescope just like a traditional mirror.

Optical telescopes have faced their own problems, due primarily to the distortions caused by the mass of larger and larger mirrors now in use. Recently, however, an ingenious solution to this problem has been found. The 10-meter (m) Keck telescope on Mauna Kea, Hawaii, for example, has a mirror made up of 36 smaller hexagonal segments, held together on a framework so that they simulate a single huge mirror (see figure). Each mirror is supported by a number of small hydraulic actuators, which correct the mirror's orientation when changes in its relative position are detected by sensors around its edge.

The Hubble Space Telescope (HST) offers an alternative approach to the problems of larger and larger telescopes. Although its mirror is only of moderate size (2.4 m), this satellite telescope is uniquely well positioned outside Earth's atmosphere and beyond the distorting ripples caused by air circulation. The HST's mirror focuses light into several different instruments, including optical and infrared cameras, and spectrographs.

Other recent developments in optical telescopy apply techniques from other fields of astronomy. For instance, the Very Large Telescope at Cerro Paranal, Chile (the largest in the world), uses interferometry to combine the data from four 8.2-m telescopes into an image equivalent to that from a 16.4-m instrument.

Issues and Debate

The world's increasing population has put a strain on ideal telescope conditions. To detect faint traces of light from elsewhere in the universe, astronomers require dark, clear skies, but these are becoming increasingly difficult to obtain. Artificial light pollution has become a serious problem with the rapid increase in street lighting over the past few decades, and many telescopes built on what were once dark, remote sites now find cities and roads encroaching. Light pollution is the combined effect of polluting particles in the lower atmosphere and poorly designed, inefficient street lighting that often lets a significant proportion of the light disappear upward. Amateur and professional astronomers are campaigning vigorously for more consideration from development planners, and of course more efficient forms of light would also save energy.

A similar problem is faced by radio astronomers, who must contend with the prevalence of mobile phones and other radio and electrical appliances that may interfere with their work. Famously, one mystery signal detected regularly at a large radio telescope was eventually traced to a nearby microwave oven. Although international legislation has been passed to prevent deliberate broadcasting in the radio astronomy "window" of wavelengths, many astronomers maintain that their science is still under siege. Most major optical telescopes have already migrated to the mountaintops, above much of the Earth's atmosphere and light pollution, but radio astronomers have plans to go even further in their quest for clear and unpolluted skies, with observatories in space or even on the far side of the Moon.

Another important issue in this increasingly profit-driven global economy is why governments should devote funding to astronomy in general, and to building large telescopes in particular. There are two ways of answering this question. The first is to point out that astronomy does in fact have important practical applications; studying what's "out there" can provide new information on what's going on "down here." For instance, Mars and Venus offer important clues to the mechanisms behind global climate change. The greenhouse effect—the way the Earth traps certain gases within its atmosphere—was first recognized in the atmosphere of Venus, while Mars offers a model for a planet stripped of its protective ozone layer. Farther out in the Universe lie clues to the nature of matter and other important questions of physics. Because astronomy often provides stunning photographs and astounding facts, it is also one of the most popular sciences with the general public, which may help to secure funding for more obscure but useful projects.

Finally, some observers comment that astronomy should not have to be "useful." They note that for centuries, theologians, philosophers, and scientists have struggled to understand humanity's place in the Universe, and maintain that this is reason enough to continue support for astronomy.

—*Giles Sparrow*

RELATED TOPICS
Microwave Communication, Remote Sensing, Satellite Technology, Space-Based Remote Sensing Tools

BIBLIOGRAPHY AND FURTHER RESEARCH
BOOKS
Fischer, Daniel, and Hilmar Duerbeck. *Hubble: A New Window on the Universe.* New York: Copernicus, 1996.
Henbest, Nigel, and Michael Marten. *The New Astronomy,* 2nd ed. New York: Cambridge University Press, 1996.
Manly, Peter L. *Unusual Telescopes.* New York: Cambridge University Press, 1991.
Wall, J. V., and A. Boksenberg, eds. *Modern Technology and Its Influence on Astronomy.* New York: Cambridge University Press, 1990.
PERIODICALS
Sky and Telescope (monthly)
Astronomy (monthly)
INTERNET RESOURCES
ESO Very Large Telescope Homepage
  http://www.eso.org/projects/vlt/
Space Telescope Science Institute Homepage
  http://www.stsci.edu/top.html

TISSUE TRANSPLANTATION

The successful transfer of living tissue from one person to another is a 20th-century phenomenon. Today, several tissues can be transplanted from living donor to recipient to make up for loss or damage of tissue.

Tissue rejection, and the need for tissue matching and immunosuppressant drugs, is a major limiting factor in transplantation. Many novel approaches, including tissue engineering and cell encapsulation, are being explored to minimize or circumvent tissue rejection.

Scientific and Technological Description

A *transplant* is the transfer of living material from one part of the body to another or from one individual to another. When the donor and recipient of the material are the same, the transplant is an *autograft;* when they are different individuals of the same species, the transfer is an *allograft;* when they are of different species, the transplant is a *xenograft.* In humans, tissues are transplanted to replace or augment tissue that is dead, diseased, or otherwise absent or malfunctioning. Skin, heart valves, muscular tissue, bone fragments, blood-forming tissue such as bone marrow, and the cornea (the transparent covering at the very front of the eye) are the main tissues that have been transplanted successfully (see figure). However, about 20 additional tissues are being investigated as potential allograftable material.

A key problem in tissue allografting is the need to match tissues between donor and recipient. Tissue implanted in the body from an outside biological source normally has cell surface substances that are detected as "foreign" by the body's immune system (the system that fights off infection). The immune system protects the body against pathogenic (disease-causing) bacteria, fungi, and viruses and

Commonly Transplanted Tissues

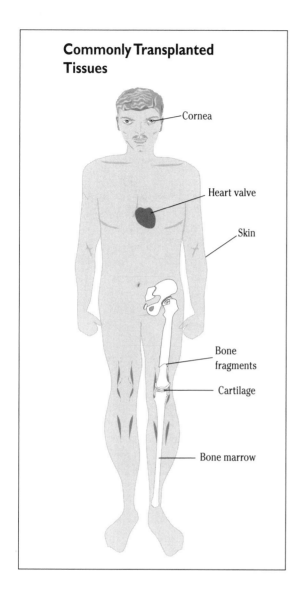

- Cornea
- Heart valve
- Skin
- Bone fragments
- Cartilage
- Bone marrow

treats a harmful invader and a potentially lifesaving tissue graft in a similar manner. The system's army of white blood cells, particularly lymphocytes, attacks the foreign cells directly or secretes chemicals called *antibodies* that do so.

If tissues are transplanted between people without any attempt to match them immunologically, tissue rejection is likely. To prevent or minimize rejection, a system of *tissue typing* is used to match donor and recipient. The most significant feature that distinguishes a recipient's tissues from those of a donor is a system of *histocompatibility antigens*. These are chemicals, usually protein in nature, found on the surface of cells. In humans, the most important of these are the *human leukocyte antigens* (HLAs). A person's complement of HLAs is inherited. As a rule, only genetically identical twins have exactly the same complement of HLAs, but closely related persons usually have many HLAs in common. Outside the close family, tissue matching is still possible. Subject to availability, surgeons seek the best match between donor and recipient. The better the match, the lower the strength of the rejection response after transplantation and the easier the control of rejection using medication.

To stave off rejection of a transplanted organ or tissue, the patient is given immunosupressant drugs that selectively limit the activity of the immune system. This means, however, that the recipient may have to take a cocktail of immunosuppressant drugs for the rest of his or her life. It also means that the recipient's immune system is potentially weakened to fend off infection by disease-causing organisms, especially parasitic fungi and protozoa. Medications are carefully administered to achieve a fine balance between the benefits of preventing rejection and unwanted side effects such as lowered immunity to infection.

Historical Development

The first well-documented procedure for successful tissue allografting was carried out in 1906. Eduard Zirn of Moravia in central Europe transplanted eye tissue from a recently deceased person to a living one in a procedure called a corneal graft. The cornea can become scarred or cloudy as a result of several quite different medical conditions, and its surgical replacement by an undamaged cornea can restore vision. The procedure is now commonplace, although its use did not become widespread until the 1950s. Corneal grafting largely circumvents one of the major problems of transplants: tissue rejection. The cornea is not in direct contact with the blood system, so the likelihood of attack by lymphocytes and their products is greatly lessened.

Tissue rejection was an overriding constraint on allografting until the 1960s. Knowledge about the nature of tissue rejection, and therefore the development of strategies to suppress or circumvent it, began to emerge as World War II was coming to an end. A British zoologist, Peter Medawar, noted that skin grafts given to burn victims failed when donor and recipient were unrelated. In experiments with rabbits, he showed that skin rejection processes were immune reactions. Macfarlane Burnet of Melbourne, Australia, drawing upon his extensive investigations into human immune response, formulated a *selective theory of immunology*. In essence, Burnet proposed that the body's immune system develops in the womb when, as a fetus, it develops the capacity to produce a wide range of specific antibodies. Later, such antibodies can be manufactured in response to the presence of specific foreign antigens.

In experiments with mice, Medawar was able to demonstrate that *activation* of the immune system—the development of the immune system's ability to distinguish between self and nonself—occurred during fetal development. "Foreign" cells given to fetal mice could result in the mice recognizing these cells as self rather than nonself in later life. Based on such findings, Medawar proposed his theory of *acquired immunological tolerance*. In applying the theory to transplants, Medawar suggested that a fetus exposed in the womb to cells of a prospective donor could acquire an immunological tolerance to the tissues of that donor. In other words, the person would be able to receive grafts from that donor without rejecting them. In 1960, Burnet and Medawar were awarded the Nobel Prize for Physiology or Medicine for their contributions to immunology. Burnet and Medawar's theories suggested ways that rejection could be countered. Today, for example, the question of how to induce immunological tolerance to transplanted tissues or organs is an active area of research.

By the 1960s, greater success with both tissue and ORGAN TRANSPLANTATION followed with the development of immunosupressant drugs. By the late 1940s, pharmacologists discovered that cortisone, a steroid hormone produced naturally by the body, depressed immune response. In the early 1950s, Medawar and others used cortisone successfully to prolong the life of allografted skin. In the 1960s, other corticosteroid drugs used to treat certain cancers were also found to suppress rejection responses. In 1962, U.S. pharmacologist George Hitchings and others developed the first specific antirejection drug, azathioprine, following screening of hundreds of related anticancer drugs. During the 1960s and 1970s it was used, in conjunction with corticosteroids, as the medication of choice for combating tissue rejection. In 1974, the immunosuppressant properties of the antibiotic cyclosporine were discovered, and by the time it was introduced in 1983, it was found to be a powerful immunosuppressant.

Current research on tissue transplantation is proceeding on many fronts. More effective immunosuppressants are being tested clinically, hopefully with fewer toxic side effects. Tissue engineers (see ARTIFICIAL TISSUE) are seeking to create donor tissues that are immunologically acceptable to the recipient. Such tissues could be grown from the recipient's own cells or derived from donor cells in which foreign antigens have been removed or masked in some way.

Another avenue of research is the transplantation of tissues from other animals. By 1999, over 10 different types of animal tissue were being tested as potential implants to treat human medical conditions ranging from Alzheimer's disease to diabetes. As early as 1994, transgenic pigs (animals containing human genes) were being reared by two biomedical companies in the United States and one in the United Kingdom with the long-term intention of offering human-compatible tissues and organs. There are many hurdles to be overcome in this area, not least because such tissue is biochemically distinct from human tissue, and even if problems of tissue rejection could be overcome, the tissue may function differently than human tissue.

A promising line of inquiry is the use of plastic membrane to encapsulate "foreign" cells. The membrane is porous and allows small molecules to pass in and out of an implant but excludes the lymphocytes that would launch an immune attack on foreign tissue. Such implants need to be very small (typically, no larger than a hair's width) if they are to be kept alive by the recipient's circulatory system with nutrients and oxygen. Many tens of thousands of microscopic tissue implants are needed to replace the biological function of damaged tissue. Currently, researchers are investigating the encapsulation of pancreas islet cells, cells that secrete insulin and are deficient in some forms of diabetes.

Uses, Effects, and Limitations

Tissue transplants are usually life-enhancing treatments, but some—notably, heart valve replacements, bone marrow transplants, and certain skin transplant procedures—are truly lifesaving. Of the tissues currently being allografted, several do not depend on functioning cells but play a structural rather than a metabolic role. The transplanted tissues provide mechanical support and so do not necessarily require their constituent cells to be alive. Transplants for which this applies include artery, bone, cartilage, tendon, and heart valve grafts.

Cornea transplants are among the most widely seen tissue transplant procedures. Corneas donated by those who have died can be cold stored for several days before use. The surgical procedure is straightforward, and because rejection is rarely a problem, cornea transplants are the most successful of all allograft operations. Worldwide, more than 50,000 corneal grafts are performed each year.

There are four heart valves—one at the exit of each heart chamber—and occasionally, one or more may be malformed from birth or may later acquire damage. Such impairment may be life threatening, because circulation is compromised and pressure from the leaking valve places an unwanted burden on muscle tissue that can trigger a heart attack. Corrective surgery may entail heart valve replacement. Of the three major types of valve replacement, one is purely synthetic, but the other two are formed from once-living tissue. Biological replacement valves contain tissue originating from the heart valves of cows or pigs, whereas homograft valves are donated by people who have died of some condition that does not damage the heart. Biological valves last seven years and more, and postoperative medication includes the use of anticoagulants (medications that prevent blood from clotting) in the short term and immunosuppressants in the longer term. Homograft valves last 10 years and more, and their use may require little or no special postoperative medication. The shortage of heart valve donors necessitates the use of artificial or biological replacements.

Until recently, permanent skin replacement meant using skin that was transferred from some other part of the patient's body. Using skin from a donor was invariably a short-lived expedient because tissue rejection rapidly ensued. However, the science of tissue engineering (see ARTIFICIAL TISSUE) has now generated skin of nonself origin that is proving successful for transplant use.

In some cases, bone marrow may be implanted into a patient to make good any loss or damage. Bone marrow—a semiliquid found at the center of some bones—is the source of many of blood's components. Bone marrow contains hematopoietic stem cells (cells that differentiate to form the different cellular components of blood: namely, white cells, red cells, and platelets). A variety of medical conditions, ranging from leukemia to non-Hodgkin's and Burkitt's lymphomas, compromise the health of bone marrow. In addition, treatments used to attack cancer, such as chemotherapy and ionizing radiation, can destroy bone marrow.

Bone marrow autografts are used to assist patients recovering from marrow-damaging cancer treatment. Prior to cancer treatment, some of the patient's marrow is withdrawn from bone sites by syringe and then stored at -190°C in preservative. It is thawed and injected into the patient after he or she has received radiation treatment or chemotherapy. The injected stem cells migrate to bone marrow sites, where they multiply and differentiate, thus compensating for any bone marrow loss. Increasingly, bone marrow allografts are considered for patients who have certain forms of leukemia or for those who have bone marrow disease or damage for reasons other than cancer treatment. A close tissue match between donor and recipient is sought. Normally, such a match is found only with a close relative: a living parent, brother, or sister. If a suitable close-match donor cannot be found, bone marrow from a more distant source is obtained.

Bone marrow allografts pose particular problems because not only recipient tissue may reject donor cells, but also the donor tissue itself contains antibody-manufacturing lymphocytes, which themselves may attack the recipient's tissues, called *graft-versus-host disease*. Various strategies are employed to combat rejection and graft-versus-host disease. Monoclonal antibodies (specific antibodies manufactured in the laboratory by fusing lymphocytes with cancerous cells) can be used to remove harmful lymphocytes selectively from the donor marrow. To combat rejection, the recipient's immune system, particularly lymphoid tissue, may be subjected to a large dose of ionizing radiation to combat the rejection response.

Bone marrow allografts may offer a promising solution to the problem of transplant organ rejection (see ORGAN TRANSPLANTATION). In certain cases, when a bone marrow allograft is administered, the combination of the recipient's bone marrow and the newly injected foreign cells seems to induce some tolerance to newly grafted organs. This is an active area of research. Bone marrow allografts carry an appreciable risk for the recipient. In the mid-1990s, the mortality rate for the procedure was between 1 and 10 percent. Procedures that use stem cells from the placenta or umbilical cord of newborn infants offer great hope as a replacement for conventional bone marrow autografts or allografts. Fetal blood recovered from these structures contains hematopoietic stem cells. These can be recovered and stored for months or years, available for use by the person later in life or available for a relative or an unrelated but tissue-matched recipient.

Issues and Debate

One of the key issues at the center of transplantation is the shortfall in supply of donors. In most cases, this problem applies less to tissue transplantation than it does to organ transplantation. Living donors can offer a small amount of tissue to a relative, whereas donating an entire organ (in the case of certain paired organs such as kidneys) is a much greater sacrifice. In the case of cornea transplants, supply is totally dependent on donations from recently deceased bodies and, as with most organ transplants, there is a serious shortfall in supply. Even when a person has given permission for his or her organs or tissues to be used for transplant purposes after death, family members traumatized by the sudden death of a close relative may override the wishes of the deceased. With more than 20 different types of transplantable tissue potentially available from a cadaver, a system for tissue recovery, backed up by a legal framework where individuals and their families must opt out of tissue donation (specifically state that they do not wish tissues to be donated) rather than opt in (state that they wish to donate), could increase donor supply substantially. As it is, many novel means of tissue or organ replacement, ranging from tissue engineering to xenotransplantation, are being pursued to make up for the shortfall of donors.

—Trevor Day

RELATED TOPICS
Artificial Organs, Artificial Tissue, Fetal Tissue Transplantation, Organ Transplantation

BIBLIOGRAPHY AND FURTHER RESEARCH
BOOKS
Caplan, Arthur L. *Am I My Brother's Keeper?* Bloomington, Ind.: Indiana University Press, 1997.
Crigger, Bette-Jane, ed. *Cases in Bioethics,* 3rd ed. New York: St. Martin's Press, 1998.
Institute of Medicine. *Xenotransplantation: Science, Ethics and Public Policy.* Washington, D.C.: National Academy Press, 1996.
Kimball, Andrew. *The Human Body Shop,* 2nd ed. Washington, D.C.: Regnery Publishing, 1997.
PERIODICALS
Antman, Karen. "When Are Bone Marrow Transplants Considered?" *Scientific American,* September 1996, 90.
Beardlsey, Tim. "Culturing Human Life." *Scientific American,* June 1998, 9.
Concar, David. "The Organ Factory of the Future?" *New Scientist,* June 18, 1994, 24.
Langer, Robert S., and Joseph P. Vacanti. "Tissue Engineering: The Challenges Ahead." *Scientific American,* April 1999, 62.
Lanza, Robert P., and Willem M. Kühtreiber. "Xenotransplantation and Cell Therapy: Progress and Controversy." *Molecular Medicine Today,* March 3, 1999, 105.
Lysaght, Michael J., and Patrick Aebischer. "Encapsulated Cells as Therapy." *Scientific American,* April 1999, 52.

Morgan, Jeffrey R., and Martin L. Yarmush. "Tissue Engineering." *Science & Medicine.* November–December 1998, 6.

Stephenson, Joan. "Terms of Engraftment: Umbilical Cord Blood Transplants Arouse Enthusiasm." *Journal of the American Medical Association,* June 21, 1995, 1813.

INTERNET RESOURCES

Bone Marrow Transplants
    http://www.aosoft.com/~cancer/ld.Bone.Marrow.html

Future Directions in Cell Transplantation
    http://www.niaid.nih.gov/Publications/transplant/future.htm

Links to Bone Marrow Transplant Sites
    http://www.sustance.com/bonemarrow/links.html

National Foundation for Transplants
    http://www.otf.org/

National Institute of Transplantation
    http://www.transplantation.com/

Pancreatic Tissue Transplants
    http://www.diabetes.com

Prosthetic Heart Valve Information
    http://www.csmc.edu/cvs/md/valve/default.htm

Tissue Banks International
    http://www.tbionline.org/newsfeatures.htm

ULTRASOUND IMAGING

Ultrasound imaging is a body scanning technique that utilizes sound waves at frequencies well above the range of human hearing. The sound waves are directed into the body and their returning echoes are detected and used to create an image of internal structures. In many developed countries, ultrasound is used routinely to monitor the fetus during pregnancy. It is also used to visualize internal organs, particularly soft tissues that do not show up well on X-ray images, and to determine the speed and direction of blood flow in the heart and blood vessels.

The simpler ultrasound techniques are low-cost methods, comparable in price to that of more traditional X-RAY IMAGING. Ultrasound images generally do not have high resolution (they do not show great detail), but most physicians consider ultrasonography to be a very safe procedure. Criticism of ultrasound as a scanning methodology tends to be centered on the possibility that it may lead to misdiagnosis. Improvements in ultrasound technology, particularly innovations in three-dimensional ultrasonography and the use of contrast agents, are likely to increase its utility and enable it in part to replace imaging techniques that use potentially harmful ionizing radiation (X-rays and gamma rays).

Scientific and Technological Description

Ultrasound imaging or *ultrasonography* uses ultrasound (high-frequency sound waves) to show structures within the body. Unlike other medical imaging techniques, ultrasound uses mechanical vibrations rather than electromagnetic energy to visualize internal structures and processes. In ultrasound imaging, high-frequency sounds are generated by tiny crystals that act as *transducers* (devices that convert energy from one form to another). When the crystals are subjected to an alternating electric voltage, they expand and contract along one axis, in phase with the alternating current. This phenomenon, the *piezoelectric effect,* is harnessed to make the crystal behave like a miniature loudspeaker, generating sound waves as it expands and contracts. As the crystal vibrates at predetermined frequencies, it emits a useful beam of ultrasound.

In most ultrasound devices, the transducer is both a sound emitter and a sound detector. Sound echoes returning from inside the body cause the crystals to vibrate and generate a tiny electric current. The time taken for return of the echo, and the echo's intensity, frequency, and other properties, provide data about the depth and nature of sound-reflective surfaces inside the body. The electrical signals from the transducers are conveyed to a computer and analyzed, and are displayed as an image on a monitor.

Sound waves travel much more slowly than do X-rays or radio waves, and they require a moderately dense medium—a liquid or solid—to travel well. Ultrasound waves are reflected off boundaries between media of markedly different density. Inside the body they are reflected from solid–air or solid–liquid boundaries and at borders between different types of tissue. The sound waves are distorted by very compact tissue such as bone. The form of the echo is a function of the relative reflectivity of the boundary layer, its depth within the body, and the nature of intervening tissues. The echo from a structure that is farther away from an ultrasound scanner will take longer to return than will the echo from a feature closer to the transducer.

In principle, the resolution of an ultrasound image (the amount of detail that can be seen) improves with increasing sound-wave frequency. However, the heating effect of ultrasound also increases at higher wave frequencies. In practice, the ultrasound frequencies used in imaging are those that are best for visualization at a given depth within the body without producing a noticeable heating effect on tissues. Other things being equal, the deeper the penetration within the body, the lower the frequency used and the lower the resolution of the resultant image. During an ultrasonic examination, the patient typically lies prone while a small hand-held probe is passed over the area of the body to be examined. A gel or lubricant is usually applied to the skin to ensure effective sound transmission between probe and body.

Dozens of different types of ultrasound scans are in use, but they generally fall within one of three categories: A-scan, B-scan or Doppler. *A-scan* equipment directs an ultrasound beam along one axis only. The technique is used to measure the depth and bulk of structures within the body, such as the thickness of the lens of the eye.

B-scan equipment effectively takes a series of A-scan images, utilizing either an array of transducers side by

Fetal Ultrasonography

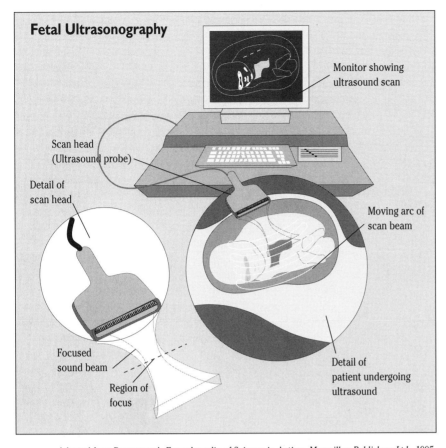

Monitor showing ultrasound scan

Scan head (Ultrasound probe)

Detail of scan head

Moving arc of scan beam

Focused sound beam

Region of focus

Detail of patient undergoing ultrasound

Adapted from Durant et al., Encyclopedia of Science in Action, *Macmillan Publishers Ltd., 1995.*

side, or a single transducer that sweeps back and forth in an arc (see figure). This yields a cross-sectional image of a specific part of the body. With all but the most basic equipment, the scans are sufficiently fast to enable the images to be viewed as a movie on a video screen. B-scans are commonly used to visualize the fetus during pregnancy. In fetal ultrasonography in middle to late pregnancy, the transducer is moved over various parts of the abdomen to scan the entire uterus (womb) and fetus from different angles. Small fetal movements can be detected. The latest generation of ultrasound *scanners* produces three-dimensional images by combining the images from many cross sections.

Doppler ultrasonography is used to measure blood flow. The technique relies on the Doppler principle or Doppler effect. In essence, sound waves of a given pitch, when coming from an approaching object, are perceived as having a higher frequency (higher pitch), while those produced by a receding object are perceived as having a lower frequency. This effect is noticeable in everyday life when a train passes by and sounds its whistle or horn. Pitch rises as the train approaches and then falls as the train passes by and moves away. The frequency (pitch) shift is proportional to the speed of the object in the line of hearing. In Doppler ultrasonography, the rate of blood flow in veins

and arteries is determined by reflecting sound off the moving red blood cells and analyzing the pitch shift of echoes returned. This gives a measurement of rate of blood flow.

Color Doppler ultrasonography generates grayscale background images on which blood flow patterns are superimposed in color. Red, blue, yellow, cyan, and white color codes are used to denote speed and direction of flow. For example, red indicates positive flow (toward the transducer) and blue indicates negative flow (away from the transducer). Depth of color shows speed of movement. Transducers are now small enough to be mounted on the tip of a fine catheter and introduced into a blood vessel. The short distance between the transducer and its target means that high-frequency sound waves can be used and image resolution is that much greater.

Historical Development

Ultrasound imaging for medical use was developed in the 1950s. It drew upon the technology of *sonar* (sound navigation and ranging), the echo-sounding technique used by Allied navies in World War II to detect U-boats and other submerged objects. Ultrasound pioneers often used decommissioned radar equipment and modified it to detect ultrasound waves.

The leading pioneer in fetal ultrasonography was Ian Donald, professor of midwifery at the University of Glasgow, Scotland. Like other ultrasound pioneers, in the United States, Sweden, and elsewhere, he employed the technique in the early 1950s to visualize abdominal tumors. Building on this success, by the late 1950s he was using the technique to diagnose disorders of the fetus in late pregnancy. By the early 1960s, he was using ultrasound to visualize the early fetus (as young as 9 weeks) to confirm the fact of pregnancy itself. Japanese researchers Shigeo Satomura and Yasuharu Numura applied the Doppler principle to ultrasound as early as 1955 to measure the speed of blood traveling through the heart. It was not until much later, with the advent of computerized imaging techniques, that the method would be used clinically more routinely.

Ultrasound pioneers, including Donald, were concerned that, like X-rays, ultrasound might be harmful to the fetus. However, with increasing use of the technique in the 1960s, and no major reports of problems, these concerns were largely discounted. Meanwhile, the introduction of computers in the 1970s to convert sound signals into visual images dramatically improved ultrasound picture quality. Today, a fetus can be visualized in sufficient detail to show skeletal abnormalities and sometimes, heart or circulatory problems.

Uses, Effects, and Limitations

Ultrasound imaging yields a poorer resolution (shows less detail) than that of most other medical imaging techniques, but it is also less potentially dangerous than other techniques and is very affordable. Ultrasound imaging is particularly suitable for examining fluid-filled organs or soft-solid structures, in the absence of airspaces or very dense material. Because of poor sound transmission through airspaces and distortion of sound waves by dense matter, ultrasonography is not normally suitable for imaging the lungs, the skeleton, or the bone-enclosed brain. Large fat deposits within the body also limit the usefulness of the technique.

Ultrasound scans of the womb during pregnancy are favored above other scanning methods because they employ no harmful ionizing radiation. Such scans can confirm that the fetus is alive, is correctly positioned, and that the placenta is in place. The scans give a reasonable estimate of fetal size and age. They also check for certain abnormalities, such as spina bifida (a disorder that can result in paralysis, bowel and bladder dysfunction, and mental retardation). A scan confirms whether there is only one baby, or more, and in late pregnancy it can be used to determine the gender of the fetus. Doppler techniques can monitor the baby's heartbeat.

In children and adults, fluid-filled organs that can be visualized by ultrasound include the bladder, the gallbladder, and the interior of the eye. Ultrasound is particularly useful for detecting cysts or tumors in soft-solid structures such as the liver, spleen, and kidneys. In scanning these organs, ultrasonography has largely replaced X-ray imaging. Doppler ultrasonography is particularly useful for evaluating the flow of blood through major blood vessels such as the carotid arteries, which carry oxygenated blood to the brain. Arteriosclerosis (partial blockage of arteries by fat deposits) raises local blood pressure and can damage smaller blood vessels nearby. This is particularly a cause for concern in the brain, where high blood pressure can cause strokes. Ultrasound scanning of the carotid artery is one way of detecting the potential problem. Color Doppler ultrasonography is especially adept at visualizing abnormal blood flow within the heart and thus in diagnosing heart valve defects. Using ultrasound is preferable to the invasive X-ray imaging techniques that use *contrast agents,* X-ray-opaque chemicals administered to the patient.

Many radiologists (physicians who specialize in imaging techniques) believe that ultrasound imaging will increase in importance in the 21st century as researchers find new alternatives to the use of ionizing radiation in medical imaging. Increasingly, ultrasound imaging is being used to guide instruments in "keyhole" surgery. Ultrasound probes have been miniaturized to a diameter of 1 millimeter and less. They are mounted in *catheters* (guide tubes) for use in conjunction with *endoscopes* (instruments that use a fiber-optic tube to explore the body). Such probes can be inserted into coronary arteries. Modern portable ultrasound machines can be moved to the operating room, the bedside, the clinic, and even, with the help of TELEMEDICINE (relaying images by telephone lines or via satellite) to a crash site or battlefield.

Contrast agents are currently being tested that will produce sharper ultrasound images. These chemical agents are introduced into the bloodstream, where they produce minute bubbles. In ultrasonography, the boundaries of the gas-filled bubbles form hard echoes that clearly visualize the blood. This creates the potential for ultrasound images of the heart or circulatory system with improved resolution but at lower cost than for other imaging techniques currently in use.

Issues and Debate

Ultrasound is widely believed to be completely harmless to patients at the frequencies, duration, and intensities normally used. It is common practice in the United States, United Kingdom, and other developed countries to undertake routine diagnostic ultrasound of the womb and fetus on one or two occasions between the 16th and 33rd weeks of pregnancy, although some physicians advocate using fetal ultrasound only when a potential problem is suspected. Since little is known about the dangers of ultrasound, they say, it is best to avoid routine screenings that may be harmful. The official stance of the U.S. Bureau of Radiological Health is as follows: "Although the body of current evidence does not indicate that diagnostic ultrasound represents an acute risk to human health, it is insufficient to

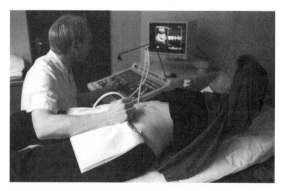

A technician performs an ultrasound on a pregnant woman (Corbis/Owen Franken).

justify an unqualified acceptance of safety." Long-term follow-up studies of babies that have been visualized by ultrasound within the womb have not shown, on balance, any developmental differences from those who were not scanned. Such studies continue.

Some consumer groups suggest that pregnant mothers might be wise to avoid fetal ultrasound tests unless they have a high-risk pregnancy and until additional long-term studies on safety have been completed. In fact, Doppler scanning of a fetus or placenta to check for circulatory abnormalities does place the fetus at potential risk of receiving higher-than-preferable levels of ultrasound, which could cause localized heating and potential tissue damage. Doppler ultrasound of the fetus or womb is used only with caution and where the benefits of conducting an investigation outweigh the risks of not doing so.

However, performing ultrasound studies requires training, skill, and experience. Being a portable, relatively affordable scanning method, it is more likely than other scanning techniques to be used in emergency situations and by inadequately trained personnel, which can lead to misdiagnosis. Ultrasound is often the only imaging technique available in the developing world. Even in the best of conditions, making accurate diagnoses from ultrasound images can be problematic. Ultrasound scans are representations of boundaries within the body and require interpretation by trained specialists for accurate diagnosis.

—Trevor Day

Related Topics

Magnetic Resonance Imaging, Scintillation Techniques, X-ray Imaging

Bibliography and Further Research

Books

Armstrong, Peter, and Martin L. Wastie. *Diagnostic Imaging,* 4th ed. Oxford: Blackwell, 1998.

Erkonen, William E. *Radiology 101: The Basics and Fundamentals of Imaging.* Philadelphia: Lippincott-Raven, 1998.

Kevles, Bettyann H. *Naked to the Bone: Medical Imaging in the Twentieth Century.* New Brunswick, N.J.: Rutgers University Press, 1997.

Lisle, David A. *Imaging for Students.* New York: Oxford University Press, 1997.

McCormick, A. K., and A. T. Eliot. *Health Physics.* Cambridge: Cambridge University Press, 1996.

Meire, Hylton B., and Pat Farrant. *Basic Ultrasound.* Chichester, West Sussex, England: Wiley, 1995.

Sanders, Roger C., ed. *Clinical Sonography: A Practical Guide,* 3rd ed. Philadelphia: Lippincott-Raven, 1998.

Weir, Jamie, and Peter H. Abrahams. *Imaging Atlas of Human Anatomy,* 2nd ed. London: Mosby-Wolfe, 1997.

Periodicals

Coghlan, Andy. "Bubbles Give a Sharper Picture." *New Scientist,* June 1, 1996, 20.

Dixon, Adrian K. "Evidence-Based Diagnostic Radiology." *Lancet,* August 16, 1997, 509.

Gibbs, W. Wayt. "Ultrasound's New Phase." *Scientific American,* June 1996, 32.

Lentle, Brian, and John Aldrich. "Radiological Sciences, Past and Present." *Lancet,* July 26, 1997, 280.

Motluk, Alison. "Brain in the Round." *New Scientist,* August 30, 1997, 7.

Paneth, Nigel. "Prenatal Sonography: Safe or Sinister?" *Lancet,* July 4, 1998, 5.

Internet Resources

Bioethics Resources
    http://adminweb.georgetown.edu/nrcbl/

A Guided Tour of the Visible Human
    http://www.madsci.org/~lynn/VH/

Teaching Aids on Medical Imaging
    http://agora.leeds.ac.uk/comir/resources/links_c.html#teaching

Three-Dimensional Ultrasound Imaging
    http://www.cs.uwa.edu.au/~bernard/us3d.html

Ultrasound Review
    http://www.usreview.com.au

The Visible Embryo
    http://www.visembryo.com

The Visible Human Project
    http://www.nlm.nih.gov/research/visible/visible_human.html

Virtual Reality

Virtual reality (usually known as VR) is a way of using computers to construct elaborate models of the world that people can interact with as though they were the real thing. The earliest (and perhaps best known) example of virtual reality is the flight simulator, a realistic model of an airplane cockpit used to train pilots. But as it has evolved, VR has found a wide range of applications in entertainment and other fields.

Virtual reality systems become more convincing—more like "real reality"—every day, and this has prompted concerns that people may eventually try to avoid the harshness of the real world by spending their time entirely in computer-generated virtual worlds. Yet supporters of VR say that the technology has the potential to vastly expand human experience.

Scientific and Technological Description

VR relies on efforts to fool people into thinking that a virtual (computer-generated) world is the real thing. As philosophers have argued for some time, what people take to be "the real thing" is itself a construction of the human brain based on sensory perception (information fed in by the senses). So VR involves replacing the real senses with virtual ones; in practice this involves "immersing" people in a three-dimensional computer-graphic scene and playing them sound effects. The scene includes parts of the person's own body, such as computer images of their hands. As they move their real hands in real space, the computer-generated hands appear to move in virtual space. VR makes it possible for someone to feel as though they are walking on a computer generated moon using their real legs, or petting a virtual dog by making movements of their real hand.

There are essentially two types of VR. The simplest involves running simulations (models) of aircraft flight or other virtual scenes on standard personal computers (PCs) or arcade consoles. Because these suspend users partway

Drinking a Virtual Cup of Coffee

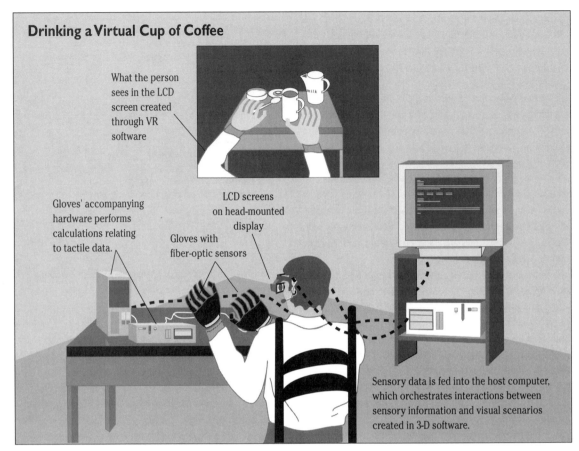

What the person sees in the LCD screen created through VR software

Gloves' accompanying hardware performs calculations relating to tactile data.

Gloves with fiber-optic sensors

LCD screens on head-mounted display

Sensory data is fed into the host computer, which orchestrates interactions between sensory information and visual scenarios created in 3-D software.

Virtual reality replaces our real senses with virtual (computer-generated) ones. Liquid-crystal display (LCD) screens replace our vision, stereo speakers replace our hearing, and special gloves replace our sense of touch. The gloves can also detect hand movements and feed them back to the VR computer. If the user reaches out with the glove, the virtual hand in the LCD image appears to reach out too. The user can grasp the virtual cup of coffee, but it won't taste of anything; VR taste has yet to be invented.

between the real and virtual worlds, they are known as *non-immersive VR*. By contrast, *fully immersive VR* aims to replace the real world entirely by getting people to wear a headset (an enlarged visor that fully covers both eyes) known as a *head-mounted display* (HMD), or sometimes even a full body suit. An HMD contains two small liquid-crystal display (LCD) screens, which provide a slightly different image to each eye to give the illusion of a three-dimensional picture. Some HMDs also contain stereo speakers that fit over the ears (see WEARABLE COMPUTERS).

Another way of looking at VR is as a radically different way for people to interface (communicate) with computers. Instead of a traditional output unit such as a monitor, the computer feeds its responses directly into a person's eyes or ears using devices known as *effectors*. Instead of an input unit such as a keyboard, the computer responds to the person's movements using delicate electronic sensors.

One of the challenges of VR is to make effectors and sensors good enough to fool the brain. Apart from HMDs, which contain sensors for detecting head and eye movements, the most common sensors are fabric gloves, which typically use FIBER OPTICS to detect finger and hand move-

ments. Some gloves contain both sensors and effectors: They can feed back tactile (touch) sensations to the user's fingers; the most advanced gloves can even stiffen to simulate touching solid objects.

Interactive virtual reality is much more complex than showing someone a film of what they might see through their eyes: Rather, the picture changes as viewers would expect in response to the movements they make. To make this happen in real time (instantaneously), the various sensors and effectors must be coordinated with a detailed model of the virtual world. Enormous computer power is needed to do this, which is why VR systems often run on SUPERCOMPUTERS.

Historical Development

Virtual reality can trace its origins back to a flight simulator developed by Edwin Link in the late 1920s at Binghamton, New York. Link's family were piano makers, and his machines used techniques and actual parts from player pianos. His machines were immediately recognized as a valuable method of training pilots, and by 1941 they were being used by 35 air forces worldwide.

The development of flight simulators was one of the main driving forces behind what is now known as virtual reality, but there were two other significant historical strands at play. In 1950, James J. Gibson, a psychologist working for the U.S. Air Force, published an influential book called *The Perception of the Visual World,* which explored how the environment appears to "wash" around people in an *optic flow* as they move through it. Subsequent research into perception provided the theoretical underpinning for VR.

The other important strand was the development of devices that made computers easier to use, a field now known as *human–computer interaction* (HCI). From the late 19th century until as recently as the 1970s, computers were programmed by reading information encoded in tiny holes punched into long streams of paper or stacks of cards. Computers were complex and aloof—the preserve of rocket scientists and professors.

That began to change in the 1960s with the work of HCI pioneers such as Ivan Sutherland. In a computer world of paper tapes and printed output, Sutherland invented a number of other devices that made computers more intuitive to use. These included the light pen (a pointing device that could be touched against a computer screen) in 1962, and in 1965, the first head-mounted display. This featured a system for tracking the position of the wearer's head and graphics that were updated in real time. HMDs were adopted by the U.S. military for training fighter pilots in the late 1970s and refined at NASA's Ames Research Laboratory from 1985 onward.

Although these various strands of research seemed to be converging, it wasn't entirely clear where they were heading until 1984, when the term *virtual reality* was coined by computer scientist and composer Jaron Lanier. His company, VPL Research, produced the first commercial virtual reality glove (the DataGlove) in 1987, initially for playing music synthesizers. Instead of the performer's hand touching the keys on a real keyboard, synthesizers could now be played by flexing the fingers inside a DataGlove. Thus performers could play virtual keyboards, move their fingers along virtual flutes, or play music in a variety of other ways without ever actually touching a real instrument. A simpler and cheaper version of the DataGlove was produced by the Mattel toy company in 1989 and became a popular device for controlling computer games.

During the 1990s, VR moved out of the computer lab and into society. This was due partly to the 1991 publication of a best-selling book by Howard Rheingold called *Virtual Reality,* and partly to the emergence of VR into popular culture. Many people's first experience with VR came when they saw a 1992 film called *The Lawnmower Man,* whose central character, a modern-day Dr. Frankenstein known as Dr. Larry Angelo, attempts to demonstrate that VR technologies could mark the next stage of evolution by using them to turn an unlikely test subject into a genius.

Meanwhile, a popular game known as Virtuality, which involved wearing a conspicuous head-mounted display, began to appear in arcades and theme parks and attracted considerable media attention.

Today, virtual reality means that sensitive gloves can be used to conduct computerized orchestras, manipulate the flow of air around virtual aircraft in wind tunnels, or guide virtual scalpels into virtual brains. VR is the most natural method of interacting with computers yet invented; it is almost like stepping into them and walking around inside.

Uses, Effects, and Limitations

Pioneers of virtual reality argue that there is no limit to its potential; provided that computer systems can fool the senses convincingly, proponents maintain, virtual reality can copy or even enhance any experience from the real world. Simulators and training aids, which were the first use of virtual reality, are still one of its biggest applications. Flight simulators enable pilots to test new aircraft designs without putting passengers at risk. In the same way, VR medical simulations enable doctors to learn high-risk operations, such as brain and heart surgery, without endangering the life of a patient.

Although VR has often been used to approximate reality as closely as possible, the same techniques can be used to produce what might be described as alternative realities, which do not exist in the real world. For example, archaeologists have used supercomputers to produce walk-through reconstructions of long-lost archaeological sites, such as the lost city of Pompeii. Just as VR can be used to recreate the past, so it can be used to speculate about the future. VR is now used routinely by architects to show clients, planners, and members of local communities how new buildings will fit into their surroundings. Rocket scientists at NASA use it to try out new SPACE SHUTTLE designs in virtual wind tunnels before they build costly prototypes. VR can also be used to give people a taste of extraordinary real-world experiences that are not routinely available, such as walking through a rain forest or taking a trip into space.

The rapid growth of the Internet during the mid-1990s led to interest in using virtual reality to enhance sites on the World Wide Web. This prompted the development of a simple, standard programming language, similar to the Web's HyperText Markup Language (HTML), which is known as Virtual Reality Markup Language (VRML). It is specially designed for producing nonimmersive VR scenes over the Internet, such as three-dimensional shopping malls that can be used for ELECTRONIC COMMERCE (Internet trading).

VR has clear benefits: Computers can make it simpler and safer to train people in dangerous jobs without putting others at risk, and VR can provide access to extraordinary experiences that ordinary people might not otherwise get the chance to explore. This can all be accomplished in a person's own home, perhaps even over the Internet. But compared to reality, VR is still very crude. Walking through a VR rain forest is nothing like walking through the teeming, steaming rain forests of Brazil. VR is still largely confined to vision and sound of poorer quality than in real life. Another limitation is the expense of providing a convincing VR experience: Although the cost of VR is constantly falling, the best HMDs and VR gloves still cost up to $100,000. More problematic than this, VR has been shown to be addictive and causes "simulation sickness" when people spend too long in virtual worlds.

Issues and Debate

If virtual reality is one of the most exciting computer-based technologies, it is also one of the most controversial. ARTIFICIAL INTELLIGENCE has proved contentious, because it aims to develop computers that can think like people. But virtual reality goes even further: It aims to replace parts of the real world (including real people) with a computer-generated alternative. While artificial intelligence is trying to make computers more like humans, virtual reality appears to be making humans more like computers.

The distinction between the human world and the computer world has become increasingly blurred. Direct human friendships are being replaced by long-distance relationships, conducted through Internet chat rooms and e-mail, between people who have never even met. Where surgeons once practiced on human cadavers or live subjects, they now train using virtual hearts and brains. Shopping, which used to mean a trip to a crowded mall and interaction with fellow shoppers, increasingly means adding virtual books, food, and other goods to a virtual shopping cart that is "pushed" around a Web site in cyberspace (see ELECTRONIC COMMERCE). Meanwhile, architects and archaeologists can walk us through imaginary buildings from the future and the past, and military strategists calculate the risks and benefits of waging war using virtual battlegrounds. For some people, virtual reality heralds exciting new possibilities; for others, VR is an assault on all that it means to be human.

Commentators, such as University of California lecturer Mark Slouka, argue that VR represents nothing less than an all-out war on the real world that undermines peoples' identity and community, and therefore their place in reality. According to Slouka, one of the most disturbing aspects of VR is that it provides an easy escape from the problems of the real world. There are no famines in cyberspace, he argues; wars can be started and stopped by clicking a mouse; global warming and ecological crisis are nonexistent. Further, there is no morality in virtual worlds: Lying, cheating, cruelty, and killing may all pass unnoticed. One concern is that violent or immoral behavior, such as that required to succeed in many computer games, may leak out into the real world.

But supporters of VR point to obvious benefits such as training brain surgeons and fighter pilots. They claim that VR is no more dangerous an escape from the real world than television, radio, the movies, or books, and unlike these pas-sive forms of entertainment, virtual reality can be highly interactive. They also point out that VR can help serve an educational and archival cause. With or without VR, species will become extinct, ancient tribes and cultures will disappear, and parts of the Earth will be degraded and destroyed. All the more reason, supporters argue, that aspects of society should be preserved using virtual reality. What is the difference, they contend, between peering through glass at a stuffed dodo in a museum and stroking a "living" dodo in a VR simulation? Some enthusiasts have gone even further. VR pioneer Jaron Lanier describes VR as an electronic drug that can extend the human spirit like no other creative medium.

Opinions are usually divided over the adoption of wide-ranging, cutting-edge technologies—and VR is no exception. Critics argue that overreliance on VR experiences could turn reality into a desolate ghetto. Supporters say that it adds a whole new dimension to human experience. The debate will no doubt continue, but it could be several decades before we know whether one of these positions will finally dominate, or whether VR and other computer technologies will change in ways that will make the debate obsolete.

—Chris Woodford

RELATED TOPICS
Computer Modeling, Electronic Commerce, Internet and World Wide Web, Supercomputers

BIBLIOGRAPHY AND FURTHER RESEARCH
BOOKS
Baker, Robin. *Designing the Future: The Computer Transformation of Reality*. London: Thames & Hudson, 1993.
Cotton, Bob, and Richard Oliver. *The Cyberspace Lexicon*. London: Phaidon, 1994.
Eddings, Joshua. *How Virtual Reality Works*. Emeryville, Calif.: Ziff-Davis, 1994.
Forte, Maurizio, and Alberto Siliotti, eds. *Virtual Archaeology: Great Discoveries Brought to Life Through Virtual Reality*. New York: Harry N. Abrams, 1997.
Heim, Michael. *The Metaphysics of Virtual Reality*. New York: Oxford University Press, 1993.
Pimentel, Ken, and Kevin Teixeira. *Virtual Reality: Through the New Looking Glass.* New York: McGraw-Hill, 1995.
Rheingold, Howard. *Virtual Reality.* New York: Summit Books, 1991.
Slouka, Mark. *War of the Worlds: Cyberspace and the High-Tech Assault on Reality*. New York: Basic Books, 1995.
Waterworth, John. *Multimedia Interaction with Computers: Human Factors Issues*. New York: Ellis Horwood, 1992.
Woolley, Benjamin. *Virtual Worlds*. Cambridge, Mass.: Blackwell, 1992.
PERIODICALS
Brown, Ed. "The Virtual Career of a Virtual Reality Pioneer: Jaron Lanier." *Fortune,* March 2, 1998, 194.
Chinnock, Chris. "Virtual Reality Goes to Work." *BYTE,* March 1996, 26.
Cosentino, Victor. "Virtual Legality." *BYTE,* March 1994, 278.
Hecht, Jeff. "Virtual Reality Leaves You Virtually Reeling." *New Scientist,* February 27, 1999, 21.
Hodgkins, Jessica. "Animating Human Motion." *Scientific American,* March 1998, 64.
Laurel, Brenda. "Virtual Reality." *Scientific American,* September 1995, 70.

McCosh, Dan. "Virtual Building." *Popular Science,* May 1999, 45.

Pountain, Dick. "VR Meets Reality." *BYTE,* July 1996, 93.

"This Week in Science: 3D Displays" (editorial). *Science,* August 30, 1996, 1149.

Vacca, John. "The Net's Next Big Thing: Virtual Reality." *BYTE,* April 1995, 28.

INTERNET RESOURCES

About.com: Focus on VRML
    http://vrml.about.com/

Active Worlds—Welcome to the 3-D Internet
    http://www.activeworlds.com

All About Virtual Reality
    http://www.allaboutvr.com

Intel's Home Computing: Virtual Stonehenge—Explore 3D Worlds
    http://www.intel.co.il/cpc/explore/stonehenge/

NASA Ames Simulation Laboratories
    http://scott.arc.nasa.gov/

University of Washington Human Interface Technology Lab
    http://www.hitl.washington.edu/

Virtual Reality Applications for Museums and Cultural Heritage
    http://kb.hitl.washington.edu/old-museumapps.html

The Visible Human Project
    http://www.nlm.nih.gov/research/visible/visible_human.html

Web-Based Surgical Simulators and Medical Education Tools
    http://synaptic.mvc.mcc.ac.uk/simulators.html

WAVE AND OCEAN ENERGY

The oceans are probably the single largest energy reserve on the Earth. Every day, tidal forces cause them to rise and fall by several meters, while the Earth's rotation drives deep ocean currents, and wind across the surface whips up powerful waves. All of these movements create mechanical energy, which can be tapped in a variety of ways and converted into electricity.

In addition, it is possible to generate electricity by utilizing the temperature differences between different parts of the ocean. However, the power and unpredictability of the sea present significant obstacles to attempts to harvest its energy.

Scientific and Technological Description

Of all the phenomena associated with the sea, ocean waves contain the most energy. Waves are generated by winds across the water's surface, interacting with currents. Because a wave moves large amounts of water at high speeds, it contains a large amount of kinetic energy, at least some of which can be converted into more useful forms. Similarly, the tides cause huge masses of water to rise and fall relative to the surface of the Earth, tugged this way and that by the pull of the Moon and the Sun. As with waves, the strength of tides varies depending on location, and some places experience far more dramatic rises and falls than others. The third major source of ocean energy is the temperature difference, or *thermal gradient,* between different depths of the sea. This is a different type of energy, but tapping it can produce useful by-products (see below).

Altogether, the power of the oceans is estimated to be around 2 terawatts (2 trillion watts or 2 million megawatts).

The majority of this energy is locked up in waves and thermal gradients—various constraints make tapping more than a small proportion of ocean energy impractical. Nevertheless, in theory, the oceans still contain more than enough energy to supply the world's current or projected future requirements. This energy is clean, requires no fuel source, and is self-renewing. The only problem is how to convert it into electricity. A wide variety of techniques have been developed to do this.

In the case of wave and tidal energy, the problem is a mechanical one. Energy from the water must be transmitted to spin a turbine and produce current. However, this energy often has to be concentrated and multiplied to produce a powerful enough force. Tidal energy plants use a *tidal barrage,* a barrier across the mouth of a bay, with sluice gates that can be opened to allow water in as the tide rises and then closed to trap it as the tide falls away. In this way the barrage builds up a "head" of water, which is allowed to escape only by flowing through electricity-generating turbines. The basic principle here is similar to that used in a traditional hydroelectric power plant with a dam.

One of the most practical methods for generating electricity from waves works on the same principle. *Focusing devices* are artificially created tapering channels that gather the waves approaching a long section of coastline, and focus them in a relatively small area. This has the effect of amplifying the wave heights, allowing them to wash over a wall and into an enclosed reservoir. The water then flows back out of this reservoir and returns to sea level through hydroelectric turbines.

Other methods for tapping wave energy are more unusual. One technique that has been used successfully on a small scale is the *oscillating water column* (OWC). An OWC is effectively a chimney surrounded by a curtain that dips down below the wave level, creating an isolated column of water. Wave movements drive this column of water up and down, forcing air out of the chimney and then drawing it back in. As the air rises and falls, it spins a turbine positioned halfway up the chimney, and creates electricity.

A third type of device is the *surface follower,* found in many different designs that all obey the same basic principle. These use the relative movement between a floating component riding the waves and a fixed component (such as a platform anchored deeper in the water) to create a pumping motion in a hydraulic fluid and drive a turbine.

Ocean thermal energy conversion (OTEC) plants use the temperature differences between the sea's surface and the ocean depths to produce electricity. They do this by pumping warm surface water as a fine spray into a partial vacuum. The sudden pressure drop makes the water boil below normal boiling point, even without external heat being applied. The steam generated is used to drive a turbine, before being condensed back into water as it comes into contact with the surface of pipes full of cold water drawn up from depths of

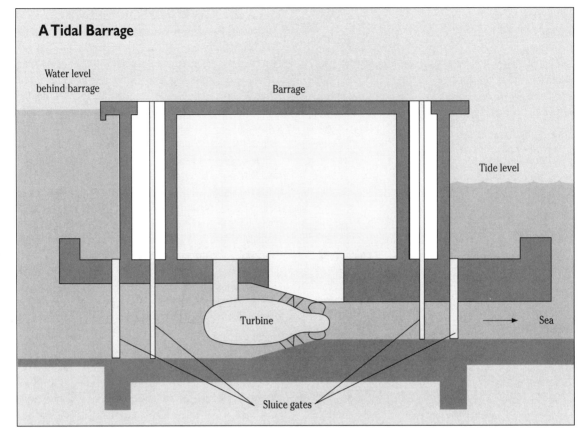

A Tidal Barrage

Water level behind barrage

Barrage

Tide level

Turbine

Sea

Sluice gates

A tidal barrage is a barrier across the mouth of a bay, with sluice gates that can be opened to allow water in as the tide barrage builds up a "head" of water, which is allowed to escape only by flowing through electricity-generated turbines. (Adapted from David J. Cuff and William J. Young, United States Energy Atlas, *2d ed., Macmillan Publishing Company, 1986; Free Press, 1980.)*

around about ½ mile. The pressure drop caused by the sudden condensation is used to create the initial vacuum, and the condensed water is free of salt, so that OTEC generators can function as desalination plants as well.

Historical Development

The potential of the oceans for energy generation has been recognized and tapped for centuries—the first devices to be powered by the sea were tidal mills used in medieval Europe as early as the 11th century. In 1582, German engineer Peter Moritz installed a mill on England's River Thames at London Bridge to drive a pump supplying water to the city's homes. But it was not until the late 19th century that scientists and inventors first began considering the potential for converting the energy of the oceans into other useful forms. The first serious investigation of wave power as a source of energy was carried out by Australian scientist R. S. Deverell around 1875, and in 1919 a commission of French scientists studied the requirements for a tidal power station.

Although it is arguably the most complicated of all ocean energy systems, OTEC, the system that exploits the ocean's temperature differences, was the first to be put into effect. The principle behind the technology was estab-

lished as early as 1881 by French physicist Jacques Arsene d'Arsonval, and a working plant was established in Cuba by 1930. Most of the devices in use for tapping ocean energy today have been developed since the 1950s. For instance, the world's first full-scale tidal power station was constructed on the Rance estuary in northern France and completed in 1966. The Rance project generates 240 megawatts of electricity at peak times and was intended to be the first of many such plants, but the French energy program switched its emphasis to nuclear fuels, and tidal power was sidelined. Today, Rance remains the largest tidal power station in the world.

Meanwhile, in Japan, naval officer Yoshio Masuda developed the OWC generator in the mid-1960s. At first, OWCs were simply built into navigational buoys to provide a source of power for their lights and beacons. However, the idea proved so successful that Japan launched an intensive study of the OWC's potential, with a full-sized prototype tested in 1979. That led to two small working stations that remain in operation today.

A variety of surface-following devices have been invented since the 1970s, but probably the most familiar is the *Salter Duck*, developed by Stephen Salter of the

University of Edinburgh, Scotland. This device consists of a shaped float, linked to others by a concrete spine. The float pitches up and down on the waves, but it contains a series of pumps that push up and down against a camshaft suspended in the middle of the float. This camshaft contains a number of gyroscopes, which generate forces that prevent it rocking back and forth with the remainder of the float, so it acts as a fixed object against which the pumps can push. The pumps drive a hydraulic fluid that spins a small turbine generator, and the electricity produced is passed back along the concrete spine to the shore. Salter Ducks and other surface-following devices are in quite widespread use today, but only in small-scale projects.

The first operating wave-focusing system was built on the North Sea coast of Norway and came into operation in 1986, supplying 350 kilowatts of power. Although only a small-scale power plant, the Norwegian project has proved highly successful, and similar, megawatt-scale plants are under construction in Australia and Indonesia.

Uses, Effects, and Limitations

Although the principles behind the various types of ocean energy devices are simple and well established, they are in use today only at a limited number of locations, primarily those where unique environmental factors make ocean energy particularly practical, or other forms of energy particularly impractical. At present, the high initial construction costs make ocean energy an uneconomical alternative in most other situations (see *Issues and Debate*), although the eventual exhaustion of fossil fuels might lead to a shift in this balance.

However, in any ocean energy project there are a number of important practical barriers to effectiveness. These must be overcome before a project can be considered. Perhaps the most important factor is that every system requires very specific conditions to work effectively. *Wave power,* for instance, is effective only in geographical regions where winds are strongest and therefore waves reach significant heights. The average power of incoming waves must be about 65 megawatts per mile of coastline for exploitation to be economical. Suitable areas tend to be at midnorthern and southern latitudes and on west-facing coasts. Near the equator, the winds are too weak, and near the poles, most of the sea is frozen.

Tidal power, on the other hand, can be harnessed cost-effectively only where there is a significant difference between low and high tides (usually, more than 5 meters). Tidal heights are governed by geographical features and are greatest in enclosed bays (the highest tides in the world, at the Bay of Fundy in southeastern Canada, can reach up to 20 meters).

Thermal energy, meanwhile, requires a temperature difference of around 20°C to work, and the sources of warm and cold water must be close together for the pumping up of the cold water to be economical. In these systems, the more energy that has to be recycled into obtaining the water sup-ply, the less net power an OTEC plant produces. Therefore, OTEC plants must be sited on coastlines with steeply shelving seabeds and warm tropical surface waters—hence the siting of the first experimental plant in Cuba.

All of these factors combine with the question of coastal suitability for construction of a large-scale power plant. Many areas with the sea conditions required are unsuitable because of their ruggedness or remoteness from civilization. The foregoing factors reduce the practical energy yields from the oceans dramatically in comparison with the potential harvestable energy cited in the global figures given above.

An essential problem that is likely to prevent ocean energy from replacing more traditional power forms completely is its reliance on specific geographical locations and its relationship to time cycles. Like solar power, tidal power cannot be generated at a constant rate—power generation is possible only after each high tide, creating a 12-hour cycle of peaks and lulls. Wave energy is practical only on certain coastlines, and OTEC requires specific offshore conditions to function efficiently. The net result is a power source that is useful for providing energy at specific locations, or for some of the time, but lacks the transportability and constant controllable output offered by fossil and nuclear fuels.

Issues and Debate

The major question surrounding all these forms of energy must be why they have, so far, made so little impression on global electricity generation. The major factors are, unfortunately, a reluctance to devote time and investment resources to new technologies while the supply of traditional fuels is still plentiful and relatively cheap, and the high capital costs of constructing tidal, wave, and OTEC power plants.

Although once a plant is in operation the electricity is effectively free (all of these systems have very low maintenance costs), the initial costs of setting up a station are prohibitive, especially as any coastal facility must be built to withstand storm conditions with waves up to 10 times more powerful than average. For example, plans to build a tidal barrage across the River Severn in the southwestern United Kingdom—a project that could have generated 10 percent of the country's electricity—were rendered uneconomical by the $15 billion construction costs.

Finally, although there is still significant potential for expanding the use of energy from the sea, the possible environmental changes that this could bring about are still poorly understood. Tidal power stations are known to lower the level of tides slightly, which might encourage the growth of algae around the high-tide mark. Some types of wave energy devices, such as chains of surface followers or focusing devices, can divert or suppress waves, altering the patterns by which sedimentation is laid down and slowly changing the coastline, although any ecological effects of this are difficult to predict. Similarly, to leave the ocean's delicate temperature balance undisturbed, OTEC plants must be carefully designed and controlled. Deep, cold water used in OTEC must not simply be dumped back onto the sea surface.

Although energy from the oceans has yet to make a significant impact on global power consumption, it seems likely that, in its various forms, it has a bright future. However, this future is more likely to involve use for small-scale local power projects than a wholesale replacement of fossil energy.

—*Giles Sparrow*

RELATED TOPICS

Biomass Energy, Geothermal Energy, Hydroelectric Power, Nuclear Energy, Solar Energy, Wind Energy

BIBLIOGRAPHY AND FURTHER RESEARCH

BOOKS

Ford, Glyn, Chris Niblett, and Lindsay Walker. *The Future for Ocean Technology*. London: Pinter Publishing, 1992.

McCormick, Michael E., and Young C. Kim. *Utilization of Ocean Waves: Wave to Energy Conversion*. New York: American Society of Civil Engineers, 1987.

Ross, David. *Power from the Waves*. New York: Oxford University Press, 1995.

Seymour, Richard J., ed. *Ocean Energy Recovery: The State of the Art*. New York: American Society of Civil Engineers, 1992

INTERNET RESOURCES

International Council for Local Environmental Initiatives
  Energy Fact Sheets
  http://www.iclei.org/efacts

U.S. Department of Energy: Energy Efficiency and Renewable Energy Network
  http://www.eren.doe.gov/

WEARABLE COMPUTERS

A wearable computer is designed to be highly usable even when the person wearing the computer is walking around. In most wearable computers, the traditional monitor is replaced by an eyepiece, while a battery and processor (the "brain" of a computer) are usually carried in a knapsack, attached to a vest, or hung on a belt.

Although many designers and scientists have experimented with making computers more portable over the past few decades, wearable computers did not become commercially available until the mid-1990s. Because they are such a new technology, their use has been limited, and their impact on society has yet to be fully determined. Wearable computers have, however, found users in heavy industry and among the disabled, and they have generated considerable debate about the potential role of computers in society.

Scientific and Technological Description

The basic design of commercially available wearable computers is in some respects quite similar to that of traditional computers. Wearable computers have a small but relatively powerful processor—those currently on the market offer 32 megabytes or more of RAM (random access memory), which stores temporary data. But a wearable computer's

processor is a bit different from that of a traditional computer. The hard drive is encased in shock-absorbing gel so that the user's movement won't damage it. Although a wearable computer may have ports to allow users to connect to peripheral devices such as a CD-ROM drive, the drives are too sensitive to motion to be contained inside the computer.

In a wearable computer, the processor is connected to a monitor and to an input device by cables. Both monitor and the input device differ wildly from those of traditional computers. The monitor is extremely small and is attached to a headset. The headset usually has an earphone, so the processor, using vocalization software, can talk to the person who is wearing it when the wearer does not want to use the monitor. The traditional input devices for a computer are a keyboard and a mouse. Neither works well in a mobile environment, so the input devices for wearable computers are quite different. Many wearable computers come with special keyboards that are very small and have many fewer keys than does a normal keyboard. To make all 26 letters in the alphabet, the user presses down more than one key at a time. Another popular input device for wearables is a microphone, which is used along with voice-recognition software to enable the user to give voice commands to the computer.

Historical Development

Although computers have been around in one form or another since the 1940s, wearable computers became commercially available only in the mid-1990s, and they have yet to find widespread acceptance. The idea of having computers that could travel with people and help them out has been around for much longer, albeit as the stuff of fantasy. In 1960 science fiction writers Manfred Clynes and Nathan Kline coined the term *cyborg* in the story "Cyborgs and Space" to describe people who used technological attachments to help them survive.

The first wearable computer was invented a year later for a much less noble purpose—cheating at roulette. Ed Thorp and Claude Shannon, who kept their invention a secret until 1966 for obvious reasons, created a small, not very powerful computer that one person could use to record the speed of the roulette wheel. The information was conveyed to a roulette-playing accomplice wearing a special hearing aid. With knowledge of the speed of the wheel, the players could confidently place bets on when and where it would stop.

As if to compensate for the illegal use found by Thorp and Shannon, in 1967 Hubert Upton invented a wearable computer with an eyeglass-mounted display to help deaf people read lips. Upton's work reflects a common theme in the technology behind wearables: A number of devices (such as one-handed miniature keyboards and voice-recognition software) that have made wearable computers possible were invented or perfected to assist disabled people.

Another trend that made wearable computers possible was the growing power of very small computers.

Following the release of the first commercial microprocessor in 1971, computers have shrunk immensely in physical size, yet have become more and more powerful. Notebook (laptop) computers made today are more powerful than the room-sized computers of the 1940s and 1950s. The 1970s also saw the advent of the first popular wearable piece of electronic equipment, the release of the Sony Walkman personal stereo in 1979. The phenomenal success of the Walkman led companies to search for ways to make portable versions of other previously immobile electronic items, such as phones and computers.

The final innovation that made computers truly wearable came when the company Reflection Technology released a head-mounted monitor in 1989. The screen was only a little over an inch wide, but because it was held very close to the eye, it was readable. The next year a group of researchers at Columbia University combined the monitor and a specially designed notebook computer to create a wearable computer that was comparable in power to a desktop computer.

Although wearable computers were no longer the stuff of science fiction in the early 1990s, they remained the stuff of academia. In 1993, Thad Starner of the Massachusetts Institute of Technology's Media Lab began wearing a computer during all his waking hours. He was joined by his colleague Steve Mann, who began transmitting images from a camera mounted on his head to a World Wide Web site in 1994. Although Starner and Mann's decision to connect to computers more or less permanently may seem peculiar, it both raised the technology to public notice and allowed researchers to improve the reliability and portability of the machines.

By 1996 a handful of companies—such as wearable manufacturers Xybernaut and ViA, and the airplane manufacturer Boeing—had begun selling, using, and holding conferences on wearable computers. Publicity increased when a traveling wearable computer fashion show, a collaboration between faculty and students at MIT and several design schools, was launched in 1997. But the technology remains in its infancy, with small sales and no widespread commercial applications.

Uses, Effects, and Limitations

As with many emerging technologies, no one is quite sure what wearable computers will ultimately be used for. So far, their most promising uses have been in certain types of industries and in assisting disabled people. Industries that require employees to operate complex equipment with both hands have shown a strong interest in wearable computers. Currently, a mechanic working on, for example, an airplane, who has a question about what he is doing, must stop working, find a manual, look up the information, and then return to work. The same mechanic wearing a computer containing the information found in the manual can simply ask his com-

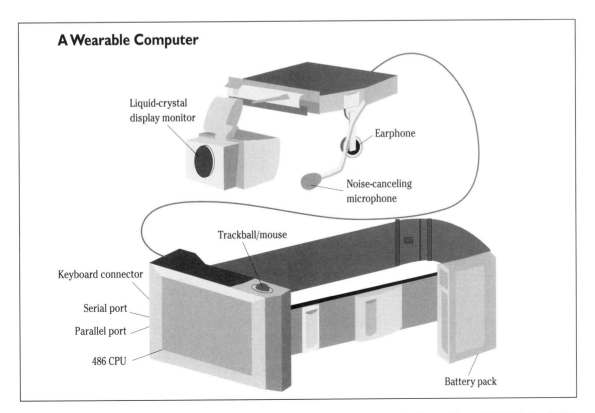

A Wearable Computer

Liquid-crystal display monitor

Earphone

Noise-canceling microphone

Trackball/mouse

Keyboard connector

Serial port

Parallel port

486 CPU

Battery pack

Adapted from Nick Baran, "Get Smart—Wear a PC," BYTE, March 1996.

Thad Starner, left, talks with Josh Weaver in May 1998. Both men, shown wearing their computers, are members of the "wearables" computer research group at MIT (AP/Victoria Arocho).

puter to look up the information, then read it off his eyepiece. If the mechanic needs to record that a particular repair was made, he can use his wearable computer to do that as well, all without leaving the bowels of the airplane.

Disabled people are also a natural market for wearable computers. There are already many different types of computer software and hardware designed to help disabled people communicate and work. Wearable computers add the very useful feature of portability to these programs and devices. For example, a person with a muscular or neurological disorder that makes it difficult to read and write might use specialized software at work to dictate material to a computer and have it read memos and e-mail to him or her. But if the person relies on a desktop computer, when they leave work, they leave a valuable tool behind. A wearable computer, in contrast, would be available whenever and wherever the person wanted to write a letter or read the newspaper.

But there are also many limits on wearable computers. Wearable computers are expensive, costing several thousand dollars. They are also not always reliable, especially in bad weather, although manufacturers have recently managed to alleviate this problem somewhat. The small size of wearable computers means that they are simply not as powerful as similarly priced desktop computers, and because they are often in motion, a wearable computer cannot be augmented by internal CD-ROM drives and similar features that are sensitive to shaking.

To augment the storage capacity of wearable computers, researchers are working on computers that would store data on remote servers through a wireless Internet connection. Some have proposed wearable computers that would not have a processor at all but would store all data on

remote servers. That would eliminate the need for the user to carry an unwieldy processor. Bulky batteries are another issue, especially given that the batteries run down fairly quickly, so the user often has to carry extras. Some researchers are exploring the idea of harnessing the energy created when people walk by putting special material into shoes that converts motion into electricity. That electricity would then be used to power wearable computers.

Another downside is that the cables that connect the various parts of the wearable computer to each other can catch on machinery, making the computer clumsy and even dangerous. Researchers are working on such cordless connections as metal embroidered into clothing or even using the body's own electromagnetic field. Another goal is to improve the monitor headset so that it does not make the user look so unusual. Many have complained that the headsets make people look like robots or creatures from a science fiction novel. Although still in an experimental stage, researchers have developed traditional-looking glasses that have a monitor embedded in the lens. Considering that workers in many industries must already wear safety goggles, such glasses would presumably eliminate aesthetic barriers to acceptance, at least at work.

Of course, hardware is only as good as the software that runs on it. Voice-recognition software, although improving, is still far from perfect (see LANGUAGE RECOGNITION SOFTWARE). Even with special noise-dampening microphones, wearable computers cannot currently use this software in noisy places, which limits the technology's use in some industries and in the military.

Issues and Debate

Proponents of wearable computers feel that as the technology gets cheaper and more reliable, it should move out of its current niches and into general use. Wearable computers, they point out, would provide users with all the benefits of traditional desktop computers, with the added bonus that these benefits would be available on demand at all times. A person wearing a computer could tap into its powerful memory, its Internet connection, and its superior calculating ability whenever they wanted—while jogging, while shopping, or while socializing—instead of having to sit at a desk in a room, at a keyboard.

But many of the proposed uses for wearable computers raise perplexing questions about privacy. A wearable computer complete with camera and Internet connectivity could make a person feel safer when they walk down a dark alley, since friends could monitor the walk on their computers. But what if that same person used the computer to record and broadcast what others thought was a private conversation? Mann's colleagues at MIT requested that he add a red light to his headset that would let them know when he was recording for his Web site. But what if such a light was sabotaged?

The privacy issue cuts both ways. When Mann wears a computer with Internet access, he can be physically alone, but he cannot truly experience solitude. Everyone can reach him, including people he may not want to hear from. When his camera is on and linked to the Web, everyone can see where he is and what he is doing. Of course, it is Mann's choice to broadcast his whereabouts, and there is no reason to think that others who decide to wear computers would choose to do the same.

Another issue facing the makers of wearable computers is that many people find normal computers annoying and frustrating to use. Researchers have viewed this as a challenge to make wearable computers more user-friendly than normal computers. One proposed way of doing this would be to make a wearable computer that was very inconspicuous and simply a part of a person's clothing, shoes, and glasses. Another, more futuristic concept would be to give the computer sensors that would help it judge a person's mood—if the user's muscles were tense and her heart was beating quickly, the computer would know she was under stress, and it could adjust its interactions with her accordingly.

But a computer that has such intimate contact with and knowledge of its user also raises questions. If someone uses a computer to assist his memory, help his thinking, and even improve his moods, where does the person stop and the computer begin? Could the computer be used as a surveillance tool, alerting authorities to a person's thoughts and feelings? These questions are intriguing but perhaps somewhat premature. The future of wearable computers no doubt will contain many surprises, with the technology being used in ways that cannot now be imagined. Wearable computers may remain a specialized technology, a useful tool for a small group of professionals and a life-enhancing device for people with certain disabilities. Or they may become so ubiquitous that today's questions about their impact on privacy or identity will quickly become obsolete.

—Mary Barr Sisson

RELATED TOPICS
Intelligent Agents, Internet and World Wide Web, Language Recognition Software

BIBLIOGRAPHY AND FURTHER RESEARCH

BOOKS
Gershenfeld, Neil. *When Things Start to Think.* New York: Henry Holt, 1999.
Picard, Rosalind W. *Affective Computing.* Cambridge, Mass.: MIT Press, 1997.

PERIODICALS
Baran, Nick. "Get Smart—Wear a PC." *BYTE,* March 1996, 36.
Byrne, Jason. "Hands On: Add a PC to Your Wardrobe." *Washington Post,* May 27, 1999, E5.
Chmielewski, Dawn C. "Wired Clothes Might Make Us Smarter." *Ventura County Star,* September 28, 1998, E3.
Clark, Michael, and Akwell Parker. "Nerd Novelties Go Mainstream." *Norfolk Virginia-Pilot,* May 31, 1999, D1.
Grossman, Wendy M. "Wearing Your Computer." *Scientific American,* January 1998, 46.
Ojeda-Zapata, Julio. "The Ultimate Accessory: Small Wearable Computers May Soon Be Part of Wardrobe." *Fort Worth Star-Telegram,* May 10, 1999, 21.
Parker, Akwell. "If the Computer Fits, Wear It." *Norfolk Virginia-Pilot,* May 31, 1999, D1.
Weeks, Linton. "Software, Hardware and Ready-to-Wear: State-of-the-Art Computer Accessories for the Smart Dresser." *Washington Post,* June 16, 1999, C1.
Weiser, Mark. "The Computer for the 21st Century." *Scientific American,* September 1991, 94–103.

INTERNET RESOURCES
MIT Wearable Computing Web Page
 http://www.media.mit.edu/wearables/
University of Oregon Computer and Information Science Wearable Computing Research Group
 http://www.cs.uoregon.edu/research/wearables/
Wearable Computers, Wearable Computing, Wearable Cameras
 http://www.wearcomp.org/
Wired for Wear
 http://www.ibm.com/news/ls/1998/09/jp_3.phtml

WEB TV

WebTV is a consumer device and associated network service, sold by WebTV Networks, that allows consumers to connect to the Internet using a standard television set. WebTV allows consumers to surf the World Wide Web, send and read e-mail, participate in chat conversations, and shop at online merchants. The user interface is designed to be as simple as possible, requiring very little computer knowledge to get started. Since its introduction in 1995, WebTV has continued to gain in popularity, especially with consumers over 40 years old. WebTV is now seen as one of the leading contenders in the race for "convergence"—in which, many industry observers contend, television broadcasting and the Internet will eventually become indistinguishable.

Scientific and Technological Description

WebTV devices are not general-purpose computers, but machines specifically designed for surfing the World Wide Web. The WebTV box itself resembles a standard video cassette recorder (VCR); in fact, it was designed this way deliberately so that it would be familiar to consumers. The WebTV unit connects to a television set via standard cables and has a phone jack for its internal modem in the back. WebTV is currently offered in two versions: Classic and Plus. The two versions have nearly identical interfaces and support most of the programs that people need in order to download and view materials from the Internet, including Javascript 1.2, Secure Sockets Layer (SSL), and several audio formats.

The Classic system consists of a 112-megahertz (MHz) processor, 2 megabytes (MB) of random access memory (RAM, which stores temporary data), and a 33-kilobyte per second (kbps) modem. There is no hard disk drive on the Classic. The Plus version is more powerful, featuring a 167-

Web TV

Catapult Entertainment, developed the idea of a specialized information appliance that could be used by anyone to access the World Wide Web. Realizing that 98 percent of American homes have televisions Perlman sought to create a device that could work with a standard television set and be cheaper and easier to use than a computer. After developing a prototype in just three days, Perlman and two associates formed WebTV Networks in April 1995.

In October 1996, the company introduced its first product, the WebTV-based Internet terminal. This product, coupled with the WebTV Classic service, allowed consumers for the first time to access the Internet through a standard television set. One year later, WebTV introduced its second-generation Internet receiver. This device, and its associated WebTV Plus service, further integrates television and Internet services by adding interactive television (ITV) links and limited picture-in-picture (PIP) display. ITV links allow a user, while watching a television show, to click on an icon that is linked to related content on the World Wide Web. For example, users might click on a link displayed during *Wheel of Fortune* that would allow them to play along with the game in progress. The PIP display adds the ability to view a television show and Web page on the screen simultaneously.

In August 1997, Microsoft Corp. acquired WebTV Networks for approximately $425 million in cash and stock. Under the terms of the agreement, WebTV Networks operates as a subsidiary of Microsoft and continues to be based in Silicon Valley, in northern California. Although some industry analysts were surprised at the marriage of software giant Microsoft and fledgling startup WebTV, industry insiders saw this as Microsoft's attempt to reach U.S. families that might never own computers. When Microsoft bought WebTV, the young company had cumulative losses of $63 million, fewer than 100,000 subscribers, and was struggling for shelf space in retail electronics stores. As of 1999, however, WebTV has grown steadily from Internet-on-TV into an enhanced television service, and

MHz processor, 8 MB of RAM, and a 56-kbps modem. There is also a 1-gigabyte (GB) hard disk on the Plus. However, the disk drive is not generally accessible to the end user—instead, it is used for internal functionality.

Both systems are generally available for under $200 plus a nominal monthly subscription fee to the WebTV service. WebTV Networks does not manufacture the WebTV hardware itself. Instead, it works with licensees such as Sony, Philips, Mitsubishi, and Samsung to build the devices. WebTV Networks also works with companies such as Hewlett-Packard and Canon to provide low-cost peripheral devices, such as printers, for WebTV users.

A new version of WebTV, released in 1999 and the result of a deal with business partner and satellite television provider EchoStar communications, will bring satellite television programming to the WebTV service. The new service will be supported by a more powerful WebTV device, containing an 8-GB hard drive capable of storing up to 8 hours of digital video. It will retail for approximately $500. Initially, the device will connect to the Internet via its internal modem. However, ultimately the device will allow much faster broadband Internet access through the satellite service.

Historical Development

In early 1995, while Internet browser software was in its infancy and the Internet was still largely unavailable to the general public, Steve Perlman, then an executive at gained subscribers through Microsoft's marketing might. WebTV is now among the 10 largest Internet service providers (ISPs) and continues to gain more subscribers. In fact, WebTV gained more than 200,000 new members during the 1998 Christmas season alone.

Uses, Effects, and Limitations

WebTV is designed primarily for simplicity and therefore is not as flexible or expandable as a general purpose computer. There are two primary limitations to the physical device. First, WebTV uses a television screen for display and therefore is limited to television display quality. The WebTV browser can display images that are 560 pixels wide and 420 pixels high. This produces images of much lower quality than those seen on most computer monitors (which can display 800 x 600 pixels or more). WebTV also uses larger typefaces (since a television viewer sits farther away from the set than a computer user would) and scrolls vertically only. These limitations mean that many Web pages do not format properly on WebTV. Furthermore, since only a small portion of most Web pages can be displayed at any time, WebTV users can become disoriented and have trouble navigating a large site effectively.

On the other hand, WebTV has better color support than that of most computer monitors, so images often look better than they would on a computer constrained by the capabilities of its video card (a video card is a circuit board that contains the necessary memory to provide a video display; some video cards can display only 256 simultaneous colors whereas a television can display millions of colors). WebTV also incorporates technology to minimize flickering and tends to give text a more polished look than a traditional browser would provide. To take advantage of the special display capabilities of the television screen, WebTV Networks added some features to the way its Web pages can be viewed that work only with a WebTV browser. These additions were achieved by changing certain attributes of the standard Hypertext Markup Language (HTML), the language in which Web pages are written. WebTV's transparency attribute, for example, can be applied to allow the background to show through an image. The gradcolor and gradangle attributes can be used to create professional-looking color gradients that would be much more difficult to achieve using standard HTML.

The second major limitation of WebTV is its lack of a standard keyboard and mouse. WebTV comes with a remote control with buttons for moving around the screen. Althoug basically functional, some users find it much more difficult to use than a mouse or other pointing device. For accessing e-mail, there is an on-screen keyboard than can be operated using the remote control. Most users, however, opt for the available wireless keyboard at an additional cost.

WebTV is a proprietary computer appliance and therefore cannot immediately take advantage of new advances on the Internet. For example, newly available media formats (e.g., *streaming* music and video—sound or video that can be listened to or viewed in real time as it is downloaded from the Internet, rather than being downloaded, stored, and played later) cannot be supported automatically, as on a PC, by downloading new software to the WebTV device. Instead, the WebTV network must work with third-party vendors to license special versions of their software that will work with WebTV. The new features must then be integrated with the existing features of the WebTV browser software. This process can take a significant amount of time. WebTV users are accustomed to waiting six months to a year for new features. By the time a new technology is introduced to WebTV, there may already be a newer version available to general-purpose computer users.

One feature of WebTV that standard computers lack is the ability to contact the WebTV network automatically to check for upgrades to the system. When they are available, the WebTV unit downloads upgrades in the background and incorporates the changes into its software. Therefore, users benefit from new features and improvements in previous software versions without having to do any work, such as contacting a Web site and following instructions to download a new version of software. This technology works so well that it was one of the company's selling points when Microsoft was investigating the purchase of WebTV.

Issues and Debate

Debate surrounding WebTV centers primarily around Microsoft's control of the company. One example is the issue of support for the Java programming language. Although WebTV had planned to offer support for this increasingly popular platform, after the Microsoft acquisition the company quietly announced that it would drop this support. Although the company claimed that the WebTV platform was not designed for such advanced uses, Microsoft opponents saw this as another tactic to promote Windows-based or proprietary standards over so-called open standards, such as Java, that can run on any operating system.

In fact, Microsoft is working to replace the core WebTV operating system with its own Windows CE operating system in an effort to spur the widespread adoption of Windows CE in other consumer devices, such as digital video (DVD) players and PERSONAL DIGITAL ASSISTANTS. With WebTV incorporating Windows CE, the manufacturers who build WebTV hardware, such as Sony and Philips, will be encouraged to use CE as a general-purpose operating system for their other products. Some people fear that ultimately Microsoft will end up controlling the standard operating system for everything from hand-held computers and telephones to DVD players (see DIGITAL VIDEO TECHNOLOGY) and cable television set-top boxes, in addition to their currently overwhelming dominance in desktop computer operating systems. The outcome of Microsoft's court battle with the U.S. Justice Department and 19 states over its alleged monopolistic business practices may ultimately address that issue.

—*Kevin Manley*

RELATED TOPICS
High-Definition Television, Internet and World Wide Web, Internet Search Engines and Portals, Personal Digital Assistants

BIBLIOGRAPHY AND FURTHER RESEARCH
BOOKS
Freeze, Jill T., and Wayne S. Freeze. *Introducing WebTV*. Redmond, Wash.: Microsoft Press, 1997.
Hill, Brad. *WebTV for Dummies,* 2nd ed. Indianapolis, Ind.: IDG Books Worldwide, 1998.
Miller, Michael. *Complete Idiot's Guide to Surfing the Internet With WebTV*. Indianapolis, Ind.: Que, 1999.
Vince, John and Rae A. Earnshaw, eds. *Digital Convergence : The Information Revolution*. New York: Springer-Verlag, 1999.

INTERNET RESOURCES
LA Times Report: Microsoft Antitrust
    http://www.latimes.com/HOME/NEWS/REPORTS/MICROSOFT/
"NBCi Teams with ValueVision for New Site, TV Channel," by Elizabeth Clampet
    http://www.internetnews.com/ec-news/article/0,1087,4_201521,00.html
Ruel.Net Set-Top Page
    http://ruel.net/top/box.links.htm
"Tech Convergence Time? Electronics Giants Say They're Ready to Merge PCs, CDs and TVs," by Jerry Kronenberg, CNN Financial Network, January 9, 1997
    http://www.cnnfn.com/digitaljam/9701/09/ces/
Total WebTV
    http://www.geocities.com/SiliconValley/Park/9319/index.html
"TV Channels on the Web," by Cameron Crouch, *PC World News*
    http://www.pcworld.com/cgi-bin/pcwtoday?ID=12812
WebTV Networks
    http://www.webtv.com

WIND ENERGY

The kinetic energy of the wind can be applied to many uses, including generating electricity through the use of an aerogenerator. Wind flows over the blades of an aerogenerator and forces the blades to turn a turbine, which powers a generator that creates electricity. Wind power is the fastest-growing energy source in the world, although it remains overshadowed by fossil fuels and nuclear power.

Scientific and Technological Description

Wind machines that specifically generate electricity are called *aerogenerators*. There are many aerogenerator designs, but they work on the same general principles. Most aerogenerators consist of a wind turbine with a rotor and blades, a powershaft, and a generator. Wind is a form of SOLAR ENERGY, because it is created by differences in atmospheric temperatures produced by the sun. When breezes blow over a rotor, the rotor spins and the kinetic energy is channeled to a generator that creates electricity. Rotor controls adjust how quickly the rotor spins. The amount of energy produced is determined by the speed of the wind and the diameter of the rotor. Faster winds and greater diameters result in more power. In fact, simply doubling the diameter of a rotor can increase its power capacity four times. The minimal speed of wind needed to spin a rotor is called the *cut-in speed,* but if the speed becomes too great, most turbines also respond to a *cut-out speed* at which the rotors stop spinning.

There are two principal types of wind turbines: horizontal-axis and vertical-axis. In the *horizontal-axis turbine,* the shaft to which the blades are attached runs parallel to the ground, whereas in the *vertical-axis turbine,* the shaft stands perpendicular to the ground. Horizontal-axis turbines are mounted on towers to take advantage of increased speeds of wind at greater heights. The towers range from 33 to 200 feet (ft.) high. When the wind changes direction, the rotor assembly moves or yaws to keep the blades perpendicular to the wind.

Vertical-axis turbines are not mounted on towers but instead keep their rotor assemblies lower to the ground. Their heights range generally from 30 to 82.5 ft., although the largest is 310 ft. high. The Darrieus turbine, with its egg-beater shape, is the most common. Because of their design, vertical-axis turbines do not yaw, which lessens the stress on the machines, but they also produce less power because wind speeds near the ground are lower than at the heights reached by horizontal-axis turbines.

Wind turbines generally have two or three blades, although some designs call for six. The blades are twisted slightly and must be strong enough to endure changes in wind speed and direction, vibrations, and turbulence. Blades have been made from many substances, including wood, plastic reinforced with fiberglass, aluminum, and steel. The choice of material depends largely on the size of the aerogenerator's blades (wood being used in smaller machines and steel being used in the largest). Aluminum is the most conventional.

There are three types of wind-power systems. Grid-connected power plants connect with wind turbines and feed their energy into the local distribution power grid for commercial use. Dispersed grid-connected systems produce electricity for homes, businesses, and farms that rely on wind power primarily but also use the utility grid when necessary. Remote stand-alone systems describe small wind turbines in areas too far from the utility grid. These turbines can provide energy for water pumping and other uses for a limited number of people.

Historical Development

Over 2000 years ago, people in Persia, China, and India began using windmills to mill crops and perform other tasks. By the late 19th century, windmills were used throughout Europe and in North America, and from 1880 to 1930, more than 6.5 million windmills were in use in the

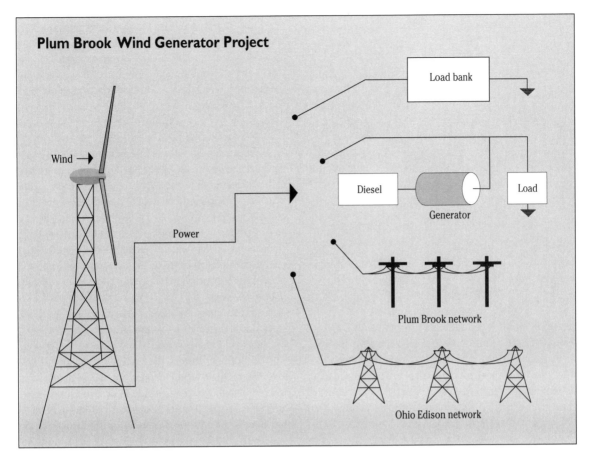

Plum Brook Wind Generator Project

Load bank

Wind →

Diesel

Generator

Load

Power

Plum Brook network

Ohio Edison network

NASA and the National Science Foundation developed the first large wind generator project in the early 1970s. Above is the Mod-O installation at Plum Brook, Ohio. It became operational in 1975. The aerogenerator feeds four outlets. Much like a conscientious car owner starts the engine of an infrequently used vehicle every once in a while, the load bank runs the generator periodically to ensure that the generator is working properly. The generator, in turn, serves as a back-up to the aerogenerator and will power important parts of the installation itself and feeds into the larger grid of Ohio. (Adapted from Tom Kovarik et al., Wind Energy, *Domus Books, 1979.)*

United States. In 1890, Danish scientists proposed the first design to produce electricity from a wind turbine, and within the next 10 years, the country built about 2500 windmills that produced electricity, most of which was for local or private use.

The United States followed Denmark's example, but the appeal of aerogenerators splintered in 1936 when the Rural Electrification Administration delivered electricity via the power grid to areas served previously by windmills and other means. In the 1940s, fossil-fuel plants superseded all other forms of energy production, and the wind energy field did not see a resurgence for another 30 years.

The resurgence resulted from a combination of governmental interest and the oil crisis in the 1970s. When members of OPEC (the Organization of Petroleum Exporting Countries) refused to sell oil to the United States in 1973, and hiked oil prices dramatically in the years that followed, the federal government began to devote more resources to examining alternative energy sources that would decrease the nation's dependence both on oil and on other countries. The National Science Foundation and the National

Aeronautics and Space Administration (NASA) had actually formed a Solar Energy Panel a year before the OPEC embargo, in 1972; its members recommended that wind deserved more study as a source of power. That study was conducted by the Wind Energy Conversion (WEC) program, which is administered under the federal Energy Department with the cooperation of NASA. The program developed the first wind farms, clusters of wind turbines for energy production, in the United States in the late 1970s.

In the early 1980s, three aerogenerators opened near Goldendale, Washington, and these were the first to produce electricity for commercial sale in the United States. To further develop renewable energy sources, the federal government, under the Public Utility Regulatory Policies Act, passed in 1978, offered attractive investment and energy tax credits for entrepreneurs who developed them. The credits equaled 25 percent of total investments, and the energy field surged. Some states offered additional credits. In California, for example, renewable energy companies received a 25 percent credit in addition to the federal credit, which created a virtual wind energy boom in the state. From 1982 to 1986, wind

In the United States, wind energy generates about 3 billion kilowatt-hours of electricity annually. More than 90 percent of that total comes from three wind farms in California, one of which is the Palm Springs wind farm shown here (C. T. Ellin Moschovitis).

power for electricity in California increased from 6 million kilowatt-hours (kWh) to more than 1.2 billion kWh. By 1985, an estimated 13,370 wind generators functioned in the state. The heady days ended during Ronald Reagan's second term as president, when the federal government cut funding for renewable energy study and reduced tax credits. The California state legislature also phased out the credits.

Today, more than 20,000 wind turbines function in the world, with the majority located in California and Denmark. The power source still provides less than 1 percent of the energy sources used for electricity in the United States, but researchers continue to refine and develop new turbine designs. In the United States, the Office of Photovoltaic and Wind Energy Technologies of the Energy Department's Office of Energy Efficiency and Renewable Energy manages the National Wind Energy Program. The program helps to foster research agreements between national laboratories and utility companies to improve wind power technologies and boasts a multimillion-dollar budget. The Energy Department also oversees the world's largest scientific institution dedicated to developing renewable energy technologies in Golden, Colorado. Named the National Renewable Energy Laboratory (NREL), the institution operates the National Wind Technology Center.

Uses, Effects, and Limitations

Wind turbines are used for water pumping, irrigation, drying grain, and heating water, but their zero-pollution electrical production makes the energy source attractive to general consumers. In the United States, wind farms and single turbines produce about 3 billion kilowatt-hours of electricity annually. More than 90 percent comes from three wind farms in California: in Altamont Pass, Tehachapi, and Palm Springs.

Wind power is attractive because, unlike fossil fuels or NUCLEAR ENERGY, wind produces no pollution or waste that

has to be stored. Wind turbines are based on principles known to human beings for thousands of years, and so long as there is an atmosphere on Earth and a sun to warm it, there will be wind. But wind power has definite limits, and the most obvious of these is that electrical production stops when the wind stops. As a result, some wind turbines are used in conjunction with other power sources, such as HYDROELECTRIC POWER, to ensure consistent energy levels.

The unreliability of available winds makes incorporating wind energy into the existing electrical grid challenging and remains a stumbling block to utilities. Energy consumption varies, and utilities must provide enough energy on demand. They rely on baseload power plants, such as coal- and nuclear-powered plants, that produce large amounts of energy without interruption. Backing these plants are intermediate plants that shut down at night and peak plants that kick in when demand is greatest. Wind cannot be relied upon fully for any one of these categories. To alleviate the variability problem, engineers are designing storage materials, such as batteries, that could supply energy when the wind dies down.

To further ensure the reliability of windpower, researchers spend much time studying site selection. For a wind turbine to work consistently, it must be located in an area with persistent winds that travel more than 10 miles per hour (mph). There can be no obstructions looming upwind that could impede wind speed. The best areas for aerogenerators are plains, open shorelines, the tops of rounded hills with gentle slopes, and the narrow gaps between mountains. These locales are not always near existing transformers and transmission lines, so sometimes costly extensions must be made over unforgiving terrain to connect a turbine to the electrical grid. Site selection is critical to the success of wind power, and some experts have calculated that an error of as little as 10 feet can result in a reduction of power by 10 percent.

To find the best locations in the United States for wind farms, researchers at the Pacific Northwest Laboratory in Richland, Washington, studied the wind energy potential of the 48 contiguous states in 1990 and created seven categories. Class 1 states have winds that travel less than 12.5 mph on average, while Class 7 states' winds bluster at more than 19.7 mph or more. Class 5 states or above are considered good candidates for wind power production; Texas and North Dakota are ranked as the best. This system is used by wind energy researchers when trying to determine the feasibility of the energy source for national consumption.

Issues and Debate

Despite being one of the few nonpolluting forms of energy, wind power has created controversy among wildlife advocates and environmentalists for several reasons. The first reason is aesthetic: Aerogenerators can be big, and many consider them unsightly. The vertical-axis turbine at Cape

Chat, Quebec, has a 310-ft.-high rotor that is 210 ft. in diameter. A horizontal-axis turbine in Oahu, Hawaii has a rotor, with a diameter of 320 ft. mounted on a 200-ft. tower. Most aerogenerators are smaller than these, but many are still large enough to be spotted a distance away, and they are not always alone. A commercial wind farm usually features many turbines that must be spaced far enough apart to avoid interference in each other's wind draw. At the Altamont Pass facility in California, for example, 7000 turbines whir. Clusters of huge turbines can blight large tracts of land. Advocates of wind power counter that much of the land can still be used for ranching, farming, forestry, and other activities. In the meantime, researchers are trying to develop smaller but equally effective turbines.

Second, turbines may interfere with wildlife: Birds that fly into the blades of turbines remain a great concern for wildlife experts and government officials. A study in California conducted from 1989 to 1991 found that most of the 182 birds found dead at a wind farm had collided with turbines. Nearly 120 of those 182 birds were birds of prey, including golden eagles, which are protected by the Endangered Species Act. A similar study in Spain identified dead birds from 13 protected species. To address the problem in the United States, the NREL has coordinated research among public and private groups. In 1994, the National Avian-Wind Power Planning Meeting was held and was followed by a second meeting in 1995. As a result, several research projects are under way.

Environmental issues aside, the quality of the design of aerogenerators has been an issue in the industry until quite recently. Large aerogenerators are expensive to build and maintain, and the generous tax credits of the 1970s and 1980s created a rush to produce hundreds of wind turbines quickly. These turbines were often built on principles of construction better suited to airplanes and helicopters that were misplaced in the wind power industry. Early investors found themselves having to repair machines that were just a few years old, and they could not always afford the costs.

Today, systems have been refined enough that the cost of wind power has decreased from 40 cents per kilowatt-hour in 1980 to about 5 cents today. Additionally, the Energy Department plans to invest $1.2 million in 10 wind turbine testing projects in 10 different states. The funding will support the design and installation of new small wind turbines for field testing. Another federal program, Wind Powering America, plans to double wind energy capacity in the country by 2005, and double it again by 2010 to create enough energy to fulfill the annual energy needs of 3 million households.

But the success of such programs depends on the growing acceptance of wind energy by utilities and consumers. Wind industrialists are hinging their hopes on the production of cheaper turbines and persuasive marketing to domestic and foreign utilities. Experts at the Energy Department predict that the next generation of turbines will decrease in cost by 20 percent, which should help programs such as Wind Powering America. Researchers continue trying to develop turbines that can take advantage of moderate and not just strong winds, and U.S. wind industry firms have expanded their interest to foreign markets. For more than a decade, they have sold complete wind power plants or specific components to such countries as Canada, the Netherlands, and Mexico, and they hope to progress further into less industrialized nations.

Of course, utilities will offer wind energy as an electricity source only if consumers will buy. Currently, wind energy provides less than 1 percent of the U.S. electrical supply, but with the ongoing deregulation of the electrical industry, Americans are freer than ever to choose their electricity source. Some can be counted on to choose nonpolluting wind energy over environmentally undesirable fossil-fuel burning, and it is upon these consumers that wind energy industrialists are pinning their hopes that the field will grow.

—*Christina Roache*

RELATED TOPICS
Biomass Energy, Hydroelectric Power, Nuclear Energy, Solar Energy

BIBLIOGRAPHY AND FURTHER RESEARCH

BOOKS

Flavin, Christopher, and Nicholas Lenssen. *Power Surge: Guide to the Coming Energy Revolution*. New York: W.W. Norton, 1994.

Glasstone, Samuel. *Energy Deskbook*. New York: Van Nostrand Reinhold, 1983.

Golob, Richard, and Eric Brus. *The Almanac of Renewable Energy: The Complete Guide to Emerging Technologies*. New York: Henry Holt & Company, 1993.

Kuecken, John A. *Alternative Energy: Projects for the 1990s*. Blue Ridge Summit, Pa.: TAB Books, 1991.

March, Frederic, et al. *Wind Power for the Electric-Utility Industry: Policy Incentives for Fuel Conservation*. Lexington, Mass.: Lexington Books, 1982.

Torrey, Volta. *Wind-Catchers: American Windmills of Yesterday and Tomorrow*. Brattleboro, Vt.: Stephen Greene Press, 1976.

PERIODICALS

Anderson, Christopher. "Renewable Energy: The Future Is Now (Again)." *Nature,* December 5, 1991, 344–345.

Annin, Peter. "Power on the Prairie: In Minnesota, They're Harvesting the Wind." *Newsweek,* October 26, 1998, 66.

"Audit Confirms Big Is Not Always Beautiful." *Nature,* February 3, 1994, 400.

Denniston, Derek. "Second Wind." *World Watch,* March–April 1993, 33–35.

Flavin, Christopher. "Bull Market in Wind Energy." *World Watch,* March–April 1999, 24–27.

"Get Wind of This." *Geographical,* September 1997, 6.

Gibbs, W. Wayt. "Change in the Wind: Utilities Are Starting to Offer Renewable Energy—for a Price." *Scientific American,* October 1997, 46.

Holmes, Hannah. "Unplugged: Homes Completely Powered with Solar and Wind Energy." *Sierra,* September–October 1993, 23–24.

McGowan, Jon. "America Reaps the Wind Harvest." *New Scientist,* August 21, 1993, 30–33.

Nadis, Steve. "Financial Blow Slows Wind-Power Projects." *Nature,* June 27, 1996, 721.

Webb, Jeremy. "Can We Learn to Love the Wind?" *New Scientist,* July 16, 1994, 12–15.

INTERNET RESOURCES
American Wind Energy Association
    http://www.awea.org
EcoNet
    http://www.igc.org/energy/wind.html
Energy Efficiency and Renewable Energy Network
    http://www.eren.doe.gov/wind
National Wind Technology Center
    http://www.nrel.gov/wind
Sandia National Laboratory Wind Energy Technology Projects
    http://www.sandia.gov/Renewable_Energy/wind_energy/
    homepage.html

X-RAY IMAGING

X-ray imaging, in a medical context, refers to the use of X-rays to produce images of internal parts of the human body for the purpose of medical diagnosis or for monitoring the progression of a disease or treatment. Medical X-ray images may be still or moving, analog or digital.

Since its inception in the late 19th century, X-ray imaging has offered great medical benefits through its capacity to visualize the interior of the human body. Concern about the health hazards posed by X-rays has led to refinements in the technology itself and in its use. X-ray imaging nevertheless continues to be the most common body-scanning method used in medicine.

Scientific and Technological Description

X-rays are electromagnetic radiation of a wavelength shorter than visible light but longer than that of gamma rays. X-rays are highly penetrating and pass through many materials, but are absorbed by dense materials such as metal and bone. For medical use, X-rays are produced by bombarding a tungsten anode (a positive electrode) with high-energy electrons from a cathode (a negative electrode). The resulting X-rays are directed by lead shielding and emerge from the X-ray machine as a beam.

In *conventional radiography,* the X-ray beam is directed through the patient's body and the emergent rays expose a photographic film (radiograph). Where rays travel through the body relatively unimpeded, as through the chest region with its air-filled lungs, the radiographic film becomes heavily exposed and appears black when the film is developed. Where bone or other dense material absorbs the X-ray beam, film exposure is reduced and that region appears pale. Between these two extremes, soft tissues of varying density and composition produce a gray-scale image on the film. Modern radiographic film contains photographic film sandwiched between fluorescent material that glows when struck by X-rays. This increases the sensitivity of the film manyfold.

Excluding its use in pure research, X-ray imaging is employed to diagnose a suspected medical condition, evaluate its extent, and in follow-up studies, to monitor the condition's progression or its response to treatment. X-ray imaging is also used in screening programs, where members of the general public are routinely checked to determine whether or not they have a specific medical condition. Examples of such mass screening programs include the use of chest X-rays to detect tuberculosis and mammograms (X-rays of women's breasts) to reveal breast cancer. In such cases, early detection of the medical condition in high-risk groups may be lifesaving or lead may to more effective and affordable treatment.

Digital radiography uses the same principles as those of conventional radiography, but the X-ray film is replaced by digital capture technology, usually an image intensifier, a screen containing fluorescent material that is backed by a photocathode. The X-ray image on the screen is transferred to a smaller screen that in turn is viewed by a camera. The digitized data are then manipulated by a computer and the image is visualized on a monitor. A PACS (picture archival and communication system) stores, labels, retrieves, and transmits the digital information within or outside the hospital. Digital radiography is set to largely replace conventional radiography in developed countries within the next decade. Digitally captured X-ray images may be recorded as a series of still images taken in rapid succession or as a moving film or videotape (*cineradiography*). Cineradiography provides time-based sequences of dynamic processes, such as the flow of liquid through blood vessels or through the gastrointestinal tract.

Tomography refers to the process of making an image that represents a cross section or slice through the body. *Computed tomography* (CT) involves capturing many X-ray image slices and then using a computer to stack them to create a three-dimensional representation of that part of the interior.

While X-ray imaging techniques are commonly noninvasive (nothing is swallowed or injected into the body), they sometimes employ *contrast media.* These are dyes or X-ray-opaque substances that fill or outline parts of the body, making them visible in X-ray images. The contrast medium is usually delivered by swallowing (as in a barium meal that reveals the inner outline of the gastrointestinal tract) or by injection into an artery or vein (as in angiography, which enables the visualization of fluid within blood vessels).

Historical Development

The history of X-ray imaging can be traced back, quite precisely, to Friday, November 8, 1895, when German physicist Wilhelm Roentgen discovered a strange glow emanating from a chemical screen on his laboratory benchtop. He had been setting up an experiment using a cathode-ray tube (a device similar to the tube of a television set). In the days that followed he came to realize that powerful rays from the cathode-ray tube produced the glow. They were interacting with barium platinocyanide and causing the chemical screen to fluoresce. Roentgen called the "unknown" rays X-rays. When he

Conventional and Digital Radiography

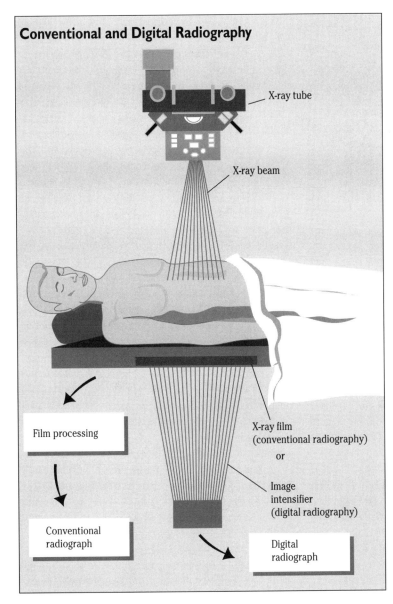

X-ray tube

X-ray beam

Film processing

Conventional radiograph

X-ray film (conventional radiography)

or

Image intensifier (digital radiography)

Digital radiograph

er models. Much early radiography (image production using X-rays) was performed by physicists or photographers. Diagnosing medical conditions from early X-ray pictures was far from easy. The profession of radiologist (a fully qualified clinician trained in radiology, the science and application of penetrating radiations) emerged only gradually during the early 20th century. By the late 1920s, contrast agents such as barium sulfate and iodine were being marketed for use with X-ray imaging to outline blood vessels and hollow organs.

In the first half-century since Roentgen's discovery, scientists began to understand the hazardous effects of X-rays and other forms of high-energy radiation. Several of the early pioneers of X-ray investigation succumbed to lethal doses of X-radiation themselves. However, as late as the 1950s, routine X-ray imaging of fetuses was still taking place in Europe and North America. The potentially damaging effect of even relatively low doses of X-rays was not appreciated, as evidenced by the fact that X-ray machines for imaging feet were common in shoestores.

In 1971, Godfrey Hounsfield and colleagues, working at the EMI laboratories in Hounslow, England, combined computing power with sensitive X-ray imaging technology to produce computed tomography, X-ray slices through the human body that were compiled to produce three-dimensional representations. In 1979, Hounsfield received the Nobel Prize for Physiology or Medicine for his work.

Recent developments seek to minimize X-ray dosages while maximizing the precision and sensitivity of the diagnostic technique. Non-X-ray techniques such as ULTRASOUND IMAGING, positron emission tomography (PET), and MAGNETIC RESONANCE IMAGING (MRI) have been added to the radiologist's repertoire. X-ray imaging continues, however, to be the most widely used diagnostic imaging technique in medicine. Apart from its passive role in diagnosis, X-ray imaging is now used in conjunction with interventional techniques such as *keyhole surgery* (surgery that utilizes a laparascope, a fiber-optic telescope) to guide instruments inside the body.

passed his hand between the tube and the fluorescing screen, the rays revealed the shadows of his bones on the screen.

X-ray imaging is one of the few scientific discoveries to be accepted rapidly and wholeheartedly by the popular media and to have its applications quickly put into effect. Roentgen's discovery was very widely reported, and by February 1896, X-rays were being used to evaluate bone and joint disorders and to find foreign objects inside the body. Roentgen chose not to patent his discovery, and so made possible rapid development of the technology by others. In so doing, he gave up his claim to a potential fortune in royalties. In 1901, Roentgen was awarded the first Nobel Prize in Physics for his discovery.

Between the late 1890s and 1920s, a legion of scientists, inventors, and engineers (among them Thomas A. Edison) improved on the basic design of the X-ray machine, producing equipment that was both safer and more powerful than earli-

Uses, Effects, and Limitations

Conventional radiography *is* less expensive than other forms of X-ray imaging. In many cases it is perfectly ade-

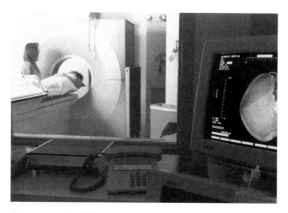

Since the early 1970s, computed tomography (CT) scanning has revolutionized imaging of the brain, allowing small tumors and hemorrhages to be detected. CT is best known in its common form, computed axial tomography (CAT), which refers to CT sections taken across the long axis of the body. Here, a CAT scanner is shown in use at a hospital (Corbis/Owen Franken).

quate for assessing skeletal damage or diagnosing medical conditions centered on the chest or abdomen. It is sometimes used in conjunction with contrast agents.

Digital radiography is gradually replacing conventional radiography in many applications. Although digital radiography equipment is expensive and startup costs associated with it are high, digital technology has several significant advantages over traditional X-ray technology. With the new technology, patients may receive a lower X-ray dose and the digital image can be computer enhanced to improve contrast. Digital images can be identified, stored, retrieved, and transmitted with greater ease than with traditional hard-copy images. Two digital images can readily be displayed side by side on a screen, or overlaid, for purposes of comparison.

In cineradiography, multiple images mean a higher X-ray dosage, which can be justified only if the additional data are clinically necessary. With cineradiography, the passage of a substance or object through the body, as in the case of a barium meal, or an endoscope inserted during keyhole surgery, can be monitored minute by minute or second by second.

Conventional, digital, and fluoroscopic techniques suffer from the limitation that body parts are superimposed one upon the other on a two-dimensional image. Computed tomography (CT), however, combines many cross sections of the body to generate a three-dimensional image of the body's interior. With CT, sensitivity to differences in X-ray absorbance is that much greater, which means that certain abnormalities in soft tissues can be better visualized than with conventional or digital radiography. CT is best known in its common form, computed axial tomography (CAT), which refers to CT sections taken across the long axis of the body.

Since the early 1970s, CT has revolutionized imaging of the brain, allowing small tumors and hemorrhages to be

detected. CT is also used to discriminate between small differences in tissue density within the chest and abdomen and can quickly determine the precise depth of a feature. These advantages come at a cost. CT, with its hundreds of X-ray scans, remains an expensive option. Also, the X-ray dosage to the patient is usually higher than with conventional or digital radiography, although the dosage is less than some other techniques previously used to examine brain damage.

The use of contrast agents is vital in certain X-ray imaging applications, such as outlining the inside of the gastrointestinal tract, the urinary system, or blood vessels. A few patients are allergic to contrast media, and resuscitation equipment and appropriate drugs need to be on hand for rare, severe adverse reactions. Some modern contrast media are less likely to cause adverse reactions but are much more expensive.

Ionizing radiations used in medicine are the greatest source of radiation exposure for most people. The hazardous effects of X-rays must be balanced against their benefits to the patient. X-rays ionize molecules within the cell, particularly DNA (the genetic material) and enzymes (proteins that catalyze reactions). In these and other ways, X-rays can disrupt cell function, although low-level damage can be corrected by the cell's repair mechanisms. The harmful effects of X-rays are threefold:

1. High levels of X-rays can cause direct tissue damage, killing cells (this property is used in radiation therapy to kill cancerous cells).

2. Radiation can increase the likelihood that cancers will develop (it is assumed that likelihood increases with increasing dosage, but see *Issues and Debate*, below); the carcinogenic properties of X-rays are highest for actively dividing cells, which is why precautions are taken to avoid irradiating an embryo in a pregnant woman.

3. Radiation damage may cause genetic mutations that are passed on to offspring; hence, precautions are taken to avoid irradiating the sex organs of children or fertile adults.

Within the next few decades, imaging technologies that do not utilize ionizing radiation, such as ultrasound (US) and magnetic resonance imaging (MRI), are likely to replace some, if not many, of the more advanced X-ray technologies.

Issues and Debate

There is little doubt that X-ray imaging, as a diagnostic tool for potentially life-threatening medical conditions, has saved many more lives than it has claimed. However, some physicians and medical consumer groups remain convinced that some X-ray scans are being performed unnecessarily, with attendant risks that have yet to be reliably quantified.

Best medical practice adopts the *alara principle*: The radiation dose that is administered is "as low as reasonably achievable." However, surveys conducted in the United Kingdom and the United States in the late 1980s and early 1990s suggested that X-ray doses given to patients in some hospitals were 20 to 30 times higher than necessary, particularly where antiquated or poorly serviced X-ray equipment was being used. As a matter of course, hospital staff should check whether a patient has been X-rayed before, to avoid duplication of effort and to minimize the possibility of cumulative radiation doses reaching dangerously high levels. Radiological agencies in the United Kingdom and the United States continue to work to institute sound and safe X-ray practice in all medical establishments.

Whether an X-ray is taken, and in what form, is a clinical judgment, and as such, must take into account numerous factors, both medical and economic. Those factors include the reliability of the diagnostic technique, the medical value of the extra information to be gained, the cost of the service, and whether treatment is feasible even if the suspected condition is confirmed. Because of economic considerations, access to the more expensive forms of X-ray scans may be limited.

A key problem in assessing radiation damage is that long-term studies have yet to demonstrate convincingly the level of danger posed by small doses of radiation. Figures quoted are derived by extrapolation from much higher doses, such as those given during radiotherapy to treat malignant tumors, or those experienced by the survivors of the Hiroshima and Nagasaki nuclear explosions. The risk estimates assume there is a more or less linear relationship between radiation dose and probability of damage. However, this relationship has yet to be convincingly confirmed or refuted.

Regarding the efficacy of X-ray screening for breast cancer, there is a broad spectrum of opinion among medical researchers. Breast cancer is likely to affect as many as one woman in ten in the developed world, as reported by the *New Scientist*. Many researchers and clinicians advocate twice-a-year mammograms for women over age 50. As the sensitivity of X-ray techniques has improved, and as imaging early cancers in the denser breast tissue of younger women is now feasible, pressure is building for routine breast screening to be introduced for women under age 50. However, the implementation of such screening procedures is controversial because they may expose young women to unnecessary radiation. In 1995, John W. Gofman of the University of California at Berkeley made the provocative assertion that most cases of breast cancer among women in the United States were, in fact, caused by previous exposure to X-rays.

One of the problems posed by new X-ray technology is that clinicians and the public wish to utilize innovations before solid, long-term studies of their efficacy have been carried out. Writing in *Lancet* in 1997, Adrian K. Dixon argued that X-ray imaging and other radiological techniques may be more justifiable and effective when used early in diagnosis, to investigate an unexplained problem, than when used much later to confirm a diagnosis or to monitor the success of treatment.

—*Trevor Day*

RELATED TOPICS
Magnetic Resonance Imaging, Scintillation Techniques, Ultrasound Imaging

BIBLIOGRAPHY AND FURTHER RESEARCH

BOOKS
Armstrong, Peter, and Martin L. Wastie. *Diagnostic Imaging,* 4th ed. Oxford: Blackwell, 1998.
Kevles, Bettyann H. *Naked to the Bone: Medical Imaging in the Twentieth Century.* New Brunswick, N.J.: Rutgers University Press, 1997.
Lisle, David A. *Imaging for Students.* New York: Oxford University Press, 1997.
Michette, Alan, and Slavka Pfauntsh, eds. *X-rays: The First Hundred Years.* Chichester, West Sussex, England: Wiley, 1996.
Patel, P. R. *Lecture Notes on Radiology.* Oxford: Blackwell, 1998.
Weir, Jamie, and Peter H. Abrahams. *Imaging Atlas of Human Anatomy,* 2nd ed. London: Mosby-Wolfe, 1997.

PERIODICALS
Dixon, Adrian K. "Evidence-Based Diagnostic Radiology." *Lancet,* August 16, 1997, 509.
Edwards, Rob. "Radiation Roulette." *New Scientist,* October 11, 1997, 37.
Evens, Ronald G. "Röntgen Retrospective: One Hundred Years of a Revolutionary Technology." *Journal of the American Medical Association,* September 20, 1995, 912.
Frankel, Richard I. "Centennial of Röntgen's Discovery of X-rays." *Western Journal of Medicine,* June 1996, 497.
Lentle, Brian, and John Aldrich. "Radiological Sciences, Past and Present." *Lancet,* July 26, 1997, 280.
Mackenzie, Debora. "Catch Them Younger." *New Scientist,* January 3, 1998, 61.
Skolneck, Andrew A. "Claim That Medical X-rays Caused Most U.S. Breast Cancers Found Incredible." *Journal of the American Medical Association,* August 2, 1995, 367.

INTERNET RESOURCES
Bioethics Resources
  http://adminweb.georgetown.edu/nrcbl/
A Guided Tour of the Visible Human
  http://www.madsci.org/~lynn/VH/
Library of X-ray Images
  http://www.netmedicine.com/xray/xrx.htm
Teaching Aids on Medical Imaging
  http://agora.leeds.ac.uk/comir/resources/links_c.html#teaching
The Visible Human Project
  http://www.nlm.nih.gov/research/visible/visible_human.html

INDEX

Page numbers in **boldface** indicate article titles. Those in *italics* indicate illustrations or photographs.